# Plant Biotechnology and Crop Improvement

# Plant Biotechnology and Crop Improvement

Editors

**Ranjith Pathirana**
**Francesco Carimi**

MDPI

Basel • Beijing • Wuhan • Barcelona • Belgrade • Novi Sad • Cluj • Manchester

*Editors*

Ranjith Pathirana
University of Adelaide
Adelaide, SA
Australia

Francesco Carimi
National Research Council
Palermo
Italy

*Editorial Office*
MDPI
St. Alban-Anlage 66
4052 Basel, Switzerland

This is a reprint of articles from the Special Issue published online in the open access journal *Plants* (ISSN 2223-7747) (available at: https://www.mdpi.com/journal/plants/special_issues/Biotechnology_Crop_Improvement).

For citation purposes, cite each article independently as indicated on the article page online and as indicated below:

Lastname, A.A.; Lastname, B.B. Article Title. *Journal Name* **Year**, *Volume Number*, Page Range.

**ISBN 978-3-7258-1307-0 (Hbk)**
**ISBN 978-3-7258-1308-7 (PDF)**
**doi.org/10.3390/books978-3-7258-1308-7**

# Contents

# About the Editors

**Ranjith Pathirana**

Ranjith Pathirana began his career as a research officer in the Department of Agriculture, Sri Lanka, where he initiated the National Coordinated Oilseeds Program. He then joined the University of Ruhuna, Sri Lanka, where he was the chair and professor of agricultural biology. He spent sabbaticals at the University of Guelph, Canada, and Massey University in New Zealand, and then joined the New Zealand Institute for Plant and Food Research where he initiated plant preservation research and pioneered cryopreservation of horticultural species and established the cryo-genebank in Palmerston North. Dr Pathirana has served as a FAO/IAEA expert in plant mutation breeding in several countries and as a consultant in plant cryopreservation. Currently he is an affiliate of the University of Adelaide and Royal Melbourne Institute of Technology (RMIT). His principal interests are in applying in vitro cell and plant biotechnologies for crop improvement, conservation, and the elucidation of biochemical pathways, particularly in plant stress response.

**Francesco Carimi**

Francesco Carimi, senior researcher at the National Research Council (CNR) in Italy, has over 35 years of experience in plant science research including the training of young researchers. He began his career at the University of Palermo. Subsequently, in 1990, he worked as a fellow in Francesco Salamini's department at the Max-Planck-Institut für Züchtungsforschung, Cologne, on the development of plant DNA markers. In 1994, he was appointed as a researcher at the CNR in Palermo. From 2001 to 2005, he held a research position and supervised a research team in the CRIBI Laboratory at the University of Padua, Italy. For a period of over 20 years, he held the role of head of the Section of Palermo of the Institute of Plant Genetics and subsequently of the Institute of Biosciences and BioResources of CNR, leading a team of researchers conducting molecular and cell biology research of woody and herbaceous plants. He has published over 110 full papers in international journals and 16 contributions to book chapters and has served as a referee for several scientific journals. He coordinated 24 national and international projects, managing, under his scientific responsibility and supervision, funds amounting to approximately EUR 28 million. He supervised PhD research of affiliated staff at the University of Palermo and Padua. His main work is in germplasm conservation, genetics, breeding, and molecular markers and biotechnologies for genetic and sanitary improvement in different plant species. He has held the role of expert in several FAO/IAEA research and training projects carried out in several developing countries.

# Preface

Traditional plant breeding has significantly boosted food production over the past half-century, enabling countries to outpace population growth. Yet, sustaining these gains and adapting to climate change are pressing concerns today. Yield increases in major cereals have plateaued over the last two decades. Meeting contemporary agricultural and horticultural demands for sustainable, nutritious crops that are environmentally resilient is challenging. Plant biotechnology emerges as a promising solution to address these challenges.

Advances in biotechnology have revolutionized our approach to food production and environmental sustainability. Techniques such as induced mutation, gene editing, and synthetic biology enable the development of crops with enhanced resilience to environmental stresses, disease resistance, and higher yields. These innovations not only boost food, feed, and fiber production but also enhance product quality while reducing environmental impact. The economic and environmental benefits of over 200 million hectares of biotech crops cultivated by 17 million farmers across 27 countries underscore the pivotal role of biotechnology in addressing global food security challenges.

This Special Issue of *Plants* gathers a diverse range of studies and reviews showcasing the latest breakthroughs in plant biotechnology. Contributions to this issue span diverse crop groups and continents, from well-domesticated species like wheat and barley to underexplored crops like prickly pear and tef. Articles cover a spectrum of biotechnological applications, including novel gene-editing methods and transcriptome analysis, emphasizing the importance of ongoing research and collaboration in advancing plant biotechnology for food security. Furthermore, these articles delve into the broader implications of biotechnologies, including ethical, socio-economic, and environmental impacts.

We hope this collection serves as a valuable resource for researchers, students, policymakers, and industry professionals, offering insights into the latest developments in plant biotechnology while fostering meaningful discussions on its wider implications. We extend our gratitude to the authors for their contributions, the anonymous reviewers for their time and expertise, and the *Plants* Editorial Office staff for their support in making this publication possible.

**Ranjith Pathirana and Francesco Carimi**
*Editors*

*plants*

MDPI

*Editorial*

# Plant Biotechnology—An Indispensable Tool for Crop Improvement

**Ranjith Pathirana [1,]* and Francesco Carimi [2]**

1 School of Agriculture, Food and Wine, Waite Campus, University of Adelaide, Urrbra, SA 5064, Australia
2 Istituto di Bioscienze e BioRisorse (IBBR), Consiglio Nazionale delle Ricerche, Via Ugo la Malfa, 153, 90146 Palermo, Italy; francesco.carimi@ibbr.cnr.it
* Correspondence: ranjith.pathirana@adelaide.edu.au or ranpathnz@gmail.com

## 1. Introduction

Traditional plant breeding has helped to increase food production dramatically over the past five decades, and many countries have managed to produce enough food for the growing population, particularly in the developing world. Sustaining these gains in crop productivity and adapting to climate change are becoming urgent concerns in modern times. In fact, yield increases in our major cereals have slowed down in the past 20 years. Global hunger is still above pre-pandemic levels, with around 690–783 million people faced with hunger in 2022, and meeting Sustainable Development Goal 2 of ending hunger by 2030 has become a daunting task [1]. Although increased yields through the Green Revolution helped to cultivate an additional 18–27 million hectares, this increased food production was accompanied by environmental degradation and micronutrient deficiencies across populations [2,3]. Developing crop cultivars that meet the present-day requirements of agriculture and horticulture is challenging, as they need to provide sustainable food and healthful nutrition for populations, and, at the same time, must be environmentally friendly and resilient to climate change. The global community is projected to face increasing food crises due to changing dietary styles and the rising population, which is set to reach almost 10 billion people by 2050 [4]). The challenge ahead is determining how to reduce the use of limiting resources (water, energy, and agricultural land) for intensive agriculture, ensuring sufficient production of food (Figure 1). Taking even the most conservative estimates, food production needs to double in the coming 30 years to meet the basic demands of the growing population [5]. Despite these challenges, there is growing evidence that food security and adequate nutrition for the global population can be achieved using climate-smart, sustainable agricultural practices, while reducing the negative impacts of agriculture on the environment, particularly greenhouse gas emissions [6].

Plant biotechnology is seen as the breakthrough technology that can help to meet this challenge in this next phase of plant breeding. Plant biotechnologies that aid in developing new varieties and individual traits within existing plant varieties include cell and tissue manipulation, marker-assisted selection, transgenic technologies, genomics, and molecular breeding. Cell and tissue culture technologies provide a range of applications in the creation, conservation, and utilization of the genetic variability in crops, such as in vitro pollination and embryo rescue for distant hybridization, the production of haploids and doubled haploids, polyploid breeding, in vitro mutagenesis, somaclonal variation, in vitro selection, germplasm preservation (in vitro for medium-term and cryopreservation for long-term), protoplast fusion for producing somatic hybrids, and gene manipulation for producing transgenic crops or the newly emerging techniques that allow for the generation of gene-edited plants.

**Citation:** Pathirana, R.; Carimi, F. Plant Biotechnology—An Indispensable Tool for Crop Improvement. *Plants* **2024**, *13*, 1133. https://doi.org/10.3390/plants13081133

Received: 7 April 2024
Accepted: 15 April 2024
Published: 18 April 2024

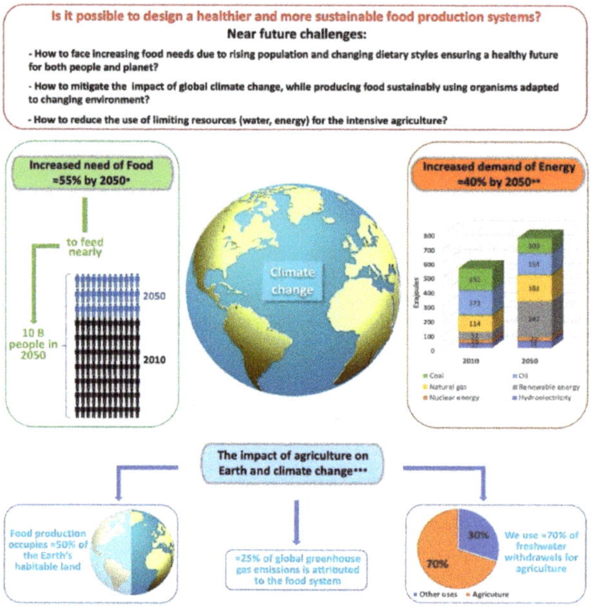


**Figure 1.** Food systems are increasingly vulnerable due to human pressures on natural ecosystems and the climate: The challenge ahead. * World Resource Institute (WRI)'s 2023 report. Available online: https://research.wri.org/wrr-food (accessed on 22 January 2024); ** Statista Energy Consumption Worldwide from 2000 to 2019, with a forecast until 2050, by Energy Source. Available online: https://www.statista.com/statistics/222066/projected-global-energy-consumption-by-source/ (accessed on 22 January 2024); *** Mahpul IN, Mohamad AH, Mazalan MF, Razak A, Rasyidee A (2021) Population, food security, nutrition and sustainable development. Available online: https://www.un.org/development/desa/dpad/publication/un-desa-policy-brief-102-population-food-security-nutrition-and-sustainable-development/ (accessed on 22 January 2024) [4].

High-resolution genetic analysis has allowed physical mapping and positional gene cloning for traits of interest, while molecular markers allow for the characterization of germplasm and finding duplicates and gaps in collections [2]. They are becoming indispensable in some breeding programs when used for the early culling of unwanted material in perennial crops such as in the case of culling male vines early in hybrid populations, screening in kiwifruit for marker-assisted selection [7,8], the development of saturated linkage maps, and pyramiding genes in introgressive breeding [9,10]. Despite the strict laws governing genetically modified crops, transgenic varieties of maize, soybean, rapeseed, cotton, tomato, potato, papaya, etc., occupy over 190 million hectares across 26 countries, grown by 17 million farmers, bringing in both economic and environmental benefits and, at the same time, some social controversy [6]. The many tools that plant biotechnology provides for crop improvement for developing resilient food systems while conserving the environment are shown in Figure 2. These aspects have been addressed in the 17 papers published in this Special Issue titled 'Plant Biotechnology and Crop Improvement'. There have been four general review papers covering different biotechnologies and thirteen original research contributions focusing on different crop groups, including tropical and temperate cereal, legume, root and tuber, fruit, ornamental, and industrial crops. With 44,000 views and 86 citations at the time of writing, this Special Issue has attracted much attention across the scientific community as expected, considering the relevance of the topic to the current challenges in global food production.

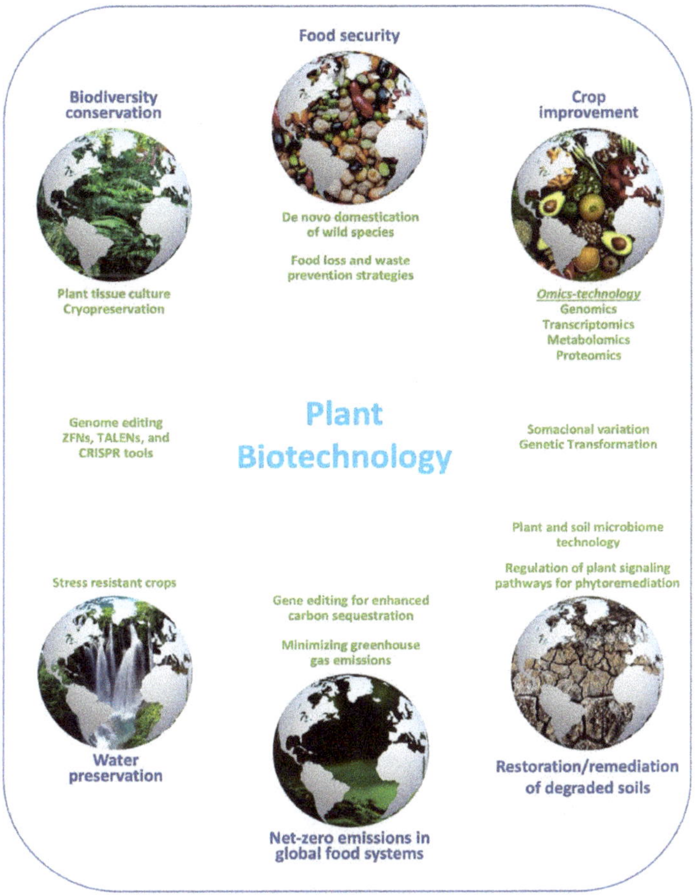


**Figure 2.** The many aspects of the contribution of biotechnology towards crop improvement for resilient food systems while also contributing to environmental protection.

This Special Issue also contains methodology development for plant genetic transformation through the use of an easy selection marker for discriminating transformed plants from escapes. The reviews in this Special Issue look at specific trait improvements such as stress resistance using gene-editing technologies and manipulation of phytohormone metabolism, as well as exploring if a second "quantum-leap" in food production is possible using these technologies. Another review describes the contribution of genomics to our understanding of crop evolution. With a wide array of applications of plant biotechnology to crop improvement available, the research papers addressed several of these technologies in extensively cultivated crops such as wheat, barley, bean, and potato, as well as in underutilized crops with high potential such as tef and prickly pear. The technologies applied to improve crops in these papers include somaclonal variation, induced mutagenesis, in vitro polyploidization, embryo rescue, and gene editing. Several papers describe the use of genomics, transcriptome analysis, molecular markers, and metabolic profiling to assist in selection in breeding programs and monitoring of transgenic plants.

The possibility of modifying phytohormone metabolism and signaling is a promising direction of research aimed at the improvement of crop productivity and stress tolerance. In her review, Nowicka (contribution 1) summarizes the state-of-the-art research concerning the modulation of phytohormone content aimed at the stimulation of plant growth and the improvement of stress tolerance. In particular, the roles of auxins, cytokinins, gibberellins, brassinosteroids, abscisic acid, ethylene, jasmonic acid, and their derivatives are analyzed. The author hypothesizes that modification of this signaling at various levels, from elements of signaling cascades, through transcription factors to miRNAs, is a very promising direction of genetic engineering of crop plants aimed at improving the resilience of plants.

Remarkable progress in genome-editing technologies has been achieved over the past 10 years and have begun to show extraordinary utility to develop crop varieties with superior qualities, or those that can tolerate adverse environmental conditions. In their review, Hamdan et al. (contribution 2) provide a detailed analysis of the genome-editing technologies that have been expertly applied to improve important agronomic traits, especially yield, quality, and stress resistance of the most important crops. In particular, the review focuses on the Clustered Regularly Interspaced Palindromic Repeats (CRISPR/Cas) system, which has been the focus in recent years as a revolutionary genome-editing tool used for various crops. The authors discuss the current developments and future applications of genome-editing technologies for developing crops that can help in mitigating the impacts of climate change on agriculture with notes on future perspectives. A bibliographic analysis is also presented covering CRISPR-related papers published from 2012 to 2021 (10 years) to identify trends and potential in CRISPR/Cas-related plant research. The authors conclude that combining conventional and more innovative technologies in agriculture would be the key to optimizing crop improvement beyond the limitations of traditional agricultural practices. A more pessimistic view is provided in a review carried out by Buzdin et al. (contribution 3), reporting that, according to estimates, global crop yields must double by 2050 to adequately feed an increasing global population without a large expansion of crop area. To achieve this "quantum-leap" in improvements in crop yield, we must respect environmental constraints and, at the same time, reduce the impact of agriculture on the environment. The authors support the long-debated idea that new technologies are unlikely to provide a rapidly growing population with significantly increased crop yield. Finally, in their review, Zhao et al. (contribution 4) analyze how recent advances in genomics have revolutionized our understanding of crop domestication. The authors summarize cutting-edge research on crop domestication by presenting the main methodologies and analyze the prospects for both targeted re-domestication and de novo domestication of wild species.

## 2. Cereal Crops

Dramatic increases in rice and wheat yields were achieved during the 'Green Revolution', where dwarfing genes were transferred to adapted cultivars through crossbreeding. The 'Green Revolution', with its high-input and technology-dependent approach, has been able to feed the growing world population in recent decades. It ensured food security, particularly in developing nations. However, long-term impacts are now evident: degraded soils, reduced groundwater levels, contaminated and salinized water bodies, and reduced biodiversity. Furthermore, high crop yields cannot be sustained without increased fertilizer use [6,11–13]. Traditional crossbreeding is straightforward when selecting for morphological traits that are easy to observe in field, such as height, grain size, color, and leaf shape, etc. The main change in rice and wheat achieved during the "Green Revolution" is dwarfing, resulting in greater partitioning of photosynthates in grains and better fertilizer response, without lodging. Hence, it was not difficult to identify dwarf plants in the segregating populations. However, traits such as nutritional quality, disease, and abiotic stress resistance are not easy to select visually in segregating populations under field conditions where breeders encounter many variables. Lab-based approaches to increase genetic variability or to genetically modify and select desirable genotypes are therefore required.

Over the past few decades, biotechnology has made significant contributions to cereal crop improvement by enhancing yield, nutritional content, biotic and abiotic stress tolerance, herbicide tolerance, and many other valuable traits. It has also played a crucial role in promoting environmental sustainability and has had positive economic impacts on agriculture. For example, the introduction of perennial cereals can alleviate many problems of annual monocultures [6,11–13]. *Thinopyrum* spp. is the most sought-after perennial grain in hybridization programs with wheat as it hybridizes freely with *Triticum*, producing fertile progeny [13], and perennial selections have outperformed the standard wheat cultivars in grain protein and mineral nutrient contents [14]. Yet, with genomic tools, selection for perennial growth and other quality traits would be easier and faster [13]. Thus, intermediate wheatgrass has been used in sequencing and marker-assisted recurrent selection [15], and a high-quality genetic map is now available online [16]. With these developments, breeding perennial wheat for large-scale cultivation will be possible.

Similarly, perennial rice (PR) will be the start of a second 'Green Revolution' as the data from 15,333 ha of perennial rice grown by 44,752 small holder farmers in southern China demonstrate [17]. The parents for the breeding program to develop PR were 'RD23', a cultivar of *Oryza sativa* ssp. *indica*, and a rhizomatous and perennial African species, *O. longistaminata*. Embryo rescue (a tissue-culture-based biotechnological intervention) of $F_1$ facilitated overcoming incompatibility and resulted in the foundation material for developing the commercialized PR. PR produced similar yields to annual rice over a period of four years, with eight harvests from a single planting. Farmers prefer PR due to 58.1% labor savings and 49.2% savings on inputs every growth cycle. Higher organic carbon and nitrogen accumulation in soils and improved soil water retention are other advantages [17]. Attempts to develop perennial rye using perennial wild rye *Secale montanum* L. [13] and perennial maize using tetraploid maize (*Zea mays* 2n = 4x = 40), tetraploid *Tripsacum dactyloides* (2n = 4x = 72), and tetraploid *Z. perennis* (2n = 4x = 40) [18] are underway.

Many other biotechnological interventions are possible in the development and selection of climate-resilient cereals. For example, Kruglova and Zinatullina [19] describe many examples of in vitro selection for drought, simulating water deficiency in culture media. They suggest using embryos at a certain developmental stage, when they are autonomous. In vitro selection for iron toxicity [20], aluminum toxicity [21], nickel, and NaCl toxicity tolerance [22] has been demonstrated in cereal crops. In vitro mutation induction and selection have also been demonstrated in many cereals [23]. More targeted mutations can be used in crop improvement thanks to the development of techniques such as Targeting Induced Local Lesions in Genomes (TILLING), as well as the latest gene-editing techniques. For example, Acevedo-Garcia et al. [24] developed bread wheat cultivars resistant to powdery mildew by TILLING. The first genome-editing tools were Zinc Finger Nucleases (ZFN) and Transcription Activator-Like Effector Nucleases (TALENs), but, later, Clustered Regularly Interspaced Short Palindromic Repeats (CRISPR) and Crispr associated protein (Cas) became the most widely used genome-editing tool due to its high editing efficiency, multiplex capability, and ease of use. Gene editing has enabled researchers to increase grain number and size in rice, and grain weight and yield in wheat. Powdery mildew resistance in wheat and resistance to *Xanthomonas* in rice have also been achieved using gene-editing technologies [25].

Wheat, barley, and tef are among the cereal crops covered in this Special Issue. In wheat breeding, crested wheatgrass *Agropyron cristatum* is considered a potential donor of valuable traits for abiotic (cold, drought, and salinity) and biotic (leaf rust, stripe rust, and powdery mildew) resistance. Crested wheatgrass belongs to the tribe Triticeae to which wheat also belongs and similar to *Triticum* has polyploid series with a basic chromosome number x = 7, but with the basic genome *P. Agropyron* is in the tertiary gene pool in the context of wheat breeding, and phylogenetically more distantly related than those in the primary and secondary gene pools of wheat. Fortunately, previous genetic studies have revealed that synteny is conserved between wheat and the P genome. Being a perennial species widely used in temperate regions for grazing beef and dairy cattle, it is also a candidate for transforming wheat into a perennial crop in futuristic sustainable agricultural systems [26]. In hybrids of these two species, chromosome recombination is key to transferring beneficial alleles from crested wheatgrass to wheat. Using in situ hybridization, a technique used to locate specific genomic DNA sequences within chromosomes, Prieto et al. (contribution 5) analyzed chromosome associations during meiosis in *Triticum aestivum* lines carrying chromosome introgressions of breeding interest (5P and 6P) in two sets of progenies: those with and without the *Ph1* locus located in the long arm of chromosome 5B of wheat, known to genetically control chromosome pairing and recombination. The authors did not find homoeologous chromosome pairing either in the presence or absence of the *Ph1* locus, indicating that this locus does not influence chromosome pairing between the two species.

The second cereal featured in this Special Issue is barley. Co-evolution of *Hordeum vulgare* and the fungus *Blumeria graminis* (D.C.) Golovin ex Speer, f. sp. *hordei* Em. Marchal (*Bgh*), causing powdery mildew is well studied and recorded, with more than 70 resistance genes. Most of the cultivated winter barley varieties carry one or more of these genes in different combinations, and this information can be used to authenticate accessions in a collection. Using sets of five single-plant progenies (SPPs) per accession from 172 winter barley accessions belonging to the core collection of Czeck gene bank, Dreiseitl and Nesvadba (contribution 6) tested 53 isolates of the pathogen for virulence/avirulence. While the majority of the accessions showed a single phenotype for resistance in their five SPPs, 78 (45.7%) accessions had more than one phenotype indicating heterogeneity in their seed stocks. With defined powdery mildew resistance genes in the SSPs, these accessions can be used with confidence in barley breeding for powdery mildew resistance. The third cereal featured in this Special Issue is the ancient grain tef (*Eragrostis tef* (Zucc.) Trotter), an underutilized cereal from Ethiopian highlands with outstanding nutritional value and more resilient than traditional cereals under marginal conditions. With no gluten epitopes, it is recommended for people suffering from celiac disease; hence, it is gaining increased attention around the globe. In this Special Issue, Numan et al. (contribution 7) describe the use of in vitro culture and mutagenesis to improve disease and lodging resistance, as well as the use of molecular markers for selection in tef. They conclude by discussing the potential of genome-editing technologies in tef improvement.

## 3. Pulse Crops

Pulses constitute an integral part of cropping systems and provide low-cost proteins in diets as well as essential micronutrients. They are the primary source of proteins in vegetarian and vegan diets as well as in the diets of the majority of the population in many developing countries where protein malnutrition is widespread. They improve soil through biological nitrogen fixation, helping to reduce nitrogen fertilizer requirements of the pulse crop, as well as for the next non-legume crop in cropping systems. The value of pulses was highlighted by declaring 2016 as the year of pulses at the 68th United Nations General Assembly (UNGA) with Food and Agriculture Organization (FAO) facilitating its implementation with the participation of Governments and various other stakeholders [27]. Recognizing the potential of pulses to achieve the 2030 Agenda for Sustainable Development, the UNGA designated 10 February 2023 as World Pulse Day [28]. Additionally, legumes are an important component of animal feed.

Conventional breeding of leguminous crops has been based on the selection for agronomic traits in the vegetative and reproductive phases that have distinct heritability values. One of the main features for mechanized cultivation of legumes is their transformation from an indeterminate growth habit to a determinate growth habit, facilitating synchronous flowering, pod maturation, and resistance to lodging. Soybean yields have increased globally from around 1130 kg ha$^{-1}$ in the early 1960s to the current 2800 kg ha$^{-1}$, with the yields in the three top soybean-producing countries (USA, Brazil, and Argentina) recording 3200–3300 kg ha$^{-1}$ [29]. The breeding of determinate cultivars is a major factor for such yield increases and the expansion of the production area through mechanization. The determinate trait is recessive and monogenically inherited, with the heterozygous individuals showing semi-determinate growth [30,31]. Determinate growth habits have been bred into many other leguminous crops used for seeds, such as pea, chickpea, pigeon pea [31,32], mung bean, black gram [33], grass pea, and cowpea [34]. Many of the first determinate cultivars were bred by mutation induction [35] and not through traditional crossbreeding.

Pulses were regarded as 'orphan crops' until recently due to lesser attention given to them compared to cereals. However, many of the pulse crops have now become 'mainstream crops', with draft genomes of many of them completed in the past decade [36–39] improving the efficiency of breeding efforts. Next-generation sequencing technologies have enabled the deployment of modern genomic tools, including a range of molecular markers associated with many agronomic traits, and disease and abiotic stress tolerances [40].

Beans, chickpeas, and peas are the most well-known and widely consumed pulses in the world [28], and two of these are featured in this Special Issue. Common beans (*Phaseolus vulgaris*) were introduced to Ethiopia in the 16th century, and farmers have selected varieties adapted to the local climate and soils over centuries. Their wide genetic diversity, particularly their tolerance to biotic and abiotic stress, has been incorporated in selections developed by the National Common Bean Improvement Program in Ethiopia. Tigist et al. (contribution 8) used 144 genotypes in a multilocation study to understand the variation in 15 agro-morphological traits. Multivariate analysis revealed six principal components. Based on agro-morphological traits, the clustering patterns were according to seed size with considerable genetic variation for the studied characters. The study revealed several accessions with distinct advantages in terms of agro-morphological traits and adaptability suitable for further improvement in the breeding program.

Chickpea (*Cicer arietinum*) is the second most consumed pulse after dry beans, and Australia is a major producer and exporter of this pulse. Among all the continents, Australia is the second driest continent after Antarctica; hence, the drought resistance of crops is a top priority in breeding programs. In both Australia and India (the largest producer of chickpea), chickpea is sown on residual summer moisture and left to grow in progressively depleting soil water, finally maturing under terminal drought. There are many traits associated with drought tolerance, such as root biomass and some leaf anatomical and physiological features. Early maturity is a drought escape strategy in crops such as chickpea sown on residual moisture. There is intensive ongoing work in identifying molecular markers for marker-assisted selection for drought tolerance in chickpea, and a quantitative trait loci (QTL) hotspot region for this trait has been found. The variety 'Geletu' with a high yield and drought tolerance was released in 2019 through a backcrossing program to introgress drought tolerance from accession ICC4958 to a high yielding Indian cultivar 'JG11', using this hotspot as a selection marker in a backcross breeding program [41]. However, other more innovative methods for screening populations for drought tolerance in the early growth stages would further accelerate breeding. In their paper in this Special Issue, Purdy et al. (contribution 9) went a step further and identified metabolites in young, watered seedlings of chickpea that can be prognostically used to predict seed numbers in mature plants under terminal drought. Among the yield components of annual crops, it is the seed number, not the seed size (weight) that is sacrificed under abiotic stress, drought in particular (contribution 9) [42,43]. Hence, identifying metabolites that can be used as indicators of seed number under terminal drought later in the life cycle would help in selecting drought-tolerant segregants early on in breeding populations. In chickpea, pinitol, sucrose (negative correlation with seed number), and gamma-aminobutyric acid (positive correlation) can be used to predict high or low seed numbers under these conditions (contribution 9). This is the first instance where a predictive marker was identified for screening drought tolerance that could be used by breeders to identify genotypes that perform well under adverse conditions, without having to expose them to drought.

## 4. Root and Tuber Crops

Almost all root and tuber crops are traditionally propagated vegetatively, and most are either sterile or partially sterile (cassava and yam); moreover, flowering is irregular and asynchronous (cassava), or crops do not flower at all (aroids such as *Colocasia*) [44]. Therefore, these crops are ideal candidates for improvement through in vitro-based biotechnological approaches. Potato [45] and sweet potato [46] have been improved through hybridization and selection; therefore, modern genomics tools are invaluable in improving the efficiency of breeding.

As tuber and root crops are an important source of carbohydrates in many impoverished communities around the world, attention has been focused on improving their mineral and vitamin contents because hidden hunger resulting from their deficiencies is prevalent in these communities, with an estimated two billion people affected [47]. About 800 million people use cassava (*Manihot esculenta* Crantz—Euphorbiaceae) as their staple food, and one third of the sub-Saharan population depends on cassava for over 50% of their caloric intake [48]. Breeding for increased mineral nutrition in cassava is hampered by the lack of genetic variation for these traits [49]; hence, transgenic approaches have been tested. For example, the overexpression of a gene for vacuolar iron sequestration, *AtV1T1*, resulted in altered partitioning of iron, with an iron content that was three to seven times higher in storage roots in transgenic plants compared to the wild type in field trials. The coexpression of a mutant *Arabidopsis thaliana* iron transporter *IRT1* and *A. thaliana* ferritin (*FER1*) produced transgenic cassava plants that accumulated iron levels that were 7–18 times higher and zinc levels that were 3–10 times higher, providing 40–50% of estimated average requirements (EAR) of iron and 60–70% of EAR of zinc for 1–6-year-old children and nonlactating, nonpregnant West African women [50]. In recent developments in the genomics of cassava, a haplotype-resolved diploid genome of an African landrace cassava ('TMEB 117') has been sequenced to a high level of accuracy providing valuable insights into the heterozygous genome of cassava and its resistance to African cassava mosaic virus [51].

Cassava mosaic disease (CMD), caused by a group of at least eight geminiviruses transmitted by white fly *Bemisia tabaci* and through infected planting material, is the most devastating disease of cassava in Africa and the Indian subcontinent. With an annual estimated economic loss of USD1.9–2.7 billion, it is considered the most damaging plant virus disease in the world [52]. The newly emerged cassava brown streak disease (CBSD) caused by two species of ipomoviruses, Cassava brown streak virus (CBSV) and Ugandan cassava brown streak virus (UCBSV), also transmitted by white fly, has become a serious threat to millions of subsistence farmers in Eastern and Central Africa. RNAi-based technology can be deployed for the simultaneous management of multiple viruses using hairpin probes with sequences from several viruses. This approach was used by Beyene et al. [53] to develop transgenic plants of the popular African cultivar 'TME204', expressing an inverted repeat construct derived from coat protein sequences from CBSV and UCBSV fused in tandem. The resulting transgenic plants showed robust resistance to both viruses while retaining the desirable agronomic characteristics of the cultivar preferred by Ugandan farmers, 'TME204' [53]. CBSD-tolerant GM cassava was approved for cultivation in Kenya in 2020 [54]. Although the disease resistance and safety of the cultivar has been tested, the release is still surrounded by skepticism and criticism [55]. A non-GM approach to mutation induction has also been attempted to develop mutants with a tolerance for CMD and CBSD. Field trials conducted in different agro-ecological regions in Kenya have revealed that the three mutants have better tolerance to these diseases than their respective parents [56]. On the other hand, an attempt to increase β carotene content by co-expression of transgenes for deoxy-d-xylulose-5-phosphate synthase and bacterial phytoene synthase in cassava resulted in reduced dry matter and starch content, despite a 15–20-fold increase in carotenoids [57].

Sweet potato (*Ipomoea batatas* (L.) Lam—Convolvulaceae) is the other most important root and tuber crop cultivated worldwide, ranking seventh overall in terms of production [57] and having considerable potential to reduce the Global Hunger Index, particularly in sub-Saharan Africa, the Pacific Islands, and parts of Asia [58]. Hexaploidy and self- and cross-incompatibility in sweet potato introduce difficulties in using both traditional breeding and genomic approaches for their improvement. Nevertheless, in recent times, next-generation sequencing, high-throughput genotyping, and phenotyping technologies have been applied to this crop, providing genomic tools and resources for its genetic improvement. The available genomic resources, databases, bioinformatic tools, and the current reference genome of sweet potato were recently reviewed by Yan et al. [46]. The improvement of sweet potato can now be fast-tracked thanks to the availability of efficient Agrobacterium transformation systems based on embryogenic suspension cultures [59] and via direct organogenesis using petiole explants [60], enabling, for example, the development of transgenic sweet potato with herbicide tolerance [61]. Biolistic transformation has also been successfully developed for this species [62]. Sweet potato feathery mottle virus (SPFMV), a *Potyvirus* in the family *Potyviridae*, is a devastating virus for sweet potato growers worldwide. Using the electroporation method of transformation, Okada et al. [63] introduced an expression vector harboring the coat protein of SPFMV and hygromycin phosphotransferase genes driven by cauliflower mosaic virus 35 S promoter into a popular sweet potato variety, 'Chikei 682-11'. Greenhouse testing of three independent transformants showed resistance to both primary and secondary infection by the virus, confirming the possibility of using coat-protein-mediated resistance to SPFMV [63].

Potato, cassava, and sweet potato, the three most important root and tuber crops worldwide, are featured in this Special Issue. Traditional hybridization-based potato breeding is cumbersome due to the tetraploid nature of cultivated potato (*Solanum tuberosum*) and its narrow genetic base. Starch content on a fresh and dry weight basis is an important breeding objective in potato breeding. Despite the common use of in vitro-produced microtubers in commercial production and germplasm conservation of potato, in vitro techniques remain in limited use as research tools for understanding the biochemical and molecular bases of the physiology of tubers or in breeding. Traits such as dormancy [64,65], cold-induced sweetening [66], and salinity tolerance [67] have been shown to be amenable to examination when using this system. In their paper published in this Special Issue, Adley et al. (contribution 10) used callus cultures to induce somaclonal variation in the variety 'Lady Rosetta' and screened 105 regenerants for starch content. They isolated a somaclonal variant with 42% and 61% higher fresh and dry weights, respectively. This somaclone had a 10% and 75% higher starch content based on the dry weight and average content per plant, respectively, compared to 'Lady Rosetta'. Molecular analysis using real-time PCR of the new variant named 'Ros 119' demonstrated upregulation of six starch-synthesis genes (contribution 10).

Cassava (*Manihot esculenta* Crantz) is the third largest source of food carbohydrates in the tropics after rice and maize. It is one of the most drought-tolerant food crops; hence, it is the staple crop in the poorest and most remote areas in Africa. With great variation in climatic conditions in the tropics, particularly with regard to rainfall and temperature, cultivars with stable high yields across environments are required, especially in large countries with varying climates. In this Special Issue, Amelework et al. (contribution 11) report the results of testing 11 advanced selections of cassava in six sites across South Africa. They analyzed of genotype and environment, and the effects of their interaction on fresh root yield (FRY) and dry matter content (DMC). The results revealed that the variation in percentage due to genotype x environment interaction was highest for FRY, whereas genotypic variation was the main contributor to the total variation in DMC. The authors identified two genotypes providing high DMC and FRY across all environments, and three sites that are representative of the variation in climatic conditions, suitable for variety evaluation and breeding.

Sweet potato (*Ipomoea batatas* (L.) Lam.) is a valuable source of carbohydrates, vitamins, fibers, and minerals, and is considered one of the most important crops in both tropical and subtropical climates. *Cylas formicarius* (Fabricius) and West Indian sweet potato weevil (*Euscepes postfasciatus* (Fairmaire)) are the most damaging pests of sweet potato in many continents including Central and South America, the South Pacific, and Japan. In the first comprehensive gene expression analysis during weevil infection in the resistant 'Kyushu No 166' cultivar published in this Special Issue, Nokihara et al. (contribution 12) show that genes related to phosphorylation, metabolic, and cellular processes, as well as terpenoid-related genes responsible for producing plant-derived juvenile hormones, are upregulated.

## 5. Industrial Crops

The only plantation crop that sustainably supplies natural rubber for aviation and other industries, as well as domestic uses, is the Pará rubber tree (*Hevea brasiliensis*); this tree originated in the Amazon, but was domesticated in Asia. As a result of domestication in a distant continent, Pará rubber tree populations have a very narrow genetic base in cultivation and are prone to many diseases. With 3–4 years from seed planting to flowering, 6–7 years to start tapping for rubber, and another 5–10 years required to assess yield, traditional breeding is a difficult and prolonged process. It takes, on average, three decades to complete the entire cycle of selection and release of new clones for planting. Therefore, marker-assisted selection and genetic transformation can accelerate breeding. Somatic embryo-based transformation has been developed for *Hevea*, and the advances made in this area have been discussed in detail in a recent review by Wang et al. [68]. The first draft of the *H. brasiliensis* genome was reported by Rahman et al. [69] in 2013. Their results indicated that 78% of the genome comprised repetitive DNA and 12.7% of the gene models unique to *Hevea*. Key genes associated with rubber biosynthesis, disease resistance, and allergenicity were identified [69]. Genome assembly of the popular rubber clone 'RRIM 600' revealed an expansion in the number of rubber-biosynthesis-related genes and their high expression in latex, explaining its high rubber yield [70]. This was further confirmed in the report by Tang et al. [71], who demonstrated the expansion of the *REF/SRPP* (rubber elongation factor/small rubber particle protein) gene family and its divergence. Using a high-density single-nucleotide polymorphism (SNP)-based map, Pootakham et al. [72] were able to anchor about two thirds of protein-coding genes into 18 linkage groups of the *H. brasiliensis* 'BPM 24' clone. Comparative analysis of the intragenomic homeologous synteny provided evidence for the presence of paleotetraploidy in the species. Chao et al. [73] demonstrated the relationship of increase in rubber yields during the domestication process with the increase in the number of laticifer rings and its high correlation with *HbPSK5* encoding the small-peptide hormone phytosulfokine—a key domestication gene of rubber. Thus, through genomic studies, our understanding of the expression of different traits of agronomic interest in rubber trees has increased. In a recent review, Priyadarshan [74] discussed the possible application of molecular markers to rubber plants in their juvenile phase to select for traits expressed after maturity using genomic selection. These studies will no doubt accelerate the breeding of new rubber clones with desired traits and improve the efficiency of breeding as well.

The three main diseases affecting rubber plantations worldwide are caused by *Phytophthora* spp. (causing shoot rot, abnormal leaf fall, patch canker, and black stripe diseases), *Corynespora cassiicola* (causing *Corynospora* leaf fall disease), and *Colletotrichum* spp. (causing *Colletotrichum* leaf fall disease). All these diseases reduce plant growth and latex yield, and are controlled using fungicides. Breeding for resistance using traditional hybridization and selection is practically impossible because of the high degree of heterozygosity in Pará rubber clones, thus requiring several backcrosses to introgress genes controlling disease resistance in this species with a long breeding cycle and the large land area required for screening such populations. Thus, early screening of breeding populations at the seedling stage can revolutionize breeding of this valuable species. Polymerase chain reaction (PCR) is a simple and rapid method that can detect nucleotide polymorphisms and sequence variation. When PCR reactions are conducted competitively in the presence of allele-specific primers to preferentially amplify only certain alleles, the variant is called allele-specific PCR (AS-PCR). Kompetitive Allele-Specific PCR (KASP) is a variant of AS-PCR modified for fluorescence-based detection of amplification results. In this Special Issue, Roy et al. (contribution 13) report the identification of 12 single nucleotide polymorphisms (SNPs) significantly associated with resistance against *Phytophthora*, *Corynespora*, and *Colletotrichum* in six linkage groups using an integrated linkage map of a $F_1$ progeny in an interspecific cross between *H. brasiliensis* ('RRII 105'—susceptible parent) and *H. benthamiana* ('F4542'—resistant parent) using 23,978 markers. To demonstrate the possible application of these findings in marker-assisted breeding of rubber for resistance to these diseases, the authors used KASP assays for all 12 SNPs that showed significant associations with the disease traits. When the KASP assays were applied to 178 'RRII' 105 × 'F4542' $F_1$ progeny, the genotypes could be clearly separated on the basis of resistance. Four $F_1$ plants were found to carry favorable alleles from *H. benthamiana* for all the three disease traits. They also predicted 41 key genes within proximity to those SNPs that were previously reported to be associated with disease resistance. This is the first report of the development of molecular markers for the three diseases, and this work has the potential to fast-track the breeding of disease-resistant Pará rubber.

## 6. New Crops for Arid Regions

The impact of climate change on the agro-forestry systems and the adaptive capacity of plants and animals will be of strategic importance in the immediate future to ensure food security. Numerous evidence suggests that reduced water availability and rising temperatures associated with global warming will have a significant impact on agriculture in the future [75]. Water is an essential component of agricultural production. According to UN and FAO data, approximately 3000–5000 L of water are needed to meet the daily food requirement of a person [76]. Furthermore, in the Global Risks Report of the World Economic Forum, water crises are stated as the third most important global risk in terms of impact on humanity [77].

Climate change has caused an increase in average temperatures and an ever-increasing demand for water. Furthermore, given that the demand for food production is likely to increase in the future [78], the challenge of sustainably producing food and non-food resources with organisms adapted to new environmental conditions will become of strategic interest. The application of biotechnology to drought-resistant crops would be a long-term solution for the production of more food with less water in increasingly warmer environments. An important contribution to achieving this objective comes from the use of cacti, known for their minimum water requirement; they have been grown extensively in arid lands, for food, feeds, and medicinal and therapeutic uses [79]. Cacti utilize Crassulacean Acid Metabolism (CAM) for photosynthesis, a unique process in desert plants that opens their stomata only at night when the plant is relatively cool, so that less moisture is lost through transpiration. Among the most interesting species, *Opuntia ficus-indica* (commonly known as prickly pear) represents an archetypal constitutive CAM species. In this Special Issue, Carra et al. (contribution 14) describe the use of the in vitro rescue of zygotic embryos for the genetic improvement of *O. ficus-indica*. Prickly pear cactus is an important forage and food source in arid and semiarid ecosystems, and is the most important cactus species cultivated globally. Both fruits and seeds have shown important antioxidant and nutritional properties, and can be a potential source of functional and nutraceutical ingredients. This crop is one of the most promising in the face of new environmental conditions due to climate change which will increasingly reduce the availability of water. In fact, it is able to produce fruits even in conditions where other crops cannot survive. The high degree of apomixis in the species is a hindrance in plant breeding programs where genetic segregation is sought for the selection of superior genotypes. Therefore, the protocols described for in vitro embryo rescue open a pathway to increase the availability of zygotic seedlings in *O. ficus-indica* breeding programs through in ovule embryo culture.

## 7. Ornamental Crops

The economic importance of ornamental plants has been increasing significantly in many countries with international demand expanding rapidly, providing many benefits to nature and humans both in urban and peri-urban areas. Ornamental plants, including cut flowers, foliage, and live plants, showed a positive trend in export growth, which led to an aggregate world value of around EUR 18 billion in 2020 [80]. Ornamental plants, cultivated both in indoor and outdoor environments, can contribute to human health and wellbeing, and can ensure essential environmental services (Figure 3), including the mitigation of the climate change, reduction in air and soil pollution, and providing food for habitants [81,82].

**Figure 3.** Ecosystem services and benefits obtained from ornamental plants in an urban and peri-urban area.

Ornamentals are a hugely diverse group of commercially significant plants that are grown and traded usually for decorative purposes, either as whole plants or for their parts. Among the ornamental plants, orchids have a special place due to their stunning displays of color and the shape of the flowers. Apart from their scientific fascination due to many unusual biological features, orchids account for a great part of the global floriculture trade. These are either traded as whole plants or cut flowers. Novelty is of great importance in the ornamental plant industry and many biotechnological approaches have applications in developing such novelties. In this Special Issue, such applications in two valuable orchid species of commercial importance have been described.

Polyploidy is much more pronounced in the plant kingdom than in the animal kingdom and has played a key role in plant evolution, species adaptation, and spread. Polyploids are more frequent among agricultural crops than in nature as they have many agronomic benefits such as larger size of organs, higher concentration of secondary metabolites, and better adaptation. Although polyploidization occurs in nature sporadically, in plant breeding it is artificially induced at higher frequencies. The development of new polyploid orchids often results in superior ornamental characteristics compared to their diploid counterparts. Orchids of the genus *Cattleya* are commercialized as hybrids. Although there are protocols for the polyploidization of *Cattelya* spp., there are no protocols for their interspecific or intergeneric hybrids, the widely commercialized types. In the current Special Issue, Vilcherrez-Atoche et al. (contribution 15) describe the use of in vitro cultured protocorms and seeds to induce polyploidy in *Brassolaeliocattleya,* a cross between three genera: *Brassavola*, *Laelia*, and *Cattleya* [83], using colchicine—an inhibitor of microtubule formation in the chromatic spindle resulting in the nondisjunction of chromosomes; this, in turn, results in the duplication of chromosomes within the nucleus. They report higher rates of polyploidization in protocorms and use flowcytometry to confirm the ploidy level in regenerants.

Dendrobium orchids are traded both as cut flowers and potted plants. They are among the top ten orchid taxa of commercial importance, with a wide variety of choices in flower color, texture, and shape, and a good vase life. However, two varieties account for 70% of the world trade, indicating a limited choice of varieties suited for the export market. Hence, there is an enormous potential for the right cultivars to break into the export market. However, compatibility barriers in intersectional crossing, negative genetic linkages in promising traits and prolonged juvenile phase and high mortality at hardening stages of in vitro cultures are barriers to improvement and commercialization of novel Dendrobiums. Induced mutations offer unique opportunities to improve elite cultivars by rectifying a defect such as late flowering, disease susceptibility or flower size [23]. The Dendrobium hybrid 'Emma White' is popular, but has a long juvenility after micropropagation. With the objective of developing an early flowering mutant of 'Emma White', Sherpa et al. (contribution 16) used gamma irradiation on protocorm-like bodies of this mutant and studied growth responses at different dosages and found optimal dose levels for producing high mutation rates with low mortality. By screening the mutant population, they were able to isolate a mutant with early flowering (294 days vs. 959 days in "Emma White'), demonstrating the value of in vitro mutagenesis in improving orchids.

## 8. Development of New Methodologies in Plant Biotechnology

Genetic transformation is a breakthrough biotechnology that has transformed agriculture in recent times, with 29 countries growing over 190.5 million ha of biotech crops. Although the number of countries is small, the impact of biotech crops is high, with most of North and South America, China, India, Australia, Indonesia, Vietnam, Myanmar, Pakistan, Bangladesh, the Philippines, and large countries in Africa (South Africa, Nigeria, Ethiopia, Kenya, Sudan, etc.) growing these crops. Thanks to improved traits such as insect resistance, growing these crops is more environmentally friendly, and the additional income from these crops is estimated at USD 225 billion for the 20-year period since 1996 thanks to the production of an additional 824 Mt of food, feed, and fiber [6]. Stable gene transformation systems and strong positive selection markers are imperative for developing transgenic plants. Co-cultivation of the host plant tissue in vitro along with Agrobacterium carrying the desired gene construct is the traditional method for transformation, and antibiotic- or herbicide-resistant genes inserted along with the desired gene/s are used as positive selection markers. Due to environmental and health concerns with such genes in plants, other markers such as β-glucuronidase or fluorescent protein markers are used, but they require destructive staining for former or expensive equipment to detect fluorescent cells for the latter option. Therefore, more robust and simple selection marker development for crop transformation is important. In this Special Issue, Lim et al. (contribution 17) report the development of a simple system for the selection of transgenic plants. They report the use of the R2R3 MYB transcription factor gene *CaAN2* from chili pepper (*Capsicum annuum*) for use as a visible selection marker with successful selection in both transient assays and in stable transformation, using tobacco as the model system. Transgenic tobacco plants harboring the chili pepper *CaAN2* readily promoted the accumulation of anthocyanin throughout the plant, allowing easy selection at the plant regeneration stage of the transformation experiment without the involvement of additional steps to identify the transgenic plants. The method has the potential to dramatically improve the efficiency of selection in plant genetic transformation, a key biotechnological approach for crop improvement.

**Author Contributions:** Conceptualization, R.P. and F.C.; methodology, R.P. and F.C.; software, R.P. and F.C.; resources, R.P. and F.C.; writing—original draft preparation, R.P. and F.C.; writing—review and editing, R.P. and F.C.; visualization, R.P. and F.C. All authors have read and agreed to the published version of the manuscript.

**Conflicts of Interest:** The author declares no conflicts of interest.

## Abbreviations

| Abbreviations | Full Name |
|---|---|
| AS-PCR | allele-specific PCR |
| Cas | Crispr associated protein |
| CAM | Crassulacean Acid Metabolism |
| CBSV | Cassava Brown Streak Virus; CMD—Cassava Mosaic Disease |
| CRISPR | Clustered Regularly Interspaced Short Palindromic Repeats |
| DMC | Dry Matter Content |
| DXS | Deoxy-d-xylulose-5-phosphate synthase |
| FAO | Food and Agriculture Organization of the United Nations |
| FRY | Fresh Root Yield |
| KASP | Kompetitive Allele-Specific PCR |
| PCR | Polymerase Chain Reaction |
| PR | Perennial Rice |
| QTL | Quantitative Trait Loci |
| SNP | Single-nucleotide Polymorphism |
| SPP | Single Plant Progeny |
| TALENs | Transcription Activator-Like Effector Nucleases TILLING—Targeting Induced Local Lesions in Genomes |
| UNGA | United Nations General Assembly |
| ZFN | Zinc Finger Nucleases |

**List of Contributions**

1. Nowicka, B. Modifications of Phytohormone Metabolism Aimed at Stimulation of Plant Growth, Improving Their Productivity and Tolerance to Abiotic and Biotic Stress Factors. *Plants* **2022**, *11*, 3430.
2. Hamdan, M.F.; Karlson, C.K.S.; Teoh, E.Y.; Lau, S.-E.; Tan, B.C. Genome Editing for Sustainable Crop Improvement and Mitigation of Biotic and Abiotic Stresses. *Plants* **2022**, *11*, 2625.
3. Buzdin, A.V.; Patrushev, M.V.; Sverdlov, E.D. Will Plant Genome Editing Play a Decisive Role in "Quantum-Leap" Improvements in Crop Yield to Feed an Increasing Global Human Population? *Plants* **2021**, *10*, 1667.
4. Zhao, Y.; Feng, M.; Paudel, D.; Islam, T.; Momotaz, A.; Luo, Z.; Zhao, Z.; Wei, N.; Li, S.; Xia, Q.; et al. Advances in Genomics Approaches Shed Light on Crop Domestication. *Plants* **2021**, *10*, 1571.
5. Prieto, P.; Palomino, C.; Cifuentes, Z.; Cabrera, A. Analysis of chromosome associations during early meiosis in wheat lines carrying chromosome introgressions from *Agropyron cristatum*. *Plants* **2021**, *10*, 2292, https://doi.org/10.3390/plants10112292.
6. Dreiseitl, A.; Nesvadba, Z. Powdery mildew resistance genes in single-plant progenies derived from accessions of a winter barley core collection. *Plants* **2021**, *10*, 1988, https://doi.org/10.3390/plants10101988.

7. Numan, M.; Khan, A.L.; Asaf, S.; Salehin, M.; Beyene, G.; Tadele, Z.; Ligaba-Osena, A. From Traditional Breeding to Genome Editing for Boosting Productivity of the Ancient Grain Tef [Eragrostis tef (Zucc.) Trotter]. *Plants* **2021**, *10*, 628.
8. Tigist, S.G.; Sibiya, J.; Amelework, A.; Keneni, G. Agromorphological and physiological performance of Ethiopian common bean (*Phaseolus vulgaris* L.) Genotypes Under Different Agroecological Conditions. *Plants* **2023**, *12*, 2342, https://doi.org/10.3390/plants12122342.
9. Purdy, S.J.; Fuentes, D.; Ramamoorthy, P.; Nunn, C.; Kaiser, B.N.; Merchant, A. The metabolic profile of young, watered chickpea plants can be used as a biomarker to predict seed number under terminal drought. *Plants* **2023**, *12*, 2172.
10. Adly, W.M.R.M.; Niedbała, G.; EL-Denary, M.E.; Mohamed, M.A.; Piekutowska, M.; Wojciechowski, T.; Abd El-Salam, E.-S.T.; Fouad, A.S. Somaclonal variation for genetic improvement of starch accumulation in potato (*Solanum tuberosum*) tubers. *Plants* **2023**, *12*, 232.
11. Amelework, A.B.; Bairu, M.W.; Marx, R.; Laing, M.; Venter, S.L. Genotype x Environment Interaction and stability analysis of selected cassava cultivars in South Africa. *Plants* **2023**, *12*, 2490.
12. Nokihara, K.; Okada, Y.; Ohata, S.; Monden, Y. Transcriptome analysis reveals key genes involved in weevil resistance in the hexaploid sweetpotato. *Plants* **2021**, *10*, 1535.
13. Roy, C.B.; Goonetilleke, S.N.; Joseph, L.; Krishnan, A.; Saha, T.; Kilian, A.; Mather, D.E. Analysis of genetic diversity and resistance to foliar pathogens based on genotyping-by-sequencing of a para rubber diversity panel and progeny of an interspecific cross. *Plants* **2022**, *11*, 3418.
14. Carra, A.; Catalano, C.; Pathirana, R.; Sajeva, M.; Inglese, P.; Motisi, A.; Carimi, F. Increased Zygote-Derived Plantlet Formation through In Vitro Rescue of Immature Embryos of Highly Apomictic *Opuntia ficus-indica* (Cactaceae). *Plants* **2023**, *12*, 2758.
15. Vilcherrez-Atoche, J.A.; Silva, J.C.; Clarindo, W.R.; Mondin, M.; Cardoso, J.C. In Vitro Polyploidization of Brassolaeliocattleya Hybrid Orchid. *Plants* **2023**, *12*, 281.
16. Sherpa, R.; Devadas, R.; Bolbhat, S.N.; Nikam, T.D.; Penna, S. Gamma Radiation Induced In-Vitro Mutagenesis and Isolation of Mutants for Early Flowering and Phytomorphological Variations in Dendrobium 'Emma White'. *Plants* **2022**, *11*, 3168.
17. Lim, S.-H.; Kim, D.-H.; Cho, M.-C.; Lee, J.-Y. Chili pepper *AN2* (*CaAN2*): A visible selection marker for nondestructive monitoring of transgenic plants. *Plants* **2022**, *11*, 820.

## References


1. FAO; IFAD; UNICEF; WFP; WHO. *The State of Food Security and Nutrition in the World 2023*; FAO: Rome, Italy, 2023.
2. Pathirana, R.; Carimi, F. Management and utilization of plant genetic resources for a sustainable agriculture. *Plants* **2022**, *11*, 2038. [CrossRef] [PubMed]
3. Pingali, P.L. Green Revolution: Impacts, limits, and the path ahead. *Proc. Natl. Acad. Sci. USA* **2012**, *109*, 12302–12308. [CrossRef] [PubMed]
4. UNDESA. UN/DESA Policy Brief #102: Population, Food Security, Nutrition and Sustainable Development. Available online: https://www.un.org/development/desa/dpad/publication/un-desa-policy-brief-102-population-food-security-nutrition-and-sustainable-development/ (accessed on 19 March 2024).
5. Hemathilake, D.M.K.S.; Gunathilake, D.M.C.C. Chapter 31—Agricultural productivity and food supply to meet increased demands. In *Future Foods*; Bhat, R., Ed.; Academic Press: Cambridge, MA, USA, 2022; pp. 539–553.
6. Wijerathna-Yapa, A.; Pathirana, R. Sustainable agro-food systems for addressing climate change and food security. *Agriculture* **2022**, *12*, 1554. [CrossRef]
7. Gill, G.P.; Harvey, C.F.; Gardner, R.C.; Fraser, L.G. Development of sex-linked PCR markers for gender identification in *Actinidia*. *Theor. Appl. Genet.* **1998**, *97*, 439–445. [CrossRef]
8. Guo, D.; Wang, R.; Fang, J.; Zhong, Y.; Qi, X. Development of sex-linked markers for gender identification of Actinidia arguta. *Sci. Rep.* **2023**, *13*, 12780. [CrossRef] [PubMed]
9. Bharadwaj, C.; Jorben, J.; Rao, A.; Roorkiwal, M.; Patil, B.S.; Jayalakshmi; Ahammed, S.K.; Saxena, D.R.; Yasin, M.; Jahagirdar, J.E.; et al. Development of high yielding fusarium wilt resistant cultivar by pyramiding of "genes" through marker-assisted backcrossing in chickpea (*Cicer arietinum* L.). *Front. Genet.* **2022**, *13*, 924287. [CrossRef] [PubMed]
10. Mei, J.; Shao, C.; Yang, R.; Feng, Y.; Gao, Y.; Ding, Y.; Li, J.; Qian, W. Introgression and pyramiding of genetic loci from wild *Brassica oleracea* into *B. napus* for improving Sclerotinia resistance of rapeseed. *Theor. Appl. Genet.* **2020**, *133*, 1313–1319. [CrossRef]
11. Benitez-Alfonso, Y.; Soanes, B.K.; Zimba, S.; Sinanaj, B.; German, L.; Sharma, V.; Bohra, A.; Kolesnikova, A.; Dunn, J.A.; Martin, A.C.; et al. Enhancing climate change resilience in agricultural crops. *Curr. Biol.* **2023**, *33*, R1246–R1261. [CrossRef]
12. Crews, T.E.; Carton, W.; Olsson, L. Is the future of agriculture perennial? Imperatives and opportunities to reinvent agriculture by shifting from annual monocultures to perennial polycultures. *Glob. Sustain.* **2018**, *1*, e11. [CrossRef]
13. Soto-Gómez, D.; Pérez-Rodríguez, P. Sustainable agriculture through perennial grains: Wheat, rice, maize, and other species. A review. *Agric. Ecosyst. Environ.* **2022**, *325*, 107747. [CrossRef]

14. Baik, B.-K.; Hoagland, L.A.; Jones, S.S.; Murphy, K.M.; Reeves, P.G. Nutritional and quality characteristics expressed in 31 perennial wheat breeding lines. *Renew. Agric. Food Syst.* **2009**, *24*, 285–292. [CrossRef]
15. Zhang, X.; Sallam, A.; Gao, L.; Kantarski, T.; Poland, J.; DeHaan, L.R.; Wyse, D.L.; Anderson, J.A. Establishment and optimization of genomic selection to accelerate the domestication and improvement of intermediate wheatgrass. *Plant Genome* **2016**, *9*, 1–18. [CrossRef]
16. Dorn, K.M.; Poland, J. *Thinopyrum intermedium* v2.1. Available online: https://phytozome-next.jgi.doe.gov/info/Tintermedium_v2_1 (accessed on 23 January 2024).
17. Zhang, S.; Huang, G.; Zhang, Y.; Lv, X.; Wan, K.; Liang, J.; Feng, Y.; Dao, J.; Wu, S.; Zhang, L.; et al. Sustained productivity and agronomic potential of perennial rice. *Nat. Sustain.* **2023**, *6*, 28–38. [CrossRef]
18. Yan, X.; Cheng, M.; Li, Y.; Wu, Z.; Li, Y.; Li, X.; He, R.; Yang, C.; Zhao, Y.; Li, H.; et al. Tripsazea, a Novel Trihybrid of Zea mays, Tripsacum dactyloides, and Zeaperennis. *G3 Genes | Genomes | Genet.* **2020**, *10*, 839–848. [CrossRef]
19. Kruglova, N.N.; Zinatullina, A.E. In Vitro Culture of Autonomous Embryos as a Model System for the Study of Plant Stress Tolerance to Abiotic Factors (on the Example of Cereals). *Biol. Bull. Rev.* **2022**, *12*, 201–211. [CrossRef]
20. Pathirana, R.; Wijithawarna, W.A.; Jagoda, K.; Ranawaka, A.L. Selection of rice for iron toxicity tolerance through irradiated caryopsis culture. *Plant Cell Tissue Organ Cult.* **2002**, *70*, 83–90.
21. Abate, E.; Hussien, S.; Laing, M.; Mengistu, F. Aluminium toxicity tolerance in cereals: Mechanisms, genetic control and breeding methods. *Afr. J. Agric. Res.* **2013**, *8*, 711–722. [CrossRef]
22. Krishania, S. *Abiotic Stress Tolerance in Eleusine coracana (L.) Gaertn by Manipulation of Tissue Culture Media*; IIS University: Jaipur, India, 2010.
23. Pathirana, R. Mutations in plant evolution, crop domestication and breeding. *Trop. Agric. Res. Ext.* **2021**, *24*, 124–157. [CrossRef]
24. Acevedo-Garcia, J.; Spencer, D.; Thieron, H.; Reinstädler, A.; Hammond-Kosack, K.; Phillips, A.L.; Panstruga, R. *mlo*-based powdery mildew resistance in hexaploid bread wheat generated by a non-transgenic TILLING approach. *Plant Biotechnol. J.* **2017**, *15*, 367–378. [CrossRef]
25. Bigini, V.; Camerlengo, F.; Botticella, E.; Sestili, F.; Savatin, D.V. Biotechnological Resources to Increase Disease-Resistance by Improving Plant Immunity: A Sustainable Approach to Save Cereal Crop Production. *Plants* **2021**, *10*, 1146. [CrossRef]
26. Ceoloni, C.; Kuzmanovic, L.; Forte, P.; Virili, M.E.; Bitti, A. Wheat-perennial Triticeae introgressions: Major achievements and prospects. In *Alien Introgression in Wheat: Cytogenetics, Molecular Biology, and Genomics*; Molnár-Láng, M., Ceoloni, C., Doležel, J., Eds.; Springer International Publishing: Cham, Switzerland, 2015; pp. 273–313.
27. FAO. International Year of Pulses 2016. Available online: https://www.fao.org/pulses-2016/en/ (accessed on 22 November 2023).
28. FAO. World Pulse Day/10 February. Available online: https://www.fao.org/world-pulses-day/en/ (accessed on 22 November 2023).
29. FAO. *FAOSTAT*; Food and Agriculture Organisation of the United Nations: Rome, Italy, 2020.
30. Bernard, R.L. Two genes affecting stem termination in soybeans. *Crop Sci.* **1972**, *12*, 235–239. [CrossRef]
31. Krylova, E.A.; Khlestkina, E.K.; Burlyaeva, M.O.; Vishnyakova, M.A. Determinate growth habit of grain legumes: Role in domestication and selection, genetic control. *EcoGen* **2020**, *18*, 43–58. [CrossRef]
32. Sharma, K.K.; Dumbala, S.R.; Bhatnagar-Mathur, P. Biotech Approaches for Crop Improvement in the Semi-arid Tropics. In *Plant Biotechnology: Experience and Future Prospects*; Ricroch, A., Chopra, S., Fleischer, S.J., Eds.; Springer International Publishing: Cham, Switzerland, 2014; pp. 193–207.
33. Raman, R.B.; Sinhamahapatra, S.P. A dwarf determinate plant type for achieving higher and stable yield in blackgram (*Vigna mungo* L. Hepper). *Bioscan* **2014**, *9*, 497–500.
34. Duc, G.; Agrama, H.; Bao, S.; Berger, J.; Bourion, V.; De Ron, A.M.; Gowda, C.L.L.; Mikic, A.; Millot, D.; Singh, K.B.; et al. Breeding annual grain legumes for sustainable agriculture: New methods to approach complex traits and target new cultivar ideotypes. *Crit. Rev. Plant Sci.* **2015**, *34*, 381–411. [CrossRef]
35. IAEA. International Atomic Energy Agency, Mutant Database. Available online: https://nucleus.iaea.org/sites/mvd/SitePages/Home.aspx (accessed on 20 January 2024).
36. Bohra, A.; Singh, N.P. Whole genome sequences in pulse crops: A global community resource to expedite translational genomics and knowledge-based crop improvement. *Biotechnol. Lett.* **2015**, *37*, 1529–1539. [CrossRef] [PubMed]
37. Jacob, C.; Carrasco, B.; Schwember, A.R. Advances in breeding and biotechnology of legume crops. *Plant Cell Tiss Organ Cult* **2016**, *127*, 561–584. [CrossRef]
38. Jegadeesan, S.; Raizada, A.; Dhanasekar, P.; Suprasanna, P. Draft genome sequence of the pulse crop blackgram [*Vigna mungo* (L.) Hepper] reveals potential R-genes. *Sci. Rep.* **2021**, *11*, 11247. [CrossRef] [PubMed]
39. Varshney, R.K.; Pandey, M.K.; Bohra, A.; Singh, V.K.; Thudi, M.; Saxena, R.K. Toward the sequence-based breeding in legumes in the post-genome sequencing era. *Theor. Appl. Genet.* **2019**, *132*, 797–816. [CrossRef] [PubMed]
40. Kumar, A.; Bohra, A.; Mir, R.R.; Sharma, R.; Tiwari, A.; Khan, M.A.; Varshney, R.K. Next generation breeding in pulses: Present status and future directions. *Crop Breed. Appl. Biotechnol.* **2021**, *21*, 1–22. [CrossRef]
41. Asati, R.; Tripathi, M.K.; Tiwari, S.; Yadav, R.K.; Tripathi, N. Molecular breeding and drought tolerance in chickpea. *Life* **2022**, *12*, 1846. [CrossRef]

42. Diaz, L.M.; Ricaurte, J.; Tovar, E.; Cajiao, C.; Teran, H.; Grajales, M.; Polania, J.; Rao, I.; Beebe, S.; Raatz, B. QTL analyses for tolerance to abiotic stresses in a common bean (*Phaseolus vulgaris* L.) population. *PLoS ONE* **2018**, *13*, e0202342. [CrossRef]
43. Sultana, R.; Choudhary, A.K.; Pal, A.K.; Saxena, K.B.; Prasad, B.D.; Singh, R. Abiotic stresses in major pulses: Current status and strategies. In *Approaches to Plant Stress and their Management*; Gaur, R.K., Sharma, P., Eds.; Springer: New Delhi, India, 2014; pp. 173–190.
44. Koundinya, A.V.V.; Das, A.; Hegde, V. Mutation breeding in tropical root and ruber crops. In *Mutation Breeding for Sustainable Food Production and Climate Resilience*; Penna, S., Jain, S.M., Eds.; Springer Nature: Singapore, 2023; pp. 779–809.
45. Bradshaw, J.E. A Brief History of the Impact of Potato Genetics on the Breeding of Tetraploid Potato Cultivars for Tuber Propagation. *Potato Res.* **2022**, *65*, 461–501. [CrossRef]
46. Yan, M.; Nie, H.; Wang, Y.; Wang, X.; Jarret, R.; Zhao, J.; Wang, H.; Yang, J. Exploring and exploiting genetics and genomics for sweetpotato improvement: Status and perspectives. *Plant Commun.* **2022**, *3*, 100332. [CrossRef] [PubMed]
47. FAO. What is Hidden Hunger. In Proceedings of the Second International Confefrence on Nutrition, Rome, Italy, 19–21 November 2014. Available online: https://www.fao.org/about/meetings/icn2/news-archive/news-detail/en/c/265240 (accessed on 22 January 2024).
48. Howeler, R.; Lutaladio, N.; Thomas, G. *Save and Grow: Cassava. A Guide to Sustainable Production Intensification*; FAO: Rome, Italy, 2013.
49. Chavez, A.L.; Bedoya, J.; Sánchez, T.; Iglesias, C.; Ceballos, H.; Roca, W. Iron, carotene, and ascorbic acid in cassava roots and leaves. *Food Nutr. Bull.* **2000**, *21*, 410–413.
50. Narayanan, N.; Beyene, G.; Chauhan, R.D.; Gaitán-Solís, E.; Gehan, J.; Butts, P.; Siritunga, D.; Okwuonu, I.; Woll, A.; Jiménez-Aguilar, D.M.; et al. Biofortification of field-grown cassava by engineering expression of an iron transporter and ferritin. *Nat. Biotechnol.* **2019**, *37*, 144–151. [CrossRef] [PubMed]
51. Landi, M.; Shah, T.; Falquet, L.; Niazi, A.; Stavolone, L.; Bongcam-Rudloff, E.; Gisel, A. Haplotype-resolved genome of heterozygous African cassava cultivar TMEB117 (*Manihot esculenta*). *Sci. Data* **2023**, *10*, 887. [CrossRef]
52. Patil, B.L.; Fauquet, C.M. Cassava mosaic geminiviruses: Actual knowledge and perspectives. *Mol. Plant Pathol.* **2009**, *10*, 685–701. [CrossRef]
53. Beyene, G.; Chauhan, R.D.; Ilyas, M.; Wagaba, H.; Fauquet, C.M.; Miano, D.; Alicai, T.; Taylor, N.J. A virus-derived stacked RNAi construct confers robust resistance to cassava brown streak disease. *Front. Plant Sci.* **2017**, *7*, 240414.
54. Mmbando, G.S. The legal aspect of the current use of genetically modified organisms in Kenya, Tanzania, and Uganda. *GM Crops Food* **2023**, *14*, 1–12. [CrossRef]
55. ACB. Push Back against Risky and Unsafe RNAi GM Cassava Cultivation in Kenya. Available online: https://acbio.org.za/gm-biosafety/push-back-against-risky-unsafe-rnai-gm-cassava-cultivation-kenya/ (accessed on 14 February 2024).
56. Kimno, L.C.; Kinyua, M.G.; Bimpong, I.K.; Were, J.O. Field screening of elite cassava (*Manihot esculenta*) mutant lines for their response to mosaic and brown streak viruses. *J. Exp. Agric. Int.* **2023**, *45*, 205–215. [CrossRef]
57. Sapakhova, Z.; Raissova, N.; Daurov, D.; Zhapar, K.; Daurova, A.; Zhigailov, A.; Zhambakin, K.; Shamekova, M. Sweet potato as a key crop for food security under the conditions of global climate change: A Review. *Plants* **2023**, *12*, 2516. [CrossRef] [PubMed]
58. Anil, S.R.; Sheela, M.N.; Visalakshi Chandra, C.; Krishna Radhika, N.; Asha, K.I. Status of biofortification in tropical root and tuber crops. *Curr. Sci.* **2023**, *120*, 169–175.
59. Yang, J.; Bi, H.P.; Fan, W.J.; Zhang, M.; Wang, H.X.; Zhang, P. Efficient embryogenic suspension culturing and rapid transformation of a range of elite genotypes of sweet potato (*Ipomoea batatas* [L.] Lam.). *Plant Sci.* **2011**, *181*, 701–711. [CrossRef] [PubMed]
60. Lou, H.R.; Maria, M.S.; Benavides, J.; Zhang, D.P.; Zhang, Y.Z.; Ghislain, M. Rapid genetic transformation of sweetpotato (*Ipomoea batatas* (L.) Lam) via organogenesis. *Afr. J. Biotechnol.* **2006**, *5*, 1851–1857.
61. Choi, H.J.; Chandrasekhar, T.; Lee, H.-Y.; Kim, K.-M. Production of herbicide-resistant transgenic sweet potato plants through *Agrobacterium tumefaciens* method. *Plant Cell Tissue Organ Cult.* **2007**, *91*, 235–242. [CrossRef]
62. Yang, K.Y.; Yi, G.B.; Kim, Y.M.; Choi, G.; Shin, B.C.; Shin, Y.M.; Kim, K.M. Production of transgenic sweet potato (*Ipomoea batatas* (L.) Lam.) lines via microprojectile bombardment. *Korean J. Breed.* **2005**, *37*, 236–240.
63. Okada, Y.; Saito, A.; Nishiguchi, M.; Kimura, T.; Mori, M.; Hanada, K.; Sakai, J.; Miyazaki, C.; Matsuda, Y.; Murata, T. Virus resistance in transgenic sweetpotato [*Ipomoea batatas* L. (Lam)] expressing the coat protein gene of sweet potato feathery mottle virus. *Theor. Appl. Genet.* **2001**, *103*, 743–751. [CrossRef]
64. Leclerc, Y.; Donnelly, D.J.; Coleman, W.K.; King, R.R. Microtuber dormancy in three potato cultivars. *Am. Potato J.* **1995**, *72*, 215–223.
65. Suttle, J.C. Involvement of ethylene in potato microtuber dormancy. *Plant Physiol.* **1998**, *118*, 843–848.
66. Pathirana, R.; Harris, J.C.; McKenzie, M.J. A comparison of microtubers and field-grown tubers of potato (*Solanum tuberosum* L.) for hexoses, sucrose and their ratios following postharvest cold storage. *Postharvest Biol. Technol.* **2008**, *49*, 180–184. [CrossRef]
67. Zhang, Z.; Mao, B.; Li, H.; Zhou, W.; Takeuchi, Y.; Yoneyama, K. Effect of salinity on physiological characteristics, yield and quality of microtubers in vitro in potato. *Acta Physiol. Plant.* **2005**, *27*, 481–489.
68. Wang, X.; Gu, X.; Xu, Z.; Yin, Z.; Yang, X.; Lin, R.; Zhou, Q.; Huang, H.; Huang, T. Current achievements and future challenges of genotype-dependent somatic embryogenesis techniques in *Hevea brasiliensis*. *Forests* **2023**, *14*, 1891. [CrossRef]
69. Rahman, A.Y.A.; Usharraj, A.O.; Misra, B.B.; Thottathil, G.P.; Jayasekaran, K.; Feng, Y.; Hou, S.; Ong, S.Y.; Ng, F.L.; Lee, L.S.; et al. Draft genome sequence of the rubber tree *Hevea brasiliensis*. *BMC Genom.* **2013**, *14*, 75. [CrossRef]

70. Lau, N.-S.; Makita, Y.; Kawashima, M.; Taylor, T.D.; Kondo, S.; Othman, A.S.; Shu-Chien, A.C.; Matsui, M. The rubber tree genome shows expansion of gene family associated with rubber biosynthesis. *Sci. Rep.* **2016**, *6*, 28594. [CrossRef]
71. Tang, C.; Yang, M.; Fang, Y.; Luo, Y.; Gao, S.; Xiao, X.; An, Z.; Zhou, B.; Zhang, B.; Tan, X.; et al. The rubber tree genome reveals new insights into rubber production and species adaptation. *Nat. Plants* **2016**, *2*, 16073. [CrossRef]
72. Pootakham, W.; Sonthirod, C.; Naktang, C.; Ruang-Areerate, P.; Yoocha, T.; Sangsrakru, D.; Theerawattanasuk, K.; Rattanawong, R.; Lekawipat, N.; Tangphatsornruang, S. De novo hybrid assembly of the rubber tree genome reveals evidence of paleotetraploidy in *Hevea* species. *Sci. Rep.* **2017**, *7*, 41457. [CrossRef] [PubMed]
73. Chao, J.; Wu, S.; Shi, M.; Xu, X.; Gao, Q.; Du, H.; Gao, B.; Guo, D.; Yang, S.; Zhang, S.; et al. Genomic insight into domestication of rubber tree. *Nat. Commun.* **2023**, *14*, 4651. [CrossRef] [PubMed]
74. Priyadarshan, P.M. Molecular markers to devise predictive models for juvenile selection in *Hevea* rubber. *Plant Breed.* **2022**, *141*, 159–183. [CrossRef]
75. Hajek, O.L.; Knapp, A.K. Shifting seasonal patterns of water availability: Ecosys-tem responses to an unanticipated dimension of climate change. *New Phytol.* **2022**, *233*, 119–125. [CrossRef]
76. FAO. Food and Agriculture Organization, Water. Available online: https://www.fao.org/water/en/ (accessed on 20 January 2024).
77. WEF. *The World Economic Forum's Global Risks Report 2021. The Global Risks Report 2021*, 16th ed. 2021; ISBN: 978-2-940631-24-7. Available online: http://www3.weforum.org/docs/WEF_The_Global_Risks_Report_2021.pdf (accessed on 22 November 2023).
78. Mishra, B.K.; Kumar, P.; Saraswat, C.; Chakraborty, S.; Gautam, A. Water security in a changing environment: Concept, challenges and solutions. *Water* **2021**, *13*, 490. [CrossRef]
79. Shedbalkar, U.U.; Adki, V.S.; Jadhav, J.P.; Bapat, V.A. Opuntia and other cacti: Applications and biotechnological insights. *Trop. Plant Biol.* **2010**, *3*, 136–150.
80. Gabellini, S.; Scaramuzzi, S. Evolving consumption trends, marketing strategies, and governance settings in ornamental horticulture: A grey literature review. *Horticulturae* **2022**, *8*, 234. [CrossRef]
81. Francini, A.; Romano, D.; Toscano, S.; Ferrante, A. The contribution of ornamental plants to urban ecosystem services. *Earth* **2022**, *3*, 1258–1274. [CrossRef]
82. Rihn, A.L.; Knuth, M.J.; Behe, B.K.; Hall, C.R. Benefit information's impact on ornamental plant value. *Horticulturae* **2023**, *9*, 740. [CrossRef]
83. Cardoso, J.C.; Martinelli, A.P.; Teixeira da Silva, J.A. A novel approach for the selection of *Cattleya* hybrids for precocious and season-independent flowering. *Euphytica* **2016**, *210*, 143–150. [CrossRef]

*plants*

MDPI

*Review*

# Modifications of Phytohormone Metabolism Aimed at Stimulation of Plant Growth, Improving Their Productivity and Tolerance to Abiotic and Biotic Stress Factors

**Beatrycze Nowicka**

Department of Plant Physiology and Biochemistry, Faculty of Biochemistry, Biophysics and Biotechnology, Jagiellonian University, Gronostajowa 7, 30-387 Kraków, Poland; beatrycze.nowicka@uj.edu.pl

**Abstract:** Due to the growing human population, the increase in crop yield is an important challenge for modern agriculture. As abiotic and biotic stresses cause severe losses in agriculture, it is also crucial to obtain varieties that are more tolerant to these factors. In the past, traditional breeding methods were used to obtain new varieties displaying demanded traits. Nowadays, genetic engineering is another available tool. An important direction of the research on genetically modified plants concerns the modification of phytohormone metabolism. This review summarizes the state-of-the-art research concerning the modulation of phytohormone content aimed at the stimulation of plant growth and the improvement of stress tolerance. It aims to provide a useful basis for developing new strategies for crop yield improvement by genetic engineering of phytohormone metabolism.

**Keywords:** phytohormones; transgenic plants; biotic stress; abiotic stress; growth regulators

**Citation:** Nowicka, B. Modifications of Phytohormone Metabolism Aimed at Stimulation of Plant Growth, Improving Their Productivity and Tolerance to Abiotic and Biotic Stress Factors. *Plants* **2022**, *11*, 3430. https://doi.org/10.3390/plants11243430

Academic Editors: Ranjith Pathirana and Francesco Carimi

Received: 6 November 2022
Accepted: 6 December 2022
Published: 8 December 2022



## 1. Introduction

Due to the constantly growing human population, ensuring high crop productivity is an important challenge for 21st century agriculture. The research aimed at obtaining high-yielding varieties is being carried out [1]. Another important issue is to obtain varieties displaying enhanced tolerance to biotic and abiotic stresses that cause significant loss of yield. Among the abiotic stresses, the most important are drought, thermal stress (too high or too low temperature), light stress, salt stress, and stress caused by environmental pollution, e.g., by heavy metal ions. Due to anthropogenic climate change, an increase in abiotic-stress-evoked losses of crop yield is expected in the near future [2]. Apart from abiotic factors, the biotic ones, such as pathogens, competing plants, parasites, and herbivores, also limit plant growth and productivity.

The application of mineral fertilizers, herbicides, and pesticides, as well as growing high-yielding varieties obtained via traditional breeding methods, enabled a significant increase in crop productivity during the second half of the 20th century [3]. For example, the average cereal yield in 1951 was 1.2 t/ha, while in 1993 it was 2.3 t/ha [4]. However, this yield increase has slowed down in the 21st century. It is currently believed that for the most important crop species, further increases in their productivity obtained by traditional breeding methods are possible only to a small extent. For this reason, research based on genetic engineering became crucial for the future of agriculture [1]. In addition to experiments on transgenic organisms, extensive genome analyses of major crop species are also being carried out. Their goal is to identify quantitative trait loci (QTLs), which are genes determining quantitative traits [5].

Research on transgenic plants conducted over the past decades resulted in the development of various strategies of genetic modification aimed at obtaining lines with increased yield or improved tolerance to stress [6,7]. One of the promising research directions is associated with the modulation of phytohormone levels [2]. Phytohormones participate in the regulation of plant growth and development. They also play a role in response to environmental factors. These compounds include auxins, cytokinins, gibberellins, abscisic acid (ABA), ethylene, jasmonic acid (JA) and its derivatives, and brassinosteroids. Auxins, cytokinins, gibberellins, and brassinosteroids are considered particularly important for the regulation of plant growth and development, while JA, ABA, and ethylene play crucial roles in stress response. However, one needs to remember that growth-stimulating hormones participate in stress responses, while those primarily associated with the stress response are also involved in the regulation of various plant developmental processes, such as dormancy, fruit maturation, or senescence [8]. Plant hormones have pleiotropic effects. Furthermore, the result of their action often depends on cross-talk between various phytohormones and signaling molecules [9]. Phytohormones occur in plants at very low concentrations; their biosynthesis and degradation are strictly regulated. In some cases, reversible inactivation by conjugate formation is also possible [10].

This review presents the current state of research on the modulation of phytohormone content aimed at obtaining transgenic plants with traits favorable for the breeder, such as increased yield and improved tolerance to abiotic and biotic stress factors.

## 2. Strategies Applied in Phytohormone-Targeted Genetic Engineering

The research aimed at improving crop performance by modification of phytohormone metabolism and signaling starts with identification of the crucial genes. This is possible mainly due to the studies carried out on mutants or by comparing crop varieties displaying desirable traits with the other ones [10]. Gene and genome sequencing enables the identification of loci crucial for the observed effects. Analyses of phenotypes and detailed analyses at the biochemical level, i.e., determination of phytohormone content, enable scientists to discover gene functions. When the sequence and function of its product are known, bioinformatics provides tools to find homologues in other species. At this point, the plant transformation can be carried out to increase or decrease the level of a certain phytohormone. The increase in hormone level can be achieved by the overexpression of the gene encoding enzyme participating in the phytohormone biosynthetic pathway or silencing of the gene whose product catalyzes hormone degradation. The decrease can be achieved by silencing of the gene crucial for phytohormone biosynthesis or by overexpression of the gene whose product is involved in hormone degradation. The manipulation of the genes encoding enzymes carrying out phytohormone conjugation was also carried out [10]. Sometimes, the increase in phytohormone level may be achieved by the enhanced production of an enzyme catalyzing the formation of a metabolite that serves as a phytohormone precursor (for example, xanthophyll precursors of ABA biosynthesis) or a cofactor needed by the hormone-synthetizing enzyme (for example, molybdenum cofactor required for abscisic aldehyde oxidase activity) (see subchapter concerning ABA). The research on the engineering of phytohormone transport was also carried out (see subchapter about auxins). The significant progress in our understanding of phytohormone signaling opens a wide range of possibilities, as various elements of signaling cascades, transcription factors, and miRNAs are emerging targets for potential modifications. These strategies have been mentioned, but their detailed description is beyond the scope of the present review. In the early research, scientists used strong, constitutive promoters to provide the overexpression of desired genes. The discovery of tissue-specific, developmental-stage-specific, and stress-responsive promoters enabled the improved control of the time and site of transgene expression [10]. Furthermore, artificial promoters have been developed. Considering gene silencing, various constructs may be applied, including antisense sequences, 3′-untranslated regions, and hairpin constructs. The recent development of the CRISPR/Cas9 system paved the way for extensive genome editing (see subchapter about future perspectives).

## 3. Auxins

Auxins play a key role in the regulation of plant growth and development, therefore, modulation of their biosynthesis and signaling has been a subject of intensive research [11].

The most important auxin is indole-3-acetic acid (IAA). The substrate for its synthesis is tryptophan, but this amino acid can be transformed in various ways, all of them leading to the same final product. One of the known pathways of IAA synthesis involves two steps: tryptophan oxidation to indole-3-pyruvate, followed by oxidation of indole-3-pyruvate to IAA. The second reaction is catalyzed by monooxygenase encoded by *YUCCA* genes. Alternatively, indole-3-pyruvate can be converted to indole-3-acetaldehyde and then to IAA. Another known IAA biosynthetic pathway leads through indole-3-acetamide. There is also a pathway specific to the Brassicaceae family, for which the intermediate is indole-3-acetonitrile [12]. The concentration of auxins depends on the synthesis, degradation, and transport of this phytohormone; it is also regulated by conjugation. In the latter case, the important role is fulfilled by enzymes encoded by the *GH3* family of genes [13].

The results of the experiments carried out on transgenic lines with changed auxin content are summarized in Table 1. In terms of practical application, the modification of auxin levels in developing flowers seems to be the most promising direction. This effect is achieved by the expression of bacterial gene encoding tryptophan monooxygenase (e.g., *iaaM* from *Pseudomonas syringae pv. savastanoi*) under the control of an ovule-specific promoter. Increased auxin content stimulates the growth of generative shoots and fruits of transgenic plants. The application of tissue-specific promoters is better than the use of constitutive ones because the enhanced auxin synthesis occurring in whole plants often leads to undesirable developmental disorders (Table 1). Apart from the genes directly involved in auxin biosynthesis and degradation, the genes whose products play regulatory roles can also be targets of manipulation. It was observed that the expression of *OsIAA6* was highly induced by drought stress. Transgenic rice with overexpression of this gene under the control of a constitutive promoter displayed enhanced expression of *YUCCA* genes and was more tolerant to drought [14].

The research carried out on mutants is also important for better understanding of auxin roles. *Arabidopsis thaliana* mutant *yuc7-1D*, with the altered gene *YUCCA7*, displayed a phenotype characteristic for plants with auxin overproduction: tall stems and curled, narrow leaves. It was also more tolerant to drought when compared to control plants [15]. *A. thaliana* mutant *arf2* producing inactive Auxin Response Factor protein (ARF) developed longer and thicker flower shoots, larger and darker leaves, and larger seeds [16]. The *brachytic2* (*br2*) maize mutant with impaired auxin transport in the stem had shorter internodes. This observation may be of practical significance because dwarfism is a desirable trait of cereals as it provides a more favorable ratio of grain biomass to shoot biomass and increases lodging resistance [17]. The modification of soybean *GmPIN1* using CRISPR/Cas9 method resulted in plants displaying changed architecture [18].

**Table 1.** Summary of the results of the experiments on transgenic plants with changed auxin concentration or transport. MDA, malonyldialdehyde; RWC, relative water content.

| Protein | Promoter and Gene | Species | Phenotype of Transgenic Plants Compared to Control Lines | References |
|---|---|---|---|---|
| indole-3-pyruvate monooxygenase | *35S::AtYUCCA6* | potato (*Solanum tuberosum*) | - changed morphology: increased height, erect leaves, narrow downward-curled leaves<br>- increased longevity<br>- increased drought tolerance: far less pronounced wilting symptoms, increased water content in the leaves of stress-exposed plants (plants not watered for 18 days), ability to recover after rewatering while control plants were dying<br>- decreased tuber biomass per plant | [19] |

**Table 1.** *Cont.*

| Protein | Promoter and Gene | Species | Phenotype of Transgenic Plants Compared to Control Lines | References |
|---|---|---|---|---|
| | *SWPA2::AtYUCCA6* oxidative stress-induced promoter | poplar (*Populus alba* × *P. glandulosa*) | - rapid shoot growth, retarded main root development, increased root hair formation<br>- disturbed leaf morphology: folded leaves, elongated petioles, long internodes<br>- delayed hormone- and dark-induced senescence of detached leaves<br>- increased drought tolerance: less pronounced symptoms of wilting (plants not watered for 6 days) | [20] |
| | *SWPA2::AtYUCCA6* | sweet potato (*Ipomea batatas*) | - changed morphology: narrow, downward-curled leaves, increased height, increased node number<br>- increased drought tolerance: higher RWC and lower content of MDA in stressed plants (plants not watered for 16 days)<br>- lower storage root formation | [21] |
| tryptophan monooxygenase from bacteria | *DefH9::iaaM* ovule-specific promoter | tomato (*Lycopersicon esculentum*) | - parthenocarpic fruit development | [22] |
| | *DefH9::iaaM* | raspberry (*Rubus idaeus*) strawberry (*Fragaria* × *ananassa*) | - more inflorescences, flowers and fruits<br>- bigger fruits<br>- increased fruit biomass normalized per plant<br>- parthenocarpic fruit development | [23] |
| | *DefH9::iaaM* | grape vine (*Vitis vinifera*) | - more inflorescences per shoot<br>- increased berry number per bunch | [24] |
| | *FBP7::iaaM* flower-specific promoter | cotton (*Gossypium hirsutum*) | - enhanced fiber yield (field trials)<br>- increased fiber quality | [25] |
| | *B33::tms1* tuber-specific promoter | potato (*Solanum tuberosum*) | - enhanced tuberization | [26] |
| enzymes catalyzing auxin conjugation | *35S::OsGH3.1* | rice (*Oryza sativa*) | - dwarfism<br>- increased tolerance to fungal infection (less pronounced symptoms of infection by *Magnaporthe grisea*) | [13] |
| | *Ubi1::OsGH3-2* | rice (*Oryza sativa*) | - changed morphology: dwarfism, smaller leaves, fewer crown roots and root hairs, smaller panicles<br>- decreased tolerance to drought: earlier and more pronounced wilting symptoms, much lower survival rate (seedlings not watered for 4 days)<br>- increased tolerance to cold stress: less pronounced symptoms on leaves, significantly increased survival rate after recovery (5 days in 4 °C) | [27] |

**Table 1.** *Cont.*

| Protein | Promoter and Gene | Species | Phenotype of Transgenic Plants Compared to Control Lines | References |
|---|---|---|---|---|
| auxin transporter | *35S::ZmPIN1a* | maize (*Zea mays*) | - increased number of lateral roots and dry weight of roots<br>- lower shoots, shorter internodes<br>- increased seed biomass per plant under high-density cultivation<br>- increased drought tolerance: wilting symptoms of stress-exposed plants appeared later, the majority of transgenic plants survived drought, while most of the wild type plants died (seedlings not watered for 5 days)<br>- increased seed biomass per plant under drought conditions (field trials) | [28] |

The knowledge concerning auxin-initiated signaling pathways has intensively broadened during the last decades. The modification of these pathways allows us to obtain plants with desired traits. For example, overexpression of the gene encoding auxin-induced protein ARGOS in *A. thaliana* resulted in the stimulation of cell proliferation and an increase in organ size [29]. Overexpression of maize ARGOS1 (ZAR1) stimulated organ growth, enhanced grain yield, and drought stress tolerance in transgenic maize [30]. Overexpression of Auxin Response Factor 19 (ARF19) homolog from *Jatropha curcas* in *A. thaliana* and *J. curcas* increased seed size and yield [31]. Expression of gene *IbARF5* from sweet potato under the *35S* promoter in *A. thaliana* resulted in enhanced tolerance to drought and salinity in transgenic plants [32]. On the other hand, downregulation of Auxin Response Factor 4 (ARF4) in tomato increased tolerance to salinity and drought stress [33]. Similar results were obtained by Chen et al. [34]. Overexpression of the gene *OsAFB6*, encoding an auxin receptor, in rice resulted in increased grain yield per plant both in short day and long day conditions [35]. Another auxin receptor, *AFB3*, when overexpressed in Arabidopsis caused the increase in salt stress tolerance [36]. Transgenic maize overexpressing Auxin Binding Protein 1 (ABP1) was more resistant to sugarcane mosaic virus [37].

## 4. Cytokinins

Cytokinins are another class of phytohormones necessary for plant growth stimulation and controlling many developmental processes. They also participate in the regulation of plant senescence. Prolonging organ longevity due to cytokinin action enables longer biomass production by the plant [38]. Among the enzymes participating in cytokinin biosynthesis, the main targets of genetic engineering are as follows: isopentenyl transferase (IPT) catalyzing condensation of isoprenoid residue with adenine nucleotide, cytokinin dehydrogenase (CKX) involved in the degradation of these phytohormones, and glycosyl transferases converting cytokinins into their conjugates [38–41]. It was observed that mutations in *ckx* genes led to the increase in cytokinin content. Arabidopsis *ckx3 ckx5* double mutant formed larger inflorescences, floral meristems, and flowers and displayed increased seed yield per plant [42]. Similarly, *ckx3 ckx5* mutants of oilseed rape showed an increased cytokinin concentration that resulted in larger and more active inflorescence meristems, increased amounts of flowers and ovules and slightly increased seed yield [43]. Natural variations in soybean *GmCKX7-1* were linked to altered cytokinin profiles and yield characteristics [44].

The first attempts of genetic engineering of cytokinin metabolism were carried out in the 1990s. Gan and Amasino [45] transformed tobacco with the *IPT* gene under the control of *SAG12* promoter from *A. thaliana*, responsible for triggering gene expression in senescing leaves. The obtained transgenic lines displayed a greater number of flowers and seeds, delayed leaf senescence, and enhanced biomass production [45]. Due to the success of this strategy, promoters from the SAG family have been often used for plant genetic engineering. However, it has to be mentioned that in some cases, the delayed senescence of older leaves delayed nutrient allocation to seeds and storage organs. As a result, no yield increase and sometimes even yield reduction were observed. Furthermore, in the situation of nitrogen shortage, it was observed that old non-senescing leaves started to compete with younger leaves, which disturbed nitrogen recycling in plants. In rice, early senescing cultivars have a higher yield than those which undergo senescence later [46]. Positive effects were obtained by crossing the transgenic line of *A. thaliana* overexpressing *CKX3* under the control of root-specific *PYK10* promoter and displaying enhanced root growth with the lines displaying enhanced leaf growth [47].

Another promoter involved in the regulation of gene expression during senescence but also during stress response is senescence associated receptor protein kinase promoter (*SARK*). Expression of the *IPT* gene under the control of this promoter allowed for obtaining transgenic rice and tobacco with increased drought tolerance [48,49]. The examples of the experiments concerning the modulation of cytokinin metabolism are shown in Table 2.

The observed effects vary depending on the species and method used; not all of them are beneficial [50]. The application of inducible promoters responding to specific conditions allows better control of cytokinin content in transgenic plants. This allows us to avoid adverse effects occurring when too many of these phytohormones are synthesized in the plant [9]. Interestingly, the overexpression of *AGO2*, encoding protein belonging to the ARGONAUTE family and playing a role in the regulation of gene expression, led to the enhanced expression of cytokinin transporter BG3 and changed the pattern of cytokinin distribution in transgenic rice. This, in turn, resulted in an increase in grain length and salt tolerance [51]. Interesting results were obtained by Wang et al. [52], who used CRISPR/Cas gene editing to introduce changes into the cytokinin biosynthetic gene *OsLOG5*. The researchers managed to obtain rice lines with improved yield properties under drought stress when compared to stressed control [52].

It is noteworthy that in many cases researchers managed to obtain transgenic lines with increased biomass production or seed yield, but also more tolerant to abiotic stresses, such as drought and salinity (Table 2). However, it needs to be emphasized that there are some inconsistencies between the literature data, because the increased tolerance to some abiotic stresses was reported for plants with both increased and decreased cytokinin content. It needs to be remembered that in the experiments on transgenic plants, various species and promoters were used; there were also differences in the stress conditions applied [53].

Furthermore, cytokinins are a group of compounds, including *trans*-zeatin; *cis*-zeatin; $N^6$-isopentenyladenine; dihydrozeatin; $N^6$-benzylaminopurine; kinetin; *ortho-*, *meta-*, *para-*topolins; and ribosides of above-mentioned compounds [54]. It is already known that particular cytokinins vary in sites and timing of their production and degradation, transport routes, signaling pathways, and activity [54,55]. In *A. thaliana*, *trans*-zeatin and isopentenyladenine are the most active forms, present in higher concentrations than other cytokinins [54]. *Trans*-hydroxylated cytokinins, namely, the *trans*-zeatin-type, are synthesized in the roots and transported to the shoots in xylem sap. They are thought to play an important role as a nitrogen-supply signal in stimulation of the shoot growth. On the other hand, $N^6$-isopentenyladenine and *cis*-zeatin-types are predominant in the phloem sap of *A. thaliana*. These species are thought to participate in systemic shoot-to-root signaling in cooperation with other signaling molecules [56]. The understanding of the specificity of certain cytokinin types synthesis, transport, and signaling is crucial for the successful genetic engineering of these phytohormones.

Plant responses aimed at restoring the homeostasis of cytokinin levels and signaling were also observed in plants with changed biosynthesis or degradation of these hormones. In some of the experiments, the cytokinin content in transgenic lines was not assessed, while in some others, the methods of cytokinin measurements were questioned by other scientists [55]. The intensive research on cytokinin synthesis, degradation, transport, and signaling is being carried out, which should enable us to explain these effects in the future.

**Table 2.** Summary of the results of the experiments on transgenic plants with changed cytokinin concentration. APX, ascorbate peroxidase; CAT, catalase; Chl, chlorophyll; CKX, cytokinin dehydrogenase; IPT, isopentenyl transferase; $O_2^{\bullet -}$, superoxide; POX, peroxidase; PSII, photosystem II; ROS, reactive oxygen species; RWC, relative water content; SOD, superoxide dismutase.

| Enzyme | Promoter and Gene | Species | Phenotype of Transgenic Plants Compared to Control Lines | References |
|---|---|---|---|---|
| IPT | *Ubi::IPT* | tall fescue (*Festuca arundinacea*) | - increased tillering<br>- increased tolerance to cold stress: delayed senescence in plants grown outdoors, decreased electrolyte leakage in detached leaves kept at temperature range 0 to −28 °C | [57] |
| | *35S::IPT* | wheat (*Triticum aestivum*) | - increased tolerance to flooding: less pronounced growth inhibition during flooding and higher yield after recovery (plants flooded for 14 days) | [58] |
| | *Wild type/35S::IPT* (scion/rootstock) | tomato (*Lycopersicon esculentum*) | - reduced root growth<br>- increased tolerance to salt stress: increased fruit yield of stressed grafted plants (plants treated with 75 mM NaCl) | [59] |
| | *SAG12::IPT* | thale cress (*Arabidopsis thaliana*) | - increased tolerance to flooding: increased biomass and carbohydrate retention of waterlogged plants (plants flooded for 5 days)<br>- improved recovery from waterlogging stress and after submergence stress (stress duration 5 days) | [60] |
| | *SAG12::IPT*<br>*SAG13::IPT*<br>senescence-specific promoters | tomato (*Lycopersicon esculentum*) | - suppression of leaf senescence<br>- stem thickening, short internodal distances<br>- loss of apical dominance<br>- advanced flowering<br>- slight increase in fruit weight per plant | [61] |
| | *SAG12::IPT* | cassava (*Manihot esculenta*) | - delayed leaf senescence observed both during dark induction of senescence of detached leaves in the greenhouse and during field trials<br>- increased drought tolerance: reduced leaf senescence and wilting during water deficit (plants watered with lesser amount of water for 4 weeks) | [62] |
| | *SAG12::IPT*<br>*HSP18::IPT*<br>heat-stress-induced promoter | creeping bentgrass (*Agrostis stolonifera*) | - increased tolerance to heat stress: enhanced growth and root biomass of plants exposed to stress (plants grown at 35 °C/30 °C day/night for 10 days) | [63] |

**Table 2.** *Cont.*

| Enzyme | Promoter and Gene | Species | Phenotype of Transgenic Plants Compared to Control Lines | References |
|---|---|---|---|---|
| | *SAG12::IPT* | creeping bentgrass (*Agrostis stolonifera*) | - increased tolerance to drought (plants not watered for 21 days): more extensive root system, decreased MDA content and electrolyte leakage in roots, lower $O_2^{\bullet-}$ and $H_2O_2$ levels in roots<br>- increased antioxidant response in drought-exposed plants: increased ascorbate content, increased activity of SOD, CAT, APX, glutathione reductase, and dehydroascorbate reductase in roots | [64] |
| | *SAG12::IPT* | creeping bentgrass (*Agrostis stolonifera*) | - increased drought tolerance: suppression of drought-induced leaf senescence and root dieback, reduced wilting, lower MDA content, enhanced activity of SOD, CAT, POX (plants not watered for 2 weeks) | [65] |
| | *SAG12::IPT* | creeping bentgrass (*Agrostis stolonifera*) | - increased proline and soluble sugar content in drought-exposed plants (plants not watered till leaf RWC dropped to 47%) | [66] |
| | *SAG12::IPT* | eggplant (*Solanum melongena*) | - increased vegetative growth rate and fruit yield per plant<br>- delayed leaf senescence<br>- decreased MDA content, increased SOD and POX activity<br>- increased drought tolerance: delayed chlorosis and wilting (plants were not watered)<br>- increased cold tolerance: delayed chlorosis and wilting (plants kept in 4 °C) | [67] |
| | *SAG12::IPT*<br>*DEG::IPT*<br>dexamethasone-inducible promoter | thale cress (*Arabidopsis thaliana*) | - increased drought tolerance: faster and more vigorous recovery of stressed plants (plants not watered for 13 days) | [68] |
| | *GHCP::IPT*<br>promoter belonging to SAG family | cotton (*Gossypium hirsutum*) | - delayed leaf senescence<br>- increased lint yield, increased fiber quality (more uniform, stronger and longer fibers)<br>- increased tolerance to salt stress: increased germination percentage under salt stress (on paper moistened with 250 mM NaCl)<br>- increased dry biomass of plants exposed to salt stress (hydroponically grown seedlings exposed to 200 mM NaCl for 21 days) | [69] |
| | *SARK::IPT* | tobacco (*Nicotiana tabacum*) | - increased tolerance to drought: less severe stress symptoms, higher leaf water content and plant dry weight, improved recovery, increased seed yield of recovered plants (plants not watered for 2 weeks)<br>- improved antioxidant defense in drought-exposed plants: increased ascorbate and glutathione content, decreased $H_2O_2$ level in leaves<br>- only minimal seed yield loss of water-restricted plants (70% less watering) | [70] |

**Table 2.** *Cont.*

| Enzyme | Promoter and Gene | Species | Phenotype of Transgenic Plants Compared to Control Lines | References |
|---|---|---|---|---|
| | *SARK::IPT* | tobacco (*Nicotiana tabacum*) | - increased tolerance to water deficit: increased $CO_2$ assimilation rate (70% reduced watering for 70 days) | [48] |
| | *SARK::IPT* | peanuts (*Arachis hypogaea*) | - increased drought tolerance: increased fresh and dry biomass of shoots and roots of stressed plants (plants not watered for 15 days, then watered with $\frac{1}{4}$ of the optimal water amount for 45 days)<br>- increased shoot dry weight and seed weight per plant grown under water deficit (field trials) | [71] |
| | *SARK::IPT* | rice (*Oryza sativa*) | - increased drought tolerance: delayed wilting, increased total dry biomass and seed yield per plant in stress-exposed plants (plants not watered for 6–10 days before flowering phase or 2 weeks after flowering) | [49] |
| | *SARK::IPT* | rice (*Oryza sativa*) | - increased tolerance to drought: less pronounced stress symptoms, increased RWC and maximum quantum efficiency of PSII, no decrease in carbon and nitrogen assimilation and protein content (plants not watered for 3 days at pre anthesis)<br>- increased sucrose and starch content in flag leaf, enhanced nitrate content, higher nitrate and nitrite reductase activity, and sustained ammonium content in drought-exposed plants | [72] |
| | *SARK::IPT* | cotton (*Gossypium hirsutum*) | - delayed leaf senescence (detached leaf assay)<br>- increased drought tolerance: increased root and shoot biomass, Chl content, and photosynthetic rate under water deficit in the greenhouse; increased root and shoot biomass and cotton yield under water deficit in growth chamber (66% less watering) | [73] |
| | *SARK::IPT* | maize (*Zea mays*) | - increased drought tolerance: delayed wilting and leaf senescence, increased water content in stress-exposed plants, 30-fold higher average seed biomass per plant (plants not watered for 3 weeks) | [74] |
| | *SARK::IPT* | sweetpotato (*Ipomea batatas*) | - delayed senescence<br>- improved tolerance to drought: improved growth characteristics and leaf RWC (plants exposed to various irrigation regimes for 96 days) | [75] |
| | *HMW::IPT* seed-specific promoter | tobacco (*Nicotiana tabacum*) | - increase in seed yield<br>- increase in ethanol-insoluble carbohydrates and protein content | [76] |

**Table 2.** *Cont.*

| Enzyme | Promoter and Gene | Species | Phenotype of Transgenic Plants Compared to Control Lines | References |
|---|---|---|---|---|
| | *lectin::IPT* seed-specific promoter | tobacco (*Nicotiana tabacum*) | - increase in seed dry weight and protein content<br>- faster growth of seedlings | [77] |
| | *TP12::IPT* flower-specific promoter | narrow-leafed lupin (*Lupinus angustifolius*) | - increased branching<br>- increased total number of fruits (pods) in some lines | [78] |
| | *AtMYB32xs::IPT* developmental-process-related promoter | canola (*Brassica napus*) | - delayed leaf senescence both under controlled conditions and in the field<br>- more flowers and siliques<br>- increased yield (field trials) | [79] |
| | *AtMYB32xs-p::IPT* | wheat (*Triticum aestivum*) | - delayed leaf senescence<br>- increased yield<br>- improved drought tolerance: improved canopy green cover, lower canopy temperatures, higher leaf water potential (field trials) | [80] |
| | *rd29A::IPT* stress-induced promoter | tobacco (*Nicotiana tabacum*) | - increased tolerance to salt stress: delayed leaf senescence and decreased MDA content in stressed plants (plants exposed to 150 mM NaCl for 2 weeks) | [81] |
| | *AtCOR15a::IPT* cold-stress-induced promoter | sugarcane (*Saccharum officinarum* and *S. spontaneum* hybrids) | - increased tolerance to cold stress: less pronounced symptoms of leaf senescence (detached leaves exposed to 27 °C, 4 °C or 4 °C and then 0 °C),<br>- increased Chl content, decreased MDA content, and electrolyte leakage in cold-stressed plants (plants were exposed to decreasing temperatures for acclimation, then incubated in 0 °C for 8 h and recovered for 24 h) | [82] |
| | *AtMT::IPT* stress-induced promoter | tobacco (*Nicotiana tabacum*) | - improved drought tolerance: less severe stress symptoms (plants not watered for 3 weeks)<br>- improved tolerance to salt stress: less severe stress symptoms, faster recovery (plants watered with 100 mM NaCl for 10 days, then with 200 mM NaCl for 11 days) | [83] |
| | *PtRD26$_{pro}$::IPT* promoter of senescence and drought-inducible transcription factor | poplar (*Populus tomentosa*) | - increased height, stimulated adventitious root generation<br>- increased net $CO_2$ assimilation<br>- increased drought tolerance: less severe stress symptoms, higher levels of maximum quantum efficiency of PSII, RWC, net $CO_2$ assimilation rate, stomatal conductance, and electron transfer rate, improved survival rate (plants not watered for 10 days) | [84] |
| *Trans*-zeatin synthetase | *hsp70::tzs* heat shock induced promoter | rapeseed (*Brassica napus*) | - reduced root system<br>- increased height and branching<br>- increased seed yield per plant | [85] |

**Table 2.** *Cont.*

| Enzyme | Promoter and Gene | Species | Phenotype of Transgenic Plants Compared to Control Lines | References |
|---|---|---|---|---|
| CKX | *35S::AtCKX1*<br>*35S::AtCKX2*<br>*35S::AtCKX3*<br>*35S::AtCKX4* | tobacco (*Nicotiana tabacum*) | - retarded shoot development, dwarfed phenotype, small leaves<br>- stimulated root growth | [86] |
| | *35S::AtCKX1*<br>*35S::AtCKX2*<br>*35S::AtCKX3*<br>*35S::AtCKX4* | thale cress (*Arabidopsis thaliana*) | - reduced shoot growth<br>- stimulated root growth | [87] |
| | *35S::CKX1*<br>*35S::CKX2*<br>*35S::CKX3*<br>*35S::CKX4* | thale cress (*Arabidopsis thaliana*) | - reduced growth of some lines<br>- increased tolerance to salt stress: less pronounced stress symptoms and improved survival rate (plants exposed to 200 mM NaCl for 6 days)<br>- increased drought tolerance: less pronounced wilting symptoms and improved survival rate (plants not watered for 2 weeks) | [88] |
| | *35S::AtCKX3* | tomato (*Solanum lycopersicum*) | - smaller leaf area, decreased stomata density<br>- decreased transpiration<br>- increased drought tolerance: increased leaf water content (plants not watered for 4 days) | [89] |
| | *35S::MsCKX* | thale cress (*Arabidopsis thaliana*) | - enlarged root system<br>- increased salt tolerance: improved root growth and seedling fresh weight (seedlings exposed to 100 or 150 mM NaCl for 7 days), improved survival rate and maximum quantum yield of PSII (plants watered with increasing concentrations of NaCl for 4 days, then with 350 mM NaCl for 10 days)<br>- improved membrane properties and antioxidant defense under salt stress: decreased ion leakage, MDA content, $H_2O_2$ and $O_2^{\bullet-}$ levels, increased proline content and SOD, CAT, POX activity (plants subjected to 150 mM NaCl for 10 days) | [90] |
| | *35S::AtCKX2* | rapeseed (*Brassica napus*) | - enlarged root system, longer primary roots, increased number of lateral and adventitious roots, increased root density, enhanced root-to-shoot ratio<br>- no reduction in shoot growth<br>- increased P, Ca, Mg, S, Zn, Cu, Mo, and Mn concentration in leaves<br>- increased Chl content under Mg- and S-deficiency<br>- improved phytoremediation capacity of Cd and Zn from contaminated medium and soil | [91] |

**Table 2.** *Cont.*

| Enzyme | Promoter and Gene | Species | Phenotype of Transgenic Plants Compared to Control Lines | References |
|---|---|---|---|---|
| | *35S::PpCKX1* | *Physcomitrella patens* | - larger size of protoplasts, curved protonemal tissues<br>- delayed transition to gametophores, reduced number of spores<br>- enhanced rhizoid development<br>- improved tolerance to dehydration: increased survival rate after drying of protonemal tissues<br>- improved tolerance to salt stress: improved growth (protonemal tissues exposed to 100 or 200 mM NaCl for 30 days) | [92] |
| | *Ubi::TaCKX1* | wheat (*Triticum aestivum*) | - increased spike number and grain number<br>- lower 1000-grain weight | [93] |
| | *W6::CKX1* root-specific promoter | tobacco (*Nicotiana tabacum*) | - stimulation of root growth, increased ratio of root to shoot biomass<br>- increased drought tolerance: increased survival rate (plants not watered for 26 days) | [94] |
| | *35S::CKX1*<br>*WRKY6::CKX1* root-specific promoter | tobacco (*Nicotiana tabacum*) | - enlarged root system and dwarfism in line transformed with *35S* promoter construct<br>- improved drought tolerance: higher water potential in lower leaves, more negative osmotic potential in leaves (plants not watered for 10 days)<br>- lower leaf temperature in *35S:CKX1* line (plants exposed for 40 °C for 2 h) | [95] |
| | *RCc3::OsCKX4* root-specific promoter | rice (*Oryza sativa*) | - enhanced root development | [96] |
| | *RCc3::OsCKX4* | rice (*Oryza sativa*) | - increased Zn concentration in roots, shoots, and grains<br>- increased grain yield per plot | [97] |
| | *bGLU::AtCKX1* root-specific promoter | barley (*Hordeum vulgare*) | - stimulated lateral root growth<br>- improved drought tolerance: higher RWC, less pronounced decrease in yield (plants exposed to water deficit)<br>- faster recovery and higher RWC in some drought-exposed transgenic lines (hydroponically grown plants were deprived of the growth medium for 24 h), improved growth after stress recovery (observed 2 weeks after severe stress application to hydroponically grown plants and 4 weeks after 3 days of watering withdrawal in soil-grown plants) | [98] |
| | *EPP::CKX1*<br>*EPP::CKX2* root-specific promoter | barley (*Hordeum vulgare*) | - stimulation of root growth<br>- increased concentration of various micro- and macro-elements in leaves<br>- increased tolerance to drought: higher $CO_2$ assimilation rate in stressed plants (plants were not watered until the soil moisture level dropped to 10%, this level was maintained for the next 2 weeks) | [99] |

**Table 2.** *Cont.*

| Enzyme | Promoter and Gene | Species | Phenotype of Transgenic Plants Compared to Control Lines | References |
|---|---|---|---|---|
| | *EPP::CKX1*<br>*EPP::CKX2* | barley (*Hordeum vulgare*) | - increased Zn concentration in grains<br>- increased Fe concentration in grains of some lines | [100] |
| | *RCc3::AtCKX1* | maize (*Zea mays*) | - stimulation of root growth<br>- increased concentration of micro- and macro-elements in leaves: K, P, Mo, Zn | [101] |
| | *CaWRKY31::CaCKX6* root-specific promoter | thale cress (*Arabidopsis thaliana*) chickpea (*Cicer arietinum*) | - increase in lateral root number, root length, and root biomass in Arabidopsis and chickpea without any penalty to vegetative and reproductive growth of shoot<br>- soil-grown chickpea exhibited higher root-to-shoot biomass<br>- enhanced drought tolerance in soil-grown chickpea: increased shoot and root growth, increased $CO_2$ assimilation rate (plants not watered for 40 days)<br>- seed yield in some chickpea lines up to 25% higher with no penalty in protein content<br>- higher levels of Zn, Fe, K, and Cu in transgenic chickpea seeds | [102] |
| | *RCc3:OsCKX5* | rice (*Oryza sativa*) | - stimulated root growth: greater volume, length, projection area, higher number of tips, enhanced surface area<br>- no detrimental impact on shoot growth<br>- increased root biomass, root to shoot ratio, deeper root system in plants grown on low-fertility soil<br>- increased P, K, Ca, Mg, Zn, Fe concentration in roots<br>- increased K, Mg, Fe, Zn concentration in shoots (not in all lines) | [103] |
| | *35S::MdCKX5.2* | thale cress (*Arabidopsis thaliana*) | - longer primary root, stimulation of lateral root development<br>- increased tolerance to drought: less severe stress symptoms, significantly improved survival rate (plants not watered for 20 days)<br>- increased tolerance to salt stress: less severe stress symptoms, longer primary roots and more lateral roots, increased fresh weight and Chl content (seedlings exposed to 100 or 150 mM of NaCl for 9 days) | [104] |
| | *OsCKX2* promoter::3′-UTR of *OsCKX2* target silencing | rice (*Oryza sativa*) | - increased grain number per plant | [105] |
| | *35S::HvCKX1*<br>*35S::TaCKX1* hairpin target silencing | barley (*Hordeum vulgare*), wheat (*Triticum aestivum*) triticale | - increased seed yield, seed number per plant and 1000-grain weight in some lines<br>- increased root biomass<br>- lower shoots | [106] |

**Table 2.** *Cont.*

| Enzyme | Promoter and Gene | Species | Phenotype of Transgenic Plants Compared to Control Lines | References |
|---|---|---|---|---|
| | *35S::HvCKX2* hairpin target silencing | barley (*Hordeum vulgare*) | - increased height<br>- increased spikes number<br>- increased seed yield, seed number, and 1000-grain weight | [107] |
| | *35S::HvCKX1* hairpin target silencing | barley (*Hordeum vulgare*) | - increased root mass<br>- increased seed yield | [108] |
| | *Ubi::shRNA-CX3* *Ubi::shRNA-CX5* hairpin target (*OsCKX2*) silencing | rice (*Oryza sativa*) | - delayed senescence<br>- increased tillering<br>- increased panicle number, grain yield per plant, and 1000-grain weight (field trials) | [109] |
| | *35S::GhCKX* hairpin target silencing | cotton (*Gossypium hirsutum*) | - delayed leaf senescence<br>- more fruiting branches and bolls, increased seed size<br>- increased seed yield and lint yield of moderately suppressed lines (per 25 $m^2$ plot size) | [110] |
| | *35S::OsCKX2* antisense target silencing | rice (*Oryza sativa*) | - enhanced panicle branching, increased seed biomass per plant, increased 1000-grain weight<br>- increased tolerance to salt stress: increased water content and higher shoots of stressed plants, at the end of stress-exposure wild type plants were dying (plants watered with 200 mM NaCl for 30 days)<br>- less pronounced decrease in yield of plants exposed to salt stress at pre-flowering stage until maturity (plants watered with 100 mM NaCl) | [111] |
| | *Act1::HvCKX1* hairpin target silencing | wheat (*Triticum aestivum*) | - increased grain number per plant | [112] |
| | *Ubi1::HvCKX1* 5′ end of the ORF and 3′ UTR, target silencing *HvCKX1* knockout obtained by CRISPR/Cas9 | barley (*Hordeum vulgare*) | - increased spike number and grain number per plant<br>- decreased 1000-grain weight<br>- increased yield for $m^2$ (field trials) | [113] |
| | *OsCKX2* knockout obtained by CRISPR/Cas9 | rice (*Oryza sativa*) | - increased shoot fresh and dry weight both in normal-phosphate and low-phosphate conditions<br>- lesser leaf yellowing and increased maximum quantum efficiency of PSII under Pi deficiency<br>- increased P concentration in roots and shoots under low-Pi conditions | [114] |
| | *proAGIP::GhCKX3b* silencing construct carpel- and stamen-specific promoter | cotton (*Gossypium hirsutum*) | - increased seed number<br>- increased lint yield | [115] |

**Table 2.** *Cont.*

| Enzyme | Promoter and Gene | Species | Phenotype of Transgenic Plants Compared to Control Lines | References |
| --- | --- | --- | --- | --- |
| glucosyl-transferase | *35S::ZOG1* | tobacco (*Nicotiana tabacum*) | - primary root elongation and diminished branching | [116] |
| | *Act1::cZOGT1*<br>*Act1::cZOGT2* | rice (*Oryza sativa*) | - short shoots<br>- delayed leaf senescence | [117] |
| | *Ubi1::ZOG1* | maize (*Zea mays*) | - delayed leaf senescence<br>- shorter stature, thinner stems, narrower leaves<br>- increased root biomass and branching<br>- disturbed floral development, smaller ear | [118] |
| | *35S::UGT85A5* | tobacco (*Nicotiana tabacum*) | - increased tolerance to salt stress: increased seed germination rate on the medium containing NaCl (100–200 mM); increased total fresh weight of salt-treated seedlings (exposed to 100 or 200 mM NaCl for 4 weeks)<br>- lesser decay of Chl content in leaf discs incubated in 100–300 mM NaCl<br>- survival under strong salt-induced stress (seedlings watered with NaCl solution increasing to 300 mM for 4 weeks)<br>- increased proline content and decreased MDA content in salt-treated plants (watered with 300 mM NaCl for 1 week) | [119] |
| | *35S::OscZOG1*<br>*OsZOG1* silencing construct under *Ubi* promoter | rice (*Oryza sativa*) | - in overexpressing line: improved growth of lateral roots, decreased shoot growth and yield-associated traits, accelerated senescence<br>- in silenced line: improved crown roots growth and tillering, higher shoots, increased yield-associated traits: panicle branching, grain number per panicle, seed size, and 1000-grain weight | [120] |
| | *35S::AtUGT76C2* | thale cress (*Arabidopsis thaliana*) | - reduced tolerance to osmotic stress at postgermination stage: decreased germination rate, slower germination and primary root growth, more severe stress symptoms (seeds sown on medium with 200, 250, 300 mM mannitol)<br>- increased tolerance to drought at mature stage: less severe stress symptoms, improved survival rate (plants not watered for 7 days), decreased water loss and faster stomata closure in detached leaves | [121] |

**Table 2.** *Cont.*

| Enzyme | Promoter and Gene | Species | Phenotype of Transgenic Plants Compared to Control Lines | References |
|---|---|---|---|---|
| | *Ubi::AtUGT76C2* | rice (*Oryza sativa*) | - enhanced root growth<br>- increased sensitivity to abiotic stresses during germination and post-germination growth: slower germination and decreased seedling growth (seeds exposed to 100 mM NaCl, 7.5% PEG8000, or 150 mM mannitol)<br>- increased tolerance to salt stress: less severe stress symptoms, decreased electrolyte leakage, increased survival rate, increased proline and soluble sugar content (seedlings watered with 200 mM NaCl for 2 weeks), lower $H_2O_2$ and $O_2^{\bullet -}$ level, increased activity of SOD, CAT, and APX (plants exposed to 200 mM NaCl for 12 h)<br>- increased drought tolerance: less severe stress symptoms, increased survival rate, increased proline and soluble sugar content (seedlings not watered for 1 week), decreased water loss from detached leaves | [122] |

## 5. Gibberellins

Gibberellins are a large group of tetracyclic diterpenoids. Among them, only a few compounds participate in the regulation of growth and development of higher plants—primarily $GA_1$ and $GA_4$ [2]. Gibberellin deficiency causes dwarfism [123]. For arable crops, especially cereals, dwarfism can be an advantage, because it improves lodging resistance and changes assimilate partitioning so that more assimilates are allocated to flowers and grains. Breeding of semi-dwarf cereal varieties has been proven to be enormously successful in increasing grain yield since the advent of the "green revolution" [124]. The rice *semidwarf-1* (*sd-1*) gene, encoding gibberellin 20 oxidase, is well known as the "green revolution gene" and is considered to be the one of the most important genes deployed in modern rice breeding. It has contributed to the significant increase in crop production that occurred in the 1960s and 1970s, especially in Asia [125,126]. The genes responsible for the "green revolution" in wheat are semi-dwarfing genes *Reduced height* (*Rht*). The most important and widely used are the alleles *Rht-B1b* and *Rht-D1b* that are found in >70% of current commercial wheat cultivars. They are known to reduce stem extension by causing partial insensitivity to gibberellins due to the changed signaling process [127,128].

The crucial enzymes associated with gibberellin metabolism are GA20ox and GA3ox gibberellin oxidases, catalyzing the last two steps of the synthesis of biologically active gibberellins, as well as GA2ox oxidase, catalyzing the oxidation of these phytohormones to inactive catabolites. The three types of enzymes mentioned above are encoded by small gene families, *GA20ox*, *GA3ox,* and *GA2ox,* respectively [129]. All of them were targets of genetic engineering. The introduction of *GA20ox* and *GA3ox* or *GA2ox* genes enables us to obtain plants with either increased or decreased active gibberellin content (Table 3). Increased gibberellin content stimulates elongation growth and lignin synthesis, while a reduced level of these phytohormones results in dwarfism, stimulation of lateral shoot formation, and reduction of lignin content (Table 3). Change in gibberellin content also allows us to obtain other useful traits. Tomato fruits with *GA20ox1* gene overexpression remained firm for a longer time, which prolonged their shelf life [130]. The formation of longer xylem fibers in transgenic poplars overexpressing *AtGA20ox1* is beneficial for paper production [131]. On the other hand, reduced lignin content in transgenic switchgrass with decreased gibberellin content facilitates the bioethanol production process. However, the *GA2ox*-overexpressing lines were semi-dwarf, which is not a desired trait in plants grown for biomass production [132]. Enhanced activity of gibberellin biosynthetic enzymes may be not beneficial in terms of resistance to certain pathogens. Transgenic rise overexpressing *OsGA20ox3* was more susceptible to *Xanthomonas oryzae* pv. *oryzae* (causing bacterial blight) and *Magnaporthe oryzae* (causing rice blast), while *OsGA20ox3* knockdown lines displayed enhanced resistance to these pathogens [133].

Modification of gibberellin signaling was also taken into consideration. DELLA protein was identified as a repressor in the gibberellin signaling pathway. The above-mentioned *Rht* alleles encode DELLA proteins [128]. Canola mutant *ds-3* bearing a mutation in the gene encoding DELLA protein is semi-dwarf [134]. The overexpression of *SLR1* encoding rice DELLA protein enhanced cold tolerance in this species, while plants with overexpression of *GA20ox1* were more sensitive to this kind of stress. These results suggest that weakening of gibberellin signaling leads to the improvement in chilling tolerance [135]. Overexpression of *GoGID1* encoding gibberellin receptor in alfalfa allowed to obtain transgenic plants with increased growth rates, heights, and biomass production when compared to the control [136].

**Table 3.** Summary of the results of the experiments on transgenic plants with changed gibberellin concentration. CAT, catalase; Chl, chlorophyll; GAMT, gibberellin methyltransferase; GAox, gibberelin oxidase; MDA, malonyldialdehyde; POX, peroxidase; RWC, relative water content; SOD, superoxide dismutase; WUE, water-use efficiency.

<table>
<tr><th>Enzyme</th><th>Promoter and Gene</th><th>Species</th><th>Phenotype of Transgenic Plants Compared to Control Lines</th><th>References</th></tr>
<tr><td colspan="5">Modifications aimed at increasing of active gibberellins level</td></tr>
<tr><td rowspan="5">GA20ox</td><td>35S::AtGA20ox1</td><td>hybrid aspen (Populus tremula × P. tremuloides)</td><td>- increased growth rate<br>- increased biomass<br>- longer and more numerous xylem fibers</td><td>[131]</td></tr>
<tr><td>35S::AtGA20ox1</td><td>tobacco (Nicotiana tabacum)</td><td>- shoot growth stimulation<br>- increased biomass production<br>- increased lignin content<br>- stimulation of xylem formation</td><td>[137]</td></tr>
<tr><td>35S::CcGA20ox1</td><td>tomato (Lycopersicon esculentum)</td><td>- changed morphology: higher shoots, non-serrated leaves, some flowers had longer style<br>- delayed flowering<br>- increased fruit number and their total weight per plant, some of the fruits were parthenocarpic, which was not observed in the control plants<br>- delayed fruit ripening</td><td>[130]</td></tr>
<tr><td>Ubi1::AtGA20ox1</td><td>maize (Zea mays)</td><td>- higher and more slender stems<br>- increased vegetative biomass<br>- increased content of lignin and cellulose</td><td>[138]</td></tr>
<tr><td>35S::PdGA20-OXIDASE<br>DX15::PdGA20-OXIDASE<br>poplar xylem-specific promoter</td><td>hybrid poplar (Populus tremula L. × Populus alba)</td><td>- increased shoot growth and biomass production, accompanied by poor root growth and unstable shoots in plants with constitutive overexpression<br>- enhanced growth without changing the overall phenotype in plants with xylem-specific expression</td><td>[139]</td></tr>
</table>

**Table 3.** *Cont.*

| Enzyme | Promoter and Gene | Species | Phenotype of Transgenic Plants Compared to Control Lines | References |
|---|---|---|---|---|
| **Modifications aimed at increasing of active gibberellins level** | | | | |
| GA3ox | *35S::StGA3ox2*<br>*StLS1::StGA3ox2* leaf-specific promoter<br>*Tub1::StGA3ox2* tuber-specific promoter | potato (*Solanum tuberosum*) | - plants with constitutive overexpression or leaf-specific overexpression were higher and tuberized earlier when grown under short day conditions<br>- plants with tuber-specific overexpression did not display differences in shoot height, their tuberization was slightly delayed<br>- increased tuber biomass per plant in line with constitutive overexpression | [129] |
| **Modifications aimed at decreasing of active gibberellins level** | | | | |
| GA2ox | *Act::OsGA2ox1*<br>*D18::OsGA2ox1* promoter of a gene participating in gibberellin synthesis in rice | rice (*Oryza sativa*) | - plants with constitutive overexpression displayed changed morphology: dwarfism, darker green, broader and shorter leaves, they also failed to set grains<br>- lines with expression under D18 promoter were semi-dwarf and developed normal flowers and grains | [140] |
| | *35S::AtGA2ox8* | canola (*Brassica napus*) | - dwarfism<br>- increased anthocyanin content in leaves<br>- stimulated branching, increased number of siliques<br>- increased seed yield per experimental plot | [141] |
| | *35S::OsGA2ox5* | thale cress (*Arabidopsis thaliana*)<br>rice (*Oryza sativa*) | - dwarfism, delayed onset of generative phase<br>- increased starch granule accumulation and gravity responses<br>- increased tolerance to salt stress: improved survival rate and less reduced seedling growth (seeds exposed to 100 mM or 140 mM NaCl for 7 days); significantly improved survival rate (seedlings exposed to 170 mM NaCl) | [142] |
| | *rbcs::PtGA2ox1* leaf-specific promoter<br>*TobRB7::PtGA2ox1* root-specific promoter<br>*LMX5::PtGA2ox1* stem-specific promoter | tobacco (*Nicotiana tabacum*) | - slower growth<br>- increased drought tolerance: increased RWC, increased proline and sugar content, decreased MDA content, elevated POX, SOD, and CAT activities (plants not watered for 19 days) | [143] |
| | *Ubi1::PvGA2ox5*<br>*Ubi1::PvGA2ox9* | switchgrass (*Panicum virgatum*) | - semi-dwarfism<br>- increased tillering<br>- decreased lignin content | [132] |

**Table 3.** *Cont.*

| Enzyme | Promoter and Gene | Species | Phenotype of Transgenic Plants Compared to Control Lines | References |
|---|---|---|---|---|
| **Modifications aimed at decreasing of active gibberellins level** | | | | |
| | *Ubi::GA2ox6* mutated versions | rice (*Oryza sativa*) | - changed morphology: reduced plant height, expanded root system, enhanced tillering<br>- increased WUE and photosynthesis rate<br>- increased grain yield (field trials)<br>- increased drought tolerance: less pronounced wilting symptoms; improved leaf recovery; increased survival rate; increased proline content and activities of SOD, CAT, POX; decreased $H_2O_2$ content (plants were air-dried for 6 h then recovered for 6 days)<br>- increased tolerance to salt stress: increased survival rate (plants exposed to 200 mM NaCl for 2 days)<br>- increased tolerance to temperature (both heat and cold) stress: increased survival rate (plants exposed to 4 °C or 42 °C for 2 days)<br>- increased tolerance to biotic stress: less pronounced symptoms of infection with the bacterium *Xanthomonas oryzae* pv. *oryzae*, increased seedling weight after infection with the fungus *Pythium arrhenomanes*, limited spread of fungus *Fusarium fujikuroi* | [144] |
| | *35S::GhGA2ox1* | cotton (*Gossypium hirsutum*) | - increased drought tolerance: increased proline, Chl, and RWC in stressed plants (plants not watered for 10 days)<br>- increased tolerance to salt stress: increased proline, Chl, and RWC in stressed plants (plants watered with 200 mM NaCl for 10 days) | [145] |
| GAMT | *35S::AtGAMT1* | tomato (*Solanum lycopersicum*) | - changed morphology: reduced plant height, smaller leaves of changed shape<br>- smaller stomata<br>- increased drought tolerance: delayed wilting symptoms, higher leaf water status, reduced transpiration, improved recovery of stressed plants (plants not watered for 14 days) | [146] |
| CYP cytochrome 450 | *ProEui::PtCYP714A3* native promoter of one of *CYP* genes in rice | rice (*Oryza sativa*) | - semi-dwarfed phenotype, promoted tillering, reduced seed size<br>- increased tolerance to salt stress: less severe stress symptoms, improved survival rate (seedlings exposed to 150 mM NaCl for 12 days) | [147] |
| | *35S::CYP71D8L* | rice (*Oryza sativa*) | - dwarfed phenotype, reduced grain number per panicle<br>- increased drought tolerance: less severe stress symptoms, increased Chl content and decreased $H_2O_2$ level after 5 days of drought (plants not watered for 10 days)<br>- increased tolerance to salt stress: less severe stress symptoms, increased Chl content and decreased $H_2O_2$ level after 3 days of stress (plants watered with 150 mM NaCl for 8 days) | [148] |

## 6. Brassinosteroids

Similar to other phytohormones, brassinosteroids have a pleiotropic effect and they participate both in the regulation of development and stress response [149]. Considering their chemical structure, these compounds belong to polyhydroxy steroids and are similar to animal steroid hormones [150]. Many experiments concerning the impact of brassinosteroids on plant development and stress tolerance were carried out by spraying the plants with solutions of these phytohormones. There are also data in the literature on genetic engineering of the pathway of their biosynthesis [151]. Important determinants of yield that are regulated by brassinosteroids are plant height, leaf angle, and inflorescence architecture [152]. The results of the experiments on transgenic plants with increased brassinosteroid content are shown in Table 4. There were also experiments aiming at decreasing the level of these phytohormones. Overexpression of *AtBAT1* encoding brassinosteroid-inactivating acyltransferase in bentgrass resulted in decreased growth rate, erect leaves, prolonged leaf longevity, and improved drought resistance [153].

The rice mutant *ebisu dwarf* (*d2*) with impaired brassinosteroid biosynthesis had erect leaves, which improves light penetration into the canopy. On the other hand, it produced smaller seeds [154]. However, another rice mutant, *osdwarf4-1,* had a stature similar to the *d2* mutant, while the morphology of its flowers and seeds remained unchanged [155,156]. Barley accessions carrying a single recessive gene *uzu*, encoding the brassinosteroid receptor, are semi-dwarf. This gene has been introduced in almost all Japanese hull-less barley cultivars [157,158]. Brassinosteroid insensitive semi-dwarf barley mutants were more tolerant to drought [159].

Transgenic tomatoes overexpressing *SlBRI1* encoding brassinosteroid receptor displayed increased height, yield, and fruit number per plant. Transgene overexpression also caused an increase in the levels of carotenoids, ascorbic acid, soluble solids, and soluble sugars during fruit ripening [160]. On the other hand, it led to a decrease in drought tolerance [161]. Rice line overexpressing *SERK2*, encoding membrane protein interacting with brassinosteroid receptor, produced larger grains and was more tolerant to salt stress [162]. Overexpression of kinases participating in brassinosteroid-induced signaling, e.g., membrane localized ZmBSK1 and downstream acting ZmCCaMK, in corn, led to the improved drought tolerance [163]. Overexpression of the gene *TaBRI1* from wheat, encoding transmembrane receptor kinase, in Arabidopsis resulted in increased sensitivity to brassinosteroids, earlier flowering, and increased silique size and seed yield [164]. While the modification of signaling through BRI receptors is linked to growth arrest, the overexpression of BRL3, a vascular-enriched member of the brassinosteroid receptor family, in *A. thaliana* enhanced tolerance to drought without penalizing plant growth [165]. In recent years, there has been significant progress in the deciphering of brassinosteroid signaling, which opens the way to successful modification of crop plants [152].

**Table 4.** Summary of the results of the experiments on transgenic plants with increased brassinosteroids concentration due to overexpression of hydroxylases participating in the biosynthesis of these phytohormones. CAT, catalase; MDA, malonyldialdehyde; POX, peroxidase; RWC, relative water content; SOD, superoxide dismutase.

| Enzyme | Promoter and Gene | Species | Phenotype of Transgenic Plants Compared to Control Lines | References |
|---|---|---|---|---|
| C-22α hydroxylase | *35S::AtDWF4* | thale cress (*Arabidopsis thaliana*) | - stimulation of generative shoot branching<br>- increase in siliques and seed number | [166] |
| | expression of cDNA of CYPs from maize, *A. thaliana*, and rice under AS promoter active in stems, leaves, and roots | rice (*Oryza sativa*) | - enhanced tillering<br>- increased grain yield per plant, increased grain weight (greenhouse and field trials)<br>- increased leaf angle, more loose stature (not desired trait) | [167] |

**Table 4.** *Cont.*

| Enzyme | Promoter and Gene | Species | Phenotype of Transgenic Plants Compared to Control Lines | References |
| --- | --- | --- | --- | --- |
| | *35S::CYP724B1* | rice (*Oryza sativa*) | - increased spikelet number per panicle<br>- increased grain size and 1000-grain weight<br>- increased leaf angle, more loose stature | [168] |
| | *35S::AtDWF4* | canola (*Brassica napus*) | - longer roots, increased root biomass<br>- larger leaves<br>- stimulation of branching, increase in siliques number<br>- increased seed yield per plant<br>- increased drought tolerance: improved survival rate, increased root and shoot biomass of recovered plants (plants not watered for 12 days)<br>- increased tolerance to heat stress: improved survival rate of stressed plants (plants exposed to 45 °C for 3 or 4 h)<br>- increased resistance to necrotrophic fungal pathogens *Leptosphaeria maculans* and *Sclerotinia sclerotiorum:* less pronounced symptoms of infection | [169] |
| | *35S::SoCYP85A1* | tobacco (*Nicotiana tabacum*) | - longer primary root and more lateral roots<br>- enhanced drought tolerance: less severe stress symptoms, higher RWC, decreased water loss rate, increased proline content in one transgenic line, decreased MDA content and $H_2O_2$ level, increased activity of SOD, CAT, and POX (plants not watered for 10 days) | [170] |
| | *35S::PtCYP85A3* | tomato (*Lycopersicon esculentum*) poplar (*Populus davidiana* × *P. bolleana*) | - promoted growth and biomass production<br>- increased plant height, shoot fresh weight and fruit yield in tomato<br>- increased plant height and stem diameter, enhanced xylem formation in poplar | [171] |
| | *Ubi::OsDWF4* *Gt1::OsDWF4* seed-specific promoter | rice (*Oryza sativa*) | - enhanced tillering<br>- increased grain yield per plant, slightly increased 1000-grain weight<br>- increased leaf angle, more loose stature (not desired trait) | [172] |
| | *Ubi::ZmDWF4* | maize (*Zea mays*) | - bigger ears, improved grain yield per ear<br>- increased 1000-grain weight<br>- faster growth, increased plant height and node number<br>- increased leaf area, delayed leaf senescence | [173] |
| | *35S::PeCPD* | poplar (*Populus tomentosa*) | - increased plant height, biomass, stem diameter and xylem formation<br>- increased tolerance to salt stress: less visual injuries, lower $H_2O_2$ and $O_2^{\bullet-}$ formation, decreased MDA content, increased levels of soluble proteins and proline, increased SOD activity (plants exposed to 50 mM NaCl for 3 days and then to 100 mM NaCl for 12 days) | [174] |

**Table 4.** *Cont.*

| Enzyme | Promoter and Gene | Species | Phenotype of Transgenic Plants Compared to Control Lines | References |
|---|---|---|---|---|
| cytochrome catalyzing conversion of 6-deoxocastasterone to castasterone | *35S::DWF* | tomato (*Lycopersicon esculentum*) | - faster germination and seedling growth<br>- increased plant height and weight<br>- slender stature, leaf deformations<br>- faster ripening of fruits<br>- decrease in fruit yield per plant but increase when normalized on the cultivated area due to higher density of plants | [175] |
| | *35S::DWF* | tomato (*Lycopersicon esculentum*) | - increased tolerance to chilling stress: lesser amount of oxidized proteins, lower level of lipid peroxidation and electrolyte leakage, increased maximum quantum efficiency of PSII, increased activity of APX and enzymes participating in ascorbate and glutathione recycling (plants exposed to 4 °C for 3 days) | [176] |
| enzyme catalyzing the conversion of 6-deoxocathasterone and 3-dehydroteasterone to 6-deoxoty phasterol and typhasterol, respectively | *Ubi::TaD11-2A* | rice (*Oryza sativa*) | - increased grain length and 1000-grain weight<br>- increased starch content and decreased amylose content | [177] |

## 7. Abscisic Acid

The most important functions of ABA include regulation of dormancy, stomata opening, as well as maturation and germination of seeds. This phytohormone also participates in the response to abiotic stress, primarily drought [2]. In higher plants, carotenoids, specifically violaxanthin or neoxanthin, are substrates for ABA biosynthesis. An important enzyme necessary for the synthesis of violaxanthin (and indirectly neoxanthin) is zeaxanthin epoxidase (ZEP). In the ABA biosynthetic pathway, both xanthophylls are converted to the conformation 9-*cis*, and then 9-*cis*-epoxycarotenoid dioxygenase (NCED) catalyzes xanthoxin formation. Later, xanthoxin undergoes two-step oxidation—first to abscisic aldehyde and then to ABA. These reactions are catalyzed by short-chain alcohol dehydrogenase/reductase (SDR) and abscisic aldehyde oxidase (AAO), respectively [178]. Seo et al. [179] observed that under stress conditions AAO expression does not change, while the expression of the *LOS5/ABA3* (*LOS5*) gene is enhanced. This gene encodes an enzyme responsible for sulphation of AAO molybdenum cofactor [180]. Therefore, experiments on plants overexpressing the *LOS5* gene were also conducted.

The results of the experiments on transgenic plants with increased ABA content are collected in Table 5. They indicate that the modifications associated with this phytohormone are a very promising direction of research aimed at obtaining varieties with increased drought tolerance. On the other hand, it was reported that ABA overproducing transgenic tomato was significantly more vulnerable to xylem embolism [181]. Apart from the synthesis and degradation of this phytohormone, ABA transport also has an impact on stress tolerance. The *Lr34res* gene conferring durable resistance to multiple fungal pathogens in rice was reported to be an ABA transporter [182].

ABA-induced signaling pathways are a subject of intensive research. The expression of tomato genes encoding ABA receptors belonging to PYR/PYL/RCAR family in *A. thaliana* improved drought tolerance of transgenic plants [183]. Similarly, overexpression of native genes of subfamily III of PYR/PYL/RCAR family in *A. thaliana* resulted in increased ABA-sensitivity and enhanced drought resistance [184]. Overexpression of *OsPYL3* and *OsPYL9* in rice enhanced tolerance to cold stress and drought; and similarly, overexpression of *TaPYL4* in wheat improved drought tolerance [185]. Transgenic poplars overexpressing *PtPYRL1* or *PtPYRL5* were more tolerant to drought, cold, and osmotic stress [186]. Membrane-bound kinase OsPKR15 was shown to interact with OsPYL11, an orthologue of AtPYL9. Ectopic expression of OsPKR15 in *A. thaliana* increased its sensitivity to ABA and resulted in the enhancement of drought tolerance [187]. The potential of ABA receptors overexpression for the improvement of water-use efficiency (WUE) in crops was proposed by Mega et al. [188]. Rice overexpressing *OsPYL6* under the control of *Arabidopsis thaliana* Responsive to Dehydration 29A (*AtRD29A*) promoter displayed enhanced tolerance to dehydration. On the other hand, the reduced grain yield under non-stress conditions due to reduction in height, biomass, panicle branching, and spikelet fertility was also observed in transgenic plants [189]. The role of SNF 1-RELATED PROTEIN KINASE 2 (SnRK2), comprising a subfamily of plant-specific protein kinases, in ABA signaling and stress tolerance is being investigated [190]. Overexpression of wheat genes of *TaSnRK2s* in *A. thaliana* resulted in improved tolerance to drought, salt, and cold stress [185]. Overexpression of *ARR5*, encoding one of the SnRK2 targets, in *A. thaliana* resulted in ABA hypersensitivity and enhanced drought tolerance [191]. Transgenic *A. thaliana* expressing *TaCIPK27*, encoding a wheat kinase involved in stress response, displayed enhanced ABA-sensitivity and improved drought tolerance [192].

**Table 5.** Summary of the results of the experiments on transgenic plants with increased abscisic acid concentration. APX, ascorbate peroxidase; CAT, catalase; LOS, sulfurase of molybdenum cofactor required for abscisic aldehyde oxidase activity; MDA, malonyldialdehyde; NCED, 9-*cis*-epoxycarotenoid dioxygenase; POX, peroxidase; RWC, relative water content; SOD, superoxide dismutase; WUE, water-use efficiency; ZEP, zeaxanthin epoxidase.

| Enzyme | Promoter and Gene | Species | Phenotype of Transgenic Plants Compared to Control Lines | References |
|---|---|---|---|---|
| ZEP | *35S::AtZEP* | thale cress (*Arabidopsis thaliana*) | - increased tolerance to salt stress: increased fresh weight of stressed plants (seedlings exposed to 0–160 mM NaCl for 10 days)<br>- increased tolerance to osmotic stress: increased fresh weight of stressed plants (seedlings exposed to 0–400 mM mannitol for 10 days)<br>- increased drought tolerance: drought survival (plants not watered for 3 weeks, control plants died); reduced water loss from detached shoots | [193] |
| | *35S::MsZEP* | tobacco (*Nicotiana tabacum*) | - increased tolerance to salt stress: increased content of soluble sugars and proline, increased activity of SOD, and decreased content of MDA (plants watered with 200 mM NaCl for 2 weeks)<br>- increased drought tolerance: less pronounced wilting symptoms, increased content of soluble sugars and proline, increased activity of SOD and decreased content of MDA (plants not watered for 2 weeks) | [194] |
| | *EsABA1* under the control of artificial superpromoter | tobacco (*Nicotiana tabacum*) | - increased tolerance to salt stress: increased shoot dry weight and total root length in stressed plants (seedlings exposed to 250 mM NaCl for 4 weeks)<br>- reduced Chl degradation in leaf discs incubated in 400 and 600 mM NaCl solutions for 3 days | [195] |

**Table 5.** *Cont.*

| Enzyme | Promoter and Gene | Species | Phenotype of Transgenic Plants Compared to Control Lines | References |
|---|---|---|---|---|
| NCED | *35S::AtNCED3* | thale cress (*Arabidopsis thaliana*) | - increased drought tolerance: lower transpiration rate, less pronounced wilting symptoms in stressed plants (plants not watered for 18 days) | [196] |
| | *35S::VuNCED1* | creeping bentgrass (*Agrostis stolonifera*) | - increased tolerance to salt stress: increased fresh and dry biomass and less pronounced wilting symptoms in stressed plants, increased survival rate (plants watered with 0.2–0.8% NaCl for 10 weeks)<br>- increased drought tolerance: increased fresh and dry biomass, less pronounced wilting symptoms in plants exposed to water deficit, increased survival rate (reduced watering for 10 weeks) | [197] |
| | *35S::OsNCED3* | thale cress (*Arabidopsis thaliana*) | - delayed seed germination, slower growth, changed leaf morphology<br>- sugar oversensitivity<br>- increased drought tolerance: less severe stress symptoms (plants not watered for 9–13 days) | [198] |
| | *35S::CrNCED1* | tobacco (*Nicotiana nudicaulis*) | - increased tolerance to salt stress: reduced Chl degradation and $H_2O_2$ and $O_2^{\bullet-}$ generation in leaf discs incubated in 200 mM NaCl solution for 4 days<br>- increased drought tolerance: reduced water loss from detached leaves, higher turgor and increased RWC in stressed plants (plants not watered for 1 week), lower $H_2O_2$ and $O_2^{\bullet-}$ generation in leaves subjected to 80 min dehydration<br>- increased tolerance to oxidative stress: reduced Chl degradation in leaf discs incubated in 1% $H_2O_2$ solution for 4 days | [199] |
| | *35S::OsNCED4* | thale cress (*Arabidopsis thaliana*) | - delayed seed germination, slower growth, changed morphology<br>- sugar oversensitivity<br>- increased drought tolerance: less severe stress symptoms (plants not watered for 9–13 days) | [200] |
| | *35S::OsNCED3* | rice (*Oryza sativa*) | - promotion of leaf senescence (darkness induction protocol)<br>- increased drought tolerance: increased survival rate of stressed plants (seedlings not watered for 18 days)<br>- increased tolerance to salt stress: increased survival rate of stressed plants (seedlings exposed to 150 NaCl) | [201] |
| | *35S::VaNCED1* | grapevine (*Vitis vinifera*) | - changed leaf morphology<br>- increased drought tolerance: less severe stress symptoms (plants not watered for 50 days) | [202] |

**Table 5.** *Cont.*

| Enzyme | Promoter and Gene | Species | Phenotype of Transgenic Plants Compared to Control Lines | References |
|---|---|---|---|---|
| | *PvNCED1* under dexamethasone-inducible promoter | tobacco (*Nicotiana plumbaginifolia*) | - increased drought tolerance: reduced water loss from detached leaves with induced transgene expression; less pronounced wilting symptoms in plants with induced transgene expression (plants not watered for 10 days) | [203] |
| | *HvLea::AtNCED6* drought-responsive promoter | barley (*Hordeum vulgare*) | - improved performance under water deficit: higher RWC and $CO_2$ assimilation rate, improved WUE (plants maintained at 10% soil moisture level for 4 days, stress imposed after anthesis) | [204] |
| | *rd29A::LeNCED1* stress-responsive promoter | petunia (*Petunia hybrida*) | - increased drought tolerance: less pronounced wilting symptoms in stressed plants, reduced water loss, significantly increased survival rate (plants not watered for 2 weeks) | [205] |
| LOS | *AtLOS5* under the control of artificial superpromoter | tobacco (*Nicotiana tabacum*) | - increased drought tolerance: reduced water loss from detached leaves; less pronounced wilting symptoms, increased activity of CAT and APX and increased proline content in stressed plants (plants not watered for 6 days) | [206] |
| | *AtLOS5* under the control of artificial superpromoter | cotton (*Gossypium hirsutum*) | - increased drought tolerance: reduced water loss from detached shoots; less pronounced wilting symptoms in stressed plants (plants not watered for 5 days)<br>- increased fresh weight; SOD, POX, and APX activities; and proline content; decreased MDA content in plants exposed to reduced watering for 8 weeks | [207] |
| | *AtLOS5* under the control of artificial superpromoter | maize (*Zea mays*) | - increased drought tolerance: reduced water loss from stressed plants and increased survival rate (plants not watered for 2 weeks)<br>- increased SOD, CAT, and POX activities and proline content, decreased content of $H_2O_2$ and MDA in plants exposed to reduced watering for 5 days | [208] |
| Hydroxylase participating in ABA catabolism | RNAi-mediated suppression of the *Hv8′ hydroxylase* construct expressed under drought-responsive promoter | barley (*Hordeum vulgare*) | - improved performance under water deficit: higher RWC and $CO_2$ assimilation rate, improved WUE (plants maintained at 10% soil moisture level for 4 days, stress imposed after anthesis) | [204] |
| | RNAi-mediated suppression of the *OsABA8ox1* | rice (*Oryza sativa*) | - increased tolerance to alkalinity: increased survival rates, decreased membrane injury, MDA, $H_2O_2$, and $O_2^{\bullet -}$ content in roots (seedlings were exposed to 10, 15, 20 mM $Na_2CO_3$)<br>- increased survival rate, less severe stress symptoms, more vigorous growth, increased Chl content, increased panicle number, spikelets per panicle, percentage of filled spikelets and 1000-grain weight (seedlings were transplanted into soil of pH 7.59, 8.86, and 9.29)<br>- increased grain yield per plant under salt stress (plants grown in soil of pH = 9.29, EC = 834.4 μS $cm^{-1}$) | [209] |

Many of the ABA-responsive transcription factors have been identified to date, among them, those belonging to NAC, bZIP, AP2/ERF, and WRKY families [210]. Stress-responsive grapevine transcription factor VvNAC17 was shown to increase sensitivity to ABA and drought tolerance when its gene was overexpressed in Arabidopsis [211]. Similarly, the overexpression of soybean *GmNAC019* in *A. thaliana* led to the hypersensitivity to ABA and higher survival rate in a soil-drying assay [212]. The overexpression of drought-induced maize *ZmWRKY26* in *A. thaliana* improved its tolerance to drought and heat [213]. The positive role of *MaWRKY80* from banana in drought stress resistance was shown in the experiment with transgenic Arabidopsis. Among other effects, this transcription factor modulated the expression of genes encoding ABA biosynthetic enzymes [214]. *Capsicum annuum* ABA Induced ERF (*CaAIEF1*) expressed in *A. thaliana* enhanced drought tolerance of transgenic plants [215]. The overexpression of *VlbZIP30*, encoding a transcription factor belonging to the bZIP family in grapevine, in transgenic *A. thaliana* improved dehydration tolerance [216]. It was shown that corn transcription factor ZmbZIP33 interacts with core components of ABA signaling. Its overexpression in Arabidopsis led to the increase in ABA content and drought tolerance [217]. Arabidopsis plants overexpressing *TabZIP14-B* from wheat exhibited enhanced tolerance to salt and cold, as well as increased ABA sensitivity [218]. A maize gene *ZmMYB3R*, encoding MYB transcription factor, is known to be induced by ABA. Its overexpression in *A. thaliana* caused increased sensitivity to ABA and enhanced tolerance to drought and salt stress [219]. Increased sensitivity to ABA resulting in the enhanced tolerance to drought, salt, and osmotic stress was also observed in *A. thaliana* with overexpression of another transcription factor from maize, *ZmHDZIV14* [220]. Dehydration responsive element binding factors (DREB) belong to the AP2/ERF family. The expression of ABA-induced *AhDREB1* from peanuts in *A. thaliana* resulted in increased ABA levels and increased sensitivity to this phytohormone, as well as in improved drought tolerance [221]. ZmPTF1 transcription factor, belonging to the bHLH family, is known to be a positive regulator of ABA synthesis. Its overexpression in maize caused an increase in ABA content and enhanced drought tolerance [222]. The other examples of genetic modification of ABA receptors, ABA signaling components, and ABA-responsible transcription factors can be found in the reviews [223–225].

## 8. Ethylene

Ethylene is another phytohormone important for the regulation of the stress response. Among its other functions, the one important for farmers is the stimulation of fruit ripening [226]. Due to the simple structure of the molecule and ethylene occurrence in the gas phase, this compound is often applied exogenously. Treatment with the ethylene precursor, 1-amino-3-cyclopropane-1-carboxylic acid (ACC), was also applied [227].

As ethylene is known to be a plant growth inhibitor, many of the experiments aimed to decrease the synthesis of this phytohormone (Table 6). The main target of genetic engineering is 1-amino-3-cyclopropane-1-carboxylic acid synthase (ACS), responsible for the synthesis of the direct precursor of this phytohormone. This enzyme was discovered to be crucial for the regulation of ethylene biosynthesis. Partial silencing of the expression of ACS encoding genes in maize resulted in higher yields of transgenic lines compared to control when plants were exposed to drought [228]. Apple and melon fruits with decreased ACS activity ripened more slowly and were firmer than the fruits of non-transformed plants, which is a desirable trait if there is a need for longer storage [229,230]. Interestingly, inoculation of pea with the soil bacterium *Variovorax paradoxus* synthesizing ACC deaminase (ACC decomposing enzyme) resulted in improved growth and seed yield under drought conditions when compared to plants inoculated with the *V. paradoxus* mutant, in which ACC deaminase activity was significantly lower [231].

**Table 6.** Summary of the results of the experiments on mutant and transgenic plants with changed ethylene concentration. ACC, 1-aminocyclopropane-1-carboxylic acid; ACS, 1-aminocyclopropane-1-carboxylic acid synthase; CAT, catalase; ERF1, ethylene response factor 1; POX, peroxidase; RWC, relative water content; SOD, superoxide dismutase.

| Protein | Promoter and Gene | Species | Phenotype of Transgenic Plants Compared to Control Lines | References |
|---|---|---|---|---|
| ACS | mutant *acs7* | thale cress (*Arabidopsis thaliana*) | - slightly faster germination<br>- faster growth at the vegetative stage<br>- increased tolerance to salt stress: improved survival of salt-treated seedlings (exposed to 150 mM NaCl); germination in presence of 150 mM NaCl slowed down to a lesser extent<br>- increased tolerance to osmotic stress: germination in presence of 300 mM mannitol slowed down to a lesser extent<br>- increased tolerance to heat stress: lower percentage of chlorosis in stressed seedlings (exposed to 43 °C for 3 h) | [232] |
| | *35S::ACS* antisense target silencing | apple tree (*Malus pumila*) | - firmer fruits<br>- increased shelf-life | [229] |
| | *35S::PmACS* antisense target silencing | melon (*Cucumis melo*) | - firmer fruits<br>- slower ripening | [230] |
| | *ZmUbi1::ZM-ACS6* hairpin target silencing | maize (*Zea mays*) | - increased grain yield of some lines in locations where drought occurred (field trials)<br>- increased yield under low-nitrogen treatment (field trials) | [228] |
| | *ZmUbi1::ScACS1*<br>*ZmUbi1::ScACS2*<br>*ZmUbi1::ScACS3*<br>hairpin target silencing | sugarcane hybrid cultivar (*Saccharum officinarum* × *Saccharum spontaneum*) | - increased plant height, leaf length, and leaf area<br>- reduced carbon assimilation<br>- no reduction in Chl content or sucrose levels<br>- induction of non-enzymatic antioxidant apparatus | [233] |
| ACC deaminase | *35S::ACCD* from bacteria<br>Other promoters used were: root-specific promoter of *rolD* gene from *Agrobacterium rhizogenes*, pathogenesis-related promoter of *prb-1b* gene from tobacco | tomato (*Lycopersicon esculentum*) | - increased tolerance to flooding, especially in lines where root-specific promoter was used: increased shoot fresh and dry weight, increased Chl content (plants flooded for 9 days) | [234] |
| | *35S::ACCD* from *Pseudomonas putida*,<br>Other promoters used were: root-specific promoter of *rolD* gene from *Agrobacterium rhizogenes*, pathogenesis-related promoter of *prb-1b* gene from tobacco | canola (*Brassica napus*) | - increased tolerance to salt stress in lines where root-specific promoter was used: increased shoot and root dry weight, increased protein and Chl content (seedlings treated with 0–200 mM NaCl for 6 weeks of growth) | [235] |

**Table 6.** *Cont.*

| Protein | Promoter and Gene | Species | Phenotype of Transgenic Plants Compared to Control Lines | References |
|---|---|---|---|---|
| | *35S::TaACCD* from fungus *Trichoderma asperellum* | thale cress (*Arabidopsis thaliana*) | - improved root growth: increase in root length and root number, increase in total fresh weight and RWC (but no difference in dry weight), increase in seed number per pod<br>- improved tolerance to salt stress: less severe stress symptoms, increased root length, number, and weight, increased RWC, decreased $H_2O_2$ content, later occurrence of $O_2^{\bullet-}$ level increase and cell damage, decreased electrolyte leakage, less pronounced decrease in Chl content, increased POX activity (seedlings watered with 150 mM NaCl for 8 days) | [236] |
| | *35S::acdS* gene of *Pseudomonas veronii* | thale cress (*Arabidopsis thaliana*) | - slightly improved tolerance to salt stress: improved seedling growth (seeds exposed to 0, 50, 100 mM NaCl, seedlings grown for 25 days)<br>- improved tolerance to flooding: less severe stress symptoms (plants water-logged for 5 days) | [237] |
| | *35S::ACCD* from *Achromobacter xylosoxidans* | geranium (*Pelargonium graveolens*) | - increased tolerance to salt stress: less pronounced decrease in Chl content, higher $CO_2$ assimilation rate, lower content of $H_2O_2$, increased activity of SOD, CAT, and POX (plants exposed to 50–200 mM NaCl for 30 days)<br>- increased drought tolerance: less pronounced decrease in Chl content, higher $CO_2$ assimilation rate, lower content of $H_2O_2$, increased activity of SOD, CAT and POX (reduced watering or ceasing watering for 15 days) | [238] |

Inactivation of *ZmACO2* encoding ACC oxidase2 catalyzing the final step of ethylene biosynthesis via genome editing using CRISPR/Cas9 method led to the reduction of ethylene production in developing ears and increased grain yield per ear [239].

There were also attempts to modify ethylene-induced signaling. The ethylene response factor superfamily is known to participate in response to various environmental stresses, such as drought, salt, heat, and cold. An elegant summary of the research on ERFs, their participation in stress response, and their genetic engineering using CRISPR/Cas9 genome editing tool was written by Debbarma et al. [240]. The modification of ERF-dependent signaling turned out to be promising also in the case of improving biotic stress response. Transgenic rice overexpressing *OsERF83* was more resistant to *Magnaporthe orizae*, causing one of the most destructive diseases in rice [241]. Overexpression of *GmERF3* in tobacco resulted in increased tolerance to drought and salt stress but also enhanced resistance to the bacteria *Ralstonia solanacearum*, fungus *Alternaria alternata*, and tobacco mosaic virus [242]. Transgenic *A. thaliana* with overexpression of *MbERF12* from *Malus baccata* displayed enhanced antioxidant response and increased tolerance to low temperature and salt stress [243]. Transgenic lines of *A. thaliana* overexpressing *ERF1* were more tolerant to drought, salt, and heat stress [244]. It was shown that ectopic constitutive expression of *ERF95* and *ERF97* led to the increase in tolerance to the heat stress in *A. thaliana* [245]. Overexpression of native ERF in rubber tree resulted in the stimulation of root growth, increased dry biomass, and increased tolerance to salt stress [246]. Overexpression of *TdSHN1*, encoding cDNA of SHINE-type ERF transcription factor from durum wheat, in tobacco improved tolerance to Cd, Cu, and Zn [247]. Overexpression of *MdERF1B* from the apple tree significantly enhanced cold tolerance of *Arabidopsis thaliana* seedlings, and transgenic apple seedlings and calli [243]. Transgenic tomato overexpressing *SlERF5* was more tolerant to drought and salt stress [248]. Other examples of modulation of ethylene signaling resulting in the enhancement of the tolerance to salt stress were reviewed by Riyazuddin et al. [249]. Members of the ARGOS family are known to be negative regulators of ethylene responses. Genetic engineering targeted at ARGOS8, including both overexpression and modification by the CRISPR/Cas 9 method, was used to obtain maize with improved grain yield under drought stress conditions [250]. The regulatory role of miRNA in stress response and its connection with ethylene signaling were also elucidated. For example, salinity-induced miR319 was reported to positively regulate ethylene synthesis and increase tolerance to salt stress in switchgrass [251].

## 9. Jasmonic Acid and Its Derivatives

Jasmonates play a role in plant response to various stress factors, including biotic ones [252]. They also participate in the regulation of plant development. The exogenous application of methyl jasmonate increased the yield of soybean [253]. Application of JA or JA together with gibberellin $GA_3$ resulted in an increase in ginseng yield [254]. The administration of JA alleviated the adverse effects of salt stress on rice and barley [255,256]. However, there is inconsistency in the results reported in the literature, as improved salt tolerance was also observed in transgenic plants with enhanced JA degradation [257]. JA is also used as an elicitor in the production of various secondary metabolites [258].

The experiments on transgenic plants with changed content of JA or its methyl ester were also carried out (Table 7). Among them, interesting ones concern plants overexpressing jasmonic acid carboxyl methyltransferase (JMT), which converts JA into its methyl ester. Another strategy is obtaining plants with overexpression of 13-lipoxygenase or 12-oxophytodienoate reductase participating in the biosynthesis of this phytohormone. Such modifications resulted in an increased tolerance to selected biotic and abiotic stresses and stimulation of the growth of underground storage organs.

**Table 7.** Summary of the results of the experiments on transgenic plants with increased concentration of jasmonic acid or its methyl ester. AOC, allene oxide cyclase; JMT, jasmonic acid carboxyl methyltransferase; LOX, lipooxygenase; MDA, malonyldialdehyde; OPR, 12-oxophytodienoate reductase; SOD, superoxide dismutase.

| Enzyme | Promoter and Gene | Species | Phenotype of Transgenic Plants Compared to Control Lines | References |
|---|---|---|---|---|
| | | **Enhancement of jasmonic acid biosynthesis** | | |
| LOX | *35S::TomloxD* | tomato (*Lycopersicon esculentum*) | - increased tolerance to heat stress: less pronounced wilting symptoms and faster recovery (seedlings exposed to 50 °C for 2 h)<br>- increased resistance to fungal infection: less pronounced symptoms of infection by *Cladosporium fulvum* | [259] |
| | *35S::TomloxD* | tomato (*Lycopersicon esculentum*) | - decreased herbivore insect feeding (plants exposed to *Helicoverpa armigera*)<br>- increased resistance to necrotrophic pathogen: less severe visual symptoms of infection with *Botrytis cinerea* | [260] |

**Table 7.** *Cont.*

| Enzyme | Promoter and Gene | Species | Phenotype of Transgenic Plants Compared to Control Lines | References |
|---|---|---|---|---|
| | | **Enhancement of jasmonic acid biosynthesis** | | |
| | *35S::CmLOX10* | thale cress (*Arabidopsis thaliana*) | - increased drought tolerance: less severe wilting symptoms, increased survival rate, lower electrolyte leakage, $H_2O_2$ and MDA level (plants not watered for 10 d)<br>- decreased stomatal aperture and water loss from leaves | [261] |
| | *35S::TgLOX4*<br>*35S::TgLOX5* | thale cress (*Arabidopsis thaliana*) | - longer leaves in *TgLOX5* overexpressing line, wider leaves<br>- increased plant height<br>- stimulated branching | [262] |
| AOC | *35S::TaAOC1*<br>*Ubi:TaAOC1* | thale cress (*Arabidopsis thaliana*) wheat (*Triticum aestivum*) | - shorter roots<br>- increased activity of SOD<br>- improved tolerance to salt stress: less pronounced reduction in root growth of transgenic wheat (seedlings were treated with increasing concentrations of NaCl for 4 days then exposed to 200 mM NaCl for the next 4 days); increased survival rate of transgenic Arabidopsis (plants were treated with increasing concentrations of NaCl for 4 days, then exposed to 200 mM NaCl for the next 2 weeks)<br>- improved tolerance to osmotic stress and oxidative stress in transgenic Arabidopsis: no or very small reduction in root length (seedlings exposed to 100, 200, 300 mM mannitol or 1, 2 mM $H_2O_2$ for 10 d) | [263] |
| | *35S::TaAOS* | tobacco (*Nicotiana benthamiana*) | - enhanced tolerance to Zn: lesser decrease in Chl content in leaf discs exposed to 10 and 20 mM $ZnCl_2$ for 6 days | [264] |
| | *Ubi::AhAOC* | rice (*Oryza sativa*) | - increased plant height and root length<br>- improved tolerance to salt stress: less pronounced reduction in seedling root growth (seeds germinated in presence of 80 or 120 mM NaCl); less severe stress symptoms, less pronounced reduction in plant height, increased content of proline and soluble sugars (plants exposed to 120 mM NaCl for 2 weeks) | [265] |
| | *35S::GhAOC1* | thale cress (*Arabidopsis thaliana*) | - enhanced tolerance to Cu: higher survival rate, increased shoot fresh weight and photosynthetic efficiency, reduced cell membrane damage and lipid peroxidation (plants watered with 120 μM $CuCl_2$ for 10 days) | [266] |
| OPR | *35S::ZmOPR1* | thale cress (*Arabidopsis thaliana*) | - increased tolerance to salt and osmotic stress: improved germination in the presence of NaCl or mannitol (seeds exposed to 100–200 mM NaCl or 100–500 mM mannitol), more seedlings remained green (observations made after 7 days)<br>- no differences in survival rate when older seedlings were exposed to NaCl (300 mM NaCl for 10 days) | [267] |
| | *Ubi1::AtOPR3* | wheat (*Triticum aestivum*) | - changed timing of development: delayed germination, slower growth, late flowering, delayed senescence<br>- increased tolerance to short-term freezing: higher maximum quantum efficiency of PSII and decreased electrolyte leakage (plants were transferred to 4 °C for 24 h; their detached leaves were then subjected to decreasing temperatures to final 1 °C, −2 °C, or −5 °C and then incubated for 24 h) | [268] |

**Table 7.** *Cont.*

| Enzyme | Promoter and Gene | Species | Phenotype of Transgenic Plants Compared to Control Lines | References |
|---|---|---|---|---|
| | | **Enhancement of methyl jasmonate biosynthesis** | | |
| JMT | *35S::AtJMT* | thale cress (*Arabidopsis thaliana*) | - increased resistance to fungal infection: lack of severe infection symptoms in plants 3 days after spraying with *Botrytis cinerea* spores | [269] |
| | *35S::AtJMT* | thale cress (*Arabidopsis thaliana*) | - increased resistance to infection with bacteria *Pseudomonas syringae* and enhanced expression of defense genes | [270] |
| | *35S::BcNTR1* | soybean (*Glycine max*) | - increased drought tolerance: less pronounced wilting symptoms, plant survival (plants not watered for 6 days, all control plants died); reduced water loss from detached leaves<br>- increased tolerance to osmotic stress: increased fresh biomass of seedlings germinating in the presence of 0.3 M mannitol | [271] |
| | *35S::AtJMT* | potato (*Solanum tuberosum*) | - stimulated tuberization, increased tuber size<br>- increased tuber yield per plant | [272] |
| | *35S::AtJMT* | ginseng (*Panax ginseng*) | - root growth stimulation<br>- increased content of protopanaxadiol group of ginsenosides | [273] |

Jasmonic acid signaling is an object of intensive research [274]. For example, overexpression of TdTIFY11a, a member of TIFY protein family participating in JA signaling, from *Triticum durum* in *A. thaliana* promoted germination under salt stress [275]. Expression of *VaNAC17*, encoding *Vitis amurensis* transcription factor, known to be induced by drought stress, in *A. thaliana* resulted in enhanced JA synthesis and drought tolerance [276]. Similarly, the overexpression of *VaNAC26* improved tolerance to drought and salt stress in *A. thaliana* [277]. The overexpression of *OsbHLH034* gene encoding transcription factor acting as positive regulator in JA signaling resulted in the increased resistance to rice bacterial blight, but it also increased sensitivity to salt stress [278]. The overexpression of *OsbHLH148* improved drought tolerance in transgenic rice [279]. Heterologous overexpression of JA-responsive transcription factor from *Ipomea batatas IbMYB116* in *A. thaliana* caused upregulation of the expression of JA biosynthetic genes, promoted JA accumulation and the JA response, and improved the tolerance to drought stress [280]. Enhanced proline accumulation and increased drought tolerance were also observed in soybean overexpressing another JA-responsive transcription factor, *GmTGA15* [281]. The modulation of the expression of JAZ proteins that are negative regulators of JA signaling allow us to obtain plants more tolerant to salinity and drought [257,282].

## 10. Future Perspectives

The examples presented here indicate that modification of phytohormone metabolism and signaling is a promising direction of research aimed at the improvement of crop productivity and stress tolerance. The progress in this field is possible due to broadening of the knowledge concerning the regulation of plant growth, development, and stress response, as well as due to the improvement of the methodology used. Many genes that can be targets of genetic engineering have been identified up to date [7]. The extensive research aiming at deciphering phytohormone signaling pathways is being carried out. The modification of this signaling at various levels, from elements of signaling cascades, through transcription factors to miRNAs, is a very promising direction of genetic engineering of crop plants.

Considering the methods of genetic engineering, the most promising innovation is genome editing using the so-called CRISPR/Cas9 system [283]. The system is based on nucleases that can be relatively easily programmed to search for specific DNA sequences. Available variants of effector nucleases allow various modifications of the target region. This makes CRISPR/Cas9 a fast, effective, and precise genome editing tool [283]. It is used both to discover functions of certain genes and to obtain plants of potential application in agriculture. CRISPR/Cas9 genome editing seems especially promising in research aimed at modulation of cytokinin levels [284].

Intensive research on the regulation of gene expression led to the discovery of many promoters specific to certain tissues, organs, or stage of plant development. The application of these promoters allows better control of the time and site of transgene expression. Scientists also designed artificial promoters [285]. There are systems enabling us to combine and introduce multiple genes at once (such as the Golden Gate modular cloning box), as well as methods for the introduction of large DNA fragments into plant cells. New successful protocols of crop species transformation are being developed [7].

An important obstacle in obtaining transgenic plants with improved yield is the well-known trade-off between stress defense and plant growth. One of its reasons is the energetic cost of the development and maintenance of various protective mechanisms, both biochemical and morphological. However, the negative effect of defense induction on growth often results from antagonistic crosstalk between phytohormones rather than from an identified metabolic expenditure. Sometimes, it is caused by pleiotropic effects of certain resistance traits or is a consequence of genetic linkage [286]. Therefore, it is possible to reduce the costs of plant defense. The strategies aimed at such a reduction were summarized by Karasov et al. [286].

To date, the majority of studies on transgenic lines with altered phytohormone content or signaling have been conducted under laboratory conditions. To obtain improved varieties suitable for regular cultivation, it is necessary to carry out large-scale field tests to determine whether the modifications introduced allow us to obtain the desired phenotype under natural conditions. At the same time, care should be taken to minimize the risk of transgene leak, so that genetically modified varieties would not pose the threat of contamination to the genomes of closely related wild species.

**Funding:** Subsidy for statutory activities from Polish Ministry of Education and Science No. 19000882 is acknowledged.

**Data Availability Statement:** Not applicable.

**Conflicts of Interest:** The author declares no conflict of interest.

## References


1. Ort, D.R.; Merchant, S.S.; Alric, J.; Barkan, A.; Blankenship, R.E.; Bock, R.; Croce, R.; Hanson, M.R.; Hibberd, J.M.; Long, S.P.; et al. Redesigning photosynthesis to sustainably meet global food and bioenergy demand. *Proc. Natl. Acad. Sci. USA* **2015**, *112*, 8529–8536. [CrossRef] [PubMed]
2. Wani, S.H.; Kumar, V.; Shriram, V.; Sah, S.K. Phytohormones and their metabolic engineering for abiotic stress tolerance in crop plants. *Crop J.* **2016**, *4*, 162–176. [CrossRef]
3. Betti, M.; Bauwe, H.; Busch, F.A.; Fernie, A.R.; Keech, O.; Levey, M.; Ort, D.R.; Parry, M.A.J.; Sage, R.; Timm, S.; et al. Manipulating photorespiration to increase plant productivity: Recent advances and perspectives for crop improvement. *J. Exp. Bot.* **2016**, *67*, 2977–2988. [CrossRef] [PubMed]
4. Long, S.P.; Zhu, X.-G.; Naidu, S.L.; Ort, D.R. Can improvement in photosynthesis increase crop yields? *Plant Cell Environ.* **2006**, *29*, 315–330. [CrossRef] [PubMed]
5. Fernandez, M.G.S.; Becraft, P.W.; Yin, Y.; Lübberstedt, T. From dwarves to giants? Plant height manipulation for biomass yield. *Trends Plant Sci.* **2009**, *14*, 454–461. [CrossRef] [PubMed]
6. Nowicka, B.; Ciura, J.; Szymańska, R.; Kruk, J. Improving photosynthesis, plant productivity and abiotic stress tolerance—Current trends and future perspectives. *J. Plant Physiol.* **2018**, *231*, 415–433. [CrossRef]
7. Nowicka, B. Target genes for plant productivity improvement. *J. Biotechnol.* **2019**, *298*, 21–34. [CrossRef]
8. Kazan, K. Diverse roles of jasmonates and ethylene in abiotic stress tolerance. *Trends Plant Sci.* **2015**, *20*, 219–229. [CrossRef]

9. Peleg, Z.; Blumwald, E. Hormone balance and abiotic stress tolerance in crop plants. *Curr. Opin. Plant Biol.* **2011**, *14*, 290–295. [CrossRef]
10. Ciura, J.; Kruk, J. Phytohormones as targets for improving plant productivity and stress tolerance. *J. Plant Physiol.* **2018**, *229*, 32–40. [CrossRef]
11. Busov, V.B.; Brunner, A.M.; Strauss, S.H. Genes for control of plant stature and form. *New Phytol.* **2008**, *177*, 589–607. [CrossRef] [PubMed]
12. Mano, Y.; Nemoto, K. The pathway of auxin biosynthesis in plants. *J. Exp. Bot.* **2012**, *63*, 2853–2872. [CrossRef] [PubMed]
13. Domingo, C.; Andrés, F.; Tharreau, D.; Iglesias, D.J.; Talón, M. Constitutive expression of *OsGH3.1* reduces auxin content and enhances defense response and resistance to a fungal pathogen in rice. *Mol. Plant Microbe Interact.* **2009**, *22*, 201–210. [CrossRef] [PubMed]
14. Jung, H.; Lee, D.K.; Choi, Y.D.; Kim, J.K. *OsIAA6*, a member of the rice Aux/IAA gene family, is involved in drought tolerance and tiller outgrowth. *Plant Sci.* **2015**, *236*, 304–312. [CrossRef] [PubMed]
15. Lee, M.; Jung, J.H.; Han, D.Y.; Seo, P.J.; Park, W.J.; Park, C.M. Activation of a flavin monooxygenase gene *YUCCA7* enhances drought resistance in Arabidopsis. *Planta* **2012**, *235*, 923–938. [CrossRef] [PubMed]
16. Okushima, Y.; Mitina, I.; Quach, H.L.; Theologis, A. AUXIN RESPONSE FACTOR 2 (ARF2): A pleiotropic developmental regulator. *Plant J.* **2005**, *43*, 29–46. [CrossRef]
17. Multani, D.S.; Briggs, S.P.; Chamberlin, M.A.; Blakeslee, J.J.; Murphy, A.S.; Johal, G.S. Loss of an MDR transporter in compact stalks of maize *br2* and sorghum *dw3* mutants. *Science* **2003**, *302*, 81–84. [CrossRef]
18. Zhang, Z.; Gao, L.; Ke, M.; Gao, Z.; Tu, T.; Huang, L.; Chen, J.; Guan, Y.; Huang, X.; Chen, X. GmPIN1-mediated auxin asymmetry regulates leaf petiole angle and plant architecture in soybean. *J. Integr. Plant Biol.* **2022**, *64*, 1325–1338. [CrossRef]
19. Kim, J.I.; Baek, D.; Park, H.C.; Chun, H.J.; Oh, D.H.; Lee, M.K.; Cha, J.Y.; Kim, W.Y.; Kim, M.C.; Chung, W.S.; et al. Overexpression of Arabidopsis *YUCCA6* in potato results in high-auxin developmental phenotypes and enhance. *Mol. Plant* **2013**, *6*, 337–349. [CrossRef]
20. Ke, Q.; Wang, Z.; Ji, C.Y.; Jeong, J.C.; Lee, H.S.; Li, H.; Xu, B.; Deng, X.; Kwak, S.S. Transgenic poplar expressing Arabidopsis *YUCCA6* exhibits auxin-overproduction phenotypes and increased tolerance to abiotic stress. *Plant Physiol. Biochem.* **2015**, *94*, 19–27. [CrossRef]
21. Park, S.C.; Kim, H.S.; Lee, H.U.; Kim, Y.H.; Kwak, S.S. Overexpression of Arabidopsis *YUCCA6* enhances environment stress tolerance and inhibits storage root formation in sweetpotato. *Plant Biotechnol. Rep.* **2019**, *13*, 345–352. [CrossRef]
22. Ficcadenti, N.; Sestili, S.; Pandolfini, T.; Cirillo, C.; Rotino, G.L.; Spena, A. Genetic engineering of parthenocarpic fruit development in tomato. *Mol. Breed.* **1999**, *5*, 463–470. [CrossRef]
23. Mezzetti, B.; Landi, L.; Pandolfini, T.; Spena, A. The *defH9-iaaM* auxin-synthesizing gene increases plant fecundity and fruit production in strawberry and raspberry. *BMC Biotechnol.* **2004**, *4*, 4. [CrossRef] [PubMed]
24. Costantini, E.; Landi, L.; Silvestroni, O.; Pandolfini, T.; Spena, A.; Mezzetti, B. Auxin synthesis-encoding transgene enhances grape fecundity. *Plant Physiol.* **2007**, *143*, 1689–1694. [CrossRef] [PubMed]
25. Zhang, M.; Zheng, X.; Song, S.; Zeng, Q.; Hou, L.; Li, D.; Zhao, J.; Wei, Y.; Li, X.; Luo, M.; et al. Spatiotemporal manipulation of auxin biosynthesis in cotton ovule epidermal cells enhances fiber yield and quality. *Nat. Biotechnol.* **2011**, *29*, 453–458. [CrossRef] [PubMed]
26. Kolachevskaya, O.O.; Alekseeva, V.V.; Sergeeva, L.I.; Rukavtsova, E.B.; Getman, I.A.; Vreugdenhil, D.; Buryanov, Y.I.; Romanov, G.A. Expression of auxin synthesis gene *tms1* under control of tuber-specific promoter enhances potato tuberization in vitro. *J. Integr. Plant Biol.* **2015**, *57*, 734–744. [CrossRef]
27. Du, H.; Wu, N.; Fu, J.; Wang, S.; Li, X.; Xiao, J.; Xiong, L. A GH3 family member, *OsGH3-2*, modulates auxin and abscisic acid levels and differentially affects drought and cold tolerance in rice. *J. Exp. Bot.* **2012**, *63*, 6467–6480. [CrossRef]
28. Li, Z.; Zhang, X.; Zhao, Y.; Li, Y.; Zhang, G.; Peng, Z.; Zhang, J. Enhancing auxin accumulation in maize root tips improves root growth and dwarfs plant height. *Plant Biotechnol. J.* **2018**, *16*, 86–99. [CrossRef]
29. Hu, Y.; Xie, Q.; Chua, N.H. The Arabidopsis auxin-inducible gene *ARGOS* controls lateral organ size. *Plant Cell* **2003**, *15*, 1951–1961. [CrossRef]
30. Guo, M.; Rupe, M.A.; Wei, J.; Winkler, C.; Goncalves-Butruille, M.; Weers, B.P.; Cerwick, S.F.; Dieter, J.A.; Duncan, K.E.; Howard, R.J.; et al. Maize *ARGOS1* (*ZAR1*) transgenic alleles increase hybrid maize yield. *J. Exp. Bot.* **2014**, *65*, 249–260. [CrossRef]
31. Sun, Y.; Wang, C.; Wang, N.; Jiang, X.; Mao, H.; Zhu, C.; Wen, F.; Wang, X.; Lu, Z.; Yue, G.; et al. Manipulation of *Auxin Response Factor 19* affects seed size in the woody perennial *Jatropha curcas*. *Sci. Rep.* **2017**, *7*, 40844. [CrossRef] [PubMed]
32. Kang, C.; He, S.; Zhai, H.; Li, R.; Zhao, N.; Liu, Q. A sweetpotato auxin response factor gene (*IbARF5*) is involved in carotenoid biosynthesis and salt and drought tolerance in transgenic Arabidopsis. *Front. Plant Sci.* **2018**, *9*, 1307. [CrossRef] [PubMed]
33. Bouzroud, S.; Gasparini, K.; Hu, G.; Barbosa, M.A.M.; Rosa, B.L.; Fahr, M.; Bendaou, N.; Bouzayen, M.; Zsögön, A.; Smouni, A.; et al. Down regulation and loss of *Auxin Response Factor 4* function using CRISPR/Cas9 alters plant growth, stomatal function and improves tomato tolerance to salinity and osmotic stress. *Genes* **2020**, *11*, 272. [CrossRef] [PubMed]
34. Chen, M.; Zhu, X.; Liu, X.; Wu, C.; Yu, C.; Hu, G.; Chen, L.; Chen, R.; Bouzayen, M.; Zouine, M.; et al. Knockout of auxin response factor SlARF4 improves tomato resistance to water deficit. *Int. J. Mol. Sci.* **2021**, *22*, 3347. [CrossRef] [PubMed]

35. He, Q.; Yang, L.; Hu, W.; Zhang, J.; Xing, Y. Overexpression of an auxin receptor *OsAFB6* significantly enhanced grain yield by increasing cytokinin and decreasing auxin concentrations in rice panicle. *Sci. Rep.* **2018**, *8*, 14051. [CrossRef]
36. Garrido-Vargas, F.; Godoy, T.; Tejos, R.; O'brien, J.A. Overexpression of the auxin receptor *AFB3* in Arabidopsis results in salt stress resistance and the modulation of *NAC4* and *SZF1*. *Int. J. Mol. Sci.* **2020**, *21*, 9528. [CrossRef]
37. Leng, P.; Ji, Q.; Asp, T.; Frei, U.K.; Ingvardsen, C.R.; Xing, Y.; Studer, B.; Redinbaugh, M.; Jones, M.; Gajjar, P.; et al. Auxin binding protein 1 reinforces resistance to Sugarcane mosaic virus in maize. *Mol. Plant* **2017**, *10*, 1357–1360. [CrossRef]
38. Zalabák, D.; Pospíšilová, H.; Šmehilová, M.; Mrízová, K.; Frébort, I.; Galuszka, P. Genetic engineering of cytokinin metabolism: Prospective way to improve agricultural traits of crop plants. *Biotechnol. Adv.* **2013**, *31*, 97–117. [CrossRef]
39. Frébort, I.; Kowalska, M.; Hluska, T.; Frébortová, J.; Galuszka, P. Evolution of cytokinin biosynthesis and degradation. *J. Exp. Bot.* **2011**, *62*, 2431–2452. [CrossRef]
40. Kamínek, M.; Šolcová, B.; Trčková, M.; Motyka, V.; Daskalova, S.; Elliott, M.C. Exploitation of Mechanisms Regulating Cytokinin Levels to Improve Cereals. In *Phytohormones in Plant Biotechnology and Agriculture*; Machácková, I., Romanov, G.A., Eds.; Springer: Dordrecht, The Netherlands, 2003; pp. 109–128.
41. Schmülling, T.; Werner, T.; Riefler, M.; Krupková, E.; Bartrina, I.; Manns, Y.; Dortay, H. Cytokinin as a Regulatory Factor for Yield and Biomass Distribution in Crop Plants. In *Phytohormones in Plant Biotechnology and Agriculture*; Machácková, I., Romanov, G.A., Eds.; Springer: Dordrecht, The Netherlands, 2003; pp. 97–108.
42. Bartrina, I.; Otto, E.; Strnad, M.; Werner, T.; Schmülling, T. Cytokinin regulates the activity of reproductive meristems, flower organ size, ovule formation, and thus seed yield in *Arabidopsis thaliana*. *Plant Cell* **2011**, *23*, 69–80. [CrossRef]
43. Schwarz, I.; Scheirlinck, M.T.; Otto, E.; Bartrina, I.; Schmidt, R.C.; Schmülling, T. Cytokinin regulates the activity of the inflorescence meristem and components of seed yield in oilseed rape. *J. Exp. Bot.* **2020**, *71*, 7146–7159. [CrossRef] [PubMed]
44. Nguyen, H.N.; Kambhampati, S.; Kisiala, A.; Seegobin, M.; Emery, R.J.N. The soybean (*Glycine max* L.) cytokinin oxidase/dehydrogenase multigene family; Identification of natural variations for altered cytokinin content and seed yield. *Plant Direct* **2021**, *5*, e00308. [CrossRef] [PubMed]
45. Gan, S.; Amasino, R.M. Inhibition of leaf senescence by autoregulated production of cytokinin. *Science* **1995**, *270*, 1986–1988. [CrossRef]
46. Jameson, P.E.; Song, J. Cytokinin: A key driver of seed yield. *J. Exp. Bot.* **2016**, *67*, 593–606. [CrossRef]
47. Vercruyssen, L.; Gonzalez, N.; Werner, T.; Schmülling, T.; Inzé, D. Combining enhanced root and shoot growth reveals cross talk between pathways that control plant organ size in Arabidopsis. *Plant Physiol.* **2011**, *155*, 1339–1352. [CrossRef] [PubMed]
48. Rivero, R.M.; Shulaev, V.; Blumwald, E. Cytokinin-dependent photorespiration and the protection of photosynthesis during water deficit. *Plant Physiol.* **2009**, *150*, 1530–1540. [CrossRef]
49. Peleg, Z.; Reguera, M.; Tumimbang, E.; Walia, H.; Blumwald, E. Cytokinin-mediated source/sink modifications improve drought tolerance and increase grain yield in rice under water-stress. *Plant Biotechnol. J.* **2011**, *9*, 747–758. [CrossRef]
50. Wilkinson, S.; Kudoyarova, G.R.; Veselov, D.S.; Arkhipova, T.N.; Davies, W.J. Plant hormone interactions: Innovative targets for crop breeding and management. *J. Exp. Bot.* **2012**, *63*, 3499–3509. [CrossRef]
51. Yin, W.; Xiao, Y.; Niu, M.; Meng, W.; Li, L.; Zhang, X.; Liu, D.; Zhang, G.; Qian, Y.; Sun, Z.; et al. ARGONAUTE2 enhances grain length and salt tolerance by activating *BIG GRAIN3* to modulate cytokinin distribution in rice. *Plant Cell* **2020**, *32*, 2292–2306. [CrossRef]
52. Wang, C.; Wang, G.; Gao, Y.; Lu, G.; Habben, J.E.; Mao, G.; Chen, G.; Wang, J.; Yang, F.; Zhao, X.; et al. A cytokinin-activation enzyme-like gene improves grain yield under various field conditions in rice. *Plant Mol. Biol.* **2020**, *102*, 373–388. [CrossRef]
53. Veselov, D.S.; Kudoyarova, G.R.; Kudryakova, N.V.; Kusnetsov, V.V. Role of cytokinins in stress resistance of plants. *Russ. J. Plant Physiol.* **2017**, *64*, 15–27. [CrossRef]
54. Thu, N.B.A.; Hoang, X.L.T.; Truc, M.T.; Sulieman, S.; Thao, N.P.; Tran, L.-S.P. Cytokinin Signaling in Plant Response to Abiotic Stresses. In *Mechanism of Plant Hormone Signaling under Stress*; Pandey, G., Ed.; John Wiley & Sons, Inc.: Hoboken, NJ, USA, 2017; pp. 71–100.
55. Chen, L.; Zhao, J.; Song, J.; Jameson, P.E. Cytokinin dehydrogenase: A genetic target for yield improvement in wheat. *Plant Biotechnol. J.* **2020**, *18*, 614–630. [CrossRef] [PubMed]
56. Sakakibara, H. Cytokinin biosynthesis and transport for systemic nitrogen signaling. *Plant J.* **2021**, *105*, 421–430. [CrossRef]
57. Hu, Y.; Jia, W.; Wang, J.; Zhang, Y.; Yang, L.; Lin, Z. Transgenic tall fescue containing the *Agrobacterium tumefaciens ipt* gene shows enhanced cold tolerance. *Plant Cell Rep.* **2005**, *23*, 705–709. [CrossRef] [PubMed]
58. Tereshonok, D.V.; Stepanova, A.Y.; Dolgikh, Y.I.; Osipova, E.S.; Belyaev, D.V.; Kudoyarova, G.R.; Vysotskaya, L.B.; Vartapetian, B.B. Effect of the *ipt* gene expression on wheat tolerance to root flooding. *Russ. J. Plant Physiol.* **2011**, *58*, 799–807. [CrossRef]
59. Ghanem, M.E.; Albacete, A.; Smigocki, A.C.; Frébort, I.; Pospíilová, H.; Martínez-Andújar, C.; Acosta, M.; Sánchez-Bravo, J.; Lutts, S.; Dodd, I.C.; et al. Root-synthesized cytokinins improve shoot growth and fruit yield in salinized tomato (*Solanum lycopersicum* L.) plants. *J. Exp. Bot.* **2011**, *62*, 125–140. [CrossRef]
60. Huynh, L.N.; VanToai, T.; Streeter, J.; Banowetz, G. Regulation of flooding tolerance of *SAG12:ipt* Arabidopsis plants by cytokinin. *J. Exp. Bot.* **2005**, *56*, 1397–1407. [CrossRef]
61. Swartzberg, D.; Dai, N.; Gan, S.; Amasino, R.; Granot, D. Effects of cytokinin production under two *SAG* promoters on senescence and development of tomato plants. *Plant Biol.* **2006**, *8*, 579–586. [CrossRef]

62. Zhang, P.; Wang, W.Q.; Zhang, G.L.; Kaminek, M.; Dobrev, P.; Xu, J.; Gruissem, W. Senescence-inducible expression of isopentenyl transferase extends leaf life, increases drought stress resistance and alters cytokinin metabolism in cassava. *J. Integr. Plant Biol.* **2010**, *52*, 653–669. [CrossRef]
63. Xu, Y.; Gianfagna, T.; Huang, B. Proteomic changes associated with expression of a gene (*ipt*) controlling cytokinin synthesis for improving heat tolerance in a perennial grass species. *J. Exp. Bot.* **2010**, *61*, 3273–3289. [CrossRef]
64. Xu, Y.; Burgess, P.; Zhang, X.; Huang, B. Enhancing cytokinin synthesis by overexpressing *ipt* alleviated drought inhibition of root growth through activating ROS-scavenging systems in *Agrostis stolonifera*. *J. Exp. Bot.* **2016**, *67*, 1979–1992. [CrossRef] [PubMed]
65. Merewitz, E.B.; Gianfagna, T.; Huang, B. Protein accumulation in leaves and roots associated with improved drought tolerance in creeping bentgrass expressing an *ipt* gene for cytokinin synthesis. *J. Exp. Bot.* **2011**, *62*, 5311–5333. [CrossRef] [PubMed]
66. Merewitz, E.B.; Du, H.; Yu, W.; Liu, Y.; Gianfagna, T.; Huang, B. Elevated cytokinin content in *ipt* transgenic creeping bentgrass promotes drought tolerance through regulating metabolite accumulation. *J. Exp. Bot.* **2012**, *63*, 1315–1328. [CrossRef]
67. Xiao, X.O.; Zeng, Y.M.; Cao, B.H.; Lei, J.J.; Chen, Q.H.; Meng, C.M.; Cheng, Y.J. $P_{SAG12}$-IPT overexpression in eggplant delays leaf senescence and induces abiotic stress tolerance. *J. Hortic. Sci. Biotechnol.* **2017**, *92*, 349–357. [CrossRef]
68. Prerostova, S.; Dobrev, P.I.; Gaudinova, A.; Knirsch, V.; Körber, N.; Pieruschka, R.; Fiorani, F.; Brzobohatý, B.; Cerný, M.; Spichal, L.; et al. Cytokinins: Their impact on molecular and growth responses to drought stress and recovery in *Arabidopsis*. *Front. Plant Sci.* **2018**, *9*, 655. [CrossRef]
69. Liu, Y.D.; Yin, Z.J.; Yu, J.W.; Li, J.; Wei, H.L.; Han, X.L.; Shen, F.F. Improved salt tolerance and delayed leaf senescence in transgenic cotton expressing the *Agrobacterium* IPT gene. *Biol. Plant.* **2012**, *56*, 237–246. [CrossRef]
70. Rivero, R.M.; Kojima, M.; Gepstein, A.; Sakakibara, H.; Mittler, R.; Gepstein, S.; Blumwald, E. Delayed leaf senescence induces extreme drought tolerance in a flowering plant. *Proc. Natl. Acad. Sci. USA* **2007**, *104*, 19631–19636. [CrossRef]
71. Qin, H.; Gu, Q.; Zhang, J.; Sun, L.; Kuppu, S.; Zhang, Y.; Burow, M.; Payton, P.; Blumwald, E.; Zhang, H. Regulated expression of an isopentenyltransferase gene (*IPT*) in peanut significantly improves drought tolerance and increases yield under field conditions. *Plant Cell Physiol.* **2011**, *52*, 1904–1914. [CrossRef]
72. Reguera, M.; Peleg, Z.; Abdel-Tawab, Y.M.; Tumimbang, E.B.; Delatorre, C.A.; Blumwald, E. Stress-induced cytokinin synthesis increases drought tolerance through the coordinated regulation of carbon and nitrogen assimilation in rice. *Plant Physiol.* **2013**, *163*, 1609–1622. [CrossRef]
73. Kuppu, S.; Mishra, N.; Hu, R.; Sun, L.; Zhu, X.; Shen, G.; Blumwald, E.; Payton, P.; Zhang, H. Water-deficit inducible expression of a cytokinin biosynthetic gene *IPT* improves drought tolerance in cotton. *PLoS ONE* **2013**, *8*, e64190. [CrossRef]
74. Bedada, L.T.; Seth, M.S.; Runo, S.M.; Teffera, W.; Mugoya, C.; Masiga, C.W.; Oduor, R.O.; Blumewald, E.; Wachira, F. Drought tolerant tropical maize (*Zea mays* L.) developed through genetic transformation with isopentenyltransferase gene. *African J. Biotechnol.* **2016**, *15*, 2447–2464. [CrossRef]
75. Nawiri, S.; Oduor, R.; Mbinda, W. Isopentenyletransferase gene enhances drought tolerance in genetically engineered sweetpotato (*Ipomoea batatas* (L.) Lam). *J. Plant Biochem. Physiol.* **2018**, *6*, 3. [CrossRef]
76. Daskalova, S.; McCormac, A.; Scott, N.; Van Onckelen, H.; Elliott, M. Effect of seed-specific expression of the *ipt* gene on *Nicotiana tabacum* L. seed composition. *Plant Growth Regul.* **2007**, *51*, 217–229. [CrossRef]
77. Ma, Q.H.; Wang, X.M.; Wang, Z.M. Expression of isopentenyl transferase gene controlled by seed-specific lectin promoter in transgenic tobacco influences seed development. *J. Plant Growth Regul.* **2008**, *27*, 68–76. [CrossRef]
78. Atkins, C.A.; Emery, N.R.J.; Smith, P.M.C. Consequences of transforming narrow leafed lupin (*Lupinus angustifolius* [L.]) with an *ipt* gene under control of a flower-specific promoter. *Transgenic Res.* **2011**, *20*, 1321–1332. [CrossRef] [PubMed]
79. Kant, S.; Burch, D.; Badenhorst, P.; Palanisamy, R.; Mason, J.; Spangenberg, G. Regulated expression of a cytokinin biosynthesis gene *IPT* delays leaf senescence and improves yield under rainfed and irrigated conditions in canola (*Brassica napus* L.). *PLoS ONE* **2015**, *10*, e0116349. [CrossRef]
80. Joshi, S.; Choukimath, A.; Isenegger, D.; Panozzo, J.; Spangenberg, G.; Kant, S. Improved wheat growth and yield by delayed leaf senescence using developmentally regulated expression of a cytokinin biosynthesis gene. *Front. Plant Sci.* **2019**, *10*, 1285. [CrossRef]
81. Qiu, W.; Liu, M.; Qiao, G.; Jiang, J.; Xie, L.; Zhuo, R. An isopentyl transferase gene driven by the stress-inducible *rd29A* promoter improves salinity stress tolerance in transgenic tobacco. *Plant Mol. Biol. Report.* **2012**, *30*, 519–528. [CrossRef]
82. Belintani, N.G.; Guerzoni, J.T.S.; Moreira, R.M.P.; Vieira, L.G.E. Improving low-temperature tolerance in sugarcane by expressing the *ipt* gene under a cold inducible promoter. *Biol. Plant.* **2012**, *56*, 71–77. [CrossRef]
83. Avni, A.; Golan, Y.; Shirron, N.; Shamai, Y.; Golumbic, Y.; Danin-Poleg, Y.; Gepstein, S. From survival to productivity mode: Cytokinins allow avoiding the avoidance strategy under stress conditions. *Front. Plant Sci.* **2020**, *11*, 879. [CrossRef]
84. Wang, H.L.; Yang, Q.; Tan, S.; Wang, T.; Zhang, Y.; Yang, Y.; Yin, W.; Xia, X.; Guo, H.; Li, Z. Regulation of cytokinin biosynthesis using $PtRD26_{pro}$-IPT module improves drought tolerance through PtARR10-PtYUC4/5-mediated reactive oxygen species removal in *Populus*. *J. Integr. Plant Biol.* **2022**, *64*, 771–786. [CrossRef] [PubMed]
85. Roeckel, P.; Oancia, T.; Drevet, J.R. Phenotypic alterations and component analysis of seed yield in transgenic *Brassica napus* plants expressing the *tzs* gene. *Physiol. Plant.* **1998**, *102*, 243–249. [CrossRef]
86. Werner, T.; Motyka, V.; Strnad, M.; Schmülling, T. Regulation of plant growth by cytokinin. *Proc. Natl. Acad. Sci. USA* **2001**, *98*, 10487–10492. [CrossRef] [PubMed]

87. Werner, T.; Motyka, V.; Laucou, V.; Smets, R.; Van Onckelen, H.; Schmülling, T. Cytokinin-deficient transgenic Arabidopsis plants show multiple developmental alterations indicating opposite functions of cytokinins in the regulation of shoot and root meristem activity. *Plant Cell* **2003**, *15*, 2532–2550. [CrossRef] [PubMed]
88. Nishiyama, R.; Watanabe, Y.; Fujita, Y.; Le, D.T.; Kojima, M.; Werner, T.; Vankova, R.; Yamaguchi-Shinozaki, K.; Shinozaki, K.; Kakimoto, T.; et al. Analysis of cytokinin mutants and regulation of cytokinin metabolic genes reveals important regulatory roles of cytokinins in drought, salt and abscisic acid responses, and abscisic acid biosynthesis. *Plant Cell* **2011**, *23*, 2169–2183. [CrossRef]
89. Farber, M.; Attia, Z.; Weiss, D. Cytokinin activity increases stomatal density and transpiration rate in tomato. *J. Exp. Bot.* **2016**, *67*, 6351–6362. [CrossRef]
90. Li, S.; An, Y.; Hailati, S.; Zhang, J.; Cao, Y.; Liu, Y.; Geng, J.; Hu, T.; Yang, P. Overexpression of the cytokinin oxidase/dehydrogenase (CKX) from *Medicago sativa* enhanced salt stress tolerance of Arabidopsis. *J. Plant Biol.* **2019**, *62*, 374–386. [CrossRef]
91. Nehnevajova, E.; Ramireddy, E.; Stolz, A.; Gerdemann-Knörck, M.; Novák, O.; Strnad, M.; Schmülling, T. Root enhancement in cytokinin-deficient oilseed rape causes leaf mineral enrichment, increases the chlorophyll concentration under nutrient limitation and enhances the phytoremediation capacity. *BMC Plant Biol.* **2019**, *19*, 83. [CrossRef]
92. Hyoung, S.; Cho, S.H.; Chung, J.H.; So, W.M.; Cui, M.H.; Shin, J.S. Cytokinin oxidase PpCKX1 plays regulatory roles in development and enhances dehydration and salt tolerance in *Physcomitrella patens*. *Plant Cell Rep.* **2020**, *39*, 419–430. [CrossRef]
93. Jabłoński, B.; Ogonowska, H.; Szala, K.; Bajguz, A.; Orczyk, W.; Nadolska-Orczyk, A. Silencing of *TaCKX1* mediates expression of other *TaCKX* genes to increase yield parameters in wheat. *Int. J. Mol. Sci.* **2020**, *21*, 4809. [CrossRef]
94. Werner, T.; Nehnevajova, E.; Köllmer, I.; Novák, O.; Strnad, M.; Krämer, U.; Schmülling, T. Root-specific reduction of cytokinin causes enhanced root growth, drought tolerance, and leaf mineral enrichment in Arabidopsis and tobacco. *Plant Cell* **2010**, *22*, 3905–3920. [CrossRef] [PubMed]
95. Macková, H.; Hronková, M.; Dobrá, J.; Turečková, V.; Novák, O.; Lubovská, Z.; Motyka, V.; Haisel, D.; Hájek, T.; Prášil, I.T.; et al. Enhanced drought and heat stress tolerance of tobacco plants with ectopically enhanced cytokinin oxidase/dehydrogenase gene expression. *J. Exp. Bot.* **2013**, *64*, 2805–2815. [CrossRef] [PubMed]
96. Gao, S.; Fang, J.; Xu, F.; Wang, W.; Sun, X.; Chu, J.; Cai, B.; Feng, Y.; Chu, C. CYTOKININ OXIDASE/DEHYDROGENASE4 integrates cytokinin and auxin signaling to control rice crown root formation. *Plant Physiol.* **2014**, *165*, 1035–1046. [CrossRef] [PubMed]
97. Gao, S.; Xiao, Y.; Xu, F.; Gao, X.; Cao, S.; Zhang, F.; Wang, G.; Sanders, D.; Chu, C. Cytokinin-dependent regulatory module underlies the maintenance of zinc nutrition in rice. *New Phytol.* **2019**, *224*, 202–215. [CrossRef] [PubMed]
98. Pospíšilová, H.; Jiskrová, E.; Vojta, P.; Mrízová, K.; Kokáš, F.; Čudejková, M.M.; Bergougnoux, V.; Plíhal, O.; Klimešová, J.; Novák, O.; et al. Transgenic barley overexpressing a cytokinin dehydrogenase gene shows greater tolerance to drought stress. *N. Biotechnol.* **2016**, *33*, 692–705. [CrossRef]
99. Ramireddy, E.; Hosseini, S.A.; Eggert, K.; Gillandt, S.; Gnad, H.; von Wirén, N.; Schmülling, T. Root engineering in barley: Increasing cytokinin degradation produces a larger root system, mineral enrichment in the shoot and improved drought tolerance. *Plant Physiol.* **2018**, *177*, 1078–1095. [CrossRef]
100. Ramireddy, E.; Galuszka, P.; Schmülling, T. Zn-fortified cereal grains in field-grown barley by enhanced root cytokinin breakdown. *Plant Signal. Behav.* **2018**, *13*, e1530023. [CrossRef]
101. Ramireddy, E.; Nelissen, H.; Leuendorf, J.E.; Van Lijsebettens, M.; Inzé, D.; Schmülling, T. Root engineering in maize by increasing cytokinin degradation causes enhanced root growth and leaf mineral enrichment. *Plant Mol. Biol.* **2021**, *106*, 555–567. [CrossRef]
102. Khandal, H.; Gupta, S.K.; Dwivedi, V.; Mandal, D.; Sharma, N.K.; Vishwakarma, N.K.; Pal, L.; Choudhary, M.; Francis, A.; Malakar, P.; et al. Root-specific expression of chickpea cytokinin oxidase/dehydrogenase 6 leads to enhanced root growth, drought tolerance and yield without compromising nodulation. *Plant Biotechnol. J.* **2020**, *18*, 2225–2240. [CrossRef]
103. do Nascimento, F.C.; de Souza, A.F.F.; de Souza, V.M.; Rangel, R.P.; Zonta, E.; Fernandes, M.S.; Santos, L.A. *OsCKX5* modulates root system morphology and increases nutrient uptake in rice. *J. Plant Growth Regul.* **2022**, *41*, 2157–2170. [CrossRef]
104. Liu, Y.; Wang, X.; Wang, X.; Gao, W.; You, C. Identification and functional characterization of apple *MdCKX5.2* in root development and abiotic stress tolerance. *Horticulturae* **2022**, *8*, 62. [CrossRef]
105. Ashikari, M.; Sakakibara, H.; Lin, S.; Yamamoto, T.; Takashi, T.; Nishimura, A.; Angeles, E.R.; Qian, Q.; Kitano, H.; Matsuoka, M. Plant science: Cytokinin oxidase regulates rice grain production. *Science* **2005**, *309*, 741–745. [CrossRef] [PubMed]
106. Zalewski, W.; Galuszka, P.; Gasparis, S.; Orczyk, W.; Nadolska-Orczyk, A. Silencing of the *HvCKX1* gene decreases the cytokinin oxidase/dehydrogenase level in barley and leads to higher plant productivity. *J. Exp. Bot.* **2010**, *61*, 1839–1851. [CrossRef] [PubMed]
107. Zalewski, W.; Orczyk, W.; Gasparis, S.; Nadolska-Orczyk, A. *HvCKX2* gene silencing by biolistic or *Agrobacterium*-mediated transformation in barley leads to different phenotypes. *BMC Plant Biol.* **2012**, *12*, 206. [CrossRef]
108. Zalewski, W.; Gasparis, S.; Boczkowska, M.; Rajchel, I.K.; Kała, M.; Orczyk, W.; Nadolska-Orczyk, A. Expression patterns of *HvCKX* genes indicate their role in growth and reproductive development of barley. *PLoS ONE* **2014**, *9*, e115729. [CrossRef] [PubMed]
109. Yeh, S.Y.; Chen, H.W.; Ng, C.Y.; Lin, C.Y.; Tseng, T.H.; Li, W.H.; Ku, M.S.B. Down-regulation of cytokinin oxidase 2 expression increases tiller number and improves rice yield. *Rice* **2015**, *8*, 36. [CrossRef]

110. Zhao, J.; Bai, W.; Zeng, Q.; Song, S.; Zhang, M.; Li, X.; Hou, L.; Xiao, Y.; Luo, M.; Li, D.; et al. Moderately enhancing cytokinin level by down-regulation of *GhCKX* expression in cotton concurrently increases fiber and seed yield. *Mol. Breed.* **2015**, *35*, 60. [CrossRef]
111. Joshi, R.; Sahoo, K.K.; Tripathi, A.K.; Kumar, R.; Gupta, B.K.; Pareek, A.; Singla-Pareek, S.L. Knockdown of an inflorescence meristem-specific cytokinin oxidase—OsCKX2 in rice reduces yield penalty under salinity stress condition. *Plant Cell Environ.* **2018**, *41*, 936–946. [CrossRef]
112. Li, Y.; Song, G.; Gao, J.; Zhang, S.; Zhang, R.; Li, W.; Chen, M.; Liu, M.; Xia, X.; Risacher, T.; et al. Enhancement of grain number per spike by RNA interference of cytokinin oxidase 2 gene in bread wheat. *Hereditas* **2018**, *155*, 33. [CrossRef]
113. Holubová, K.; Hensel, G.; Vojta, P.; Tarkowski, P.; Bergougnoux, V.; Galuszka, P. Modification of barley plant productivity through regulation of cytokinin content by reverse-genetics approaches. *Front. Plant Sci.* **2018**, *9*, 1676. [CrossRef]
114. Yan, H.; Wang, Y.; Chen, B.; Wang, W.; Sun, H.; Sun, H.; Li, J.; Zhao, Q. *OsCKX2* regulates phosphate deficiency tolerance by modulating cytokinin in rice. *Plant Sci.* **2022**, *319*, 111257. [CrossRef] [PubMed]
115. Zeng, J.; Yan, X.; Bai, W.; Zhang, M.; Chen, Y.; Li, X.; Hou, L.; Zhao, J.; Ding, X.; Liu, R.; et al. Carpel-specific down-regulation of *GhCKXs* in cotton significantly enhances seed and fiber yield. *J. Exp. Bot.* **2022**, *73*, 6758–6772. [CrossRef] [PubMed]
116. Havlová, M.; Dobrev, P.I.; Motyka, V.; Štorchová, H.; Libus, J.; Dobrá, J.; Malbeck, J.; Gaudinová, A.; Vanková, R. The role of cytokinins in responses to water deficit in tobacco plants over-expressing *trans*-zeatin O-glucosyltransferase gene under *35S* or *SAG12* promoters. *Plant, Cell Environ.* **2008**, *31*, 341–353. [CrossRef]
117. Kudo, T.; Makita, N.; Kojima, M.; Tokunaga, H.; Sakakibara, H. Cytokinin activity of *cis*-zeatin and phenotypic alterations induced by overexpression of putative *cis*-zeatin-O-glucosyltransferase in rice. *Plant Physiol.* **2012**, *160*, 319–331. [CrossRef] [PubMed]
118. Pineda Rodo, A.; Brugière, N.; Vankova, R.; Malbeck, J.; Olson, J.M.; Haines, S.C.; Martin, R.C.; Habben, J.E.; Mok, D.W.S.; Mok, M.C. Over-expression of a zeatin *O*-glucosylation gene in maize leads to growth retardation and tasselseed formation. *J. Exp. Bot.* **2008**, *59*, 2673–2686. [CrossRef]
119. Sun, Y.G.; Wang, B.; Jin, S.H.; Qu, X.X.; Li, Y.J.; Hou, B.K. Ectopic expression of Arabidopsis glycosyltransferase *UGT85A5* enhances salt stress tolerance in tobacco. *PLoS ONE* **2013**, *8*, e59924. [CrossRef]
120. Shang, X.L.; Xie, R.R.; Tian, H.; Wang, Q.L.; Guo, F.Q. Putative zeatin *O*-glucosyltransferase OscZOG1 regulates root and shoot development and formation of agronomic traits in rice. *J. Integr. Plant Biol.* **2016**, *58*, 627–641. [CrossRef]
121. Li, Y.J.; Wang, B.; Dong, R.R.; Hou, B.K. AtUGT76C2, an Arabidopsis cytokinin glycosyltransferase is involved in drought stress adaptation. *Plant Sci.* **2015**, *236*, 157–167. [CrossRef]
122. Li, Y.; Liu, F.; Li, P.; Wang, T.; Zheng, C.; Hou, B. An Arabidopsis cytokinin-modifying glycosyltransferase UGT76C2 improves drought and salt tolerance in rice. *Front. Plant Sci.* **2020**, *11*, 560696. [CrossRef]
123. Golovatskaya, I.F. Effect of gibberellin on Arabidopsis growth, development, and hormonal balance under green and blue light. *Russ. J. Plant Physiol.* **2008**, *55*, 315–320. [CrossRef]
124. Liu, C.; Zheng, S.; Gui, J.; Fu, C.; Yu, H.; Song, D.; Shen, J.; Qin, P.; Liu, X.; Han, B.; et al. Shortened basal internodes encodes a gibberellin 2-Oxidase and contributes to lodging resistance in rice. *Mol. Plant* **2018**, *11*, 288–299. [CrossRef]
125. Monna, L.; Kitazawa, N.; Yoshino, R.; Suzuki, J.; Masuda, H.; Maehara, Y.; Tanji, M.; Sato, M.; Nasu, S.; Minobe, Y. Positional cloning of rice semidwarfing gene, *sd-1*: Rice "green revolution gene" encodes a mutant enzyme involved in gibberellin synthesis. *DNA Res.* **2002**, *9*, 11–17. [CrossRef]
126. Spielmeyer, W.; Ellis, M.H.; Chandler, P.M. Semidwarf (*sd-1*), "green revolution" rice, contains a defective gibberellin 20-oxidase gene. *Proc. Natl. Acad. Sci. USA* **2002**, *99*, 9043–9048. [CrossRef] [PubMed]
127. Tang, T.; Botwright Acuña, T.; Spielmeyer, W.; Richards, R.A. Effect of gibberellin-sensitive *Rht18* and gibberellin-insensitive *Rht-D1b* dwarfing genes on vegetative and reproductive growth in bread wheat. *J. Exp. Bot.* **2021**, *72*, 445–458. [CrossRef]
128. Thomas, S.G. Novel *Rht-1* dwarfing genes: Tools for wheat breeding and dissecting the function of DELLA proteins. *J. Exp. Bot.* **2017**, *68*, 354–358. [CrossRef]
129. Bou-Torrent, J.; Martínez-García, J.F.; García-Martínez, J.L.; Prat, S. Gibberellin A1 metabolism contributes to the control of photoperiod-mediated tuberization in potato. *PLoS ONE* **2011**, *6*, e24458. [CrossRef]
130. García-Hurtado, N.; Carrera, E.; Ruiz-Rivero, O.; López-Gresa, M.P.; Hedden, P.; Gong, F.; García-Martínez, J.L. The characterization of transgenic tomato overexpressing gibberellin 20-oxidase reveals induction of parthenocarpic fruit growth, higher yield, and alteration of the gibberellin biosynthetic pathway. *J. Exp. Bot.* **2012**, *63*, 5803–5813. [CrossRef] [PubMed]
131. Eriksson, M.E.; Israelsson, M.; Olsson, O.; Moritz, T. Increased gibberellin biosynthesis in transgenic trees promotes growth, biomass production and xylem fiber length. *Nat. Biotechnol.* **2000**, *18*, 784–788. [CrossRef] [PubMed]
132. Wuddineh, W.A.; Mazarei, M.; Zhang, J.; Poovaiah, C.R.; Mann, D.G.J.; Ziebell, A.; Sykes, R.W.; Davis, M.F.; Udvardi, M.K.; Stewart, C.N. Identification and overexpression of *gibberellin 2-oxidase* (*GA2ox*) in switchgrass (*Panicum virgatum* L.) for improved plant architecture and reduced biomass recalcitrance. *Plant Biotechnol. J.* **2015**, *13*, 636–647. [CrossRef]
133. Qin, X.; Liu, J.H.; Zhao, W.S.; Chen, X.J.; Guo, Z.J.; Peng, Y.L. Gibberellin 20-oxidase gene *OsGA20ox3* regulates plant stature and disease development in rice. *Mol. Plant Microbe Interact.* **2013**, *26*, 227–239. [CrossRef]
134. Zhao, B.; Li, H.; Li, J.; Wang, B.; Dai, C.; Wang, J.; Liu, K. *Brassica napus DS-3*, encoding a DELLA protein, negatively regulates stem elongation through gibberellin signaling pathway. *Theor. Appl. Genet.* **2017**, *130*, 727–741. [CrossRef]
135. Li, Z.; Wang, B.; Zhang, Z.; Luo, W.; Tang, Y.; Niu, Y.; Chong, K.; Xu, Y. OsGRF6 interacts with SLR1 to regulate *OsGA2ox1* expression for coordinating chilling tolerance and growth in rice. *J. Plant Physiol.* **2021**, *260*, 153406. [CrossRef]

136. Wang, X.; Li, J.; Ban, L.; Wu, Y.; Wu, X.; Wang, Y.; Wen, H.; Chapurin, V.; Dzyubenko, N.; Li, Z.; et al. Functional characterization of a gibberellin receptor and its application in alfalfa biomass improvement. *Sci. Rep.* **2017**, *7*, 41296. [CrossRef] [PubMed]
137. Biemelt, S.; Tschiersch, H.; Sonnewald, U. Impact of altered gibberellin metabolism on biomass accumulation, lignin biosynthesis, and photosynthesis in transgenic tobacco plants. *Plant Physiol.* **2004**, *135*, 254–265. [CrossRef] [PubMed]
138. Voorend, W.; Nelissen, H.; Vanholme, R.; De Vliegher, A.; Van Breusegem, F.; Boerjan, W.; Roldán-Ruiz, I.; Muylle, H.; Inzé, D. Overexpression of *GA20-OXIDASE1* impacts plant height, biomass allocation and saccharification efficiency in maize. *Plant Biotechnol. J.* **2016**, *14*, 997–1007. [CrossRef] [PubMed]
139. Fladung, M. Xylem-specific overexpression of the *GIBBERELLIN ACID 20 OXIDASE* gene (*GA20-OXIDASE*) from pine in hybrid poplar (*Populus tremula* L. × *P. alba* L.) revealed reliable increase in growth and biomass production just in a single-copy-line. *Gesunde Pflanz.* **2022**, *74*, 239–248. [CrossRef]
140. Sakamoto, T.; Morinaka, Y.; Ishiyama, K.; Kobayashi, M.; Itoh, H.; Kayano, T.; Iwahori, S.; Matsuoka, M.; Tanaka, H. Genetic manipulation of gibberellin metabolism in transgenic rice. *Nat. Biotechnol.* **2003**, *21*, 909–913. [CrossRef]
141. Zhou, B.; Lin, J.; Peng, W.; Peng, D.; Zhuo, Y.; Zhu, D.; Huang, X.; Tang, D.; Guo, M.; He, R.; et al. Dwarfism in *Brassica napus* L. induced by the over-expression of a gibberellin 2-oxidase gene from *Arabidopsis thaliana*. *Mol. Breed.* **2012**, *29*, 115–127. [CrossRef]
142. Shan, C.; Mei, Z.; Duan, J.; Chen, H.; Feng, H.; Cai, W. OsGA2ox5, a gibberellin metabolism enzyme, is involved in plant growth, the root gravity response and salt stress. *PLoS ONE* **2014**, *9*, e87110. [CrossRef]
143. Zhong, T.; Zhang, L.; Sun, S.; Zeng, H.; Han, L. Effect of localized reduction of gibberellins in different tobacco organs on drought stress tolerance and recovery. *Plant Biotechnol. Rep.* **2014**, *8*, 399–408. [CrossRef]
144. Lo, S.F.; Ho, T.H.D.; Liu, Y.L.; Jiang, M.J.; Hsieh, K.T.; Chen, K.T.; Yu, L.C.; Lee, M.H.; Chen, C.Y.; Huang, T.P.; et al. Ectopic expression of specific GA2 oxidase mutants promotes yield and stress tolerance in rice. *Plant Biotechnol. J.* **2017**, *15*, 850–864. [CrossRef] [PubMed]
145. Shi, J.B.; Wang, N.; Zhou, H.; Xu, Q.H.; Yan, G.T. The role of gibberellin synthase gene *GhGA2ox1* in upland cotton (*Gossypium hirsutum* L.) responses to drought and salt stress. *Biotechnol. Appl. Biochem.* **2019**, *66*, 298–308. [CrossRef] [PubMed]
146. Nir, I.D.O.; Moshelion, M.; Weiss, D. The Arabidopsis *GIBBERELLIN METHYL TRANSFERASE 1* suppresses gibberellin activity, reduces whole-plant transpiration and promotes drought tolerance in transgenic tomato. *Plant Cell Environ.* **2014**, *37*, 113–123. [CrossRef] [PubMed]
147. Wang, C.; Yang, Y.; Wang, H.; Ran, X.; Li, B.; Zhang, J.; Zhang, H. Ectopic expression of a cytochrome P450 monooxygenase gene *PtCYP714A3* from *Populus trichocarpa* reduces shoot growth and improves tolerance to salt stress in transgenic rice. *Plant Biotechnol. J.* **2016**, *14*, 1838–1851. [CrossRef] [PubMed]
148. Zhou, J.; Li, Z.; Xiao, G.; Zhai, M.; Pan, X.; Huang, R.; Zhang, H. CYP71D8L is a key regulator involved in growth and stress responses by mediating gibberellin homeostasis in rice. *J. Exp. Bot.* **2020**, *71*, 1160–1170. [CrossRef]
149. Anwar, A.; Liu, Y.; Dong, R.; Bai, L.; Yu, X.; Li, Y. The physiological and molecular mechanism of brassinosteroid in response to stress: A review. *Biol. Res.* **2018**, *51*, 46. [CrossRef]
150. Bartwal, A.; Mall, R.; Lohani, P.; Guru, S.K.; Arora, S. Role of secondary metabolites and brassinosteroids in plant defense against environmental stresses. *J. Plant Growth Regul.* **2013**, *32*, 216–232. [CrossRef]
151. Divi, U.K.; Krishna, P. Brassinosteroid: A biotechnological target for enhancing crop yield and stress tolerance. *N. Biotechnol.* **2009**, *26*, 131–136. [CrossRef]
152. Nolan, T.M.; Vukasinović, N.; Liu, D.; Russinova, E.; Yin, Y. Brassinosteroids: Multidimensional regulators of plant growth, development, and stress responses. *Plant Cell* **2020**, *32*, 295–318. [CrossRef]
153. Han, Y.J.; Kim, Y.S.; Hwang, O.J.; Roh, J.; Ganguly, K.; Kim, S.K.; Hwang, I.; Kim, J. Overexpression of *Arabidopsis thaliana* brassinosteroid-related acyltransferase 1 gene induces brassinosteroid-deficient phenotypes in creeping bentgrass. *PLoS ONE* **2017**, *12*, e0187378. [CrossRef]
154. Hong, Z.; Ueguchi-Tanaka, M.; Umemura, K.; Uozu, S.; Fujioka, S.; Takatsuto, S.; Yoshida, S.; Ashikari, M.; Kitano, H.; Matsuoka, M. A rice brassinosteroid-deficient mutant, *ebisu dwarf* (*d2*), is caused by a loss of function of a new member of cytochrome P450. *Plant Cell* **2003**, *15*, 2900–2910. [CrossRef] [PubMed]
155. Sakamoto, T. Phytohormones and rice crop yield: Strategies and opportunities for genetic improvement. *Transgenic Res.* **2006**, *15*, 399–404. [CrossRef] [PubMed]
156. Sakamoto, T.; Morinaka, Y.; Ohnishi, T.; Sunohara, H.; Fujioka, S.; Ueguchi-Tanaka, M.; Mizutani, M.; Sakata, K.; Takatsuto, S.; Yoshida, S.; et al. Erect leaves caused by brassinosteroid deficiency increase biomass production and grain yield in rice. *Nat. Biotechnol.* **2006**, *24*, 105–109. [CrossRef] [PubMed]
157. Braumann, I.; Urban, W.; PreuSS, A.; Dockter, C.; Zakhrabekova, S.; Hansson, M. Semi-dwarf barley (*Hordeum vulgare* L.) *brh2* and *ari-l* mutants are deficient in a U-box E3 ubiquitin ligase. *Plant Growth Regul.* **2018**, *86*, 223–234. [CrossRef]
158. Chono, M.; Honda, I.; Zeniya, H.; Yoneyama, K.; Saisho, D.; Takeda, K.; Takatsuto, S.; Hoshino, T.; Watanabe, Y. A semidwarf phenotype of barley *uzu* results from a nucleotide substitution in the gene encoding a putative brassinosteroid receptor. *Plant Physiol.* **2003**, *133*, 1209–1219. [CrossRef]
159. Gruszka, D.; Janeczko, A.; Puła, J.; Lepiarczyk, A.; Pociecha, E. Impact of drought exerted during spike development on tillering, yield parameters and grain chemical composition in semi-dwarf barley mutants deficient in the brassinosteroid metabolism. *Agronomy* **2020**, *10*, 1595. [CrossRef]

160. Nie, S.; Huang, S.; Wang, S.; Cheng, D.; Liu, J.; Lv, S.; Li, Q.; Wang, X. Enhancing brassinosteroid signaling via overexpression of tomato (*Solanum lycopersicum*) *SlBRI1* improves major agronomic traits. *Front. Plant Sci.* **2017**, *8*, 1386. [CrossRef]
161. Nie, S.; Huang, S.; Wang, S.; Mao, Y.; Liu, J.; Ma, R.; Wang, X. Enhanced brassinosteroid signaling intensity via *SlBRI1* overexpression negatively regulates drought resistance in a manner opposite of that via exogenous BR application in tomato. *Plant Physiol. Biochem.* **2019**, *138*, 36–47. [CrossRef]
162. Dong, N.; Yin, W.; Liu, D.; Zhang, X.; Yu, Z.; Huang, W.; Liu, J.; Yang, Y.; Meng, W.; Niu, M.; et al. Regulation of brassinosteroid signaling and salt resistance by SERK2 and potential utilization for crop improvement in rice. *Front. Plant Sci.* **2020**, *11*, 621859. [CrossRef]
163. Liu, L.; Xiang, Y.; Yan, J.; Di, P.; Li, J.; Sun, X.; Han, G.; Ni, L.; Jiang, M.; Yuan, J.; et al. BRASSINOSTEROID-SIGNALING KINASE 1 phosphorylating CALCIUM/CALMODULIN-DEPENDENT PROTEIN KINASE functions in drought tolerance in maize. *New Phytol.* **2021**, *231*, 695–712. [CrossRef]
164. Singh, A.; Breja, P.; Khurana, J.P.; Khurana, P. Wheat *Brassinosteroid-Insensitive1* (*TaBRI1*) interacts with members of *TaSERK* gene family and cause early flowering and seed yield enhancement in Arabidopsis. *PLoS ONE* **2016**, *11*, e0153273. [CrossRef] [PubMed]
165. Fàbregas, N.; Lozano-Elena, F.; Blasco-Escámez, D.; Tohge, T.; Martínez-Andújar, C.; Albacete, A.; Osorio, S.; Bustamante, M.; Riechmann, J.L.; Nomura, T.; et al. Overexpression of the vascular brassinosteroid receptor BRL3 confers drought resistance without penalizing plant growth. *Nat. Commun.* **2018**, *9*, 4680. [CrossRef] [PubMed]
166. Choe, S.; Fujioka, S.; Noguchi, T.; Takatsuto, S.; Yoshida, S.; Feldmann, K.A. Overexpression of *DWARF4* in the brassinosteroid biosynthetic pathway results in increased vegetative growth and seed yield in Arabidopsis. *Plant J.* **2001**, *26*, 573–582. [CrossRef] [PubMed]
167. Wu, C.; Trieu, A.; Radhakrishnan, P.; Kwok, S.F.; Harris, S.; Zhang, K.; Wang, J.; Wan, J.; Zhai, H.; Takatsuto, S.; et al. Brassinosteroids regulate grain filling in rice. *Plant Cell* **2008**, *20*, 2130–2145. [CrossRef] [PubMed]
168. Zhu, X.; Liang, W.; Cui, X.; Chen, M.; Yin, C.; Luo, Z.; Zhu, J.; Lucas, W.J.; Wang, Z.; Zhang, D. Brassinosteroids promote development of rice pollen grains and seeds by triggering expression of Carbon Starved Anther, a MYB domain protein. *Plant J.* **2015**, *82*, 570–581. [CrossRef] [PubMed]
169. Sahni, S.; Prasad, B.D.; Liu, Q.; Grbic, V.; Sharpe, A.; Singh, S.P.; Krishna, P. Overexpression of the brassinosteroid biosynthetic gene *DWF4* in *Brassica napus* simultaneously increases seed yield and stress tolerance. *Sci. Rep.* **2016**, *6*, 28298. [CrossRef] [PubMed]
170. Duan, F.; Ding, J.; Lee, D.; Lu, X.; Feng, Y.; Song, W. Overexpression of *SoCYP85A1*, a spinach cytochrome p450 gene in transgenic tobacco enhances root development and drought stress tolerance. *Front. Plant Sci.* **2017**, *8*, 1909. [CrossRef]
171. Jin, Y.L.; Tang, R.J.; Wang, H.H.; Jiang, C.M.; Bao, Y.; Yang, Y.; Liang, M.X.; Sun, Z.C.; Kong, F.J.; Li, B.; et al. Overexpression of *Populus trichocarpa CYP85A3* promotes growth and biomass production in transgenic trees. *Plant Biotechnol. J.* **2017**, *15*, 1309–1321. [CrossRef]
172. Li, Q.F.; Yu, J.W.; Lu, J.; Fei, H.Y.; Luo, M.; Cao, B.W.; Huang, L.C.; Zhang, C.Q.; Liu, Q.Q. Seed-specific expression of *OsDWF4*, a rate-limiting gene involved in brassinosteroids biosynthesis, improves both grain yield and quality in rice. *J. Agric. Food Chem.* **2018**, *66*, 3759–3772. [CrossRef]
173. Liu, N.; Zhao, Y.J.; Wu, J.W.; Wei, Y.M.; Ren, R.C.; Zang, J.; Zhang, W.T.; Zhang, L.; Shen, Q.; Zhang, X.S.; et al. Overexpression of *ZmDWF4* improves major agronomic traits and enhances yield in maize. *Mol. Breed.* **2020**, *40*, 71. [CrossRef]
174. Feng, Q.; Ou, Y.; Han, Y.; de Dios, V.R.; Wang, L.; Zhang, Q.; Yao, Y. The brassinosteroid biosynthesis enzyme gene *PeCPD* improves plant growth and salt tolerance in *Populus tomentosa*. *Ind. Crops Prod.* **2021**, *162*, 113218. [CrossRef]
175. Li, X.J.; Chen, X.J.; Guo, X.; Yin, L.L.; Ahammed, G.J.; Xu, C.J.; Chen, K.S.; Liu, C.C.; Xia, X.J.; Shi, K.; et al. *DWARF* overexpression induces alteration in phytohormone homeostasis, development, architecture and carotenoid accumulation in tomato. *Plant Biotechnol. J.* **2016**, *14*, 1021–1033. [CrossRef] [PubMed]
176. Xia, X.J.; Fang, P.P.; Guo, X.; Qian, X.J.; Zhou, J.; Shi, K.; Zhou, Y.H.; Yu, J.Q. Brassinosteroid-mediated apoplastic $H_2O_2$-glutaredoxin 12/14 cascade regulates antioxidant capacity in response to chilling in tomato. *Plant Cell Environ.* **2018**, *41*, 1052–1064. [CrossRef] [PubMed]
177. Xu, H.; Sun, H.; Dong, J.; Ma, C.; Li, J.; Li, Z.; Wang, Y.; Ji, J.; Hu, X.; Wu, M.; et al. The brassinosteroid biosynthesis gene *TaD11-2A* controls grain size and its elite haplotype improves wheat grain yields. *Theor. Appl. Genet.* **2022**, *135*, 2907–2923. [CrossRef] [PubMed]
178. Xiong, L.; Zhu, J.K. Regulation of abscisic acid biosynthesis. *Plant Physiol.* **2003**, *133*, 29–36. [CrossRef] [PubMed]
179. Seo, M.; Peeters, A.J.M.; Koiwai, H.; Oritani, T.; Marion-Poll, A.; Zeevaart, J.A.D.; Koornneef, M.; Kamiya, Y.; Koshiba, T. The Arabidopsis aldehyde oxidase 3 (*AAO3*) gene product catalyzes the final step in abscisic acid biosynthesis in leaves. *Proc. Natl. Acad. Sci. USA* **2000**, *97*, 12908–12913. [CrossRef]
180. Xiong, L.; Ishitani, M.; Lee, H.; Zhu, J.K. The Arabidopsis *LOS5/ABA3* locus encodes a molybdenum cofactor sulfurase and modulates cold stress- and osmotic stress-responsive gene expression. *Plant Cell* **2001**, *13*, 2063–2083. [CrossRef]
181. Lamarque, L.J.; Delzon, S.; Toups, H.; Gravel, A.I.; Corso, D.; Badel, E.; Burlett, R.; Charrier, G.; Cochard, H.; Jansen, S.; et al. Over-accumulation of abscisic acid in transgenic tomato plants increases the risk of hydraulic failure. *Plant Cell Environ.* **2020**, *43*, 548–562. [CrossRef]

182. Krattinger, S.G.; Kang, J.; Bräunlich, S.; Boni, R.; Chauhan, H.; Selter, L.L.; Robinson, M.D.; Schmid, M.W.; Wiederhold, E.; Hensel, G.; et al. Abscisic acid is a substrate of the ABC transporter encoded by the durable wheat disease resistance gene *Lr34*. *New Phytol.* **2019**, *223*, 853–866. [CrossRef]
183. Gonzalez-Guzman, M.; Rodriguez, L.; Lorenzo-Orts, L.; Pons, C.; Sarrion-Perdigones, A.; Fernandez, M.A.; Peirats-Llobet, M.; Forment, J.; Moreno-Alvero, M.; Cutler, S.R.; et al. Tomato PYR/PYL/RCAR abscisic acid receptors show high expression in root, differential sensitivity to the abscisic acid agonist quinabactin, and the capability to enhance plant drought resistance. *J. Exp. Bot.* **2014**, *65*, 4451–4464. [CrossRef]
184. Li, X.; Li, G.; Li, Y.; Kong, X.; Zhang, L.; Wang, J.; Li, X.; Yang, Y. ABA receptor subfamily III enhances abscisic acid sensitivity and improves the drought tolerance of Arabidopsis. *Int. J. Mol. Sci.* **2018**, *19*, 1938. [CrossRef] [PubMed]
185. Gietler, M.; Fidler, J.; Labudda, M.; Nykiel, M. Review abscisic acid—Enemy or savior in the response of cereals to abiotic and biotic stresses? *Int. J. Mol. Sci.* **2020**, *21*, 4607. [CrossRef] [PubMed]
186. Yu, J.; Ge, H.; Wang, X.; Tang, R.; Wang, Y.; Zhao, F.; Lan, W.; Luan, S.; Yang, L. Overexpression of pyrabactin resistance-like abscisic acid receptors enhances drought, osmotic, and cold tolerance in transgenic poplars. *Front. Plant Sci.* **2017**, *8*, 1752. [CrossRef] [PubMed]
187. Nagar, P.; Sharma, N.; Jain, M.; Sharma, G.; Prasad, M.; Mustafiz, A. OsPSKR15, a phytosulfokine receptor from rice enhances abscisic acid response and drought stress tolerance. *Physiol. Plant.* **2022**, *174*, e13569. [CrossRef]
188. Mega, R.; Tsujimoto, H.; Okamoto, M. Genetic manipulation of abscisic acid receptors enables modulation of water use efficiency. *Plant Signal. Behav.* **2019**, *14*, e1642039. [CrossRef]
189. Santosh Kumar, V.V.; Yadav, S.K.; Verma, R.K.; Shrivastava, S.; Ghimire, O.; Pushkar, S.; Rao, M.V.; Kumar, T.S.; Chinnusamy, V. The abscisic acid receptor OsPYL6 confers drought tolerance to *indica* rice through dehydration avoidance and tolerance mechanisms. *J. Exp. Bot.* **2021**, *72*, 1411–1431. [CrossRef]
190. Lou, D.; Wang, H.; Liang, G.; Yu, D. OsSAPK2 confers abscisic acid sensitivity and tolerance to drought stress in rice. *Front. Plant Sci.* **2017**, *8*, 993. [CrossRef]
191. Huang, X.; Hou, L.; Meng, J.; You, H.; Li, Z.; Gong, Z.; Yang, S.; Shi, Y. The antagonistic action of abscisic acid and cytokinin signaling mediates drought stress response in Arabidopsis. *Mol. Plant* **2018**, *11*, 970–982. [CrossRef]
192. Wang, Y.; Li, T.; John, S.J.; Chen, M.; Chang, J.; Yang, G.; He, G. A CBL-interacting protein kinase TaCIPK27 confers drought tolerance and exogenous ABA sensitivity in transgenic Arabidopsis. *Plant Physiol. Biochem.* **2018**, *123*, 103–113. [CrossRef]
193. Park, H.Y.; Seok, H.Y.; Park, B.K.; Kim, S.H.; Goh, C.H.; Lee, B.; Lee, C.H.; Moon, Y.H. Overexpression of Arabidopsis *ZEP* enhances tolerance to osmotic stress. *Biochem. Biophys. Res. Commun.* **2008**, *375*, 80–85. [CrossRef]
194. Zhang, Z.; Wang, Y.; Chang, L.; Zhang, T.; An, J.; Liu, Y.; Cao, Y.; Zhao, X.; Sha, X.; Hu, T.; et al. *MsZEP*, a novel zeaxanthin epoxidase gene from alfalfa (*Medicago sativa*), confers drought and salt tolerance in transgenic tobacco. *Plant Cell Rep.* **2016**, *35*, 439–453. [CrossRef] [PubMed]
195. Sun, Y.; Zhu, L.; Guo, J.; Zhou, C. Heterologous expression of *EsABA1* enhances salt tolerance with increased accumulation of endogenous ABA in transgenic tobacco. *Turk. J. Bot.* **2014**, *38*, 1067–1079. [CrossRef]
196. Iuchi, S.; Kobayashi, M.; Taji, T.; Naramoto, M.; Seki, M.; Kato, T.; Tabata, S.; Kakubari, Y.; Yamaguchi-Shinozaki, K.; Shinozaki, K. Regulation of drought tolerance by gene manipulation of 9-*cis*-epoxycarotenoid dioxygenase, a key enzyme in abscisic acid biosynthesis in Arabidopsis. *Plant J.* **2001**, *27*, 325–333. [CrossRef] [PubMed]
197. Aswath, C.R.; Kim, S.H.; Mo, S.Y.; Kim, D.H. Transgenic plants of creeping bent grass harboring the stress inducible gene, *9-cis-epoxycarotenoid dioxygenase*, are highly tolerant to drought and NaCl stress. *Plant Growth Regul.* **2005**, *47*, 129–139. [CrossRef]
198. Hwang, S.G.; Chen, H.C.; Huang, W.Y.; Chu, Y.C.; Shii, C.T.; Cheng, W.H. Ectopic expression of rice *OsNCED3* in Arabidopsis increases ABA level and alters leaf morphology. *Plant Sci.* **2010**, *178*, 12–22. [CrossRef]
199. Xian, L.; Sun, P.; Hu, S.; Wu, J.; Liu, J.H. Molecular cloning and characterization of *CrNCED1*, a gene encoding 9-*cis*-epoxycarotenoid dioxygenase in *Citrus reshni*, with functions in tolerance to multiple abiotic stresses. *Planta* **2014**, *239*, 61–77. [CrossRef]
200. Hwang, S.G.; Lee, C.Y.; Tseng, C.S. Heterologous expression of rice *9-cis-epoxycarotenoid dioxygenase 4* (*OsNCED4*) in Arabidopsis confers sugar oversensitivity and drought tolerance. *Bot. Stud.* **2018**, *59*, 2. [CrossRef] [PubMed]
201. Huang, Y.; Guo, Y.; Liu, Y.; Zhang, F.; Wang, Z.; Wang, H.; Wang, F.; Li, D.; Mao, D.; Luan, S.; et al. 9-*cis*-Epoxycarotenoid dioxygenase 3 regulates plant growth and enhances multi-abiotic stress tolerance in rice. *Front. Plant Sci.* **2018**, *9*, 162. [CrossRef]
202. He, R.; Zhuang, Y.; Cai, Y.; Agüero, C.B.; Liu, S.; Wu, J.; Deng, S.; Walker, M.A.; Lu, J.; Zhang, Y. Overexpression of 9-*cis*-epoxycarotenoid dioxygenase cisgene in grapevine increases drought tolerance and results in pleiotropic effects. *Front. Plant Sci.* **2018**, *9*, 970. [CrossRef]
203. Qin, X.; Zeevaart, J.A.D. Overexpression of a 9-*cis*-epoxycarotenoid dioxygenase gene in *Nicotiana plumbaginifolia* increases abscisic acid and phaseic acid levels and enhances drought tolerance. *Plant Physiol.* **2002**, *128*, 544–551. [CrossRef]
204. Seiler, C.; Harshavardhan, V.T.; Reddy, P.S.; Hensel, G.; Kumlehn, J.; Eschen-Lippold, L.; Rajesh, K.; Korzun, V.; Wobus, U.; Lee, J.; et al. Abscisic acid flux alterations result in differential abscisic acid signaling responses and impact assimilation efficiency in barley under terminal drought stress. *Plant Physiol.* **2014**, *164*, 1677–1696. [CrossRef] [PubMed]
205. Estrada-Melo, A.C.; Ma, C.; Reid, M.S.; Jiang, C.Z. Overexpression of an ABA biosynthesis gene using a stress-inducible promoter enhances drought resistance in petunia. *Hortic. Res.* **2015**, *2*, 15013. [CrossRef] [PubMed]

206. Yue, Y.; Zhang, M.; Zhang, J.; Duan, L.; Li, Z. Arabidopsis *LOS5/ABA3* overexpression in transgenic tobacco (*Nicotiana tabacum* cv. *Xanthi-nc*) results in enhanced drought tolerance. *Plant Sci.* **2011**, *181*, 405–411. [CrossRef] [PubMed]
207. Yue, Y.; Zhang, M.; Zhang, J.; Tian, X.; Duan, L.; Li, Z. Overexpression of the *AtLOS5* gene increased abscisic acid level and drought tolerance in transgenic cotton. *J. Exp. Bot.* **2012**, *63*, 3741–3748. [CrossRef] [PubMed]
208. Lu, Y.; Li, Y.; Zhang, J.; Xiao, Y.; Yue, Y.; Duan, L.; Zhang, M.; Li, Z. Overexpression of Arabidopsis molybdenum cofactor sulfurase gene confers drought tolerance in maize (*Zea mays* L.). *PLoS ONE* **2013**, *8*, e52126. [CrossRef]
209. Liu, X.; Xie, X.; Zheng, C.; Wei, L.; Li, X.; Jin, Y.; Zhang, G.; Jiang, C.J.; Liang, Z. RNAi-mediated suppression of the abscisic acid catabolism gene *OsABA8ox1* increases abscisic acid content and tolerance to saline–alkaline stress in rice (*Oryza sativa* L.). *Crop J.* **2022**, *10*, 354–367. [CrossRef]
210. Basu, S.; Rabara, R. Abscisic acid—An enigma in the abiotic stress tolerance of crop plants. *Plant Gene* **2017**, *11*, 90–98. [CrossRef]
211. Ju, Y.L.; Yue, X.F.; Min, Z.; Wang, X.H.; Fang, Y.L.; Zhang, J.X. VvNAC17, a novel stress-responsive grapevine (*Vitis vinifera* L.) NAC transcription factor, increases sensitivity to abscisic acid and enhances salinity, freezing, and drought tolerance in transgenic Arabidopsis. *Plant Physiol. Biochem.* **2020**, *146*, 98–111. [CrossRef]
212. Hoang, X.L.T.; Nguyen, N.C.; Nguyen, Y.N.H.; Watanabe, Y.; Tran, L.S.P.; Thao, N.P. The soybean GmNAC019 transcription factor mediates drought tolerance in Arabidopsis in an abscisic acid-dependent manner. *Int. J. Mol. Sci.* **2020**, *21*, 286. [CrossRef]
213. Wang, C.T.; Ru, J.N.; Liu, Y.W.; Li, M.; Zhao, D.; Yang, J.F.; Fu, J.D.; Xu, Z.S. Maize WRKY transcription factor ZmWRKY106 confers drought and heat tolerance in transgenic plants. *Int. J. Mol. Sci.* **2018**, *19*, 3046. [CrossRef]
214. Liu, G.; Li, B.; Li, X.; Wei, Y.; He, C.; Shi, H. *MaWRKY80* positively regulates plant drought stress resistance through modulation of abscisic acid and redox metabolism. *Plant Physiol. Biochem.* **2020**, *156*, 155–166. [CrossRef] [PubMed]
215. Hong, E.; Lim, C.W.; Han, S.W.; Lee, S.C. Functional analysis of the pepper ethylene-responsive transcription factor, CaAIEF1, in enhanced ABA sensitivity and drought tolerance. *Front. Plant Sci.* **2017**, *8*, 1407. [CrossRef] [PubMed]
216. Tu, M.; Wang, X.; Zhu, Y.; Wang, D.; Zhang, X.; Cui, Y.; Li, Y.; Gao, M.; Li, Z.; Wang, Y.; et al. VlbZIP30 of grapevine functions in dehydration tolerance via the abscisic acid core signaling pathway. *Hortic. Res.* **2018**, *5*, 49. [CrossRef]
217. Cao, L.; Lu, X.; Wang, G.; Zhang, Q.; Zhang, X.; Fan, Z.; Cao, Y.; Wei, L.; Wang, T.; Wang, Z. Maize *ZmbZIP33* is involved in drought resistance and recovery ability through an abscisic acid-dependent signaling pathway. *Front. Plant Sci.* **2021**, *12*, 442. [CrossRef] [PubMed]
218. Zhang, L.; Zhang, L.; Xia, C.; Gao, L.; Hao, C.; Zhao, G.; Jia, J.; Kong, X. A novel wheat C-bZIP gene, *TabZIP14-B*, participates in salt and freezing tolerance in transgenic plants. *Front. Plant Sci.* **2017**, *8*, 710. [CrossRef] [PubMed]
219. Wu, J.; Jiang, Y.; Liang, Y.; Chen, L.; Chen, W.; Cheng, B. Expression of the maize MYB transcription factor ZmMYB3R enhances drought and salt stress tolerance in transgenic plants. *Plant Physiol. Biochem.* **2019**, *137*, 179–188. [CrossRef] [PubMed]
220. Fang, P.; Yan, H.; Chen, F.; Peng, Y. Overexpression of maize ZmHDZIV14 increases abscisic acid sensitivity and mediates drought and salt stress in Arabidopsis and tobacco. *Plant Mol. Biol. Report.* **2021**, *39*, 275–287. [CrossRef]
221. Zhang, B.; Su, L.; Hu, B.; Li, L. Expression of *AhDREB1*, an AP2/ERF transcription factor gene from peanut, is affected by histone acetylation and increases abscisic acid sensitivity and tolerance to osmotic stress in Arabidopsis. *Int. J. Mol. Sci.* **2018**, *19*, 1441. [CrossRef]
222. Li, Z.; Liu, C.; Zhang, Y.; Wang, B.; Ran, Q.; Zhang, J. The bHLH family member ZmPTF1 regulates drought tolerance in maize by promoting root development and abscisic acid synthesis. *J. Exp. Bot.* **2019**, *70*, 5471–5486. [CrossRef]
223. Rehman, A.; Azhar, M.T.; Hinze, L.; Qayyum, A.; Li, H.; Peng, Z.; Qin, G.; Jia, Y.; Pan, Z.; He, S.; et al. Insight into abscisic acid perception and signaling to increase plant tolerance to abiotic stress. *J. Plant Interact.* **2021**, *16*, 222–237. [CrossRef]
224. Cardoso, A.A.; Gori, A.; Da-Silva, C.J.; Brunetti, C. Abscisic acid biosynthesis and signaling in plants: Key targets to improve water use efficiency and drought tolerance. *Appl. Sci.* **2020**, *10*, 6322. [CrossRef]
225. Seo, D.H.; Seomun, S.; Choi, Y.D.; Jang, G. Root development and stress tolerance in rice: The key to improving stress tolerance without yield penalties. *Int. J. Mol. Sci.* **2020**, *21*, 1807. [CrossRef] [PubMed]
226. Müller, M.; Munné-Bosch, S. Ethylene response factors: A key regulatory hub in hormone and stress signaling. *Plant Physiol.* **2015**, *169*, 32–41. [CrossRef]
227. Poór, P.; Nawaz, K.; Gupta, R.; Ashfaque, F.; Khan, M.I.R. Ethylene involvement in the regulation of heat stress tolerance in plants. *Plant Cell Rep.* **2022**, *41*, 675–698. [CrossRef] [PubMed]
228. Habben, J.E.; Bao, X.; Bate, N.J.; Debruin, J.L.; Dolan, D.; Hasegawa, D.; Helentjaris, T.G.; Lafitte, R.H.; Lovan, N.; Mo, H.; et al. Transgenic alteration of ethylene biosynthesis increases grain yield in maize under field drought-stress conditions. *Plant Biotechnol. J.* **2014**, *12*, 685–693. [CrossRef] [PubMed]
229. Dandekar, A.M.; Teo, G.; Defilippi, B.G.; Uratsu, S.L.; Passey, A.J.; Kader, A.A.; Stow, J.R.; Colgan, R.J.; James, D.J. Effect of down-regulation of ethylene biosynthesis on fruit flavor complex in apple fruit. *Transgenic Res.* **2004**, *13*, 373–384. [CrossRef]
230. Silva, J.A.; Da Costa, T.S.; Lucchetta, L.; Marini, L.J.; Zanuzo, M.R.; Nora, L.; Nora, F.R.; Twyman, R.M.; Rombaldi, C.V. Characterization of ripening behavior in transgenic melons expressing an antisense 1-aminocyclopropane-1-carboxylate (ACC) oxidase gene from apple. *Postharvest Biol. Technol.* **2004**, *32*, 263–268. [CrossRef]
231. Belimov, A.A.; Dodd, I.C.; Hontzeas, N.; Theobald, J.C.; Safronova, V.I.; Davies, W.J. Rhizosphere bacteria containing 1-aminocyclopropane-1-carboxylate deaminase increase yield of plants grown in drying soil via both local and systemic hormone signalling. *New Phytol.* **2009**, *181*, 413–423. [CrossRef]

232. Dong, H.; Zhen, Z.; Peng, J.; Chang, L.; Gong, Q.; Wang, N.N. Loss of *ACS7* confers abiotic stress tolerance by modulating ABA sensitivity and accumulation in Arabidopsis. *J. Exp. Bot.* **2011**, *62*, 4875–4887. [CrossRef]
233. Neris, D.; Mattiello, L.; Zuñiga, G.; Purgatto, E.; Menossi, M. Reduction of ethylene biosynthesis in sugarcane induces growth and investment in the non-enzymatic antioxidant apparatus. *Plant Cell Rep.* **2022**, *41*, 979–993. [CrossRef]
234. Grichko, V.P.; Glick, B.R. Flooding tolerance of transgenic tomato plants expressing the bacterial enzyme ACC deaminase controlled by the *35S*, *rolD* or PRB-1*b* promoter. *Plant Physiol. Biochem.* **2001**, *39*, 19–25. [CrossRef]
235. Sergeeva, E.; Shah, S.; Glick, B.R. Growth of transgenic canola (*Brassica napus* cv. Westar) expressing a bacterial 1-aminocyclopropane-1-carboxylate (ACC) deaminase gene on high concentrations of salt. *World J. Microbiol. Biotechnol.* **2006**, *22*, 277–282. [CrossRef]
236. Zhang, F.; Zhang, J.; Chen, L.; Shi, X.; Lui, Z.; Li, C. Heterologous expression of ACC deaminase from *Trichoderma asperellum* improves the growth performance of *Arabidopsis thaliana* under normal and salt stress conditions. *Plant Physiol. Biochem.* **2015**, *94*, 41–47. [CrossRef]
237. Jung, H.; Ali, S.; Kim, J.Y.; Kim, W.C. Transgenic Arabidopsis expressing *acdS* gene of *Pseudomonas veronii*-KJ alleviate the adverse effects of salt and water-logging stress. *Plant Breed. Biotechnol.* **2018**, *6*, 221–232. [CrossRef]
238. Singh, P.; Pandey, S.S.; Dubey, B.K.; Raj, R.; Barnawal, D.; Chandran, A.; Rahman, L. ur Salt and drought stress tolerance with increased biomass in transgenic *Pelargonium graveolens* through heterologous expression of ACC deaminase gene from *Achromobacter xylosoxidans*. *Plant Cell Tissue Organ Cult.* **2021**, *147*, 297–311. [CrossRef]
239. Ning, Q.; Jian, Y.; Du, Y.; Li, Y.; Shen, X.; Jia, H.; Zhao, R.; Zhan, J.; Yang, F.; Jackson, D.; et al. An ethylene biosynthesis enzyme controls quantitative variation in maize ear length and kernel yield. *Nat. Commun.* **2021**, *12*, 5832. [CrossRef] [PubMed]
240. Debbarma, J.; Sarki, Y.N.; Saikia, B.; Boruah, H.P.D.; Singha, D.L.; Chikkaputtaiah, C. Ethylene response factor (ERF) family proteins in abiotic stresses and CRISPR–Cas9 genome editing of ERFs for multiple abiotic stress tolerance in crop plants: A review. *Mol. Biotechnol.* **2019**, *61*, 153–172. [CrossRef]
241. Tezuka, D.; Kawamata, A.; Kato, H.; Saburi, W.; Mori, H.; Imai, R. The rice ethylene response factor OsERF83 positively regulates disease resistance to *Magnaporthe oryzae*. *Plant Physiol. Biochem.* **2019**, *135*, 263–271. [CrossRef]
242. Zhang, G.; Chen, M.; Li, L.; Xu, Z.; Chen, X.; Guo, J.; Ma, Y. Overexpression of the soybean *GmERF3* gene, an AP2/ERF type transcription factor for increased tolerances to salt, drought, and diseases in transgenic tobacco. *J. Exp. Bot.* **2009**, *60*, 3781–3796. [CrossRef]
243. Han, D.; Han, J.; Xu, T.; Li, X.; Yao, C.; Li, T.; Sun, X.; Wang, X.; Yang, G. Overexpression of *MbERF12*, an ERF gene from *Malus baccata* (L.) Borkh, increases cold and salt tolerance in *Arabidopsis thaliana* associated with ROS scavenging through ethylene signal transduction. *Vitr. Cell. Dev. Biol. Plant* **2021**, *57*, 760–770. [CrossRef]
244. Cheng, M.C.; Liao, P.M.; Kuo, W.W.; Lin, T.P. The Arabidopsis ETHYLENE RESPONSE FACTOR1 Regulates abiotic stress-responsive gene expression by binding to different cis-acting elements in response to different stress signals. *Plant Physiol.* **2013**, *162*, 1566–1582. [CrossRef] [PubMed]
245. Huang, J.; Zhao, X.; Bürger, M.; Wang, Y.; Chory, J. Two interacting ethylene response factors regulate heat stress response. *Plant Cell* **2021**, *33*, 338–357. [CrossRef] [PubMed]
246. Lestari, R.; Rio, M.; Martin, F.; Leclercq, J.; Woraathasin, N.; Roques, S.; Dessailly, F.; Clément-Vidal, A.; Sanier, C.; Fabre, D.; et al. Overexpression of *Hevea brasiliensis* ethylene response factor *HbERF-IXc5* enhances growth and tolerance to abiotic stress and affects laticifer differentiation. *Plant Biotechnol. J.* **2018**, *16*, 322–336. [CrossRef] [PubMed]
247. Djemal, R.; Khoudi, H. The ethylene-responsive transcription factor of durum wheat, *TdSHN1*, confers cadmium, copper, and zinc tolerance to yeast and transgenic tobacco plants. *Protoplasma* **2022**, *259*, 19–31. [CrossRef]
248. Pan, Y.; Seymour, G.B.; Lu, C.; Hu, Z.; Chen, X.; Chen, G. An ethylene response factor (ERF5) promoting adaptation to drought and salt tolerance in tomato. *Plant Cell Rep.* **2012**, *31*, 349–360. [CrossRef]
249. Riyazuddin, R.; Verma, R.; Singh, K.; Nisha, N.; Keisham, M.; Bhati, K.K.; Kim, S.T.; Gupta, R. Ethylene: A master regulator of salinity stress tolerance in plants. *Biomolecules* **2020**, *10*, 959. [CrossRef]
250. Shi, J.; Gao, H.; Wang, H.; Lafitte, H.R.; Archibald, R.L.; Yang, M.; Hakimi, S.M.; Mo, H.; Habben, J.E. *ARGOS8* variants generated by CRISPR-Cas9 improve maize grain yield under field drought stress conditions. *Plant Biotechnol. J.* **2017**, *15*, 207–216. [CrossRef]
251. Liu, Y.; Li, D.; Yan, J.; Wang, K.; Luo, H.; Zhang, W. MiR319 mediated salt tolerance by ethylene. *Plant Biotechnol. J.* **2019**, *17*, 2370–2383. [CrossRef]
252. Wasternack, C. Action of jasmonates in plant stress responses and development—Applied aspects. *Biotechnol. Adv.* **2014**, *32*, 31–39. [CrossRef]
253. Mabood, F.; Zhou, X.; Lee, K.D.; Smith, D.L. Methyl jasmonate, alone or in combination with genistein, and *Bradyrhizobium japonicum* increases soybean (*Glycine max* L.) plant dry matter production and grain yield under short season conditions. *Field Crops Res.* **2006**, *95*, 412–419. [CrossRef]
254. Kim, S.K.; Kim, J.T.; Jang, S.W.; Lee, S.C.; Lee, B.H.; Lee, I.J. Exogenous effect of gibberellins and jasmonate on tuber enlargement of *Dioscorea opposita*. *Agron. Res.* **2005**, *3*, 39–44.
255. Kang, D.J.; Seo, Y.J.; Lee, J.D.; Ishii, R.; Kim, K.U.; Shin, D.H.; Park, S.K.; Jang, S.W.; Lee, I.J. Jasmonic acid differentially affects growth, ion uptake and abscisic acid concentration in salt-tolerant and salt-sensitive rice cultivars. *J. Agron. Crop Sci.* **2005**, *191*, 273–282. [CrossRef]

256. Walia, H.; Wilson, C.; Condamine, P.; Liu, X.; Ismail, A.M.; Close, T.J. Large-scale expression profiling and physiological characterization of jasmonic acid-mediated adaptation of barley to salinity stress. *Plant Cell Environ.* **2007**, *30*, 410–421. [CrossRef] [PubMed]
257. Delgado, C.; Mora-Poblete, F.; Ahmar, S.; Chen, J.T.; Figueroa, C.R. Jasmonates and plant salt stress: Molecular players, physiological effects, and improving tolerance by using genome-associated tools. *Int. J. Mol. Sci.* **2021**, *22*, 3082. [CrossRef]
258. Zhao, J.; Davis, L.C.; Verpoorte, R. Elicitor signal transduction leading to production of plant secondary metabolites. *Biotechnol. Adv.* **2005**, *23*, 283–333. [CrossRef]
259. Hu, T.; Zeng, H.; Hu, Z.; Qv, X.; Chen, G. Overexpression of the tomato 13-lipoxygenase gene *TomloxD* increases generation of endogenous jasmonic acid and resistance to *Cladosporium fulvum* and high temperature. *Plant Mol. Biol. Report.* **2013**, *31*, 1141–1149. [CrossRef]
260. Yan, L.; Zhai, Q.; Wei, J.; Li, S.; Wang, B.; Huang, T.; Du, M.; Sun, J.; Kang, L.; Li, C.B.; et al. Role of tomato lipoxygenase D in wound-induced jasmonate biosynthesis and plant immunity to insect herbivores. *PLoS Genet.* **2013**, *9*, e1003964. [CrossRef]
261. Xing, Q.; Liao, J.; Cao, S.; Li, M.; Lv, T.; Qi, H. *CmLOX10* positively regulates drought tolerance through jasmonic acid-mediated stomatal closure in oriental melon (*Cucumis melo* var. *makuwa Makino*). *Sci. Rep.* **2020**, *10*, 17452. [CrossRef]
262. Sun, Q.; Zhang, B.; Yang, C.; Wang, W.; Xiang, L.; Wang, Y.; Chan, Z. Jasmonic acid biosynthetic genes *TgLOX4* and *TgLOX5* are involved in daughter bulb development in tulip (*Tulipa gesneriana*). *Hortic. Res.* **2022**, *9*, uhac006. [CrossRef]
263. Zhao, Y.; Dong, W.; Zhang, N.; Ai, X.; Wang, M.; Huang, Z.; Xiao, L.; Xia, G. A wheat allene oxide cyclase gene enhances salinity tolerance via jasmonate signaling. *Plant Physiol.* **2014**, *164*, 1068–1076. [CrossRef]
264. Liu, H.H.; Wang, Y.G.; Wang, S.P.; Li, H.J.; Xin, Q.G. Improved zinc tolerance of tobacco by transgenic expression of an allene oxide synthase gene from hexaploid wheat. *Acta Physiol. Plant.* **2014**, *36*, 2433–2440. [CrossRef]
265. Liu, H.H.; Wang, Y.G.; Wang, S.P.; Li, H.J. Cloning and characterization of peanut allene oxide cyclase gene involved in salt-stressed responses. *Genet. Mol. Res.* **2015**, *14*, 2331–2340. [CrossRef] [PubMed]
266. Wang, Y.; Liu, H.; Xin, Q. Improvement of copper tolerance of Arabidopsis by transgenic expression of an allene oxide cyclase gene, *GhAOC1*, in upland cotton (*Gossypium hirsutum* L.). *Crop J.* **2015**, *3*, 343–352. [CrossRef]
267. Gu, D.; Liu, X.; Wang, M.; Zheng, J.; Hou, W.; Wang, G.; Wang, J. Overexpression of *ZmOPR1* in Arabidopsis enhanced the tolerance to osmotic and salt stress during seed germination. *Plant Sci.* **2008**, *174*, 124–130. [CrossRef]
268. Pigolev, A.V.; Miroshnichenko, D.N.; Pushin, A.S.; Terentyev, V.V.; Boutanayev, A.M.; Dolgov, S.V.; Savchenko, T.V. Overexpression of Arabidopsis *OPR3* in hexaploid wheat (*Triticum aestivum* L.) alters plant development and freezing tolerance. *Int. J. Mol. Sci.* **2018**, *19*, 3989. [CrossRef]
269. Seo, H.S.; Song, J.T.; Cheong, J.J.; Lee, Y.H.; Lee, Y.W.; Hwang, I.; Lee, J.S.; Choi, Y.D. Jasmonic acid carboxyl methyltransferase: A key enzyme for jasmonate-regulated plant responses. *Proc. Natl. Acad. Sci. USA* **2001**, *98*, 4788–4793. [CrossRef]
270. Jung, C.K.; Lyou, S.H.; Koo, Y.J.; Song, J.T.; Choi, Y.D.; Cheong, J.J. Constitutive expression of defense genes in transgenic Arabidopsis overproducing methyl jasmonate. *J. Appl. Biol. Chem.* **2003**, *46*, 52–57.
271. Xue, R.G.; Zhang, B.; Xie, H.F. Overexpression of a *NTR1* in transgenic soybean confers tolerance to water stress. *Plant Cell Tissue Organ Cult.* **2007**, *89*, 177–183. [CrossRef]
272. Sohn, H.B.; Lee, H.Y.; Seo, J.S.; Jung, C.; Jeon, J.H.; Kim, J.H.; Lee, Y.W.; Lee, J.S.; Cheong, J.J.; Choi, Y.D. Overexpression of jasmonic acid carboxyl methyltransferase increases tuber yield and size in transgenic potato. *Plant Biotechnol. Rep.* **2011**, *5*, 27–34. [CrossRef]
273. Kim, Y.S.; Han, J.Y.; Lim, S.; Kim, H.J.; Lee, M.H.; Choi, Y.E. Overexpressing Arabidopsis jasmonic acid carboxyl methyltransferase (*AtJMT*) results in stimulation of root growth and ginsenoside heterogeneity in *Panax ginseng*. *Plant Omics* **2012**, *5*, 28–32.
274. Huang, H.; Liu, B.; Liu, L.; Song, S. Jasmonate action in plant growth and development. *J. Exp. Bot.* **2017**, *68*, 1349–1359. [CrossRef] [PubMed]
275. Ebel, C.; BenFeki, A.; Hanin, M.; Solano, R.; Chini, A. Characterization of wheat (*Triticum aestivum*) *TIFY* family and role of triticum durum *TdTIFY11a* in salt stress tolerance. *PLoS ONE* **2018**, *13*, e0200566. [CrossRef] [PubMed]
276. Su, L.; Fang, L.; Zhu, Z.; Zhang, L.; Sun, X.; Wang, Y.; Wang, Q.; Li, S.; Xin, H. The transcription factor VaNAC17 from grapevine (*Vitis amurensis*) enhances drought tolerance by modulating jasmonic acid biosynthesis in transgenic Arabidopsis. *Plant Cell Rep.* **2020**, *39*, 621–634. [CrossRef] [PubMed]
277. Fang, L.; Su, L.; Sun, X.; Li, X.; Sun, M.; Karungo, S.K.; Fang, S.; Chu, J.; Li, S.; Xin, H. Expression of *Vitis amurensis NAC26* in Arabidopsis enhances drought tolerance by modulating jasmonic acid synthesis. *J. Exp. Bot.* **2016**, *67*, 2829–2845. [CrossRef] [PubMed]
278. Onohata, T.; Gomi, K. Overexpression of jasmonate-responsive *OsbHLH034* in rice results in the induction of bacterial blight resistance via an increase in lignin biosynthesis. *Plant Cell Rep.* **2020**, *39*, 1175–1184. [CrossRef]
279. Seo, J.S.; Joo, J.; Kim, M.J.; Kim, Y.K.; Nahm, B.H.; Song, S.I.; Cheong, J.J.; Lee, J.S.; Kim, J.K.; Choi, Y.D. OsbHLH148, a basic helix-loop-helix protein, interacts with OsJAZ proteins in a jasmonate signaling pathway leading to drought tolerance in rice. *Plant J.* **2011**, *65*, 907–921. [CrossRef]
280. Zhou, Y.; Zhu, H.; He, S.; Zhai, H.; Zhao, N.; Xing, S.; Wei, Z.; Liu, Q. A novel sweetpotato transcription factor gene *IbMYB116* enhances drought tolerance in transgenic Arabidopsis. *Front. Plant Sci.* **2019**, *10*, 1025. [CrossRef]
281. Chen, Z.; Fang, X.; Yuan, X.; Zhang, Y.; Li, H.; Zhou, Y.; Cui, X. Overexpression of transcription factor GmTGA15 enhances drought tolerance in transgenic soybean hairy roots and arabidopsis plants. *Agronomy* **2021**, *11*, 170. [CrossRef]

282. Kim, H.; Seomun, S.; Yoon, Y.; Jang, G. Jasmonic acid in plant abiotic stress tolerance and interaction with abscisic acid. *Agronomy* **2021**, *11*, 1886. [CrossRef]
283. Piatek, A.A.; Lenaghan, S.C.; Stewart, C.N. Advanced editing of the nuclear and plastid genomes in plants. *Plant Sci.* **2018**, *273*, 42–49. [CrossRef]
284. Mandal, S.; Ghorai, M.; Anand, U.; Roy, D.; Kant, N.; Mishra, T.; Mane, A.B.; Jha, N.K.; Lal, M.K.; Tiwari, R.K.; et al. Cytokinins: A genetic target for increasing yield potential in the CRISPR era. *Front. Genet.* **2022**, *13*, 883930. [CrossRef] [PubMed]
285. Vanhaeren, H.; Inzé, D.; Gonzalez, N. Plant growth beyond limits. *Trends Plant Sci.* **2016**, *21*, 102–109. [CrossRef] [PubMed]
286. Karasov, T.L.; Chae, E.; Herman, J.J.; Bergelson, J. Mechanisms to mitigate the trade-off between growth and defense. *Plant Cell* **2017**, *29*, 666–680. [CrossRef] [PubMed]

*plants*

*Review*

# Genome Editing for Sustainable Crop Improvement and Mitigation of Biotic and Abiotic Stresses

**Mohd Fadhli Hamdan [1], Chou Khai Soong Karlson [1], Ee Yang Teoh [1], Su-Ee Lau [1,2] and Boon Chin Tan [1,*]**

1 Centre for Research in Biotechnology for Agriculture, Universiti Malaya, Kuala Lumpur 50603, Malaysia
2 Department of Cell and Molecular Biology, Faculty of Biotechnology and Biomolecular Sciences, Universiti Putra Malaysia, Serdang 43400, Malaysia
* Correspondence: boonchin@um.edu.my; Tel.: +60-3-7967-7982

**Abstract:** Climate change poses a serious threat to global agricultural activity and food production. Plant genome editing technologies have been widely used to develop crop varieties with superior qualities or can tolerate adverse environmental conditions. Unlike conventional breeding techniques (e.g., selective breeding and mutation breeding), modern genome editing tools offer more targeted and specific alterations of the plant genome and could significantly speed up the progress of developing crops with desired traits, such as higher yield and/or stronger resilience to the changing environment. In this review, we discuss the current development and future applications of genome editing technologies in mitigating the impacts of biotic and abiotic stresses on agriculture. We focus specifically on the CRISPR/Cas system, which has been the center of attention in the last few years as a revolutionary genome-editing tool in various species. We also conducted a bibliographic analysis on CRISPR-related papers published from 2012 to 2021 (10 years) to identify trends and potential in the CRISPR/Cas-related plant research. In addition, this review article outlines the current shortcomings and challenges of employing genome editing technologies in agriculture with notes on future prospective. We believe combining conventional and more innovative technologies in agriculture would be the key to optimizing crop improvement beyond the limitations of traditional agricultural practices.

**Keywords:** biotechnology; climate change; CRISPR; crop improvement; genome editing

**Citation:** Hamdan, M.F.; Karlson, C.K.S.; Teoh, E.Y.; Lau, S.-E.; Tan, B.C. Genome Editing for Sustainable Crop Improvement and Mitigation of Biotic and Abiotic Stresses. *Plants* **2022**, *11*, 2625. https://doi.org/10.3390/plants11192625

Academic Editors: Ranjith Pathirana and Francesco Carimi

Received: 17 August 2022
Accepted: 30 September 2022
Published: 6 October 2022



## 1. Introduction

Climate change, such as extreme weather or temperature, drought, increasing soil salinity, and flooding, significantly affects the food production system, posing serious threats to food security. The adverse effects of climate change on agricultural productivity have been reported in several regions, including Asia [1], sub-Saharan Africa [2], and the European Union (EU) [3]. For example, the heatwave and drought in the EU in 2018 have reduced cereal production by 8% compared to the previous five-year average [4], causing fodder shortages for livestock and increasing commodity prices. The impacts of climate change on agriculture in developing countries are more significant than in developed countries, mainly as these countries are located in tropical latitudes, which are more sensitive to climate change [5]. In addition, differences in vulnerability between these regions might be due to differences in endowments of human skills, physical infrastructure, and rapid demography growth, causing developing countries to have lower levels of resilience [6–8]. Ensuring sustainable crop production and food security has become challenging not only due to the growing environmental pressures but also the ever-increasing human population. Around 720 to 811 million people, about a tenth of the global population, still suffer from hunger. Meanwhile, more than 2 billion people are in the 'food insecure' category [9]. Another 130 million people may be added to the latter category due to the recent COVID-19 pandemic [10]. These problems will continue to worsen with the projected global population growth since the yield of grain crops, such as rice, wheat, and maize, has already reached a plateau [11]. With an estimated world population of 9.7 billion by 2050, crop productivity will need to increase by another ~70% while simultaneously reducing the environmental impacts [12]. Moreover, climate change increases the severity of biotic and biotic stresses on crops. Biotic stresses, such as pathogens, insect pests, and weeds, cause average output losses ranging from 17.2% in potatoes to 30.0% in rice [13]. Likewise, abiotic stresses, such as temperature extremes, drought, and lack of nutrient deficiency, caused the loss of 51–82% of the global crop output annually [14]. As the intensity of biotic and abiotic stresses on crops increases because of climate change, novel approaches are required to enhance plant tolerance. Given that the conventional agricultural practices are inadequate to meet current and future food demands and deal with the aggravated impacts of biotic and abiotic stresses due to climate change, developing practical and effective adaptation strategies is indispensable to enhance crop productivity and ensure food security. Ideally, the strategies driving this effort should be sustainable and environmentally friendly while minimizing adverse environmental impacts.

Crop breeding, including cross-breeding and mutation breeding, has been used to enhance crop performance under climate change scenarios. However, breeding programs can be laborious and time-consuming, even aided by marker-assisted selection. It can take 8 to 10 years [15] or 6 to 15 years [16] to produce a genetically superior cultivar for agricultural production. Plant breeders have used cross-breeding based on naturally occurring mutations [15] or mutation breeding techniques based on ionizing radiation and chemical mutagens to generate new varieties with desired agronomic traits, including improved stress-tolerance potential and biofortification [17]. Nevertheless, since cross-breeding is limited to traits present in the parental genomes, low variability in elite germplasms restrains the use of this technique. The outcomes of the mutation breeding technique are unpredictable even though lower mutation rates have been reported in essential genes compared to non-essential genes [18]. In addition, complex and tedious screening and selection procedures are required to identify the desired trait from a large population of mutagenized plants [19]. Transgenic technologies that involve transferring desired trait-coding genes into the elite cultivars are undoubtedly an alternative to counter losses in crop yield [20]. However, the time and expenses for developing a genetically modified (GM) crop with desirable traits are enormous. The major limitation of this method is the low public acceptance of GM crops and, related to this, the complex and strict safety regulatory procedures [21]. In addition, different countries have adopted different regulatory procedures. However, to date, only a few countries, such as Switzerland, strictly restricted or legally prohibited the cultivation of GMOs [22].

Given the importance of securing sustainable crop yield, the challenge now is to improve the existing technologies or develop alternative technologies/solutions to increase crop yields. Here, we discuss the possibility of using genome editing, particularly the CRISPR/Cas9 system, to alleviate the impact of environmental stress and enhance crop production. A bibliometric analysis of CRISPR-related articles published in the SCOPUS database was done to evaluate its current trend of publications from 2012 to 2021. The selected timeline represents the first decade since the discovery of CRISPR/Cas9 in 2012 for use in genome editing [23]. This content analysis allows us to identify certain 'hot spots' or themes and reveal the potential of CRISPR-related research in plants.

## 2. Genome Editing Technologies

Genome editing techniques using sequence-specific nucleases (SSNs) have become popular in plant research. They have been used to develop high-yielding crops, improve the adaptability of crops to environmental stresses or enhance their nutrition content [24]. To date, there are four SSNs, namely meganucleases, zinc finger nucleases (ZFNs), transcription activator-like effector nucleases (TALENs), and clustered regularly interspaced palindromic repeat (CRISPR)/CRISPR-associated (Cas) protein systems (Figure 1). These technologies allow precise targeting and modifying of specific DNA sequences in three common steps: (1) an exogenous engineered nuclease consisting of a recognition module and nuclease domain recognizes the target DNA sequence, (2) the engineered nuclease binds to the target DNA sequence and induces double-strand breaks (DSBs) at or in the vicinity of the target site and (3) the DSBs will then be repaired by either non-homologous end-joining (NHEJ) or homologous recombination (HR). NHEJ is an error-prone repair mechanism that often results in insertion and deletion (Indel) mutations, whereas HR results in a precise repair of DSBs [25].

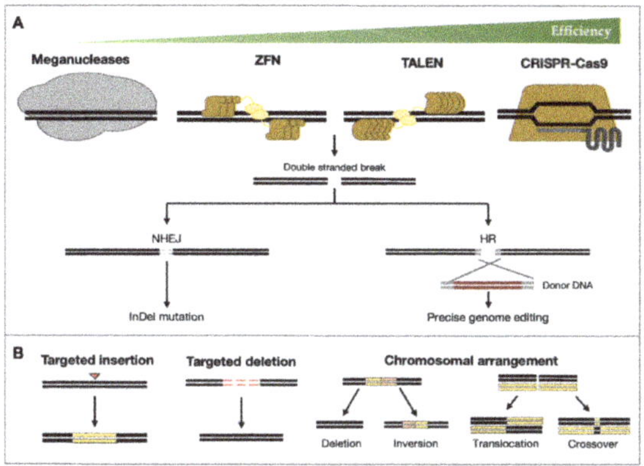


**Figure 1.** Different types of sequence-specific nucleases and types of editing. (**A**) Meganucleases, zinc finger nuclease (ZFN), transcription activator-like effector nuclease (TALEN), and CRISPR/Cas9 induces double-stranded breaks, which were corrected by non-homologous end-joining (NHEJ) and homologous recombination (HR). (**B**) Schematic diagram of target insertion, target deletion, and chromosomal arrangement through genome editing technologies. InDel, insertion-deletion.

Meganucleases were the first SSN used to create targeted DSBs in eukaryote genomes [26]. They are naturally occurring endonucleases found in prokaryotes, archaea, and unicellular eukaryotes [27]. The first meganuclease, *I-SecI*, was discovered in yeast [28]. Meganucleases are the most specific naturally occurring endonucleases as they recognize 14–40 bp long DNA sequences [29]. These enzymes have a larger recognition site than the type II restriction enzymes used in recombinant DNA technology. Due to their long recognition sequence and high specificity, meganucleases can efficiently target and modify any sequence of interest [30]. For these reasons, meganucleases have been used to create targeted DSBs in eukaryote genomes since the 1990s. In 1993, Puchta and colleagues published a landmark paper demonstrating that *I-SecI*-induced DSBs enhance HR in *Nicotiana tabacum* [31]. This marked the arrival of precise genetic engineering in plants using SSNs. Since then, several efforts have been made to introduce trait genes into plants. For example, D'Halluin et al. [32] inserted multiple trait genes, 4-hydroxyphenylpyruvate dioxygenase (*hppd*) and modified enol-pyruvylshikimate-3-phosphate synthase genes (*epsps*) into cotton using meganucleases to enhance its herbicide tolerance. Although meganucleases have been successfully applied for targeted gene editing in the plant, they have a few limitations, such as the low catalytic activity of the enzyme, prone to sequence degeneracy, and the lack of mature DNA binding structure, which hinders their wide applications [33,34].

ZFNs are fusions of the DNA recognition domain of zinc finger protein and the cleavage domain of the FokI endonuclease [35]. ZFNs act through DNA/protein recognition, and each zinc finger recognizes three base pairs (bp). As FokI must dimerize to become active, ZFNs should be designed as a pair to ensure the correct orientation and appropriate spacing for FokI dimerization [36]. To date, ZFN-mediated gene modification has been reported in various crops, such as soybean [37], maize [38], wheat [39], and rice [40]. Yet, their application as an editing tool in crops is limited because of the complexity and cost of the protein construction for each targeted site, and the potential cytotoxicity effects, presumably due to cleavage at off-target sites [41].

Similar to ZFNs, TALENs comprise transcriptional activator-like effector (TALE) repeats (comprise the DNA binding domain) and a FokI endonuclease (comprises the cleavage domain) [30]. TALEs are type III effector proteins derived from *Xanthomonas* spp. Their DNA binding ability was first reported in plants in 2007 [42,43]. In 2009, the recognition code of TALE targeting DNA sequence was also decrypted [44]. The DNA binding domain in TALE monomers contains a central repeat domain, which consists of tandem repeats of 34 amino acid residues. Each 34-amino-acid-long repeat in the central repeat domain targets only one nucleotide in the target DNA sequence. This made TALENs a better gene-editing tool compared to ZFNs as they allow flexible target design. Two hypervariable amino acid residues at the 12th and 13th positions are highly variable (termed as repeat variable di-residue [RVD]) and critical for specific nucleotide recognition.

TALENs have been demonstrated in various plant species, such as Arabidopsis, tobacco, soybean, sugarcane, maize, and wheat [45,46]. The use of TALENs in crop improvement was first reported in rice, where *OsSWEET14* (bacterial blight susceptibility gene) was disrupted, and the resulting mutant rice displayed bacterial blight resistance [47]. Other applications of TALENs in crop improvement include producing flavor in rice [48], developing powdery mildew resistant wheat [49], enhancing the nutrient content of soybean [50], and increasing anthocyanin levels in tomatoes [51]. However, despite their potential for crop improvement, several challenges of TALENs have limited their applications. One major drawback is the inefficient delivery of the TALEN system into a cell due to the large size of cDNA encoding TALEN (about 3 kb). Furthermore, the construction of TALE repeats remains a bottleneck and the efficiency of TALENs targeting a gene is variable [45].

The discovery of the CRISPR/Cas9 genome editing system has revolutionized the fields of functional genomics in animal and plant biology. Originating from bacteria and archaea as an adaptive immunity system, the CRISPR/Cas9 system has become a viable tool for targeted genome editing in prokaryotes and eukaryotes.

## 3. The CRISPR/Cas System

In 1987, CRISPR was discovered accidentally in the *Escherichia coli* genome while Ishino et al. [52] were sequencing the *iap* gene encoding alkaline phosphatase isozyme conversion enzyme. Downstream the *iap* gene, the authors discovered a unique set of tandemly repeating 29-nucleotide (nt) DNA sequences interspersed with 32-nt spacer sequences. They were unaware of the biological role of these repeats due to the lack of sequence homology with other known sequences. Later in 1993, long tandem repeats were discovered by Mojica et al. while sequencing several *Haloferax mediterranei* genome segments [53]. This marks the first finding of archaeal direct repeats. The series of interspaced repetitions was later classified as clustered regularly interspaced short palindromic repeats (CRISPR) [54]. As biological science advances towards the genomic era, CRISPR has now been identified in various bacterial and archaeal genomes.

In 2005, it was finally revealed that spacers present within CRISPRs were produced by invading phages and plasmids [55,56]. This established the CRISPR/Cas system as an adaptive immune system of bacteria and archaea that defends bacteria from bacteriophages and mobile genetic elements (MGEs) by eliminating invasive genomic elements [57]. CRISPRs can prompt the capture of invading DNA fragments to serve as a record of prior genetic aggressions [55,56]. The significance of CRISPR/Cas systems as adaptive immunity has been reinforced by subsequent experimental findings, which reiterated that new spacer sequences from the infecting phage are acquired by the bacterial CRISPR array [58–61].

The CRISPR/Cas systems were initially classified into three types (Types I, II and III) based on proteins and accessory RNAs. Type I and III systems use a complex of multi-Cas protein for target DNA recognition and cleavage, whereas the Type II system relies on a single Cas9 protein to accomplish the interference [62]. Further experimentation and analysis have further divided the classification into 2 classes, 6 types and 33 subtypes [63]. As the classification of CRISPR has been described in earlier reviews [64,65], we are not explaining it in detail in this paper.

The type II CRISPR/Cas system is the most widely used and best studied due to its straightforward constructs compared to the other systems. Type II CRISPR/Cas system employs a single protein, Cas9, and two non-coding RNAs, CRISPR RNA (crRNA) and trans-activating crRNA (tracrRNA), for target recognition and cleavage. The dualtracr-RNA:crRNA guides the Cas9 nuclease to recognize protospacer adjacent motifs (PAMs or 5′-NGG-3′) on the target DNA sequence. Cleavage of the target DNA is then performed by two Cas9 nuclease domains, the HNH domain (cleave the DNA strands complementary) and RuvC-like domain (cleave the non-complementary). Both induce a DSB three bases upstream of the PAM site of the target region.

The newly established CRISPR/Cas system replaces the dual-tracrRNA:crRNA with a single guide RNA (sgRNA) to ease the genome modifications. With this system, one can perform genome editing by simply modifying the 20-nucleotide sgRNA to be complementary to the target DNA. Overall, a CRISPR/Cas9 project involves the steps below:

(i) Target and PAM sequence identification;
(ii) Evaluate off-target effects;
(iii) sgRNA synthesis;
(iv) Cloning of the sgRNA into a suitable plant expression plasmid;
(v) Plant transformation and screening of the edited lines.

As described above, CRISPR-mediated genome editing involves the generation of Cas9-induced DSBs that are repaired by NHEJ or HR. For developing agronomic traits, precise genome modification is required. Although HR can be used to precisely repair the DSBs when DNA donor templates are supplied, it is rarely used in crop improvement because of its low efficiency in higher plants. The recent developed powerful technologies, i.e., base editing and prime editing, have partly overcome such barriers and greatly improved crop breeding opportunities.

Base editing (BE) is a technique that directly converts one target genomic DNA base into another at a targeted locus without producing a double-stranded break. Combining cytosine or adenine deaminases with CRISPR-Cas9, a range of cytosine base editors (CBEs) and adenine base editors (ABEs) has been developed in recent years [66–68]. These varieties allow exact C-to-T or A-to-G base conversions without causing a DSB. Using CBE, cytosine (C) is deaminated to create uracil (U). The uracil (U) is read as thymine during DNA replication (T). CBE consequently provides a single-base substitution from CG to TA [69]. In ABE, the inactive CRISPR–Cas9 domain is connected to adenosine deaminase, which helps convert adenine (A) to inosine, unlike cytidine deaminase in CBE. This inosine is read during DNA replication as guanine (G). Consequently, ABE generates AT to GC base substitutions [70]. Since their discovery, base editors have become valuable tools for precisely modifying the genomes of eukaryotic organisms [71–74].

Prime editing (PE) is another innovation made to the genome editing toolbox. Previous BE procedures created single base substitutions for four transitions (C > T; T > C; A > G; G > A), and newer studies included two transversions (C > G and G > C). Instead of a deaminase, PE uses an extended guide RNA (pegRNA)-guided reverse transcriptase, which allows pegRNA to install substitutions, insertions, and deletions [75]. In contrast, PE contains all 12 alterations, including the eight transversions. This increases the versatility and robustness of the gene editing strategy. Although still in its infancy, PE applications show promise in multiple cell types, organoids, and mice embryos. Data on its application in plant systems have also started to emerge. In maize, PE has introduced W542L and S621I double mutations in two ALS genes, *ZmALS1* and *ZmALS2*, which may confer resistance to several ALS-inhibiting herbicides. In rice, the level of PE efficiencies ranged from 2.22 to 31.3% [76]. In one experiment, triple amino acid substitutions (T169I, A170V, and P173S) were introduced into *OsEPSPS* [77], which may confer a higher level of glyphosate resistance [78].

In addition to the Cas9 protein, three more family proteins, namely Cas12, Cas13, and Cas14, as well as their orthologs, have been identified. Cas12 family protein is considered more advanced and versatile than Cas9 due to several characteristics, such as smaller size, lack of need for trans-activating crRNA (tracrRNA), and ability to cleave DNA via its RuvC domain. In addition, it can edit many genes from a single RNA transcript due to an intrinsic RNAse that can process its own guide RNA array [79–81]. Type IV Cas13 has an RNA-guided RNase domain that could degrade nearby single-stranded-RNA (ssRNA) molecules [82]. It has been applied for targeted RNA interference in various organisms, including animals, human viruses, and plants [83]. Cas14 is exceptional for sequence detection since it does not need a PAM and is particularly sensitive to mismatches in the center of its target region [84].

Given the simplicity, versatility, and efficacy of the CRISPR/Cas9 system, this technology shows great potential for target mutagenesis in various plant species. Despite these advantages and significant developments in the CRISPR/Cas system, continuous efforts to improve its efficiency and practicality in agriculture are still desirable. Figure 2 summarizes the general procedures involved in plant transformation and CRISPR-based plant genome editing.

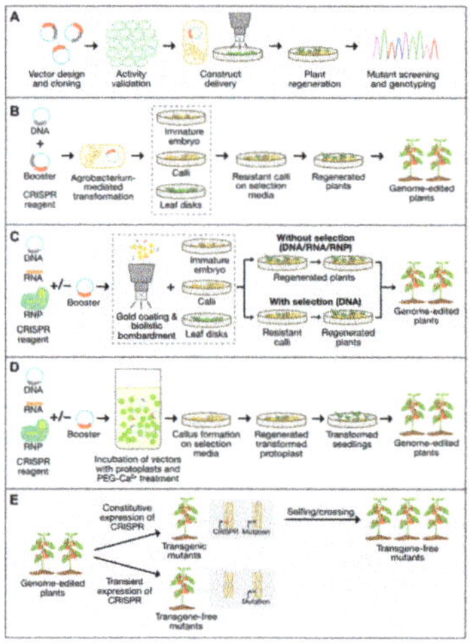


**Figure 2.** General procedures in plant transformation, delivery methods of CRISPR cargo, and transgene-free mutant development. (**A**) Major steps in plant genome editing. Once transformation vectors are designed and constructed, their activity may be validated with protoplasts before being delivered into the host plant. Protoplast transformation can also be used directly to produce transformed cells and eventually plants as described in panel D. The general procedure of transformation also usually followed by a selection process to select resistant cells and seedlings, and finally a sequencing process to confirm presence of transformed mutants. (**B**) Plant genome editing via *Agrobacterium*-mediated delivery of CRISPR DNA. *Agrobacterium* containing the vectors are transfected into plant cells in the form of calli, embryos, or leaf explants, followed by the selection process to produce genome-edited plants (**C**) Conventional and transient expression approaches for particle bombardment-mediated genome editing via CRISPR DNA, RNA, or RNP delivery. Transformation vectors-coated gold particles are bombarded into plant cells followed by the selection process (**D**) Protoplast transformation with CRISPR DNA, RNA, or RNP. Transformation vectors, protoplasts, PEG, and $Ca^{2+}$ ions are mixed before further selection processes to isolate transformed calli, seedlings, and finally genome-edited plants. (**E**) Two ways to obtain transgene-free mutants. Using the conventional method, a selection agent is used to select resistant calli and transgenic plants. Transformation vectors can be segregated out from the mutant genomes via selfing or crossing. Using the transient method, no selection agent is needed to segregate out the transformation vectors to produce transgene-free mutants. [RNP, ribonucleotide protein].

## 4. CRISPR/Cas9 for Genome Editing in Crops

The CRISPR/Cas9 system has been used in various crops to develop desirable and heritable traits, such as yield improvement, and biotic and abiotic stress management. Table 1 summarizes the applications of the CRISPR/Cas9 for crop improvement.

**Table 1.** Examples of CRISPR/Cas9 applications for crop improvement.

| Improvement | Trait | Crop | sgRNA Target Area | Type of Editing | Target Area | Result | References |
|---|---|---|---|---|---|---|---|
| Abiotic stress resistance | Drought | Chickpea | cDNA | Frameshift deletion | *Coumarate ligase (4CL)* and *Reveille 7 (RVE7)* | Enhanced tolerance | [85] |
| | Cold | Rice | cDNA | InDel mutation | *OsMYB30* | Improved tolerance | [86] |
| | Herbicide | Maize | cDNA | Base editing | *ZmALS1*, *ZmALS2* | Plants with Sulfonylurea herbicide-resistant | [87] |
| | Salinity | Tomato | DBD domain of cDNA | 49-bp deletion | *SlARF4* | Enhanced salinity tolerance | [88] |
| | Heavy metals | Rice | cDNA | Downregulation | *OsNramp5* | Decreased cadmium accumulation | [89] |
| | Heat | Tomato | cDNA | 1-bp insertion 4-bp deletion | *SlMAPK3* | Enhanced heat tolerance | [90] |
| Biotic stress resistance | Viral disease | Barley | Coding sequence | Base editing | MP, CP, Rep/Rep, IR/Virus genome | Resistant plants | [91] |
| | Fungal disease | Rice | Genome | 80-bp insert | ALB1, RSY1/ Fungal gene | Improved resistance to rice blast | [92] |
| | Bacterial disease | Tomato | JAS domain C-terminal | Deletion | *SlDMR6-1*/Host S-gene | Resistant plants | [93] |
| | Insect pest | Soybean | Coding region | 1-bp and 33-bp deletion | *GmUGT* | Enhanced resistance to *Helicoverpa armigera* and *Spodoptera litura* | [94] |
| Plant/crop quality | Crop growth | Rice | cDNA | Frameshift | *PYL1–PYL6* and *PYL12(gp-1)*, *PYL7–PYL11* and *PYL13(gp-2)* | Improved plant growth and grain productivity | [95] |
| | Crop yield | Wheat | cDNA | 10-bp deletion | *TaCKX2-1*, *TaGLW7*, *TaGW2*, and *TaGW8* | Improved grain yield | [96] |
| | Crop nutrition | Rice | Genomic Safe Harbor | 5.2kb insertion | 5.2 kb carotenoid cassette insertion | Increased β-carotene content | [97] |
| | Grain size | Rice | cDNA | InDel mutation | *OsGS3* | Increased grain size | [98] |
| | Grain number | Rice | cDNA | InDel mutation | *OsGn1a* | Increased grain number | [98,99] |
| | Fruit size | Tomato | Promoter | 85-bp deletion | *SlENO* | Enhanced fruit size | [100] |

### *4.1. Abiotic Stress*

Climate change leads to various abiotic stresses, threatening agricultural food production worldwide [101]. About 90% of all arable lands are prone to single or multiple abiotic stresses, such as water stress, extreme temperature, and salinity [102]. To survive, plants have evolved various mechanisms to respond to and cope with these stresses [103]. However, the plant stress-responsive and adaptation mechanisms are complex and governed by various genes, posing challenges to developing novel cultivars using the conventional methods [104]. As such, targeted genome editing on a single or multiple target sites through the CRISPR/Cas9 system could be a promising approach to developing abiotic stress-resilient crop varieties [25].

The CRISPR/Cas9 approach has been exploited to improve crop survival under adverse environmental stresses. For example, Zhang et al. [105] developed salinity-resistant rice through the CRISPR/Cas9 approach. By knocking out the *OsRR22* gene, the authors found that the generated rice showed better plant growth than wild-type under salinity conditions [105]. A recent study indicated that *OsNAC041* is a critical transcription factor involved in the salt stress response in rice. A targeted *osnac041* mutant obtained using the CRISPR/Cas9 method showed a higher plant height than wild-type [106]. Other studies demonstrated that members of the AP2/ERF domain containing the RAV (related to ABI3/VP1) transcription factor family are involved in salinity stress adaption [107,108]. For instance, when the rice was exposed to salt stress, the *OsRAV2* gene was activated. To determine the role of the GT-1 element in the *OsRAV2* gene, Duan et al. [109] designed a sgRNA targeting the GT-1 region of the promoter. They found that the mutant lines could not express the OsRAV2 gene under salinity conditions, confirming the importance of this gene in response to salinity stress. A similar finding has been reported by Liu et al. [110], where the CRISPR/Cas9-mediated *OsGT*$\gamma$*-2* knockout lines showed salt-hypersensitive phenotypes. Besides rice, the CRISPR/Cas9 genome editing technology has also been applied to other crops, such as wheat [111], soybean [112], maize [113], and tomato [114].

Drought stress disturbs physiological and biochemical processes in plants, limiting plant growth and yield [115]. Several genes and phytohormone signaling pathways have been shown to play critical roles in drought stress responses. Of these, abscisic acid (ABA) is a central regulator of water use and coordinates the plant's responses to drought stress. Hence, several studies have been conducted to improve drought tolerance in crops by targeting the genes involved in ABA signaling. For example, Zhang et al. [116] determined the role of *OsABA8ox2*, which encodes ABA 8′-hydroxylase, in rice drought tolerance. The authors found that the CRISPR/Cas9-mediated *OsABA8ox2* knockout lines showed increased drought-induced ABA in roots and induced root formation beneficial to drought tolerance. In contrast, overexpressing *OsABA8ox2* in rice suppressed root elongation and exhibited hypersensitivity to drought stress [116]. The ENHANCED RESPONSE TO ABA1 (ERA1), which encodes the β-subunit of the protein farnesyltransferase, was mutated in Japonica rice cv. Nipponbare using the CRISPR/Cas9 system [117]. The rice *osera1* mutant lines showed increased sensitivity to ABA and drought tolerance through stomatal regulation, suggesting that ERA1 could be a potential candidate gene for enhancing drought tolerance in crops. Another study by Yin et al. [118] showed that the *OsEPFL9* (Epidermal Patterning Factor like-9) mutants had more than an eight-fold reduction in stomatal density (SD) in the CRISPR/Cas9-edited rice plants. The reduced SD allows the edited rice lines to resist drought stress. Under optimal conditions, a significant reduction in carbon assimilation and conductance and enhanced water use efficiency (WUE) was observed when SD was reduced by 50% in barley and wheat [119,120]. Likewise, in well-watered conditions, a CRISPR-based knockout of grapevine *VvEPFL9-1* reduced SD by 60% and caused reduced carbon assimilation as compared to WT [121]. In tomatoes, *slmapk3* mutants generated through CRISPR/Cas9 showed that *SlMAPK3* is involved in drought response, and the *slmapk3* mutants showed more severe wilting symptoms and suffered cell membrane damage under drought stress [122].

Some studies used the CRISPR/Cas9 technology to reduce mineral toxicity. For example, Nieves-Cordones et al. [123] developed low cesium-containing rice plants by inactivating the K+ transporter OsHAK1 using the CRISPR/Cas9 system. In rice, knocking out OsARM1 and *OsNramp5* showed improved arsenic tolerance [124] and low cadmium accumulation [125], respectively. Another example of increasing plant stress resistance was shown by Shao et al. [126], where the authors developed a semi-dwarf variety of bananas using the CRISPR/Cas9 system to disrupt the genes responsible for the gibberellin biosynthesis. As a result, the developed bananas are more resistant to storms and heavy wind. Besides generating knockouts on the susceptible genes, genome-editing tools can also be used for knock-ins of a desirable gene. For instance, Shimatani et al. [127] used CRISPR/Cas9 to insert a maize promoter before the drought tolerance gene, *ARGOS8*. Consequently, the edited maize crops showed a greater grain yield during water stress.

These studies demonstrated that the CRISPR/Cas system could edit the plant genome, allowing us to investigate the role of genes involved in response to abiotic stresses. However, reports on targeting abiotic stress tolerance genes are scarce, primarily due to the complexity associated with abiotic stress tolerance, often involving the modulation of several genes to alter the trait of interest.

### 4.2. Biotic Stress

Plants are constantly plagued by pathogens, such as viruses, bacteria, and fungi, which can significantly reduce crop quality and yield [128]. The majority of disease-resistant crops against non-viral diseases are produced through genome editing and targeted mutagenesis of genes that negatively influence defense [129]. While few such genes are available for increasing disease resistance, many of these loci have already been successfully exploited for increased resistance.

In rice, genome editing has shown a remarkable result in combating diseases using CRISPR/Cas9. Most pathogens use the sucrose transporters that are encoded by the *SWEET* gene family in many plants [130]. In two experiments, CRISPR/Cas9 was utilized to target the promoter region of a few *OsSWEET* genes to develop resistance against bacterial leaf blight [131,132]. Knockout of the *OsERF922* gene that expresses ethylene response in the plant using CRISPR/Cas9 reduced the effect of leaf blast disease, thereby enhancing its tolerance toward it [133]. Additionally, CRISPR/Cas9 editing of the eukaryotic elongation factor, *eIF4G,* in rice resulted in plants that were immune to the rice tungro virus [134]. The infected CRISPR-edited plants contained no detectable viral proteins and produced better yields than wild-type plants.

The advancement of the CRISPR/Cas9 system has furthered the development of resistance to multiple diseases at the same time. Engineering the broad spectrum of disease resistance in staple crops on a large scale could provide a single solution to several diseases that are affecting crop production [131]. The editing of *bsr-k1*, a rice gene that binds to and increases the turnover of defense-related genes [135], is an example of this strategy. By "turning off" these critical defense genes, edited rice plants were resistant to both leaf blast and bacterial leaf blight. When challenged with rice leaf blast in the field, the transgenic lines show a greater yield of 50% more without affecting other agronomic features [135]. Likewise, the same strategy has also been applied to other crops for disease resistance. For example, broad-spectrum resistance was obtained by altering a single locus in tomatoes [136]. The *SlDMR6-1* mutations by CRISPR/Cas9 in the edited lines maintain an increased salicylic acid level in the plant with a significant reduction of disease symptoms and pathogen abundance, gaining resistance to *Pseudomonas syringae*, *Phytophthora capsici*, and *Xanthomonas* spp. [136]. In barley, CRISPR/Cas9-mediated editing of *MORC1*, a defense-related gene, increased resistance to barley powdery mildew *and Fusarium graminearum* [137]. In addition, the authors showed that the edited barley plants had lower levels of fungal DNA and fewer lesions.

In some species, targeting homologs of Mildew-resistance Locus (MLO) and other loci enhanced the resistance to these fungal infections. By concurrently targeting the three homologs of the MLO, TaMLO-A, TaMLO-B, and TaMLO-D, CRISPR/Cas9 can increase the resistance of wheat to powdery mildew [49]. Another example is the Tomelo transgene-free tomato, which is resistant to powdery mildew disease and was produced by targeting *SlMlo1* gene using CRISPR/Cas9 [138]. Zhang et al. [139] changed the three homologs of the wheat *TaEDR1* gene simultaneously using CRISPR/Cas9 to improve resistance to powdery mildew disease. In grapevine, targeting of the MLO homologs boosted the resistance to powdery mildew, whereas the edited line of grapevine had a two-fold reduction in powdery mildew sporulation [140]. In other efforts, knockout of the 14-3-3 c and 14-3-3 d protein simultaneously, a negative regulator of disease response, in cotton enhanced resistance to *Verticillium dahliae* [141]. The edited cotton showed fewer disease symptoms and lowered pathogen presence compared to control [141].

### 4.3. Yield

One of the essential keys to sustaining food production is crop yield. It is the most direct means to address the ever-rising food demand from a growing population. However, crop yield is a complex trait regulated by many factors. Therefore, much research has been done to identify the quantitative trait loci (QTLs) associated with morpho-agronomic and yield-related traits in various crop plants [142].

One way genome editing can increase crop yield is to eliminate genes that have a detrimental impact on yields, such as genes limiting grain size and weight [143,144]. In one recent example, CRISPR/Cas9 was used to individually knock out the genes of four negative yield regulators (*Gn1a*, *DEP1*, *GS3*, and *IPA1*) in the rice cultivar Zhonghua 11. Each of the individual knockout mutants, *Gn1a*, *DEP1*, and *GS3*, showed increased yield characteristics in the $T_2$ generation [145]. Similarly, Xu et al. [146] used a CRISPR/Cas9-mediated multiplex genome-editing technology to knock out three main rice negative regulators of grain weight (*GW2*, *GW5*, and *TGW6*) simultaneously, and the resulting mutants had a considerable increase in thousands of grain weights. In another study on wheat, CRISPR/Cas was used to knock out the three homoeoalleles of *GASR7*, and the mutant plant produced a much heavier kernel weight when compared to the wild-type wheat plants [147]. Besides grain, targeting a tomato cis-regulatory region in the CLAVATA-WUSCHEL stem cell circuit (CLV-WUS) using CRISPR/Cas9 resulted in an edited tomato with an increased number of locules (seed compartments) and bigger fruit size [148].

Alternatively, genome editing can also influence crop yield through other strategies. CRISPR/Cas9 technology was employed in maize to create high amylopectin variants from superior cultivars by knocking out the waxy gene [149]. The edited maize cultivars yielded 5.5 bushels per acre more than conventionally bred high amylopectin varieties. Furthermore, they could be developed in a shorter time, demonstrating the feasibility of genome editing in particular specialized applications [149]. Furthermore, reducing the ABA response of rice plants can also enhance the yield. Rice plants with simultaneous mutations of class I PYL genes (encoding receptors for ABA) using CRISPR/Cas9 had better yields than the control [95]. Under well-watered conditions, triple knockout of *PYLs 1,4,6* resulted in a 30% increase in yield [95]. It is interesting to see how these ABA-encoding PYL genes affect yield under less-optimal conditions. A recent study shows that under drought conditions, the wheat *PYL1-1B* (*TaPYL1-1B*) is responsible for increased yield and drought resistance, where it exhibited higher ABA sensitivity, photosynthetic capacity and WUE [150].

A higher yield of tomatoes can also be achieved by modifying the flower repressor gene using CRISPR/Cas9. Knockout of the flowering repressor *SELF-PRUNING 5G* (*SP5G*) gene produced tomato plants that have rapid flowering, which in turn yield earlier with compact determined growth [151]. In contrast, mutations in the *SELF PRUNING* (*SP*) gene changed the plant architecture to a bushier state with more branches [152]. The resultant mutants with two modifications had faster flowering time and earlier fruit ripening than the control lines. In another study, CRISPR-based knockout of tomato *SlAGL6* enhanced yield under heat stress. The tomato agl6 mutants displayed facultative parthenocarpy without any pleiotropic effect and produced seedless fruits of equal weight and shape to WT [153]. Under salinity stress, the CRISPR-edited soybean *gmaitr* mutants yield was much less affected than the WT in plant height, number of pods per plant, and seed weight [112]. The number of studies on plant yield and resilience improvement is expected to grow, in line with the rapid advancement of genome editing tools.

## 5. VOSviewer Bibliometric Analysis

We used 'Visualization of similarities (VOS) viewer' version 1.6.17; [154] to visualize and analyze the bibliometric network of CRISPR-related publications extracted from the SCOPUS database for the past 10 years (2012 to 2021). VOSviewer is a handy tool that allows a graphical representation and interpretation of networks representing co-authorship, journals, institutions, or co-occurring keywords based on a selected topic of interest [155].

Based on our keyword search, more than 5200 scientific papers focused on "CRISPR" OR "genome editing" AND "plants" have been published in the last ten years (2012–2021). We compiled a list of relevant publications with the co-occurrence of keywords in the title, abstract, and keywords sections from all publication types (2012–2021), including journal articles, books, and conference proceedings, to generate bibliographic maps and networks using the software VOSviewer. The criteria were set as follows: the keywords repeated at least five times were selected, singular and plural forms were standardized to singular forms to avoid redundancies, and full names and abbreviations were standardized to full names. Interchangeable keywords (e.g., 'corn' and maize') and spelling differences (e.g., 'colour' vs 'color') were also standardized in the 'thesaurus' option before running the bibliographic analysis. Based on these premises, 50 keywords were used and clustered according to their strength of association. Four clusters (sets of closely related nodes) were generated and integrated into a network overlay visualization map. The maps and networks for the analyses are presented in Figure 3. The list of 50 keywords based on their ranking is presented in Table 2.

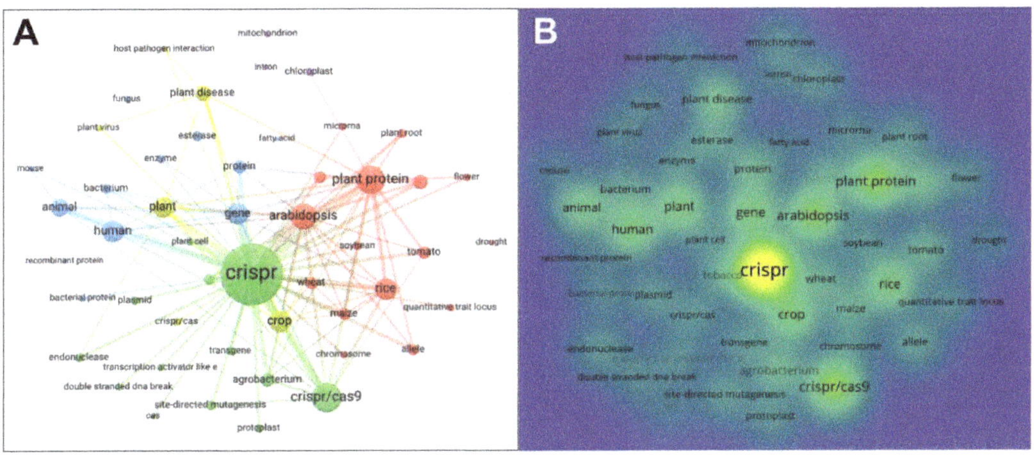


**Figure 3.** Co-occurrence network of 50 most used keywords in CRISPR-related plant research from 2012 to 2021. (**A**) Network visualization of the keywords based on total link strength. Green, yellow, red, and blue nodes represent four different clusters of keywords identified. A minimum strength of 40 was set for the lines to appear between the nodes. The relatedness of the keywords depends on the number of articles in which they occur together, which is indicated by the size of the nodes/keywords, and the length/thickness of the lines between the nodes. The bigger the nodes/keywords, the larger the weight of the nodes/keywords. The shorter and thicker the lines between the nodes, the more frequently they appear together in the publications. (**B**) Density visualization of the keywords based on occurrences. The density of a keyword depends on the number of keywords around the node. Keywords in the yellow areas indicate a more frequent occurrence in the publications while green areas indicate a less frequent appearance.

Table 2 shows that "crispr", "CRISPR/Cas9", and "plant protein" are the three most used keywords in CRISPR-related plant research publications. Of several different types of plants/crops (e.g., model plants, food crops, industrial crops, and ornamental plants) [156], only the model plants (Arabidopsis and tobacco) and food crops (rice, tomato, wheat, maize and soybean) are present in the network map. Their total number of occurrences (shown in brackets) are as follows: Arabidopsis (673), tobacco (192), rice (525), tomato (239), wheat (224), maize (213), and soybean (134).

**Table 2.** A list of 50 most frequently occurring keywords in CRISPR-related plant research publications from 2012 to 2021. The ranking is based on the number of occurrences in the publications. Total link strength indicates the number of publications in which two keywords occur together.

| Rank | Keyword | Occurrences | Total Link Strength | Rank | Keyword | Occurrences | Total Link Strength |
|---|---|---|---|---|---|---|---|
| 1 | crispr | 2386 | 6260 | 26 | chloroplast | 146 | 384 |
| 2 | CRISPR/Cas9 | 821 | 1873 | 27 | plasmid | 146 | 499 |
| 3 | plant protein | 769 | 2385 | 28 | crispr/cas | 144 | 267 |
| 4 | arabidopsis | 673 | 2111 | 29 | transgene | 141 | 542 |
| 5 | human | 535 | 1629 | 30 | protoplast | 140 | 494 |
| 6 | crop | 531 | 1450 | 31 | soybean | 134 | 453 |
| 7 | rice | 525 | 1462 | 32 | enzyme | 128 | 437 |
| 8 | gene | 514 | 1677 | 33 | flower | 124 | 427 |
| 9 | plant | 512 | 1597 | 34 | quantitative trait locus | 123 | 405 |
| 10 | animal | 409 | 1287 | 35 | transcription activator like effector nuclease | 117 | 494 |
| 11 | plant disease | 335 | 991 | 36 | chromosome | 110 | 376 |
| 12 | transcription factor | 292 | 1008 | 37 | microrna | 110 | 387 |
| 13 | agrobacterium | 282 | 1026 | 38 | mitochondrion | 108 | 285 |
| 14 | protein | 262 | 938 | 39 | double stranded dna break | 105 | 440 |
| 15 | tomato | 239 | 772 | 40 | plant cell | 105 | 375 |
| 16 | wheat | 224 | 690 | 41 | bacterial protein | 99 | 414 |
| 17 | plant leaf | 220 | 814 | 42 | plant virus | 96 | 310 |
| 18 | maize | 213 | 742 | 43 | fungus | 91 | 314 |
| 19 | allele | 195 | 708 | 44 | drought | 76 | 211 |
| 20 | esterase | 193 | 398 | 45 | intron | 75 | 156 |
| 21 | tobacco | 192 | 699 | 46 | host pathogen interaction | 72 | 267 |
| 22 | bacterium | 181 | 625 | 47 | cas | 70 | 292 |
| 23 | site-directed mutagenesis | 170 | 630 | 48 | mouse | 60 | 215 |
| 24 | endonuclease | 155 | 687 | 49 | fatty acid | 54 | 159 |
| 25 | plant root | 152 | 498 | 50 | recombinant protein | 53 | 192 |

Four CRISPR system-related keywords, "crispr" (2386), "cas" (70), "crispr/cas" (144), and "CRISPR/Cas9" (821), have been identified in the top 50 keywords during the keyword search. These keywords were not grouped in the 'thesaurus' option before the analysis since each keyword may represent a unique value. The highest cited keyword, "crispr" (2386 occurrence), may represent the investigation of CRISPR as a biological phenomenon in the bacterial immune system, which later formed the basis of "CRISPR/Cas" technology. After the discovery of the CRISPR/Cas technology, studies on CRISPR as a biological phenomenon have continued to provide knowledge to further improve the CRISPR/Cas system applications.

Further advancements in the CRISPR/Cas systems are oriented to expand its applications to other organisms and cell types and identify other alternatives to Cas9 proteins to improve CRISPR editing scope and specificity [156]. This is reflected by the presence of the keyword "cas" in the network map. The three most distinguishable variants of Cas proteins identified so far are Cas3 in type I systems, Cas9 in type II systems, and Cas10 in type III systems [157]. In addition, many other Cas proteins have also been harnessed to expand the CRISPR/Cas targeting scope, including Cas9, Cas12a, and Cas13 variants and orthologs. The expanding list of these Cas proteins and their applications has been covered extensively in recent reviews [64,158–161].

The closeness of the keyword "CRISPR/Cas9" with its surrounding keywords, such as "chromosome", "gene", "transgene", "site-directed mutagenesis", and "crop" indicated the application of the CRISPR/Cas system for the past decade as genome editing tools in crops, allowing specific and targeted changes in the gene of interest. A key approach in plant engineering for the past few decades has involved the integration of specifically assembled DNA cassettes or foreign genes into the host plant. The ability to express non-native segments of DNA in certain plants or crops resulted in novel plants with desirable traits such as herbicide resistance, pest resistance, and disease resistance [162]. It is also worth noting that transgene-free plants produced by genome editing using the CRISPR/Cas-based system (e.g., site-directed-nuclease-1 (SDN-1) type) is rapidly becoming its main selling point to avoid unnecessary regulatory issues and to gain better public perception [163,164]. These two factors are important for genome editing technology to be fully utilized and positively impact on the agricultural sector [165]. This may explain the relatedness of the keywords "crispr/cas" and "transgene" in the network map.

Genome editing reagent delivery into the host genome is a crucial topic in plant genome editing. The two most common ways of transferring a gene of interest into a host plant involved *Agrobacterium*-mediated transformation (AMT) or direct DNA transfer [166]. Both techniques aim to express the integrated transgene, silencing endogenous gene expression or modifying endogenous gene activity or function [167]. Compared to direct DNA transfer, AMT is more cost-effective and accessible to most plant researchers due to its low input (requiring low copies of DNA fragments carrying the gene of interest) and high throughput (high transformation efficiency) [168]. AMT also enables the transfer of large DNA fragments with minimal rearrangement, unlike the direct DNA transfer technique. These qualities made AMT the favored approach for plant transformation. This scenario is reflected by the network map (Figure 3), where the keyword "agrobacterium" stays in proximity to keywords, such as "crispr/cas", "transgene", and "crops".

In contrast, keywords that may be related to physically or chemically direct DNA transfer methods, such as "biolistic delivery", "gene gun", "plant bombardment", "electroporation", "microinjection", or "Polyethylene glycol (PEG)-mediated transfer" are not present in the network map. Another topic commonly associated with genome editing and CRISPR tools is using protoplasts (plant cells without cell walls) as a rapid validation system. It provides a platform to test the mutagenesis efficiency of RNA-guided endonucleases, promoters, sgRNA designs, or Cas proteins before the full-scale transformation in the host plant [169]. The popularity of this approach is reflected in the network map in which the keyword "protoplast" is present near "site-directed mutagenesis", "crispr/cas", and "transgene" nodes.

Apart from identifying the research "hot spots", the network map in Figure 3 also revealed gaps in the current state of CRISPR research. For example, the lack of connecting lines and the relatively large distance between "CRISPR" and both plant organelles, "chloroplast" and "mitochondria" indicate the lack of CRISPR application in those organelles, as compared to its already wide application in the nuclear genome in various species. This scenario is probably due to the impermeability of those plant organelles to most RNA and DNA [170]. In addition, the delivery system of CRISPR reagents into the host plant genome remains a challenge in plant transformation. For example, the use of plant bombardment to deliver CRISPR/Cas components may not require a binary vector. However, this technique has other disadvantages, such as random integration into the plant genome, less editing efficiency, and being costlier compared to AMT [171].

Given that one of the main aims of plant genome editing is to mitigate the effects of climate change, it is quite surprising to see the absence of keywords related to environmental stresses (e.g., drought, extreme weather, and elevated temperature or carbon dioxide level). Perhaps these keywords are more used in the other sections (e.g., Introduction or Conclusion sections) and less frequently elaborated in detail in the sections extracted for this analysis (i.e., Title, Abstract, and Keyword sections). In summary, it is possibly safe to assume that the first decade of CRISPR/Cas research may have focused on the 'foundation' of the CRISPR/Cas editing system by making various technical improvements in its applications (e.g., the discovery of different Cas proteins, improvements on the delivery system, and evaluation of altered DNA and possible off-target mutations).

## 6. Limitations and Challenges

Genome editing technologies have been employed to make precise changes in plant genomes. They significantly impact both fundamental research and agricultural improvement [172]. Recent modification approaches, particularly CRISPR/Cas, have increased the effectiveness and feasibility of genome editing without the need for incorporating foreign DNA. However, there are still significant obstacles to these technologies in terms of efficient and practical application in crop improvement. One prominent limitation is the off-target effect of these technologies that rely on using SSNs for targeted disruption, insertion, or replacement of selected loci [173]. While most off-targets are caused by identical sequences homologous to the targeted sequence, these effects can also occur in the region close to the target site with unrelated sequences [174]. Efforts have been made to reduce the off-target effects, especially in the CRISPR/Cas system. For instance, many different alternatives of the traditional Cas9 protein have been introduced and developed for higher efficiency and lower off-target effects [175–180]. Others, such as base editors that allow for exact nucleotide modifications, epigenome modifiers that change DNA confirmation and related expression levels, and prime editing for precision insertion of small DNA segments, are all prospective options [181–183].

Another major challenge of utilizing genome editing technologies to develop improved crops is the stringent regulatory frameworks and extensive risk assessment procedures on GM crops [184]. Most nations have biosafety frameworks in place to govern GM crops generated using recombinant DNA technology. These biosafety frameworks often draw on the fundamental concepts for food safety and the environmental risk assessment of conventional GM crops inserted with foreign gene(s) with desired characteristic(s) [185]. However, with the advent of the gene-edited crop, it is necessary to reassess the present definition of GM crops and the accompanying regulatory frameworks, because different gene editing techniques may introduce different types of alterations in the plant genome. For example, SDN-3 mutation is more similar to the conventional recombinant technique, which introduces a whole transgene into the plant genome, therefore the final product is usually considered a GMO in many nations. In contrast, the SDN-1 can introduce single base substitutions, and in certain cases without the need for introducing DSB. As a result, the genetic features of certain gene-edited crops may differ from conventional GM crops, therefore requiring a case-by-case approach to assess the risk associated with each individual product of the genome editing event [186].

Currently, there is no common regulatory approach at the international level for genome editing because of the continuous debate over the similarities, and differences between gene-edited crops and conventional GM crops. Hence, many countries do not have a clear regulatory policy on the gene-edited crops produced, which further impedes the development and implementation of these improved crops in the field. Nevertheless, the broad use of gene-editing technology poses major technological problems for regulatory bodies to identify and distinguish the regulated crop, as it can be difficult to distinguish the naturally occurring edited events in the plant genome from artificial means. Therefore, an agenda supported by various entities such as experts, associations, regulators, and researchers are needed to address these complex issues and concerns raised by gene-editing in the plant for the benefit of all [187].

## 7. Conclusions and Future Prospects

Genome editing technologies can potentially improve plant agriculture and food production to feed the world's growing population. Due to their efficiency, ease of engineering, and robustness, CRISPR/Cas systems have revolutionized plant genome engineering and globalized its applications. The current consensus is that CRISPR/Cas systems have the potential to improve plants and crops in various ways, such as increasing crop quality and yield, introducing abiotic stress-resistant traits (such as drought-, herbicide-, and insecticide-resistance), improving food safety by removing the need for an antibiotic-resistant marker, and prolonging food product shelf-life.

The main findings from the bibliographic analysis can be summarized as follows: (1) CRISPR/Cas systems are mainly used for nuclear genome editing. In addition to the nuclear genome, further development and applications of the CRISPR/Cas tools in the plant organelles (i.e., mitochondria and chloroplast genomes) are expected to increase as the technology advances, (2) most CRISPR/Cas editing so far is done on model plants (e.g., Arabidopsis and tobacco) or food crops (e.g., rice, tomato, and wheat). Discovery of novel Cas variants and orthologs and other CRISPR reagents should further expand the targeting scope of CRISPR/Cas systems to other types of plants/crops irrespective of species, (3) the issue of 'transgene' usage is one of the most widely discussed in the field of genome editing. Emerging studies on novel genome editing tools are focused on transgene-free editing, which are deemed to be more 'regulatory-friendly' and may attract improved public approval, and (4) the research publications are mostly focused on technical advancement in CRISPR systems (e.g., types of editing, targeting scope expansion, types of genomes targeted, and its delivery system) as portrayed by the frequency and relatedness of the extracted keywords in the network map (Figure 3).

Keywords related to regulatory, biosecurity, policymaking, and public acceptance issues are not present in the 50 most used keywords. Keywords related to climate change were also absent from the extracted sections. Despite this, climate change is one of the main driving forces for agricultural innovations to improve food sustainability and security. Regulatory approval and public opinion are also among the key deciding factors for genome-edited plants or crop adaption and commercialization [165,188]. The expanding gap between the fast-paced advances of CRISPR/Cas systems and the surrounding issues related to its regulation, adoption, and public acceptance should not be neglected if the potential of the technology in agriculture and food production is to be fully realized.

The present status of CRISPR technology allows for a wide range of applications aimed at increasing plant yield, disease resistance, and resilience to environmental changes. However, various technological advancements are still required, including precise editing and direct delivery of gene engineering reagents. One way to improve CRISPR delivery into the host genome is to reduce the cargo capacity so that a smaller delivery vehicle can be used to transfer CRISPR proteins through a cell. Another possibility is to use a hypercompact CRISPR CasΦ system. The CasΦ protein (~70 kD) has a molecular weight of half that of Cas9 and Cas12a enzymes [189]. Similar to Cas12a, CasΦ does not require a tracrRNA and produces staggered 5′-overhangs. It also has a minimal PAM requirement, allowing a wider range of target sites in the genome. Despite its low editing efficiency (0.85%), the possibility of using a hypercompact Cas delivery system may pave the way for the use of efficient small gene editors, further expanding the CRISPR editing toolbox.

The bibliographic analysis indicates that the trend of CRISPR/Cas research for the past decade has focused on various ways to improve the CRISPR/Cas functionality (e.g., targeting scope and delivery system). However, only recently, 'natural brakes' that could switch off the CRISPR/Cas activity when needed have been discovered. These tools are known as 'anti-CRISPR' technology, which uses phages and other mobile genetic elements that express anti-CRISPR proteins (Acrs). These proteins may nullify CRISPR/Cas activity by blocking Cas from binding or cleaving nucleic acid substrates [190]. Future improvements on these 'natural brakes' allow for more customized control of plant genome editing and expression, a needed innovation to improve the robustness of the existing CRISPR/Cas toolbox.

The recent development of biotechnologies and the production of novel crop varieties may benefit agricultural efficiency in the face of climate change. Establishing a technology adoption system across multiple farmlands is important to fully realize the potential benefits of these technologies and crop varieties. One of the issues towards adopting technology is the insufficient baseline empirical data to model the risks and benefits of sustainable farming across multiple farm types, farm sizes, and environments [191]. As technological developments are rapidly evolving, there is a constant need to deliver broad knowledge of sustainable farming to the public or industry to reduce the uncertainty about biotechnology and facilitate the adoption of agricultural biotechnology. These combined efforts will hopefully bring a paradigm shift in the farmer's perspective on sustainable farming and work towards the common goal of food security.

**Author Contributions:** Conceptualization, B.C.T. and M.F.H.; writing—original draft preparation, M.F.H., C.K.S.K., E.Y.T., S.-E.L. and B.C.T.; funding acquisition, B.C.T. All authors have read and agreed to the published version of the manuscript.

**Funding:** This work was supported by the Fundamental Research Grant Scheme (FRGS/1/2018/STG03/UM/02/2) from the Ministry of Higher Education Malaysia, Royal Society-Newton Advanced Fellowship (NA170200; IF004-2018) from the Academy of Sciences Malaysia, and RU Fund (ST003-2021; RU004A-2020) from the Universiti Malaya.

**Institutional Review Board Statement:** Not applicable.

**Informed Consent Statement:** Not applicable.

**Data Availability Statement:** Not applicable.

**Conflicts of Interest:** The authors declare no conflict of interest.

## References

1. Ozdemir, D. The impact of climate change on agricultural productivity in Asian countries: A heterogeneous panel data approach. *Environ. Sci. Pollut. Res.* **2022**, *29*, 8205–8217. [CrossRef] [PubMed]
2. Kuyah, S.; Sileshi, G.; Libère, N.; Chirinda, N.; Ndayisaba, P.; Dimobe, K.; Oborn, I. Innovative agronomic practices for sustainable intensification in sub-Saharan Africa. A review. *Agron. Sustain. Dev.* **2021**, *41*, 16. [CrossRef]
3. Brás, T.; Seixas, J.; Carvalhais, N.; Jägermeyr, J. Severity of drought and heatwave crop losses tripled over the last five decades in Europe. *Environ. Res. Lett.* **2021**, *16*, 065012. [CrossRef]
4. EC. Short-Term Outlook for EU Agricultural Markets in 2018 and 2019. Available online: https://agriculture.ec.europa.eu/data-and-analysis/markets/outlook/short-term_en (accessed on 1 May 2022).
5. Yim, H. Incorporating climate change adaptation into sustainable development. *J. Int. Dev. Coop.* **2017**, *2017*, 139–171. [CrossRef]
6. Galluzzi, G.; Seyoum, A.; Halewood, M.; Lopez Noriega, I.; Welch, E.W. The role of genetic resources in breeding for climate change: The case of public breeding programmes in eighteen developing countries. *Plants* **2020**, *9*, 1129. [CrossRef]
7. Bandt, O.d.; Jacolin, L.; Lemaire, T. *Climate Change in Developing Countries: Global Warming Effects, Transmission Channels and Adaptation Policies*; Banque de France: Paris, France, 2021; pp. 1–65.
8. Ludwig, F.; Terwisscha van Scheltinga, C.; Verhagen, J.; Kruijt, B.; van Ierland, E.; Dellink, R.; de Bruin, H.; Kabat, P. *Climate Change Impacts on Developing Countries—EU Accountability*; Think Tank Publications: Bulgaria, The Netherlands; pp. 1–45.
9. FAO; IFAD; UNICEF; WFP; WHO. *The State of Food Security and Nutrition in the World 2021: Transforming Food Systems for Food Security, Improved Nutrition and Affordable Healthy Diets for All*; World Health Organization: Rome, Italy, 2021; pp. 1–240.
10. UN. *Policy Brief: The Impact of COVID-19 on Food Security and Nutrition*; United Nations: Rome, Italy, 2020; pp. 1–22.
11. Grassini, P.; Eskridge, K.M.; Cassman, K.G. Distinguishing between yield advances and yield plateaus in historical crop production trends. *Nat. Commun.* **2013**, *4*, 2918. [CrossRef] [PubMed]
12. Lau, S.-E.; Teo, W.F.A.; Teoh, E.Y.; Tan, B.C. Microbiome engineering and plant biostimulants for sustainable crop improvement and mitigation of biotic and abiotic stresses. *Discov. Food* **2022**, *2*, 9. [CrossRef]
13. Savary, S.; Willocquet, L.; Pethybridge, S.J.; Esker, P.; McRoberts, N.; Nelson, A. The global burden of pathogens and pests on major food crops. *Nat. Ecol. Evol.* **2019**, *3*, 430–439. [CrossRef]
14. Oshunsanya, S.O.; Nwosu, N.J.; Li, Y. Abiotic stress in agricultural crops under climatic conditions. In *Sustainable Agriculture, Forest and Environmental Management*; Jhariya, M.K., Banerjee, A., Meena, R.S., Yadav, D.K., Eds.; Springer Singapore: Singapore, 2019; pp. 71–100.
15. Gao, C. Genome engineering for crop improvement and future agriculture. *Cell* **2021**, *184*, 1621–1635. [CrossRef]
16. Lyzenga, W.J.; Pozniak, C.J.; Kagale, S. Advanced domestication: Harnessing the precision of gene editing in crop breeding. *Plant Biotech. J.* **2021**, *19*, 660–670. [CrossRef]
17. Chaudhary, J.; Deshmukh, R.; Sonah, H. Mutagenesis approaches and their role in crop improvement. *Plants* **2019**, *8*, 467. [CrossRef] [PubMed]
18. Zhang, J. Important genomic regions mutate less often than do other regions. *Nature* **2022**, *602*, 38–39. [CrossRef] [PubMed]
19. Ma, L.; Kong, F.; Sun, K.; Wang, T.; Guo, T. From classical radiation to modern radiation: Past, present, and future of radiation mutation breeding. *Front. Public Health* **2021**, *9*, 11. [CrossRef]
20. Sedeek, K.E.M.; Mahas, A.; Mahfouz, M. Plant genome engineering for targeted improvement of crop traits. *Front. Plant Sci.* **2019**, *10*, 16. [CrossRef]
21. Herman, R.A.; Fedorova, M.; Storer, N.P. Will following the regulatory script for GMOs promote public acceptance of gene-edited crops? *Trends Biotechnol.* **2019**, *37*, 1272–1273. [CrossRef] [PubMed]

22. Turnbull, C.; Lillemo, M.; Hvoslef-Eide, T.A.K. Global regulation of genetically modified crops amid the gene edited crop boom—A review. *Front. Plant Sci.* **2021**, *12*, 19. [CrossRef]
23. Jinek, M.; Chylinski, K.; Fonfara, I.; Hauer, M.; Doudna, J.A.; Charpentier, E. A programmable dual-RNA-guided DNA endonuclease in adaptive bacterial immunity. *Science* **2012**, *337*, 816–821. [CrossRef]
24. Chen, K.; Wang, Y.; Zhang, R.; Zhang, H.; Gao, C. CRISPR/Cas genome editing and precision plant breeding in agriculture. *Annu. Rev. Plant Biol.* **2019**, *70*, 667–697. [CrossRef]
25. Wada, N.; Ueta, R.; Osakabe, Y.; Osakabe, K. Precision genome editing in plants: State-of-the-art in CRISPR/Cas9-based genome engineering. *BMC Plant Biol.* **2020**, *20*, 234. [CrossRef]
26. Kashtwari, M.; Mansoor, S.; Wani, A.A.; Najar, M.A.; Deshmukh, R.K.; Baloch, F.S.; Abidi, I.; Zargar, S.M. Random mutagenesis in vegetatively propagated crops: Opportunities, challenges and genome editing prospects. *Mol. Biol. Rep.* **2021**, *49*, 5729–5749. [CrossRef]
27. Silva, G.; Poirot, L.; Galetto, R.; Smith, J.; Montoya, G.; Duchateau, P.; Paques, F. Meganucleases and other tools for targeted genome engineering: Perspectives and challenges for gene therapy. *Curr. Gene Ther.* **2011**, *11*, 11–27. [CrossRef] [PubMed]
28. Faye, G.; Dennebouy, N.; Kujawa, C.; Jacq, C. Inserted sequence in the mitochondrial 23s ribosomal RNA gene of the yeast *Saccharomyces cerevisiae*. *Mol. Genet. Genomics* **1979**, *168*, 101–109. [CrossRef] [PubMed]
29. Bhambhani, S.; Kondhare, K.; Giri, A. Advanced genome editing strategies for manipulation of plant specialized metabolites pertaining to biofortification. *Phytochem. Rev.* **2022**, *21*, 81–99. [CrossRef]
30. Iqbal, Z.; Iqbal, M.S.; Ahmad, A.; Memon, A.G.; Ansari, M.I. New prospects on the horizon: Genome editing to engineer plants for desirable traits. *Curr. Plant Biol.* **2020**, *24*, 100171. [CrossRef]
31. Puchta, H.; Dujon, B.; Hohn, B. Homologous recombination in plant cells is enhanced by in vivo induction of double strand breaks into DNA by a site-specific endonuclease. *Nucleic Acids Res.* **1993**, *21*, 5034–5040. [CrossRef] [PubMed]
32. D'Halluin, K.; Vanderstraeten, C.; Van Hulle, J.; Rosolowska, J.; Van Den Brande, I.; Pennewaert, A.; D'Hont, K.; Bossut, M.; Jantz, D.; Ruiter, R.; et al. Targeted molecular trait stacking in cotton through targeted double-strand break induction. *Plant Biotechnol. J.* **2013**, *11*, 933–941. [CrossRef] [PubMed]
33. Bijlani, S.; Pang, K.M.; Sivanandam, V.; Singh, A.; Chatterjee, S. The role of recombinant AAV in precise genome editing. *Front. Genom. Editing* **2021**, *3*, 799722. [CrossRef]
34. Beyer, H.M.; Iwai, H. Structural basis for the propagation of homing endonuclease-associated inteins. *Front. Mol. Biosci.* **2022**, *9*, 14. [CrossRef]
35. Shen, H.; Li, Z. DNA double-strand break repairs and their application in plant DNA integration. *Genes* **2022**, *13*, 322. [CrossRef]
36. Lee, J.G.; Sung, Y.H.; Baek, I.J. Generation of genetically-engineered animals using engineered endonucleases. *Arch. Pharm. Res.* **2018**, *41*, 885–897. [CrossRef]
37. Bonawitz, N.D.; Ainley, W.M.; Itaya, A.; Chennareddy, S.R.; Cicak, T.; Effinger, K.; Jiang, K.; Mall, T.K.; Marri, P.R.; Samuel, J.P.; et al. Zinc finger nuclease-mediated targeting of multiple transgenes to an endogenous soybean genomic locus via non-homologous end joining. *Plant Biotechnol. J.* **2019**, *17*, 750–761. [CrossRef] [PubMed]
38. Shukla, V.K.; Doyon, Y.; Miller, J.C.; DeKelver, R.C.; Moehle, E.A.; Worden, S.E.; Mitchell, J.C.; Arnold, N.L.; Gopalan, S.; Meng, X.; et al. Precise genome modification in the crop species *Zea mays* using zinc-finger nucleases. *Nature* **2009**, *459*, 437–441. [CrossRef] [PubMed]
39. Ran, Y.; Patron, N.; Kay, P.; Wong, D.; Buchanan, M.; Cao, Y.Y.; Sawbridge, T.; Davies, J.P.; Mason, J.; Webb, S.R.; et al. Zinc finger nuclease-mediated precision genome editing of an endogenous gene in hexaploid bread wheat (*Triticum aestivum*) using a DNA repair template. *Plant Biotechnol. J.* **2018**, *16*, 2088–2101. [CrossRef] [PubMed]
40. Jung, Y.-J.; Nogoy, F.M.; Lee, S.-K.; Cho, Y.-G.; Kang, K. Application of ZFN for site directed mutagenesis of rice *SSIVa* gene. *Biotechnol. Bioproc. E* **2018**, *23*, 108–115. [CrossRef]
41. Rasheed, A.; Gill, R.A.; Hassan, M.U.; Mahmood, A.; Qari, S.; Zaman, Q.U.; Ilyas, M.; Aamer, M.; Batool, M.; Li, H.; et al. A critical review: Recent advancements in the use of CRISPR/Cas9 technology to enhance crops and alleviate global food crises. *Curr. Issues Mol. Bio.* **2021**, *43*, 1950–1976. [CrossRef]
42. Romer, P.; Hahn, S.; Jordan, T.; Strauss, T.; Bonas, U.; Lahaye, T. Plant pathogen recognition mediated by promoter activation of the pepper *Bs3* resistance gene. *Science* **2007**, *318*, 645–648. [CrossRef] [PubMed]
43. Kay, S.; Hahn, S.; Marois, E.; Hause, G.; Bonas, U. A bacterial effector acts as a plant transcription factor and induces a cell size regulator. *Science* **2007**, *318*, 648–651. [CrossRef]
44. Boch, J.; Scholze, H.; Schornack, S.; Landgraf, A.; Hahn, S.; Kay, S.; Lahaye, T.; Nickstadt, A.; Bonas, U. Breaking the code of DNA binding specificity of TAL-type III effectors. *Science* **2009**, *326*, 1509–1512. [CrossRef]
45. Zhang, Y.; Massel, K.; Godwin, I.D.; Gao, C. Applications and potential of genome editing in crop improvement. *Genom. Biol.* **2018**, *19*, 210. [CrossRef]
46. Martinez-Fortun, J.; Phillips, D.W.; Jones, H.D. Potential impact of genome editing in world agriculture. *Emerg. Top. Life Sci.* **2017**, *1*, 117–133. [CrossRef]
47. Li, T.; Liu, B.; Spalding, M.H.; Weeks, D.P.; Yang, B. High-efficiency TALEN-based gene editing produces disease-resistant rice. *Nat. Biotechnol.* **2012**, *30*, 390–392. [CrossRef] [PubMed]
48. Shan, Q.; Zhang, Y.; Chen, K.; Zhang, K.; Gao, C. Creation of fragrant rice by targeted knockout of the *OsBADH2* gene using TALEN technology. *Plant Biotechnol. J.* **2015**, *13*, 791–800. [CrossRef] [PubMed]

49. Wang, Y.; Cheng, X.; Shan, Q.; Zhang, Y.; Liu, J.; Gao, C.; Qiu, J.-L. Simultaneous editing of three homoeoalleles in hexaploid bread wheat confers heritable resistance to powdery mildew. *Nat. Biotechnol.* **2014**, *32*, 947–951. [CrossRef]
50. Demorest, Z.L.; Coffman, A.; Baltes, N.J.; Stoddard, T.J.; Clasen, B.M.; Luo, S.; Retterath, A.; Yabandith, A.; Gamo, M.E.; Bissen, J.; et al. Direct stacking of sequence-specific nuclease-induced mutations to produce high oleic and low linolenic soybean oil. *BMC Plant Biol.* **2016**, *16*, 225. [CrossRef] [PubMed]
51. Čermák, T.; Baltes, N.J.; Čegan, R.; Zhang, Y.; Voytas, D.F. High-frequency, precise modification of the tomato genome. *Genome Biol.* **2015**, *16*, 232. [CrossRef] [PubMed]
52. Ishino, Y.; Shinagawa, H.; Makino, K.; Amemura, M.; Nakata, A. Nucleotide sequence of the *iap* gene, responsible for alkaline phosphatase isozyme conversion in *Escherichia coli*, and identification of the gene product. *J. Bacteriol.* **1987**, *169*, 5429–5433. [CrossRef]
53. Mojica, F.J.; Ferrer, C.; Juez, G.; Rodriguez-Valera, F. Long stretches of short tandem repeats are present in the largest replicons of the *Archaea haloferax* mediterranei and *Haloferax volcanii* and could be involved in replicon partitioning. *Mol. Microbiol.* **1995**, *17*, 85–93. [CrossRef]
54. Jansen, R.; Embden, J.D.; Gaastra, W.; Schouls, L.M. Identification of genes that are associated with DNA repeats in prokaryotes. *Mol. Microbiol.* **2002**, *43*, 1565–1575. [CrossRef]
55. Mojica, F.J.; Diez-Villasenor, C.; Garcia-Martinez, J.; Soria, E. Intervening sequences of regularly spaced prokaryotic repeats derive from foreign genetic elements. *J. Mol. Evol.* **2005**, *60*, 174–182. [CrossRef]
56. Pourcel, C.; Salvignol, G.; Vergnaud, G. CRISPR elements in *Yersinia pestis* acquire new repeats by preferential uptake of bacteriophage DNA, and provide additional tools for evolutionary studies. *Microbiology (Reading)* **2005**, *151*, 653–663. [CrossRef]
57. Sorek, R.; Lawrence, C.M.; Wiedenheft, B. CRISPR-mediated adaptive immune systems in bacteria and archaea. *Annu. Rev. Biochem.* **2013**, *82*, 237–266. [CrossRef] [PubMed]
58. Makarova, K.S.; Grishin, N.V.; Shabalina, S.A.; Wolf, Y.I.; Koonin, E.V. A putative RNA-interference-based immune system in prokaryotes: Computational analysis of the predicted enzymatic machinery, functional analogies with eukaryotic RNAi, and hypothetical mechanisms of action. *Biol. Direct* **2006**, *1*, 7. [CrossRef] [PubMed]
59. Barrangou, R.; Fremaux, C.; Deveau, H.; Richards, M.; Boyaval, P.; Moineau, S.; Romero, D.A.; Horvath, P. CRISPR provides acquired resistance against viruses in prokaryotes. *Science* **2007**, *315*, 1709–1712. [CrossRef] [PubMed]
60. Deveau, H.; Barrangou, R.; Garneau, J.E.; Labonte, J.; Fremaux, C.; Boyaval, P.; Romero, D.A.; Horvath, P.; Moineau, S. Phage response to CRISPR-encoded resistance in *Streptococcus thermophilus*. *J. Bacteriol.* **2008**, *190*, 1390–1400. [CrossRef]
61. Horvath, P.; Romero, D.A.; Coute-Monvoisin, A.C.; Richards, M.; Deveau, H.; Moineau, S.; Boyaval, P.; Fremaux, C.; Barrangou, R. Diversity, activity, and evolution of CRISPR loci in *Streptococcus thermophilus*. *J. Bacteriol.* **2008**, *190*, 1401–1412. [CrossRef]
62. Karlson, C.K.S.; Mohd-Noor, S.N.; Nolte, N.; Tan, B.C. CRISPR/dCas9-based systems: Mechanisms and applications in plant sciences. *Plants* **2021**, *10*, 2055. [CrossRef]
63. Makarova, K.S.; Wolf, Y.I.; Iranzo, J.; Shmakov, S.A.; Alkhnbashi, O.S.; Brouns, S.J.J.; Charpentier, E.; Cheng, D.; Haft, D.H.; Horvath, P.; et al. Evolutionary classification of CRISPR-Cas systems: A burst of class 2 and derived variants. *Nat. Rev. Microbiol.* **2020**, *18*, 67–83. [CrossRef]
64. Zhang, Y.; Malzahn, A.A.; Sretenovic, S.; Qi, Y. The emerging and uncultivated potential of CRISPR technology in plant science. *Nat. Plant.* **2019**, *5*, 778–794. [CrossRef]
65. Chaudhuri, A.; Halder, K.; Datta, A. Classification of CRISPR/Cas system and its application in tomato breeding. *Theor. Appl. Genet.* **2022**, *135*, 367–387. [CrossRef]
66. Talas, A.; Simon, D.A.; Kulcsar, P.I.; Varga, E.; Krausz, S.L.; Welker, E. Bear reveals that increased fidelity variants can successfully reduce the mismatch tolerance of adenine but not cytosine base editors. *Nat. Commun.* **2021**, *12*, 6353. [CrossRef]
67. Wu, S.; Li, L.; Li, M.; Sun, S.; Zhao, Y.; Xue, X.; Chen, F.; Zhong, J.; Guo, J.; Qu, Q.; et al. Two compact Cas9 ortholog-based cytosine base editors expand the DNA targeting scope and applications in vitro and in vivo. *Front. Cell. Dev. Biol.* **2022**, *10*, 809922. [CrossRef]
68. Wang, G.; Xu, Z.; Wang, F.; Huang, Y.; Xin, Y.; Liang, S.; Li, B.; Si, H.; Sun, L.; Wang, Q.; et al. Development of an efficient and precise adenine base editor (ABE) with expanded target range in allotetraploid cotton (*Gossypium hirsutum*). *BMC Biol.* **2022**, *20*, 45. [CrossRef]
69. Komor, A.C.; Kim, Y.B.; Packer, M.S.; Zuris, J.A.; Liu, D.R. Programmable editing of a target base in genomic DNA without double-stranded DNA cleavage. *Nature* **2016**, *533*, 420–424. [CrossRef] [PubMed]
70. Gaudelli, N.M.; Komor, A.C.; Rees, H.A.; Packer, M.S.; Badran, A.H.; Bryson, D.I.; Liu, D.R. Programmable base editing of a • t to g • c in genomic DNA without DNA cleavage. *Nature* **2017**, *551*, 464–471. [CrossRef] [PubMed]
71. Molla, K.A.; Yang, Y. CRISPR/Cas-mediated base editing: Technical considerations and practical applications. *Trends Biotechnol.* **2019**, *37*, 1121–1142. [CrossRef] [PubMed]
72. Li, Q.; Li, Y.; Yang, S.; Huang, S.; Yan, M.; Ding, Y.; Tang, W.; Lou, X.; Yin, Q.; Sun, Z.; et al. CRISPR-Cas9-mediated base-editing screening in mice identifies DND1 amino acids that are critical for primordial germ cell development. *Nat. Cell Biol.* **2018**, *20*, 1315–1325. [CrossRef]
73. Molla, K.A.; Sretenovic, S.; Bansal, K.C.; Qi, Y. Precise plant genome editing using base editors and prime editors. *Nat. Plants* **2021**, *7*, 1166–1187. [CrossRef]

74. Kang, B.C.; Yun, J.Y.; Kim, S.T.; Shin, Y.; Ryu, J.; Choi, M.; Woo, J.W.; Kim, J.S. Precision genome engineering through adenine base editing in plants. *Nat. Plants* **2018**, *4*, 427–431. [CrossRef]
75. Scholefield, J.; Harrison, P.T. Prime editing—An update on the field. *Gene Ther.* **2021**, *28*, 396–401. [CrossRef]
76. Jiang, Y.Y.; Chai, Y.P.; Lu, M.H.; Han, X.L.; Lin, Q.; Zhang, Y.; Zhang, Q.; Zhou, Y.; Wang, X.C.; Gao, C.; et al. Prime editing efficiently generates W542L and S621I double mutations in two ALS genes in maize. *Genome Biol.* **2020**, *21*, 257. [CrossRef]
77. Li, H.; Li, J.; Chen, J.; Yan, L.; Xia, L. Precise modifications of both exogenous and endogenous genes in rice by prime editing. *Mol. Plant* **2020**, *13*, 671–674. [CrossRef] [PubMed]
78. Perotti, V.E.; Larran, A.S.; Palmieri, V.E.; Martinatto, A.K.; Alvarez, C.E.; Tuesca, D.; Permingeat, H.R. A novel triple amino acid substitution in the EPSPS found in a high-level glyphosate-resistant *Amaranthus hybridus* population from Argentina. *Pest Manag. Sci.* **2019**, *75*, 1242–1251. [CrossRef] [PubMed]
79. Ha, D.I.; Lee, J.M.; Lee, N.E.; Kim, D.; Ko, J.H.; Kim, Y.S. Highly efficient and safe genome editing by CRISPR-Cas12a using CRISPR RNA with a ribosyl-2′-o-methylated uridinylate-rich 3′-overhang in mouse zygotes. *Exp. Mol. Med.* **2020**, *52*, 1823–1830. [CrossRef] [PubMed]
80. Wang, Y.; Wang, M.; Zheng, T.; Hou, Y.; Zhang, P.; Tang, T.; Wei, J.; Du, Q. Specificity profiling of CRISPR system reveals greatly enhanced off-target gene editing. *Sci. Rep.* **2020**, *10*, 2269. [CrossRef] [PubMed]
81. Swarts, D.C.; van der Oost, J.; Jinek, M. Structural basis for guide RNA processing and seed-dependent DNA targeting by CRISPR-Cas12a. *Mol. Cell* **2017**, *66*, 221–233.e4. [CrossRef]
82. Mahas, A.; Marsic, T.; Lopez-Portillo Masson, M.; Wang, Q.; Aman, R.; Zheng, C.; Ali, Z.; Alsanea, M.; Al-Qahtani, A.; Ghanem, B.; et al. Characterization of a thermostable CAS13 enzyme for one-pot detection of SARS-COV-2. *Proc. Natl. Acad. Sci. USA* **2022**, *119*, e2118260119. [CrossRef]
83. Bayoumi, M.; Munir, M. Potential use of CRISPR/cas13 machinery in understanding virus–host interaction. *Front. Microbiol.* **2021**, *12*, 3580. [CrossRef]
84. Savage, D.F. Cas14: Big advances from small CRISPR proteins. *Biochemistry* **2019**, *58*, 1024–1025. [CrossRef]
85. Badhan, S.; Ball, A.S.; Mantri, N. First report of CRISPR/Cas9 mediated DNA-free editing of *4CL* and *RVE7* genes in chickpea protoplasts. *Int. J. Mol. Sci.* **2021**, *22*, 396. [CrossRef]
86. Zeng, Y.; Wen, J.; Zhao, W.; Wang, Q.; Huang, W. Rational improvement of rice yield and cold tolerance by editing the three genes *OsPIN5b*, *GS3*, and *OsMYB30* with the CRISPR-Cas9 system. *Front. Plant Sci.* **2019**, *10*, 1663. [CrossRef]
87. Li, Y.; Zhu, J.; Wu, H.; Liu, C.; Huang, C.; Lan, J.; Zhao, Y.; Xie, C. Precise base editing of non-allelic acetolactate synthase genes confers sulfonylurea herbicide resistance in maize. *Crop J.* **2020**, *8*, 449–456. [CrossRef]
88. Bouzroud, S.; Gasparini, K.; Hu, G.; Antonia, M.; Barbosa, M.; Luan Rosa, B.; Fahr, M.; Bendaou, N.; Bouzayen, M.; Zsögön, A.; et al. Down regulation and loss of auxin response factor 4 function using CRISPR/Cas9 alters plant growth, stomatal function and improves tomato tolerance to salinity and osmotic stress. *Genes* **2020**, *11*, 272. [CrossRef] [PubMed]
89. Chen, Q.; Tang, W.; Zeng, G.; Sheng, H.; Shi, W.; Xiao, Y. Reduction of cadmium accumulation in the grains of male sterile rice Chuang-5S carrying *Pi48* or *Pi49* through marker-assisted selection. *3 Biotech* **2020**, *10*, 539. [CrossRef]
90. Yu, W.; Wang, L.; Zhao, R.; Sheng, J.; Zhang, S.; Li, R.; Shen, L. Knockout of *SlMAPK3* enhances tolerance to heat stress involving ROS homeostasis in tomato plants. *BMC Plant Biol.* **2019**, *19*, 354. [CrossRef]
91. Kis, A.; Hamar, E.; Tholt, G.; Ban, R.; Havelda, Z. Creating highly efficient resistance against wheat dwarf virus in barley by employing CRISPR/Cas9 system. *Plant Biotechnol. J.* **2019**, *17*, 1004–1006. [CrossRef]
92. Foster, A.J.; Martin-Urdiroz, M.; Yan, X.; Wright, H.S.; Soanes, D.M.; Talbot, N.J. CRISPR-Cas9 ribonucleoprotein-mediated co-editing and counterselection in the rice blast fungus. *Sci. Rep.* **2018**, *8*, 14355. [CrossRef] [PubMed]
93. Ortigosa, A.; Gimenez-Ibanez, S.; Leonhardt, N.; Solano, R. Design of a bacterial speck resistant tomato by CRISPR/Cas9-mediated editing of *SlJAZ2*. *Plant Biotechnol. J.* **2019**, *17*, 665–673. [CrossRef] [PubMed]
94. Zhang, Y.; Guo, W.; Chen, L.; Shen, X.; Yang, H.; Fang, Y.; Ouyang, W.; Mai, S.; Chen, H.; Chen, S.; et al. CRISPR/Cas9-mediated targeted mutagenesis of *GmUGT* enhanced soybean resistance against leaf-chewing insects through flavonoids biosynthesis. *Front. Plant Sci.* **2022**, *13*, 802716. [CrossRef]
95. Miao, C.; Xiao, L.; Hua, K.; Zou, C.; Zhao, Y.; Bressan, R.A.; Zhu, J.-K. Mutations in a subfamily of abscisic acid receptor genes promote rice growth and productivity. *Proc. Natl. Acad. Sci. USA* **2018**, *115*, 6058–6063. [CrossRef]
96. Zhang, Z.; Hua, L.; Gupta, A.; Tricoli, D.; Edwards, K.J.; Yang, B.; Li, W. Development of an *Agrobacterium*-delivered CRISPR/Cas9 system for wheat genome editing. *Plant Biotechnol. J.* **2019**, *17*, 1623–1635. [CrossRef]
97. Dong, O.X.; Yu, S.; Jain, R.; Zhang, N.; Duong, P.Q.; Butler, C.; Li, Y.; Lipzen, A.; Martin, J.A.; Barry, K.W.; et al. Marker-free carotenoid-enriched rice generated through targeted gene insertion using CRISPR-Cas9. *Nat. Commun.* **2020**, *11*, 1178. [CrossRef] [PubMed]
98. Zhou, J.; Xin, X.; He, Y.; Chen, H.; Li, Q.; Tang, X.; Zhong, Z.; Deng, K.; Zheng, X.; Akher, S.A.; et al. Multiplex QTL editing of grain-related genes improves yield in elite rice varieties. *Plant Cell Rep.* **2019**, *38*, 475–485. [CrossRef] [PubMed]
99. Yuyu, C.; Aike, Z.; Pao, X.; Xiaoxia, W.; Yongrun, C.; Beifang, W.; Yue, Z.; Liaqat, S.; Shihua, C.; Liyong, C.; et al. Effects of *GS3* and *GL3.1* for grain size editing by CRISPR/Cas9 in rice. *Rice Sci.* **2020**, *27*, 405–413. [CrossRef]
100. Yuste-Lisbona, F.J.; Fernandez-Lozano, A.; Pineda, B.; Bretones, S.; Ortiz-Atienza, A.; Garcia-Sogo, B.; Muller, N.A.; Angosto, T.; Capel, J.; Moreno, V.; et al. ENO regulates tomato fruit size through the floral meristem development network. *Proc. Natl. Acad. Sci. USA* **2020**, *117*, 8187–8195. [CrossRef] [PubMed]

101. Neupane, D.; Adhikari, P.; Bhattarai, D.; Rana, B.; Ahmed, Z.; Sharma, U.; Adhikari, D. Does climate change affect the yield of the top three cereals and food security in the world? *Earth* **2022**, *3*, 45–71. [CrossRef]
102. Lau, S.E.; Hamdan, M.F.; Pua, T.L.; Saidi, N.B.; Tan, B.C. Plant nitric oxide signaling under drought stress. *Plants* **2021**, *10*, 360. [CrossRef]
103. Mohd Amnan, M.A.; Pua, T.L.; Lau, S.E.; Tan, B.C.; Yamaguchi, H.; Hitachi, K.; Tsuchida, K.; Komatsu, S. Osmotic stress in banana is relieved by exogenous nitric oxide. *PeerJ* **2021**, *9*, e10879. [CrossRef]
104. Vats, S.; Kumawat, S.; Kumar, V.; Patil, G.B.; Joshi, T.; Sonah, H.; Sharma, T.R.; Deshmukh, R. Genome editing in plants: Exploration of technological advancements and challenges. *Cell* **2019**, *8*, 1386. [CrossRef]
105. Zhang, A.; Liu, Y.; Wang, F.; Li, T.; Chen, Z.; Kong, D.; Bi, J.; Zhang, F.; Luo, X.; Wang, J. Enhanced rice salinity tolerance via CRISPR/Cas9-targeted mutagenesis of the *OsRR22* gene. *Mol. Breed.* **2019**, *39*, 47. [CrossRef]
106. Bo, W.; Zhaohui, Z.; Huanhuan, Z.; Xia, W.; Binglin, L.; Lijia, Y.; Xiangyan, H.; Deshui, Y.; Xuelian, Z.; Chunguo, W.; et al. Targeted mutagenesis of NAC transcription factor gene, *OsNAC041*, leading to salt sensitivity in rice. *Rice Sci.* **2019**, *26*, 98–108. [CrossRef]
107. Xie, Z.; Nolan, T.M.; Jiang, H.; Yin, Y. AP2/ERF transcription factor regulatory networks in hormone and abiotic stress responses in *Arabidopsis*. *Front. Plant Sci.* **2019**, *10*, 228. [CrossRef] [PubMed]
108. Faraji, S.; Filiz, E.; Kazemitabar, S.K.; Vannozzi, A.; Palumbo, F.; Barcaccia, G.; Heidari, P. The AP2/ERF gene family in *Triticum durum*: Genome-wide identification and expression analysis under drought and salinity stresses. *Genes* **2020**, *11*, 1464. [CrossRef] [PubMed]
109. Duan, Y.B.; Li, J.; Qin, R.Y.; Xu, R.F.; Li, H.; Yang, Y.C.; Ma, H.; Li, L.; Wei, P.C.; Yang, J.B. Identification of a regulatory element responsible for salt induction of rice *OsRAV2* through *ex situ* and *in situ* promoter analysis. *Plant Mol. Bio.* **2016**, *90*, 49–62. [CrossRef] [PubMed]
110. Liu, X.; Wu, D.; Shan, T.; Xu, S.; Qin, R.; Li, H.; Negm, M.; Wu, D.; Li, J. The trihelix transcription factor *OsGT* $\gamma$*-2* is involved adaption to salt stress in rice. *Plant Mol. Bio.* **2020**, *103*, 545–560. [CrossRef]
111. Nazir, R.; Mandal, S.; Mitra, S.; Ghorai, M.; Das, N.; Jha, N.K.; Majumder, M.; Pandey, D.K.; Dey, A. Clustered regularly interspaced short palindromic repeats (CRISPR)/CRISPR-associated genome-editing toolkit to enhance salt stress tolerance in rice and wheat. *Physiol. Plant.* **2022**, *174*, 12. [CrossRef]
112. Wang, T.; Xun, H.; Wang, W.; Ding, X.; Tian, H.; Hussain, S.; Dong, Q.; Li, Y.; Cheng, Y.; Wang, C.; et al. Mutation of *GmAITR* genes by CRISPR/Cas9 genome editing results in enhanced salinity stress tolerance in soybean. *Front. Plant Sci.* **2021**, *12*, 779598. [CrossRef]
113. Zhang, M.; Cao, Y.; Wang, Z.; Wang, Z.Q.; Shi, J.; Liang, X.; Song, W.; Chen, Q.; Lai, J.; Jiang, C. A retrotransposon in an HKT1 family sodium transporter causes variation of leaf Na(+) exclusion and salt tolerance in maize. *New Phytol.* **2018**, *217*, 1161–1176. [CrossRef]
114. Tran, M.T.; Doan, D.T.H.; Kim, J.; Song, Y.J.; Sung, Y.W.; Das, S.; Kim, E.J.; Son, G.H.; Kim, S.H.; Van Vu, T.; et al. CRISPR/Cas9-based precise excision of *SlHyPRP1* domain(s) to obtain salt stress-tolerant tomato. *Plant Cell Rep.* **2021**, *40*, 999–1011. [CrossRef]
115. Amnan, M.A.M.; Aizat, W.M.; Khaidizar, F.D.; Tan, B.C. Drought stress induces morpho-physiological and proteome changes of *Pandanus amaryllifolius*. *Plants* **2022**, *11*, 221. [CrossRef]
116. Zhang, Y.; Wang, X.; Luo, Y.; Zhang, L.; Yao, Y.; Han, L.; Chen, Z.; Wang, L.; Li, Y. *OsABA8ox2*, an ABA catabolic gene, suppresses root elongation of rice seedlings and contributes to drought response. *Crop J.* **2020**, *8*, 480–491. [CrossRef]
117. Ogata, T.; Ishizaki, T.; Fujita, M.; Fujita, Y. CRISPR/Cas9-targeted mutagenesis of *osera1* confers enhanced responses to abscisic acid and drought stress and increased primary root growth under nonstressed conditions in rice. *PLoS ONE* **2020**, *15*, e0243376. [CrossRef] [PubMed]
118. Yin, X.; Biswal, A.K.; Dionora, J.; Perdigon, K.M.; Balahadia, C.P.; Mazumdar, S.; Chater, C.; Lin, H.C.; Coe, R.A.; Kretzschmar, T.; et al. CRISPR-Cas9 and CRISPR-Cpf1 mediated targeting of a stomatal developmental gene *EPFL9* in rice. *Plant Cell Rep.* **2017**, *36*, 745–757. [CrossRef] [PubMed]
119. Caine, R.S.; Yin, X.; Sloan, J.; Harrison, E.L.; Mohammed, U.; Fulton, T.; Biswal, A.K.; Dionora, J.; Chater, C.C.; Coe, R.A.; et al. Rice with reduced stomatal density conserves water and has improved drought tolerance under future climate conditions. *New Phytol.* **2019**, *221*, 371–384. [CrossRef] [PubMed]
120. Hughes, J.; Hepworth, C.; Dutton, C.; Dunn, J.A.; Hunt, L.; Stephens, J.; Waugh, R.; Cameron, D.D.; Gray, J.E. Reducing stomatal density in barley improves drought tolerance without impacting on yield. *Plant Physiol.* **2017**, *174*, 776–787. [CrossRef]
121. Clemens, M.; Faralli, M.; Lagreze, J.; Bontempo, L.; Piazza, S.; Varotto, C.; Malnoy, M.; Oechel, W.; Rizzoli, A.; Dalla Costa, L. VvEPFL9-1 knock-out via CRISPR/Cas9 reduces stomatal density in grapevine. *Front. Plant Sci.* **2022**, *13*, 878001. [CrossRef]
122. Wang, L.; Chen, L.; Li, R.; Zhao, R.; Yang, M.; Sheng, J.; Shen, L. Reduced drought tolerance by CRISPR/Cas9-mediated SlMAPK3 mutagenesis in tomato plants. *J. Agric. Food Chem.* **2017**, *65*, 8674–8682. [CrossRef]
123. Nieves-Cordones, M.; Mohamed, S.; Tanoi, K.; Kobayashi, N.I.; Takagi, K.; Vernet, A.; Guiderdoni, E.; Perin, C.; Sentenac, H.; Very, A.A. Production of low-Cs+ rice plants by inactivation of the k+ transporter *OsHAK1* with the CRISPR-Cas system. *Plant J.* **2017**, *92*, 43–56. [CrossRef]
124. Wang, F.Z.; Chen, M.X.; Yu, L.J.; Xie, L.J.; Yuan, L.B.; Qi, H.; Xiao, M.; Guo, W.; Chen, Z.; Yi, K.; et al. *OsARM1*, an R2R3 MYB transcription factor, is involved in regulation of the response to arsenic stress in rice. *Front. Plant Sci.* **2017**, *8*, 1868. [CrossRef]
125. Tang, L.; Mao, B.; Li, Y.; Lv, Q.; Zhang, L.; Chen, C.; He, H.; Wang, W.; Zeng, X.; Shao, Y.; et al. Knockout of OsNramp5 using the CRISPR/Cas9 system produces low Cd-accumulating indica rice without compromising yield. *Sci Rep.* **2017**, *7*, 14438. [CrossRef]

126. Shao, X.; Wu, S.; Dou, T.; Zhu, H.; Hu, C.; Huo, H.; He, W.; Deng, G.; Sheng, O.; Bi, F.; et al. Using CRISPR/Cas9 genome editing system to create MaGA20ox2 gene-modified semi-dwarf banana. *Plant Biotechnol. J.* **2020**, *18*, 17–19. [CrossRef]
127. Shimatani, Z.; Kashojiya, S.; Takayama, M.; Terada, R.; Arazoe, T.; Ishii, H.; Teramura, H.; Yamamoto, T.; Komatsu, H.; Miura, K.; et al. Targeted base editing in rice and tomato using a CRISPR-Cas9 cytidine deaminase fusion. *Nat. Biotechnol.* **2017**, *35*, 441–443. [CrossRef]
128. Talakayala, A.; Ankanagari, S.; Garladinne, M. CRISPR-Cas genome editing system: A versatile tool for developing disease resistant crops. *Plant Stress* **2022**, *3*, 100056. [CrossRef]
129. Yin, K.; Qiu, J.L. Genome editing for plant disease resistance: Applications and perspectives. *Philos. Trans. R. Soc. Lond. B Biol. Sci.* **2019**, *374*, 20180322. [CrossRef] [PubMed]
130. Jiang, W.; Zhou, H.; Bi, H.; Fromm, M.; Yang, B.; Weeks, D.P. Demonstration of CRISPR/Cas9/sgRNA-mediated targeted gene modification in Arabidopsis, tobacco, sorghum and rice. *Nucleic Acids Res.* **2013**, *41*, e188. [CrossRef]
131. Xu, Z.; Xu, X.; Gong, Q.; Li, Z.; Li, Y.; Wang, S.; Yang, Y.; Ma, W.; Liu, L.; Zhu, B.; et al. Engineering broad-spectrum bacterial blight resistance by simultaneously disrupting variable tale-binding elements of multiple susceptibility genes in rice. *Mol. Plant.* **2019**, *12*, 1434–1446. [CrossRef]
132. Oliva, R.; Ji, C.; Atienza-Grande, G.; Huguet-Tapia, J.C.; Perez-Quintero, A.; Li, T.; Eom, J.-S.; Li, C.; Nguyen, H.; Liu, B.; et al. Broad-spectrum resistance to bacterial blight in rice using genome editing. *Nat. Biotechnol.* **2019**, *37*, 1344–1350. [CrossRef] [PubMed]
133. Wang, F.; Wang, C.; Liu, P.; Lei, C.; Hao, W.; Gao, Y.; Liu, Y.G.; Zhao, K. Enhanced rice blast resistance by CRISPR/Cas9-targeted mutagenesis of the ERF transcription factor gene *OsERF922*. *PLoS ONE* **2016**, *11*, e0154027. [CrossRef] [PubMed]
134. Macovei, A.; Sevilla, N.R.; Cantos, C.; Jonson, G.B.; Slamet-Loedin, I.; Čermák, T.; Voytas, D.F.; Choi, I.R.; Chadha-Mohanty, P. Novel alleles of rice eif4g generated by CRISPR/Cas9-targeted mutagenesis confer resistance to rice tungro spherical virus. *Plant Biotechnol. J.* **2018**, *16*, 1918–1927. [CrossRef] [PubMed]
135. Zhou, X.; Liao, H.; Chern, M.; Yin, J.; Chen, Y.; Wang, J.; Zhu, X.; Chen, Z.; Yuan, C.; Zhao, W.; et al. Loss of function of a rice TPR-domain RNA-binding protein confers broad-spectrum disease resistance. *Proc. Natl. Acad. Sci. USA* **2018**, *115*, 3174–3179. [CrossRef]
136. Thomazella, D.P.d.T.; Seong, K.; Mackelprang, R.; Dahlbeck, D.; Geng, Y.; Gill, U.S.; Qi, T.; Pham, J.; Giuseppe, P.; Lee, C.Y.; et al. Loss of function of a dmr6 ortholog in tomato confers broad-spectrum disease resistance. *Proc. Natl. Acad. Sci. USA* **2021**, *118*, e2026152118. [CrossRef]
137. Kumar, N.; Galli, M.; Ordon, J.; Stuttmann, J.; Kogel, K.-H.; Imani, J. Further analysis of barley MORC1 using a highly efficient RNA-guided Cas9 gene-editing system. *Plant Biotechnol. J.* **2018**, *16*, 1892–1903. [CrossRef] [PubMed]
138. Nekrasov, V.; Wang, C.; Win, J.; Lanz, C.; Weigel, D.; Kamoun, S. Rapid generation of a transgene-free powdery mildew resistant tomato by genome deletion. *Sci. Rep.* **2017**, *7*, 482. [CrossRef]
139. Zhang, Y.; Bai, Y.; Wu, G.; Zou, S.; Chen, Y.; Gao, C.; Tang, D. Simultaneous modification of three homoeologs of TaEDR1 by genome editing enhances powdery mildew resistance in wheat. *Plant J.* **2017**, *91*, 714–724. [CrossRef] [PubMed]
140. Wan, D.-Y.; Guo, Y.; Cheng, Y.; Hu, Y.; Xiao, S.; Wang, Y.; Wen, Y.-Q. CRISPR/Cas9-mediated mutagenesis of *VvMLO3* results in enhanced resistance to powdery mildew in grapevine (*Vitis vinifera*). *Hortic. Res.* **2020**, *7*, 116. [CrossRef]
141. Wang, Y.Y.; Zhang, Y.J.; Zhou, R.; Dossa, K.; Yu, J.Y.; Li, D.H.; Liu, A.L.; Mmadi, M.A.; Zhang, X.R.; You, J. Identification and characterization of the bzip transcription factor family and its expression in response to abiotic stresses in sesame. *PLoS ONE* **2018**, *13*, 21. [CrossRef] [PubMed]
142. Wei, J.; Fang, Y.; Jiang, H.; Wu, X.T.; Zuo, J.H.; Xia, X.C.; Li, J.Q.; Stich, B.; Cao, H.; Liu, Y.X. Combining QTL mapping and gene co-expression network analysis for prediction of candidate genes and molecular network related to yield in wheat. *BMC Plant Biol.* **2022**, *22*, 288. [CrossRef] [PubMed]
143. Song, G.; Jia, M.; Chen, K.; Kong, X.; Khattak, B.; Xie, C.; Li, A.; Mao, L. CRISPR/Cas9: A powerful tool for crop genome editing. *Crop J.* **2016**, *4*, 75–82. [CrossRef]
144. Ma, X.; Zhu, Q.; Chen, Y.; Liu, Y.-G. CRISPR/Cas9 platforms for genome editing in plants: Developments and applications. *Mol. Plant* **2016**, *9*, 961–974. [CrossRef]
145. Li, M.; Li, X.; Zhou, Z.; Wu, P.; Fang, M.; Pan, X.; Lin, Q.; Luo, W.; Wu, G.; Li, H. Reassessment of the four yield-related genes *Gn1a*, *DEP1*, *GS3*, and *IPA1* in rice using a CRISPR/Cas9 system. *Front. Plant Sci.* **2016**, *7*, 377. [CrossRef]
146. Xu, R.; Yang, Y.; Qin, R.; Li, H.; Qiu, C.; Li, L.; Wei, P.; Yang, J. Rapid improvement of grain weight via highly efficient CRISPR/Cas9-mediated multiplex genome editing in rice. *J. Genet. Genomics* **2016**, *43*, 529–532. [CrossRef]
147. Zhang, Y.; Liang, Z.; Zong, Y.; Wang, Y.; Liu, J.; Chen, K.; Qiu, J.-L.; Gao, C. Efficient and transgene-free genome editing in wheat through transient expression of CRISPR/Cas9 DNA or RNA. *Nat. Commun.* **2016**, *7*, 12617. [CrossRef] [PubMed]
148. Rodriguez-Leal, D.; Lemmon, Z.H.; Man, J.; Bartlett, M.E.; Lippman, Z.B. Engineering quantitative trait variation for crop improvement by genome editing. *Cell* **2017**, *171*, 470–480. [CrossRef] [PubMed]
149. Gao, H.; Gadlage, M.J.; Lafitte, H.R.; Lenderts, B.; Yang, M.; Schroder, M.; Farrell, J.; Snopek, K.; Peterson, D.; Feigenbutz, L.; et al. Superior field performance of waxy corn engineered using CRISPR-Cas9. *Nat. Biotechnol.* **2020**, *38*, 579–581. [CrossRef]
150. Mao, H.; Jian, C.; Cheng, X.; Chen, B.; Mei, F.; Li, F.; Zhang, Y.; Li, S.; Du, L.; Li, T.; et al. The wheat ABA receptor gene TaPYL1-1b contributes to drought tolerance and grain yield by increasing water-use efficiency. *Plant Biotechnol. J.* **2022**, *20*, 846–861. [CrossRef] [PubMed]

151. Soyk, S.; Muller, N.A.; Park, S.J.; Schmalenbach, I.; Jiang, K.; Hayama, R.; Zhang, L.; Van Eck, J.; Jimenez-Gomez, J.M.; Lippman, Z.B. Variation in the flowering gene self pruning 5g promotes day-neutrality and early yield in tomato. *Nat. Genet.* **2017**, *49*, 162–168. [CrossRef]
152. Carmel-Goren, L.; Liu, Y.S.; Lifschitz, E.; Zamir, D. The self-pruning gene family in tomato. *Plant Mol. Biol.* **2003**, *52*, 1215–1222. [CrossRef]
153. Klap, C.; Yeshayahou, E.; Bolger, A.M.; Arazi, T.; Gupta, S.K.; Shabtai, S.; Usadel, B.; Salts, Y.; Barg, R. Tomato facultative parthenocarpy results from slagamous-like 6 loss of function. *Plant Biotechnol. J.* **2017**, *15*, 634–647. [CrossRef]
154. van Eck, N.J.; Waltman, L. Software survey: VOSviewer, a computer program for bibliometric mapping. *Scientometrics* **2010**, *84*, 523–538. [CrossRef]
155. Romero, L.; Portillo-Salido, E. Trends in Sigma-1 receptor research: A 25-year bibliometric analysis. *Front. Pharmacol.* **2019**, *10*, 564. [CrossRef]
156. Hamdan, M.F.; Mohd Noor, S.N.; Abd-Aziz, N.; Pua, T.-L.; Tan, B.C. Green revolution to gene revolution: Technological advances in agriculture to feed the world. *Plants* **2022**, *11*, 1297. [CrossRef]
157. Liu, Z.; Dong, H.; Cui, Y.; Cong, L.; Zhang, D. Application of different types of CRISPR/Cas-based systems in bacteria. *Microb. Cell Fact.* **2020**, *19*, 172. [CrossRef] [PubMed]
158. Zhang, Y.; Ren, Q.; Tang, X.; Liu, S.; Malzahn, A.A.; Zhou, J.; Wang, J.; Yin, D.; Pan, C.; Yuan, M.; et al. Expanding the scope of plant genome engineering with Cas12a orthologs and highly multiplexable editing systems. *Nat. Commun.* **2021**, *12*, 1944. [CrossRef] [PubMed]
159. Huang, T.K.; Puchta, H. Novel CRISPR/Cas applications in plants: From prime editing to chromosome engineering. *Transgenic Res.* **2021**, *30*, 529–549. [CrossRef]
160. Wada, N.; Osakabe, K.; Osakabe, Y. Expanding the plant genome editing toolbox with recently developed CRISPR-Cas systems. *Plant Physiol.* **2022**, *188*, 1825–1837. [CrossRef]
161. Zess, E.; Begemann, M. CRISPR-Cas9 and beyond: What's next in plant genome engineering. *In Vitro Cell. Dev. Biol. Plant* **2021**, *57*, 584–594. [CrossRef]
162. Budeguer, F.; Enrique, R.; Perera, M.F.; Racedo, J.; Castagnaro, A.P.; Noguera, A.S.; Welin, B. Genetic transformation of sugarcane, current status and future prospects. *Front. Plant Sci.* **2021**, *12*, 20. [CrossRef]
163. Lassoued, R.; Macall, D.M.; Hesseln, H.; Phillips, P.W.B.; Smyth, S.J. Benefits of genome-edited crops: Expert opinion. *Transgenic Res.* **2019**, *28*, 247–256. [CrossRef] [PubMed]
164. Ahmad, S.; Tang, L.; Shahzad, R.; Mawia, A.M.; Rao, G.S.; Jamil, S.; Wei, C.; Sheng, Z.; Shao, G.; Wei, X.; et al. CRISPR-based crop improvements: A way forward to achieve zero hunger. *J. Agric. Food Chem.* **2021**, *69*, 8307–8323. [CrossRef]
165. Entine, J.; Felipe, M.S.S.; Groenewald, J.H.; Kershen, D.L.; Lema, M.; McHughen, A.; Nepomuceno, A.L.; Ohsawa, R.; Ordonio, R.L.; Parrott, W.A.; et al. Regulatory approaches for genome edited agricultural plants in select countries and jurisdictions around the world. *Transgenic Res.* **2021**, *30*, 551–584. [CrossRef]
166. Lobato-Gomez, M.; Hewitt, S.; Capell, T.; Christou, P.; Dhingra, A.; Giron-Calva, P.S. Transgenic and genome-edited fruits: Background, constraints, benefits, and commercial opportunities. *Hortic. Res.* **2021**, *8*, 166. [CrossRef]
167. Anjanappa, R.B.; Gruissem, W. Current progress and challenges in crop genetic transformation. *J. Plant Physiol.* **2021**, *261*, 153411. [CrossRef] [PubMed]
168. Zhao, H.; Jia, Y.; Cao, Y.; Wang, Y. Improving T-DNA transfer to *Tamarix hispida* by adding chemical compounds during *Agrobacterium tumefaciens* culture. *Front. Plant Sci.* **2020**, *11*, 501358. [CrossRef] [PubMed]
169. Yue, J.J.; Yuan, J.L.; Wu, F.H.; Yuan, Y.H.; Cheng, Q.W.; Hsu, C.T.; Lin, C.S. Protoplasts: From isolation to CRISPR/Cas genome editing application. *Front. Genom. Ed.* **2021**, *3*, 717017. [CrossRef] [PubMed]
170. Yoo, B.C.; Yadav, N.S.; Orozco, E.M.; Sakai, H. Cas9/gRNA-mediated genome editing of yeast mitochondria and *Chlamydomonas* chloroplasts. *PeerJ* **2020**, *8*, 29. [CrossRef] [PubMed]
171. Sandhya, D.; Jogam, P.; Allini, V.R.; Abbagani, S.; Alok, A. The present and potential future methods for delivering CRISPR/Cas9 components in plants. *J. Genet. Eng. Biotechnol.* **2020**, *18*, 25. [CrossRef] [PubMed]
172. Hua, K.; Zhang, J.; Botella, J.; Ma, C.; Kong, F.; Liu, B.; Zhu, J. Perspectives on the application of genome-editing technologies in crop breeding. *Mol. Plant* **2019**, *12*, 1047–1059. [CrossRef]
173. Chen, S.-J. Minimizing off-target effects in CRISPR-Cas9 genome editing. *Cell Biol. Toxicol.* **2019**, *35*, 399–401. [CrossRef]
174. Kawall, K.; Cotter, J.; Then, C. Broadening the GMO risk assessment in the EU for genome editing technologies in agriculture. *Environ. Sci. Eur.* **2020**, *32*, 106. [CrossRef]
175. Muller, M.; Lee, C.M.; Gasiunas, G.; Davis, T.H.; Cradick, T.J.; Siksnys, V.; Bao, G.; Cathomen, T.; Mussolino, C. *Streptococcus thermophilus* CRISPR-Cas9 systems enable specific editing of the human genome. *Mol. Ther.* **2016**, *24*, 636–644. [CrossRef]
176. Nishimasu, H.; Shi, X.; Ishiguro, S.; Gao, L.; Hirano, S.; Okazaki, S.; Noda, T.; Abudayyeh, O.O.; Gootenberg, J.S.; Mori, H.; et al. Engineered CRISPR-Cas9 nuclease with expanded targeting space. *Science* **2018**, *361*, 1259–1262. [CrossRef]
177. Chen, J.S.; Dagdas, Y.S.; Kleinstiver, B.P.; Welch, M.M.; Sousa, A.A.; Harrington, L.B.; Sternberg, S.H.; Joung, J.K.; Yildiz, A.; Doudna, J.A. Enhanced proofreading governs CRISPR–Cas9 targeting accuracy. *Nature* **2017**, *550*, 407–410. [CrossRef] [PubMed]
178. Slaymaker, I.M.; Gao, L.Y.; Zetsche, B.; Scott, D.A.; Yan, W.X.; Zhang, F. Rationally engineered Cas9 nucleases with improved specificity. *Science* **2016**, *351*, 84–88. [CrossRef] [PubMed]

179. Kleinstiver, B.P.; Pattanayak, V.; Prew, M.S.; Tsai, S.Q.; Nguyen, N.T.; Zheng, Z.; Joung, J.K. High-fidelity CRISPR–Cas9 nucleases with no detectable genome-wide off-target effects. *Nature* **2016**, *529*, 490–495. [CrossRef] [PubMed]
180. Dugar, G.; Leenay, R.T.; Eisenbart, S.K.; Bischler, T.; Aul, B.U.; Beisel, C.L.; Sharma, C.M. CRISPR RNA-dependent binding and cleavage of endogenous RNAs by the *Campylobacter jejuni* Cas9. *Mol. Cell* **2018**, *69*, 893–905. [CrossRef]
181. Nakamura, M.; Gao, Y.; Dominguez, A.A.; Qi, L.S. CRISPR technologies for precise epigenome editing. *Nat. Cell Biol.* **2021**, *23*, 11–22. [CrossRef]
182. Galonska, C.; Charlton, J.; Mattei, A.L.; Donaghey, J.; Clement, K.; Gu, H.; Mohammad, A.W.; Stamenova, E.K.; Cacchiarelli, D.; Klages, S.; et al. Genome-wide tracking of dCas9-methyltransferase footprints. *Nat. Commun.* **2018**, *9*, 597. [CrossRef]
183. Lin, Q.; Zong, Y.; Xue, C.; Wang, S.; Jin, S.; Zhu, Z.; Wang, Y.; Anzalone, A.V.; Raguram, A.; Doman, J.L.; et al. Prime genome editing in rice and wheat. *Nat. Biotechnol.* **2020**, *38*, 582–585. [CrossRef]
184. Saeglitz, C.; Bartsch, D. Regulatory and associated political issues with respect to Bt transgenic maize in the European Union. *J. Invertebr. Pathol.* **2003**, *83*, 107–109. [CrossRef]
185. Jones, H.D. Challenging regulations: Managing risks in crop biotechnology. *Food Energy Secur.* **2015**, *4*, 87–91. [CrossRef]
186. Eckerstorfer, M.F.; Grabowski, M.; Lener, M.; Engelhard, M.; Simon, S.; Dolezel, M.; Heissenberger, A.; Luthi, C. Biosafety of genome editing applications in plant breeding: Considerations for a focused case-specific risk assessment in the EU. *BioTech* **2021**, *10*, 10. [CrossRef]
187. Agapito-Tenfen, S.Z.; Okoli, A.S.; Bernstein, M.J.; Wikmark, O.G.; Myhr, A.I. Revisiting risk governance of GM plants: The need to consider new and emerging gene-editing techniques. *Front. Plant Sci.* **2018**, *9*, 1874. [CrossRef]
188. Busch, G.; Ryan, E.; von Keyserlingk, M.A.G.; Weary, D.M. Citizen views on genome editing: Effects of species and purpose. *Agric. Human Values* **2022**, *39*, 151–164. [CrossRef]
189. Pausch, P.; Al-Shayeb, B.; Bisom-Rapp, E.; Tsuchida, C.A.; Li, Z.; Cress, B.F.; Knott, G.J.; Jacobsen, S.E.; Banfield, J.F.; Doudna, J.A. CRISPR-casphi from huge phages is a hypercompact genome editor. *Science* **2020**, *369*, 333–337. [CrossRef] [PubMed]
190. Calvache, C.; Vazquez-Vilar, M.; Selma, S.; Uranga, M.; Fernandez-Del-Carmen, A.; Daros, J.A.; Orzaez, D. Strong and tunable anti-CRISPR/Cas activities in plants. *Plant Biotechnol. J.* **2022**, *20*, 399–408. [CrossRef] [PubMed]
191. Garrett, R.; Niles, M.T.; Gil, J.D.; Gaudin, A.; Chaplin-Kramer, R.; Assmann, A.; Assmann, T.S.; Brewer, K.; de Faccio Carvalho, P.C.; Cortner, O.J.A.S. Social and ecological analysis of commercial integrated crop livestock systems: Current knowledge and remaining uncertainty. *Agric. Syst.* **2017**, *155*, 136–146. [CrossRef]

plants

MDPI

*Review*

# Will Plant Genome Editing Play a Decisive Role in "Quantum-Leap" Improvements in Crop Yield to Feed an Increasing Global Human Population?

**Anton V. Buzdin [1,2,3], Maxim V. Patrushev [4] and Eugene D. Sverdlov [4,5,*]**

1 The Laboratory of Clinical and Genomic Bioinformatics, I.M. Sechenov First Moscow State Medical University, 119991 Moscow, Russia; buzdin@oncobox.com
2 Moscow Institute of Physics and Technology, Dolgoprudny, Moscow Region, 141701 Moscow, Russia
3 Shemyakin-Ovchinnikov Institute of Bioorganic Chemistry, Russian Academy of Sciences, 117997 Moscow, Russia
4 Kurchatov Center for Genome Research, National Research Center Kurchatov Institute, 123182 Moscow, Russia; maxpatrushev@yandex.ru
5 Institute of Molecular Genetics, National Research Center Kurchatov Institute, 123182 Moscow, Russia
* Correspondence: edsverd@gmail.com

**Abstract:** Growing scientific evidence demonstrates unprecedented planetary-scale human impacts on the Earth's system with a predicted threat to the existence of the terrestrial biosphere due to population increase, resource depletion, and pollution. Food systems account for 21–34% of global carbon dioxide ($CO_2$) emissions. Over the past half-century, water and land-use changes have significantly impacted ecosystems, biogeochemical cycles, biodiversity, and climate. At the same time, food production is falling behind consumption, and global grain reserves are shrinking. Some predictions suggest that crop yields must approximately double by 2050 to adequately feed an increasing global population without a large expansion of crop area. To achieve this, "quantum-leap" improvements in crop cultivar productivity are needed within very narrow planetary boundaries of permissible environmental perturbations. Strategies for such a "quantum-leap" include mutation breeding and genetic engineering of known crop genome sequences. Synthetic biology makes it possible to synthesize DNA fragments of any desired sequence, and modern bioinformatics tools may hopefully provide an efficient way to identify targets for directed modification of selected genes responsible for known important agronomic traits. CRISPR/Cas9 is a new technology for incorporating seamless directed modifications into genomes; it is being widely investigated for its potential to enhance the efficiency of crop production. We consider the optimism associated with the new genetic technologies in terms of the complexity of most agronomic traits, especially crop yield potential (Yp) limits. We also discuss the possible directions of overcoming these limits and alternative ways of providing humanity with food without transgressing planetary boundaries. In conclusion, we support the long-debated idea that new technologies are unlikely to provide a rapidly growing population with significantly increased crop yield. Instead, we suggest that delicately balanced humane measures to limit its growth and the amount of food consumed per capita are highly desirable for the foreseeable future.

**Keywords:** CRISPR/Cas9; crop yield potential (Yp) limits; quantum leap; synthetic biology

**Citation:** Buzdin, A.V.; Patrushev, M.V.; Sverdlov, E.D. Will Plant Genome Editing Play a Decisive Role in "Quantum-Leap" Improvements in Crop Yield to Feed an Increasing Global Human Population? *Plants* **2021**, *10*, 1667. https://doi.org/10.3390/plants10081667

Academic Editors: Ranjith Pathirana and Francesco Carimi

Received: 12 July 2021
Accepted: 7 August 2021
Published: 13 August 2021



## 1. Introduction

Growing scientific evidence demonstrates unprecedented planetary-scale human impacts on the Earth's system [1]. In 1971, John Harte published *"Patient Earth"* [2], which discusses rising problems in the nascent field of environmental science, including human population growth, resource scarcity, and nuclear contamination. In 1972, Donella Meadows and her colleagues published a landmark book entitled *"The Limits to Growth"* [3], the message of which is that the resources of the Earth probably cannot maintain the current rate of economic and population growth well after 2100 even with the use of advanced technologies, which is still in dispute today. The major challenges facing civilization have become undoubtedly evident [4]. Rising atmospheric $CO_2$ levels were predicted by Svante Arrhenius in 1896 [1], and a strong correspondence between the temperature and the concentration of carbon dioxide in the atmosphere observed during the glacial cycles of the past several hundred thousand years can be seen at the site of NOAA's National Centers for Environmental Information [5].

The warming atmosphere has influenced global wind and precipitation patterns and increased the intensity of extreme weather [6]. Catastrophic fires observed in different areas of the planet are evident expressions of this change. Atmospheric chemistry is also affected by human activities [7]. Food systems account for 21–34% of global emissions. Over the past half-century, land-use changes have significantly impacted ecosystems, biogeochemical cycles, biodiversity, and climate. The global area equipped for irrigation has doubled since the 1960s; agriculture now represents 70% of freshwater withdrawals around the world and global fertilizer use has quadrupled, leading to increased nutrient runoff into inland waters and coastal seas [8]. Expanding agricultural land use is a significant contributor to rising atmospheric $CO_2$ levels and biodiversity loss due to deforestation and the draining of wetlands [9]. The list of influences could be continued. The causes of slowing global grain production and shrinking reserves [10] are yet to be answered [11]. Some predictions suggest that crop yields must double by 2050 to adequately feed an increasing global population without a large expansion of crop area, although this is a hotly debated issue [1,9,11–17]. Doubling agricultural yield within the next 30 years requires an annual increase of ~2.2%, which exceeds the average annual increase witnessed over the past 50 years [18]. We do not discuss the highly important problems of poor food distribution and wastage of one-third of the world's food talked over in this review [19].

"Quantum-leap" improvements in crop cultivar productivity are needed to achieve sufficient annual yields within a narrow window of permissible environmental perturbations. In 2009, "the planetary boundaries framework" was put forward defining a "safe operating space for humanity" [20] (see also [21]). It was argued that a set of nine "planetary boundaries" must not be crossed by humanity at the cost of its own peril through scientifically defined targets of the maximum allowed human interference with processes that regulate the state of the planet. The nine processes are climate change; biogeochemical (nitrogen and phosphorus) flows; land-system change; freshwater use; aerosol loading; ozone depletion; ocean acidification; the loss of biosphere integrity, including functional and genetic biodiversity; and the introduction of novel entities, such as toxic chemicals and plastics. The concept of planetary boundaries has generated significant academic debate and policy recommendations worldwide [21–23]. However, it is a very useful reminder for people to be extremely cautious in transforming nature.

"Green Revolution" technologies of the post-war years have resulted in cereal production increases of 30% per capita over the last 50 years [8]. However, the revolution had great unintended consequences [24]. Greater than 90% losses in crop genetic diversity have occurred over the 20th century due to agricultural practices relying upon only three plant species: rice (*Oryza sativa*), maize (*Zea mays*), and wheat (*Triticum aestivum*) towards supplying nearly 60% of the world's plant-based food [8]. Improvement in crop diversity is necessary to achieve resilience to abiotic and biotic stress [25]. Enhancement in crop yield and stability requires a systems approach by combining agronomic and technological management with the implementation of new crop cultivars [11,12,25,26]. Ideally, new crop varieties should have genetic combinations that alleviate losses from the multiple environmental and pest constraints encountered during the crop lifecycle in a farmer's field. The integration of mechanistic understanding, genetic variation, and genome-scale breeding will be essential for technological solutions to manage shortfalls in agriculture yield and stability in a growing worldwide population [11,26,27]. This is also true for forage legumes [28], which play a crucial role as feedstock in the global production of meat [29] and occupy comparable planting space as food crops [30].

There are three strategies for significantly improving crop yield and stability: mutation breeding, genetic engineering, and directed modification. Mutation breeding involves exposing seeds to chemicals or radiation to generate mutants with desirable traits to be bred with other cultivars. Genetic engineering of crops with known genome sequences allows the creation of agronomically relevant variations [31], and synthetic biology techniques allow the synthesis of extended DNA fragments with desired sequences [32,33]. Currently, the most widely adopted genetically modified trait in crops is resistance to herbicides and insects, which has been incorporated into *Zea mays*, *Glycine max* (soybean), *Gossypium hirsutum* (cotton), and *Brassica napus* (canola). They are usually monogenic traits. Directed modification involves targeting specific DNA regions or selected genes that are known as important agronomic traits for DNA editing. New technologies such as "clustered regularly interspaced short palindromic repeats/CRISPR-associated protein 9" (CRISPR/Cas9) are being widely investigated for their potential to enhance crop production efficiency. These expectations are very high [15,16]. Bioinformatics plays an important role in the selection of targets for directed modification by relying on existing information about DNA, RNA, and protein sequences contained in databases such as GenBank, Ensembl, or UniProt as well as their functional activities contained in databases such as gene ontology (GO). A database Gramene for plants is being actively developed (see below, Section 12, page 15). However, most GO annotations are incomplete and imperfect [34,35]. Therefore, predicting the associations between genes and phenotypes is rather problematic, as is the identification of adequate targets for modification. It is suggested that a better understanding of the mechanisms controlling yields in variable environments is required for necessary crop improvement.

Here, we assess the validity of the optimism associated with new genetic technologies in terms of the complexity of most agronomic traits, especially crop yield potential (Yp) and its theoretical limits. We discuss possible directions for overcoming these limits and suggest alternative ways of providing humanity with sufficient food without transgressing planetary security boundaries. The suggestions vary from innovative urban agriculture development [36–38] to the development of crops tolerant to poor soil [39]. Due to the limited space, we do not consider the problems associated with the second and third green revolutions since these terms are heterogeneously interpreted by different authors.

## 2. The Green Revolution and Its Genes

Modern agriculture has its roots in the green revolution that began with the introduction of high-yielding wheat and rice cultivars in the 1960s [24,38,40–42]. Norman Borlaug, the father of the green revolution, bred wheat to favour shorter, stronger stalks that better support larger seed heads. In 1953, he crossed a Japanese dwarf variety of wheat called Norin 10 with a high-yielding American cultivar called Brevor 14. Norin 10/Brevor 14 is a semi-dwarf cultivar that is one-half to two-thirds the height of standard varieties and produces more stalks and, thus, more heads of grain per plant. Borlaug also crossbred the semi-dwarf Norin 10/Brevor 14 cultivar with disease-resistant cultivars to produce wheat varieties that are adapted to tropical and sub-tropical climates. Borlaug's new semi-dwarf, disease-resistant varieties dramatically changed the potential yield of spring wheat. By 1963, 95% of Mexico's wheat crops used the semi-dwarf varieties developed by Borlaug and the harvest was six times than that in 1944, the year Borlaug arrived in Mexico. Mexico has become fully self-sufficient in wheat production and a net exporter of wheat.

Plant height is a major agronomic trait closely correlated with crop yield, which is controlled by multiple genes that may be optimized through breeding strategies (see "Complexity of agronomic traits" for further discussion). The genes responsible for dwarfing traits interfere with the action or production of gibberellic acid (GA) plant hormones. Two main "green revolution genes", namely *Rht (reduced height)*, which encodes a growth repressor DELLA protein of GA signaling, and *sd1* (*semi-dwarf1*), which encodes (GA) 20-oxidase [41,43–45], were cloned from wheat and rice, respectively, and are now widespread through international breeding programs. Only 3 out of the more than 21 reduced-height (*Rht*) genes reported in wheat have been used extensively in wheat breeding programs (*Rht-1* homoeoalleles *Rht-B1b* and *Rht-D1b*, *Rht8*, and *Rht12*) [42,46–49]. Remarkably, both *Rht-1* homoeoalleles originate from the same Japanese variety, Norin 10. In addition to the widely used GA-insensitive dwarfing genes *Rht-B1* and *Rht-D1*, there is a wide spectrum of loci that can be used to modulate plant height [50]. Rice *sd1* was initially defined as a semi-dwarfism gene encoding a defective enzyme in the GA biosynthetic pathway [41,51–54], and pleiotropic changes were revealed by the recovery of the wild-type cultivar following restoration of the *sd1* mutant protein to wild-type in rice [54].

Since interference with the plant's response to GA triggers adverse effects for a range of important agronomic traits, attempts have been made to identify mutants without these shortages. Recently, a major *Rht* locus on wheat chromosome 6A, *Rht24*, substantially reduced plant height alone as well as in combination with *Rht-1b* alleles. Unlike the two *Rht-1b* alleles, plants carrying *Rht24* remain sensitive to GA treatment [40]. Nowadays, as many as 61 genes (*d1* to *d61*) are known to cause dwarfism in rice [55].

In addition to GAs, brassinosteroids and strigolactones are also involved in controlling plant height. The genes involved in changing the levels of these hormones offer additional opportunities for expanding the genetic basis of semi-dwarf rice breeding [52,56]. The above-mentioned genes controlling plant height lie in a complex regulatory network, and additional dwarfing genes are involved in other pathways [40,56]. Phytohormones regulate many aspects of plant life by activating transcription factors that bind sequence-specific response elements (REs) in regulatory regions of target genes. Specific RE variants are highly conserved in core hormone response genes and regulate the magnitude and spatial profile of hormonal responses suggesting that hormone-regulated transcription factors bind a spectrum of REs, each coding for a distinct transcriptional response profile [57]. Such intricate regulation adds an additional level of trait complexity, see [58] for a detailed review.

In his 1970 Nobel lecture, Borlaug summarized the qualities of wheat that he had bred: "It is the unusual breadth of adaption combined with high genetic Yp, short straw, a strong responsiveness and high efficiency in the use of heavy doses of fertilizers, and *a broad spectrum of disease resistance* that has made the Mexican dwarf varieties a powerful catalyst that they have become in launching the green revolution" [59].

Plant diseases are responsible for substantial crop losses each year and pose a threat to global food security and agricultural sustainability. However, it is challenging to breed varieties with resistance that is effective, stable, and broad spectrum [60–63]. Plant growth and disease resistance are tightly regulated, and many negative correlations between growth and defence are the result of regulatory crosstalk [60]. There is increasing evidence that resistance to one disease involves trade-offs with responses to other bio-antagonists. For example, numerous pleiotropic effects of mildew resistance locus O (*MLO*) were recently reviewed [64]. There may occur substantial genotype-by-environment interactions in fitness costs, which makes experiments studying disease resistance in plants especially challenging [60,62–68].

Insecticide resistance mutations are widely assumed to carry fitness costs [66]. For example, an intercellular sucrose transporter was recently identified as a major susceptibility locus in blight-resistant rice *O. sativa* [69]. This transporter moves sugars from the photosynthetic tissues into the phloem for transport to tissues requiring externally supplied sugar for growth and development. A disadvantage of transporting sugars outside of the plant cell is the supply of a carbon source for endophytes [69]. Coordination between growth and disease resistance demonstrates the activity of GAs in the presence and absence of microbes [70,71]. When a microbe is detected, an immune cascade overrides the destabilising activity of GA on DELLA proteins and re-establishes DELLA-mediated suppression of growth [72].

It is suggested that genes with large effects on defence against a bio-antagonist also have large pleiotropic effects on survival, growth, reproduction, and responses to other bio-antagonists. The exceptions to this rule include many resistance genes but in many cases have little or no detectable fitness cost. Therefore, the alleles of resistance genes selected by plant breeders have extensive benefits in providing strong disease-resistant crops with yields comparable to wild-type [64].

Plants adopt various strategies to reduce the cost of mounting resistance. This largely involves the restriction of defences to particular tissues, developmental stages, and/or windows of time. Cost amelioration may involve tritrophic interactions [73,74], where the activity of another species (such as within a protective microbiome) prevents the invasion of pathogens [75].

The commonly accepted notion is that most spontaneous mutations are deleterious with negative fitness effects on the survival of the individuals who carry them, and only a small fraction is beneficial. It is also suggested that the majority of deleterious mutations have small fitness effects (1% or less on average) [76]. Meanwhile, compensatory mutations may counteract the negative effects of other deleterious mutations, although alone, they are also deleterious. A large variation in the fitness effects of deleterious mutations may be an important factor in the survival and growth of some small natural populations. *In silico* modelling of gene regulatory networks [77,78] implied that compensatory mutations are surprisingly frequent and can drive gene regulatory network evolution. Furthermore, predictions indicate that the smaller the population, the larger the effect of compensatory mutations on fitness recovery, with the compensatory effect increasing sharply with a decreasing population [76]. Future empirical studies should test this prediction, and, in any case, it should serve as a warning to researchers selecting favourable traits in experimental fields.

## 3. Theoretical Limits on Crop Productivity. A Complex System Cannot Be Predictably Modified but Can Be Replaced by a Functionally Similar Complex System

The theoretical limits of crop productivity were initially examined in 1992 by [79]. They stated that environmental limitations render crops unable to achieve their genetic yield potential even in the best field conditions. Crop growth in controlled hydroponic conditions with high $CO_2$ levels is limited by photosynthetic photon flux even at daily levels that are three times higher than maximal summer sunlight. Therefore, biomass productivity and edible yield are still well below the predicted maximal output. Photosynthesis emerges as the key remaining route to increasing the genetic Yp of the major crops, yet it has improved little and falls far short of its biological limit. A better understanding of the mechanisms of photosynthetic processes should lead to the development of strategies for Yp improvement [80].

The Yp of crop can be described by the following equation [80–82]:

$$Yp = Q \cdot \varepsilon_i \cdot \varepsilon_c \cdot \varepsilon_p$$

Q is the total solar radiation; $\varepsilon_i$ is the efficiency of light capture; $\varepsilon_c$ is the efficiency of conversion of the intercepted light into biomass; and $\varepsilon_p$ is the harvest index, the proportion of biomass partitioned into grain. In the absence of environmental stress, parameters, such as harvest index, are close to the theoretical maximum. Plant breeding brings $\varepsilon_p$ and $\varepsilon_i$ close to their theoretical maxima, leaving $\varepsilon_c$, primarily determined by photosynthesis, as the only remaining major prospect for improving Yp [82]. This is an excellent idea, and it has many active apologists.

Photosynthesis is dependent upon the interactions between chloroplast organelles and the nucleus [32,83] since chloroplast fitness relies on nuclear-encoded genes. In angiosperms, the chloroplast genome (plastome) expresses only approximately 50 protein-coding genes involved in tRNA and rRNA genes, chlorophyll synthesis, photosynthesis, and metabolic processes [84]. Meanwhile, up to 2500–3500 nuclear-encoded proteins are predicted to be chloroplast localised in *Arabidopsis thaliana* [84]. However, organelle-to-nucleus signaling coordinates the expression of nuclear genes encoding chloroplast proteins with the metabolic and developmental state of the organelle [85].

Improvement of the currently low conversion efficiency of light to biomass (~2%) has received considerable attention [86]. The dramatic environmental changes that land plants have repeatedly experienced in the course of their evolution probably resulted in the formation of a rather robust photosynthetic system, with unoptimised efficiency for particular conditions. This means that modern photosynthetic productivity is maintained at optimal rates under many adverse conditions rather than tuning efficiency to conditions used in modern farming [87]. It is predicted that improving photosynthetic efficiency will not be easy. The successful modification of photosynthesis to enhance plant growth and yield has been limited to a few cases [88,89]. This provides some support that the genetic engineering approach is an avenue worth pursuing for the improvement in Yp through the optimisation of photosynthetic processes [18,80,81,90,91].

Four major research areas for redesigning photosynthesis were suggested [32]: (i) studying natural variations in photosynthesis, (ii) coordinating photosynthesis with pathways using photosynthesis, (iii) transfer of highly efficient photosynthetic systems existing in non-host species, and (iv) engineering photosynthetic systems not existing in nature.

Overexpressing *A. thaliana* SBPase in tobacco (*Nicotiana tabacum*) was the first successful photosynthetic carbon metabolism engineering. The transgenic plant had higher SBPase activity, increased photosynthetic rate, greater accumulation of sucrose and starch, and a higher total biomass increment (for a brief recent review see [83]); we also mention a couple of other examples reviewed in [83] for clarity.

SBPase activity was increased in the transgenic rice cultivar (*Oryza sativa* L. ssp. *japonica*) by overexpressing OsSbp cDNA from the rice cultivar 9311 (*Oryza sativa* ssp. *indica*). The transgenic plants accumulated SBPase in chloroplasts and developed enhanced tolerance of transgenic rice plants to salt stress at the young seedling stage [92]. Chilling is a factor limiting growth and yield in tomato production. Genetically engineering tomato plants with an appropriate target gene could ameliorate the chilling injury. It was reported [93] that in transgenic tomato plants (*Solanum lycopersicum*) with increased SBPase activity, photosynthetic rates were increased as well as sucrose and starch accumulation. Tomato plants with increased SBPase activity were more chilling tolerant. Thus, the higher level of SBPase activity provides an advantage to photosynthesis, growth, and chilling tolerance in tomato plants. Another consequence of this work is that an individual enzyme in the Calvin cycle may be a useful target for genetic engineering to improve production and stress tolerance in crops.

The level of the SBPase in wheat has been increased through transformation and expression of a *Brachypodium distachyon* SBPase gene construct and showed enhanced leaf photosynthesis and increased total biomass and dry seed yield [94]. South et al. inserted a synthetic glycolic acid metabolic pathway in *N. tabacum* chloroplasts by expressing pumpkin (*Cucurbita maxima*) malate synthase and green alga (*Chlamydomonas reinhardtii*) glycolate dehydrogenase and demonstrated that engineering alternative glycolate metabolic pathways into crop chloroplasts while inhibiting glycolate export into the native pathway can drive increases in C3 crop yield under agricultural field conditions [95].

These examples show that challenges in Yp improvement can potentially be overcome using genetic engineering in conjunction with synthetic biology and computational modelling strategies [80]. Synthetic biology tools allow the redesign of entire processes using simpler existing systems, for example, by introducing algal/bacterial 'inventions', such as carboxysomes, into land plants [87]. Moreover, instead of changing single components, synthetic biology tools allow the engineering and redesigning of entire processes [18,83,89–91].

On the other hand, we can expect a certain success from the side of "blue" biotechnology, namely from unicellular algal biotechnology. Algae are important sources of nutrients for both humans and agricultural species [96,97]. Microalgae can produce biomass that is enriched in proteins [98–100], low saturated fatty acids [101], and antioxidants [102]. Since they have comparatively rapid growth rates with varying photosynthetic apparatus and mechanisms [103], they may be regarded as attractive targets for genetic modifications to improve their metabolic productivity. However, scepticism remains as to whether increased photosynthetic capacity may increase food crop yields [104] (see below).

**4. Complexity of Agronomic Traits**

The cause of slowing crop yield growth is yet to be determined [11]. The gap between the actual yield and yield potential may be accounted for by several variables, including genetics, environment, management practices, and socioeconomic factors. These results indicate that many agronomic traits in plants are complex systems. Complex genetic architectures include numerous interacting loci (or alleles) with small effects and interactions with the genetic background, environment, or age. Complex agronomic traits, such as plant height, harvest index, total biomass, number of productive tillers, grain number per spike, spike length (SL), number of kernels per spike, thousand seed weight, and grain weight per spike, and physiological traits, such as canopy temperature (CT), chlorophyll content, photosynthetic rate, and water-soluble carbohydrates (WSC), contribute to grain yield improvement in wheat [105]. The complexity of optimizing a particular complex trait may be visualised by a comparison with recent research demonstrating that human height is associated with at least 10,000 DNA markers [106] and missing heritability [107].

Yield depends on many intrinsic and extrinsic factors, such as plant height, biotic and abiotic stresses, the efficiency of light energy capture by photosynthetic light reactions, the efficiency of conversion of light energy into biomass, and harvest index [18]. Such complex traits are defined as quantitative traits (QT). These traits are collectively regulated by several loci (QTL) that may interact with each other and with the environment and affect the mode of gene action. Recent advances in next-generation sequencing make it possible to map QTL genomic regions and help map phenotypes for thousands of traits. This leads to the partial reconstruction of gene networks at the transcript level and explains the relationship among traits [55,108,109]. Many agronomically important traits are quantitatively inherited, especially yield and yield-contributing traits [49,55]. Crop yield is a QT [55] that is controlled by many plant genes. In wheat, for example, three main phenotypic yield components were identified: thousand kernel weight (TKW); kernel number per spike; and spike number per unit area (SN), which determines wheat yield. Correspondingly, many genetic loci related to wheat yield have been identified. Recently, 58 QTL-rich clusters were defined based on their distribution on chromosomes [110]; however, their complete genetic architecture and key genetic loci for selection remain largely unknown [110]. Current methodologies in quantitative genetics [111] can only detect the determinant genes with the strongest effects, which are unlikely to represent all of the components required to produce a phenotypic characteristic [112]. There is every reason to believe that there are constraints on the magnitude of allowable variations of regulatory genes that are typically dosage sensitive. This multifactoriality makes it difficult to identify the appropriate targets for gene editing.

Recently, the role of the environment in the variability of phenotypic traits in maize, including crop yield, was investigated [113]. The variation in observed phenotypes can be partitioned into three main factors: genotype, environment, and genotype $\times$ environment interaction (G $\times$ E), in addition to other minor factors and measurement errors. In plant breeding, G $\times$ E plays an important role as the relative performance of different genotypes in different environments influences plant breeders' recommendations of best-performing varieties for specific regions. Typically, plant breeders minimize G $\times$ E by producing cultivars that are appropriate for regions that share common environmental characteristics. With an improved understanding of specific components of genotype, environment, and G $\times$ E, breeders may use data-based approaches to enhance their ability to position a larger number of genotypes in environments to maximize productivity. Grain yield is of primary importance in breeding *Zea mays* L. and is commonly considered alongside several traits that affect it indirectly or directly, such as flowering, height, and yield-component traits. Due to their differences in heritability and sensitivity to environmental factors, these different types of traits may show different levels of G $\times$ E [113]. G $\times$ E interactions showed between 9.0 and 20.4% of the phenotypic variance with greater effects in the yield-component traits [113].

However, the ability to fine-tune the expression of a QTL rather than only utilizing what is available in wild relatives is shown as a promising way to increase yield. For example, researchers edited genes that altered the promoters in three pathways contributing to productivity in tomato plants—plant architecture, fruit size, and inflorescence [11]. The resulting plants displayed a series of previously unobserved phenotypes, including several with increased yields.

## 5. Nitrogen Input

In the current bottleneck of crop production, we should leave habitual standards and search for new approaches to the problem of human survival. The considerations of experienced sceptics should be considered, for example, T. R Sinclair et al. [104]: "It seems crucial to further elucidate the role of resource inputs other than carbon in influencing crop yield... Given the conclusion that nitrogen input to crops has been and will continue to be critical in limiting crop grain yield, there are important questions for future research targeting nitrogen availability and use in crop plants ... ".

In the above-cited Nobel lecture, Borlaug stated: "In my dream I see green, vigorous, high-yielding fields of wheat, rice, maize, sorghums, and millets, which are obtaining, free of expense, 100 kg of nitrogen per hectare from nodule-forming, nitrogen-fixing bacteria...".

In the following sections, we discuss some of the strategies directed at the solution of this critical issue.

## 6. Is It Possible to Transfer Nitrogen-Fixing Genes from Legumes to Non-Legumes?

Biological $N_2$ fixation, catalysed by the prokaryotic enzyme nitrogenase, is an attractive alternative to the use of synthetic N fertilizers. Associations with nitrogen-fixing bacteria delivering the complete nitrogen needs of the host plants are limited to a select group of species. It is tempting to try to radically improve nitrogen availability for cereal crops by transferring the symbiosis trait encoding genes from legumes to non-legumes, especially to economically important crops, such as rice, maize, and wheat (reviewed in [25,114]).

Extensive genetic and biochemical studies have identified the common core set of genes/gene products required for functional nitrogen biosynthesis [25,114–116]. Molybdenum nitrogenase is an $O_2$-labile metalloenzyme composed of NifDK and NifH proteins, which requires several *nif* gene products. The sensitivity of nitrogenase to O2 and the apparent complexity of nitrogenase biosynthesis are the main barriers identified to date. The expression of active NifH requires NifM and NifH and possibly NifU and NifS, whereas active NifDK requires NifH, NifD, NifK, NifB, NifE, and NifN and possibly NifU, NifS, NifQ, NifV, NifY, NifW, and NifZ. Plastids and mitochondria are potentially viable subcellular locations for nitrogenases since they provide the ATP and electrons required for nitrogenase activity. These organelles differ in their internal $O_2$ levels and their ability to incorporate ammonium into amino acids. The direct transfer of *nif* genes into cereals to increase cereal crop productivity is challenging due to the sensitivity of nitrogenase to $O_2$ and the apparent complexity of nitrogenase biosynthesis.

Associations with nitrogen-fixing bacteria that deliver the complete nitrogen needs of the host plants are limited to a select group of species. The ability to fix nitrogen has been found in a wide range of bacterial genera, many of which are known to be associative (residing on or near the root surface) and endophytic (residing within plant cells) rhizobacteria, including *Azospirillum, Azotobacter, Burkholderia, Gluconacetobacter, Herbaspirillum, Klebsiella, Paenibacillus*, and *Pseudomonas* [117]. Engineering biological nitrogen fixation in plants by the direct introduction of *nif* genes, as well as in the case of photosynthesis, requires elegant synthetic biology approaches to ensure highly active and stable nitrogenase activity through expression in the appropriate stoichiometry of the subunits. This, if possible, may be achieved by synthetic engineering of the nitrogenase system into mitochondria or chloroplasts since these organelles potentially provide the reducing power and ATP required for nitrogen fixation [118].

## 7. Taking Advantage of Plant–Microbe Interactions

Billions of microorganisms and macroorganisms (from viruses to nematodes) live on, inside, and near plants, both above and below ground [119]. The results of beneficial plant–microbe interactions include the direct stimulation of plant growth, the protection of plants from pathogens and insect pests through the direct production of toxins or through induced resistance in the plant, and improved resilience to environmental stress (e.g., drought, salinity). Beneficial interactions occur in the root zone (rhizosphere), leaf surfaces (phyllosphere), and internal tissues (endosphere) [11,120,121]. In leguminous plants, some *Rhizobium* species of bacteria induce the formation of root nodules in a symbiotic relationship that converts atmospheric and largely inert $N_2$ into ammonia ($NH_3$) and other molecular precursors that the plant uses in the biosynthesis of nucleotides, coenzymes, and amino acids. In many more species of plants, fungal symbionts (arbuscular mycorrhizal fungi) form hyphae that increase the ability of plant roots to access minerals (particularly phosphorus) and water [119]. The hundreds of land plant and algal genomes that are now available enable genome-wide comparisons of gene families associated with plant immunity and symbiosis. However, few plant–microbe interactions have been studied in depth, with only a few land plant lineages. Subsequent investigations may reveal new types of symbiotic or pathogenic interactions [117,122]. Synthetic biology tools may provide an opportunity to design plant–microbe associations that improve crop productivity. Such associations can be studied using gene editing in plants, microbes, or both [118]. For example, plant genes controlling nodule formation by nitrogen-fixing rhizobacteria may be expanded to non-legume crops to reduce the need for fertilizer application, and microbial consortia present in the root zone could be engineered to produce novel plant growth promoters or protectants.

However, the complexity of plant genetics, metabolism, and nitrogen fixation machinery makes this an extremely challenging task [117,122]. Early ancestral cereals were associated with nitrogen-fixing bacteria according to a recent study (reviewed in [117]). Efforts are being made by multiple research groups to enhance biological nitrogen fixation in cereal crops through (a) enabling symbiosis between plants and nitrogen-fixing bacteria using genetic engineering and (b) identifying and utilising nitrogen-fixing bacteria to supplement plant nitrogen requirements. Symbiosis of native cereal crops with root-associated nitrogen-fixing bacteria offers a potentially sustainable solution for nitrogen management on a shorter timescale [114,117,123].

Utilising the plant microbiome is a reliable approach for the next green revolution and to meet global food demand in sustainable and eco-friendly agriculture [124]. Technological developments such as next-generation sequencing, gene editing, and synthetic biology allow the manipulation of plant and microbe genotypes at an unprecedented scale. Combining the prospecting of plant and bacterial natural diversity with genetic engineering will hopefully provide a more sustainable global food source in the short and long term [123].

## 8. Plant-Mediated Strategies for Shaping the Rhizosphere Microbiome

Characteristics of interest are manipulated using two different approaches: plant breeding and genetic engineering (see [121] and references therein). Plant breeding techniques for selecting a specific microbial community aim to increase crop yield by providing plant resistance to a variety of stresses. Many studies have manipulated plants by modifying the production of key exudates, which directs the establishment of specific plant microbiomes. For example, Koyama et al. [125] reported that transgenic plants have a greater ability to secrete citrate from the roots and therefore grow better in phosphate-limited soil. Yang et al. [126] and Gevaudant et al. [127] manipulated the pH of the rhizosphere by using transgenic lines of *Arabidopsis thaliana* and *N. tabacum* plants overexpressing an $H^+$-ATPase protein, which increased $H^+$-efflux from the roots of the plant. This created a more acidic environment in the rhizosphere resulting in enhanced growth at a lower pH, increased resistance to drought stress through the expression of pyrophosphate-energised vacuolar membrane proton pump 1 (AVP1), and augmented tolerance to salinity stress in tobacco lines [128]. Similar examples can be continued. More recently, site-directed genetic engineering of DNA has used methods such as CRISPR/Cas9 and transcription activator-like effector nucleases (TALENs).

## 9. Genome Editing Technologies for "Quantum-Leap" Improvements in Yield-Limited Crops Are Ready, but Where Are the Targets?

There are two problems in targeted genome editing: (1) the identification of target(s), which should be modified to achieve desirable phenotypic change, and (2) precise targeted genome modification. The advent of precise genome-editing tools is expected to revolutionise the way we create new plant varieties. Three groups of tools currently available are classified according to their mechanism of action: programmable sequence-specific nucleases, base-editing enzymes, and oligonucleotide-directed mutagenesis [129,130]. The most commonly used today is CRISPR/Cas9, which has been implemented in more than 20 crop plants (reviewed [121,130,131]) for a variety of desired traits to improve crop yield and management of abiotic and biotic stress tolerance in plants. This genome-editing technology is adopted from the prokaryotic adaptive immunity system, which is found in several bacterial and archaeal genomes. It uses small stretches of RNA sequences coupled with nucleases Cas-, the enzymes that specifically cut the genome of invading viruses to suppress them. This system is used for the introduction of desired variations at the chosen location in the genome [132]. The technology can be applied to modify virtually any genomic sequence with the only restraint being the accessibility of the protospacer adjacent motif (PAM). PAM is a short, typically 2–6 bp sequence recognised by any compatible Cas-nuclease sequence near the target sites. Different Cas-nucleases distinguish various PAMs, and, frequently, there are even differences among orthologs. In such a way, site-directed nucleases (SDNs) are constructed and used for targeted genome editing to introduce double-stranded breaks (DSBs) at precise sites into plant genomes. This technology can be used to modify virtually any genomic sequence with the only restraint being the accessibility of the PAM near the target sites [133,134].

In the immense literature devoted to the CRISPR technology used for editing plant genomes, we found 39,100 articles with the words "CRISPR plant editing", which contained contradictory data on the efficiency and specificity of this technology. As this technology is widely used, we do not describe it in detail here since it is best described in multiple reviews [15,16,132,135–138]. CRISPR/Cas9 is economical, easy to use, highly accurate, and effective even when performing multiplex genome editing (MGE) [15,139–144]. CRISPR/Cas9, CRISPR/Cas12a, and base editing are being improved continuously [15]. A simple scheme showing that this technology may be used to insert point mutations or extended DNA fragments into defined genomic sites is shown in Figure 1.

MGE involves the simultaneous targeting of multiple related or unrelated targets. The latter is the most straightforward using the CRISPR/Cas9 system because multiple guide RNAs (gRNAs) can be delivered either as independent expression cassettes with their own promoters or as polycistronic transcripts processed into mature gRNAs by endogenous or introduced nucleases. MGE in plants initially focused on input traits, such as herbicide resistance, but has recently expanded to include hormone biosynthesis and perception, metabolic engineering, plant development, and molecular farming, with numerous simultaneous targeting events reported. Knockout mutations in all three homologs of TaMLO (*T. aestivum* mildew resistance locus o) provided resistance against powdery mildew in wheat [141]. A non-complete list of edited genomes can be found [145] (see above, "Complexity of agronomic traits").

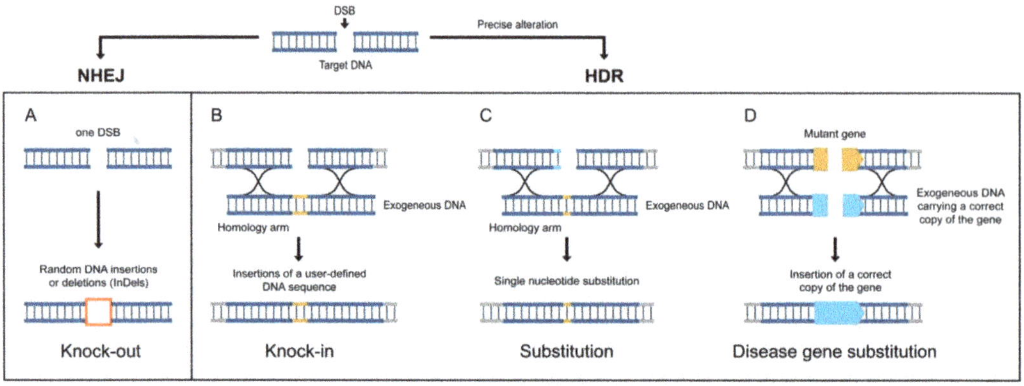


**Figure 1.** Genome-editing schemes with site-specific DNA nucleases. Double-stranded breaks (DSBs) induced by a nuclease (e.g., CRISPR/Cas9) at a specific site can be repaired by non-homologous end joining (NHEJ) or homology-directed recombination (HDR). (**A**) Repair by NHEJ usually results in the insertion or deletion of random base pairs causing gene knockout. (**B**) HDR with a donor DNA template with homologous arms can be exploited to achieve gene insertion and deletion, (**C**) to modify a gene by introducing precise nucleotide substitutions, and (**D**) inserting a correct gene copy instead of the mutant gene. DSB = double-stranded break.

Direct modification typically involves targeting protein-coding regions; however, recent examples include promoter modifications to generate mutants with varying gene expression levels [11,142,146]. CRISPR/Cas9 editing of promoters generates diverse cis-regulatory alleles that provide beneficial quantitative variation for breeding. A genetic scheme was devised to rapidly evaluate the phenotypic impact of numerous promoter variants on genes regulating three major productivity traits in tomato: fruit size, inflorescence branching, and plant architecture. This procedure allows for the immediate selection and fixation of novel alleles. It also provides an approach for testing complex relationships between gene regulatory changes and the control of QTs [147].

Researchers have paid great attention to the frequency of nonspecific mutations in CRISPR/Cas9. The data are highly contradictory, with one report stating an off-target specificity of 9.8–97.3% in *Arabidopsis* [148] to no evidence of off-target cleavage activity when specific gRNAs predicted by bioinformatics were chosen [149]. DNA breaks introduced by single-gRNA/Cas9 frequently resolved into deletions extending over many kilobases [150]. Furthermore, lesions distal to the cut site and crossover events have been identified [150]. Specificity may depend on the delivery method [151], and the problem of off-targeting may be tackled by the use of recently discovered CRISPR/Cpf12a (Cpf1), which creates a staggered double-strand break at the target site [136,152]. However, recently, Murugan et al. [153] revealed that Cas12a has multiple nicking activities against dsDNA substrates. SDN-mediated off-target changes can contribute to a small number of additional genetic variants compared to those that occur naturally in breeding populations or are introduced by induced-mutagenesis methods [154].

The second modified CRISPR "base editor" system can generate precise single-base mutations in the targeted DNA. It does not rely on DSB formation to induce targeted changes but instead uses a partly disabled nuclease with an additional protein domain. The targeting components of the nucleases are still intact, thereby allowing site-directed nucleotide changes and additional protein units to target specific genomic locations. Applications include targeted base editing with deaminase domains, transcriptional knockdown using repressors, targeted DNA methylation, and numerous other applications. In addition, the technology does not need the introduction of DSB to modify a base pair, and, consequently, the likelihood of major perturbations in the genome, such as deletion or chromosomal translocations, is considerably reduced. For example, cytosine and adenine base editors converting C to T and A to G, respectively, fuse a nickase-type Cas9 with a deaminase domain and, thus, do not induce DSBs (Figure 2). The two single-base editors (cytidine deaminase and adenosine deaminase) were fused to produce simultaneous double-base conversions (C → T and G → A) [15,154–159].

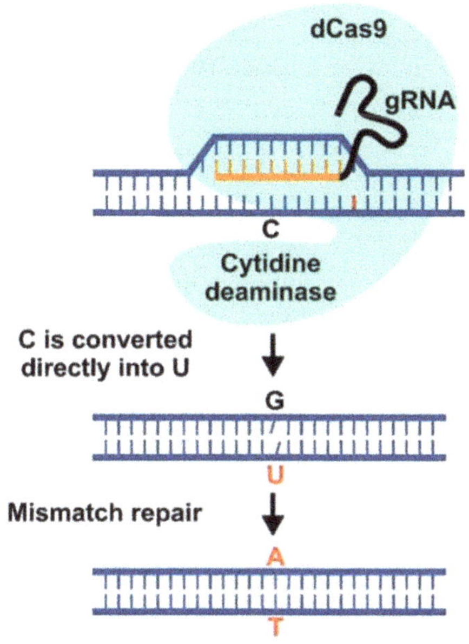


**Figure 2.** Base editing illustrated with cytidine deamination. In this type of base editing, cytidine deaminase fused with dCas9 targets the desired location. There is no DSB; C is converted directly into U on the free strand; and during mismatch repair, a C → T substitution can be created when the modified strand is used as a template [15].

Cas9 can be easily adapted to facilitate genome-scale perturbations. For example, Cas9 nuclease can be converted into an RNA-guided DNA-binding protein (dCas9) and then fused to transcription activation or repressor domains. These dCas9-activator fusions target the promoter/enhancer regions of endogenous genes to modulate gene expression [160,161].

The significance of nonspecific CRISPR-caused mutations is unclear given the natural background of mutations that constantly appear stochastically in the genome. The estimated haploid spontaneous single nucleotide mutation rate for *A. thaliana* is about $7 \times 10^{-9}$ per site per generation [162]. Approximately the same rate was reported in rice lines with 3.4-fold higher mutation rates in heterozygotes ($1.1 \times 10^{-8}$) than homozygotes [163]. The spontaneous mutation rate in *Zea mays* is 2.2 to $3.9 \times 10^{-8}$ per site per generation [164] (reviewed in [154]). A detailed analysis of various off-target effects is presented in an excellent review by Graham [154]. Other stochastic de novo mutations occur during in vitro culture referred to as somaclonal variation. The estimated mutation rate in *Arabidopsis* root explants (living cells transferred to culture medium) is between $4 \times 10^{-7}$ and $2.4 \times 10^{-6}$ mutations per nucleotide, while a mutation rate of $1.0 \times 10^{-7}$ occurs in rice plants regenerated through tissue culture [154]. During in vitro culture, many regenerated plants develop differences in appearance relative to the parental genotype, and these induced changes may include heritable genetic and epigenetic alterations [154].

When discussing the significance of off-target mutations, it should be noted that plants differ from animals in substantive ways that alter the impact of induced changes. First, unlike many animals, genetic changes in juvenile plants can be transmitted to reproductive tissues [154]. In addition, plants frequently develop multiple independent reproductive structures, with only a fraction affected by new mutations. Breeding to develop new lines for commercial release involves an intensive process of selection of individual plants with useful phenotypes while eliminating individuals with undesirable mutations or phenotypes (commonly known as "off-types") [154]. For these reasons, off-target edits in crops present fewer safety concerns than those that could arise with the therapeutic applications of genome editing.

In conclusion, it should be mentioned that to generate a wide variety of new traits in plants using CRISP/Cas genome editing [165], Agrobacterium-, biolistic-, and also virus-mediated methods were used to deliver CRISPR/Cas into plant cells [166,167]. A high-throughput gene-editing assay Automated Protoplast Transformation System was recently developed [168].

## 10. Success Stories of CRISPR/Cas9

Since the introduction of CRISPR/Cas9 for editing mammalian genomes, it has been applied to modify the genomes of several model and crop plants, including tobacco, tomato, barley, *Arabidopsis*, wheat, rice, and maize. The genes involved in the regulation of fruit size and signalling pathways have been suggested as promising targets for genome editing-based crop improvement [169]. Following the first reports of CRISPR/Cas9-based genome editing in wheat protoplasts, gene-edited plants were generated using transformation with CRISPR/Cas9 in different forms, including plasmids, linear DNA fragments, linear RNA, and ribonucleoprotein complexes (reviewed in [139–141]).

Considerable efforts have been undertaken to identify QTLs controlling yield in various crop plants [15,165]. This was achieved in a rice cultivar by individually knocking out four negative regulators of yield (*Gn1a*, *DEP1*, *GS3*, and *IPA1*) using CRISPR/Cas9. Three of the resulting knockout mutations (*Gn1a*, *DEP1*, and *GS3*) showed enhanced yield in the T2 generation with increased grain size and number and denser erect panicles with a 30–68% increase in yield per panicle (reviewed in [15,165]). The editing of two yield-regulating genes, *Gn1a* and *DEP1*, developed superior alleles in rice with even greater yields than those of the natural high-yield alleles [170]. Similarly, the simultaneous knockout of three major rice negative regulators of grain weight (GW2, GW5, and TGW6) using a CRISPR/Cas9-mediated MGE system resulted in a significant increase in the thousand-grain weight. This approach can be used for the rapid breeding of QTLs in crop varieties [144].

Several phytohormones, such as abscisic acid (ABA), control plant growth and stress responses that affect crop yields. CRISPR/Cas9 generated mutations in genes encoding the ABA receptors pyrabactin resistance 1-like 1 (PYL1), PYL4, and PYL6 and created a rice line that produced up to 31% more grains than the original variety in field tests. This work highlights the potential of modifying hormones to control growth and improve yields in rice [171] (reviewed in [15]).

The successful modification of a gene encoding a maize negative regulator of ethylene responses, *ARGOS8*, using CRISPR/Cas9 was reported (reviewed in [165]). The homology-directed repair pathway was used to insert the maize native *GOS2* promoter into the untranslated region of *ARGOS8*, resulting in drought-tolerant maize with improved yield in limited water supply. Other studies reviewed in [165] confirmed that CRISPR/Cas9 can be used to manipulate abiotic stress genes, indicating its potential for future crop improvement. However, several essential traits, such as crop yield and abiotic stress resistance, are controlled by multiple genes, and the same QTL can have highly varied and opposing effects in different backgrounds.

These positive examples are certainly impressive; however, one should expect that they are the exception rather than the rule when a complex trait is the target for modification. For complex traits, such as crop yield, the result is expected to be unpredictable. Although CRISPR/Cas9 mutagenesis has improved our options for addressing gene function, recent results suggest that compensatory mechanisms in CRISPR mutants may hide gene functions [77]. Possibly, there may be no revolutionary breakthrough in plant selection by using directed mutations or changes in the genes involved in the formation of a complex trait, such as crop yield [105]. On the other hand, CRISPR/Cas9 mutagenesis has the unique capacity of seamless mutation by directly changing certain nucleotide positions without affecting the backbone sequence. This offers new possibilities and challenges to plant biotechnology companies and society and, theoretically, can result in an accelerated evolution of genetically modified organism (GMO) crop species that cannot be identified using traditional algorithms. In turn, surpassing the regulation of "hidden" GMO consumption may result in dramatic economical, ethical, and biotechnological implications.

## 11. Reference Genomes and Assessment of Genomic Variation

A prerequisite for target identification is the availability of a comprehensive and reliable sequence of the genome and its functional map. Although complete sequences of the genomes of higher plants are not well studied in comparison with the genomes of humans and animals, there has been rapid progress in this field from a highly fragmented genome assembly with incomplete gene models to a full "pseudomolecule" reference sequence along with detailed gene model annotation. Reference sequence allows the physical anchoring of genes in complete chromosomal order and provides improved gene models that facilitate the design of transgenic constructs and primers [172,173].

Many of the crop and vegetable species that constitute a major part of the global diet now have high-quality reference genome sequences (reviewed in [154,172]). However, reference genomes have several limitations, the most apparent being that no genes or gene variants are present in any single accession. The steps required to assemble pan- and super-pangenomes were reviewed in [174–177]. It is important to note that most whole-genome sequencing studies to date have used short-read sequencing technologies. As a result, the diversity in breeding populations due to structural variations, such as differing transposable element location and abundance, presence–absence variation (PAV), and gene copy number variants (CNVs), have been difficult to measure. PAVs and CNVs typically refer to changes that include genes. Although structural variants are less common than single nucleotide polymorphisms (SNPs), they are an important source of variation. There are many examples of gene PAVs or CNVs impacting agronomic traits (reviewed in [154]). Due to the density of naturally occurring variation, intra- and interspecific crosses of plants that occur during plant breeding of millions of SNPs and thousands of PAV or CNV sequence variations appeared. More long-read sequencing technologies will allow more accurate measurements of polymorphisms in breeding populations [154].

Plant genomes are notoriously repetitive and difficult to assemble [178], though long-read sequencing technologies have been quickly adopted [178,179] allowing high-quality de novo assembly. The rapidly increasing number of long-read, whole-genome sequencing (WGS) produces an increasing number of high-quality plant genome assemblies.

## 12. The Research Bottleneck in Plant Sciences Is Shifting from Genotyping to Phenotyping

Despite the tremendous progress made with continually expanding genomic technologies to unravel and understand crop genomes, the impact of genomics data on crop improvement is still far from satisfactory. This is largely due to a lack of effective phenotypic data and problems with genome functional mapping. Our ability to collect high-quality phenotypic data lags behind the current capacity to generate high-throughput genomics data. Thus, the research bottleneck in plant sciences is shifting from genotyping to phenotyping for unlocking information coded in plant genomes. The phenomics data generated have been used to identify genes/QTLs through QTL mapping, association mapping, and genome-wide association studies (GWAS, see below) for genomic-assisted breeding for crop improvement [47,180–182].

Bioinformatics, which is known to play an important role in the selection of targets for targeted modification, relies on existing information about the DNA, RNA, and protein sequences contained in databases such as GO. In 1998, the GO consortium released the first common vocabulary describing gene function across species, thus enabling a genome-wide and comparative approach to functional genomics [183]. The current release

(10 August 2020) has 44,262 GO terms, 8,047,076 annotations, 1,556,208 gene products, and 4643 species [184].

A knowledgebase Gramene [185] based on the comparative functional analyses of genomic and pathway data for model plants and major crops contains, in the current release, 93 reference genomes over 3.9 million genes in 122,947 families with orthologous and paralogous classifications. Gramene integrates ontology-based protein structure–function annotation and information on genetic, epigenetic, expression, and phenotypic diversity; gene functional annotations extracted from this latest achievement will undoubtedly play an important role in the functional mapping of plant genomes.

Genomic databases have been powerful in integrating data from multiple studies, and international efforts are now bringing together phenotypic data alongside genotypic data (e.g., [186–189]. However, most GO annotations are incomplete and imperfect [34,35,190,191]. This is also true for the Gene Ontology Meta Annotator for Plants (GOMAP), which is an optimised, high-throughput, and reproducible pipeline for genome-scale GO annotation for plant genomes [189]. Therefore, predicting the associations between genes and phenotypes is rather problematic, as is the identification of adequate targets for modification. Several other challenges remain, with the most common for crop species being polyploidy, which is particularly evident in wheat. Due to functional redundancy, it will be necessary to knock out all homoeologs of a gene to determine its phenotypic impact [172,173,192]. In wheat, over 98% of the genome is non-coding; therefore, it is necessary to identify open chromatin regions to define non-coding but functionally important regions. Finally, it is essential to compare multiple wheat varieties to observe the effects of the same editing in different genetic backgrounds.

GWAS is an approach used in genetic research to associate specific genetic variations with phenotypic traits [34,35,190,191]. The method involves scanning the genomes of many different individuals and searching for genetic markers that can be used to predict the presence of a trait. The genetic markers can be used to understand how genes contribute to the trait and to uncover causal genetic polymorphisms in plants. This will aid breeders in developing improved plant varieties to meet the food needs of an ever-increasing world population.

Genome-wide screenings were first applied to humans a few years before the methods were leveraged for use in plants. In 2004, a publication first appeared that applied GWAS-like methods to barley (reviewed in [193]). GWAS is a good first step towards the discovery and deployment of key genes, further research is necessary to evaluate the reproducibility and transferability of GWAS results across environments and genetic backgrounds. The development of optimal experimental settings for GWAS analysis will require an interdisciplinary approach. The identification of key traits involves GWAS, proper analytical methods, using appropriate genetic resources for mapping, and choosing an adequate genotyping platform. With the arrival of rapid genotyping and next-generation sequencing technologies, GWAS has become a routine strategy for decoding genotype–phenotype associations in many species. Over the last decade, more than 1000 studies have revealed substantial genotype–phenotype associations in crops and provided opportunities to probe functional genomics [193–195]. A successful GWAS should to incorporate elements of candidate gene discovery and QTL deployment. Therefore, the following criteria for a successful GWAS were proposed: the study needs to identify true marker-trait associations with meaningful effect sizes; proximity to underlying genes for the traits of interest; and transferability within a similar population and across a reasonably broad set of environments [193].

## 13. The Use of Locally Available Resources to Adapt to Climate Variability and Change

A quite reasonable strategy for improving crop yield is the use of locally available resources to adapt to climate variability and change [196]. There are more than 50,000 edible plants, but only 15 crops are used, resulting in 90% of the world's demand, and three of them (rice, maize, and wheat) provide two-thirds of human caloric intake [197]. Ironically, more

than 70% of wild relatives of domesticated crops are threatened by extinction and in urgent need of conservation due to the expansion of agriculture into natural ecosystems [198]. Globally, gene banks and botanical gardens hold more than 7.4 million seeds or plant tissues from thousands of species [197]. These collections must be maintained, curated, and explored [11]. Among the most promising candidates are orphan crops that have either originated in a geographic location or those that have become 'indigenised' over many years of cultivation; they may offer 'new' opportunities as they are uniquely suited to harsh local environments, provide nutritional diversity, and enhance agrobiodiversity within farmer fields [199]. There is quite a lot of space to improve orphan crops, and genome editing accelerates modifications that would be problematic in a traditional breeding program. [200,201].

## 14. De Novo Domestication of Wild Plants

An attractive alternative route for future agriculture is the de novo domestication of wild plants [11,200,202,203]. This involves a multidisciplinary approach, including research from botany, archaeology, genetics, biogeography, and other disciplines [204]. The transformation of wild plants into domesticated crops usually involves modification of a common set of characteristics across different species, referred to as 'domestication syndrome' traits and previously defined as "the characteristic collection of phenotypic traits associated with the genetic change to a domesticated form of an organism from a wild progenitor form" [205,206]. Examples of these traits include loss of pod shattering/seed dehiscence, loss of seed dormancy, reduced anti-nutritional compounds, changes in growth habit, phenology, flower colour, and seed colour. Understanding the genetic control of domestication syndrome traits facilitates the efficient transfer of useful traits from wild progenitors into crops through crossing and selection [205,207]. It has become apparent in recent years that understanding the nature of the plurality of processes underlying domestication syndrome is the key to understanding the origins of domestication.

## 15. Reference, Pan-, and Super-Pan Genome Sequences Provide a Strong Basis for the Location of Domestication Syndrome Genes

In recent years, there has been a large influx of plant genome sequencing projects. The high level of genomic variation led to the realisation that single reference genomes do not represent the diversity within a species and led to the expansion of the pangenome concept. This suggests that the genomes of individuals within a population or species share a core set of genes that unifies them (the core genome) but also contains a fraction of genes that are absent from one or more individuals (the accessory or dispensable genome), which together give rise to the pangenome of such a population or species [175–177,208].

The numerous genome sequences allowed for a better understanding of the domestication processes of crops and animals and to follow some of the genetic changes that permitted domestication to occur [209]. However, the first fragmented data produced were insufficient to detect all or at least the most important genes associated with domestication syndromes [208]. However, they allow the implementation of GWAS to detect some domestication traits [210]. Since domestication reduces the genetic diversity of a taxon, often eliminating portions of the dispensable genome that contain genes involved in local adaptation, the use of wild relatives is crucial for generating a representative pangenome for a species [177]. Once a pangenome is generated, it can be used alongside whole-genome sequencing data to analyse the structural variants between and within populations, revealing novel loci involved in the development of domestication-related traits that would have remained hidden using only a single reference genome [211]. As sequencing technologies become cheaper, multiple pangenomes from different species of the same genus should eventually be combined to create a super-pangenome representing the entire genetic content available in a genus with one or more domesticated taxa, and it should include the diversity of all the wild relatives [175].

Wild plants can be regarded as reservoirs of useful genes [202], yet there is a vast array of plant species whose agricultural potential remains untapped [11]. Recent technological advances have opened a new approach to de novo domestication of wild plants as a viable solution for designing desired crops while maintaining food security and a more sustainable low-input agriculture. In these innovative fields, the potential application of CRISPR-like technologies for genome editing is very wide. The process of domesticating wild progenitors into edible crops is closely linked to the modification of developmental processes, and the steps that are needed to face the current challenges will equally require developmental modifications [212,213]. Therefore, studying the genetic basis of crop domestication is largely equivalent to studying aspects of plant development. Consequently, understanding the genetic and molecular mechanisms underlying developmental processes has great potential to further improve the performance of today's major crops and determine routes for fast-tracking domestication of less developed crops [212]. Genes controlling plant development have been studied in multiple plant systems. This has provided deep insights into conserved genetic pathways controlling core developmental processes, including meristem identity, phase transitions, determinacy, stem elongation, and branching. These pathways control plant growth patterns and are fundamentally important in crop biology and agriculture.

There are recent examples of crops that have been targeted for rapid domestication [11]. Cape gooseberry or pichuberry (*Physalis peruviana*) was chosen for its growing popularity since it is highly nutritious and can be eaten as fresh fruit or used to make juice or jam. It is native to the Andean region of South America and has many "wild" characteristics that prevent it from being easily cultivated. Knowledge of the genes related to the improvement and domestication of the tomato, a distant relative of *P. peruviana*, has motivated scientists to identify similar genes in the undomesticated pichuberry that could be targeted for domestication. Gene editing of *P. peruviana's* genetic ortholog of the tomato gene CLAVATA1 (SlCLV1), which controls meristem proliferation, gives rise to plants with narrow leaves and flowers with more organs. This offers a proof of principle for rapid, targeted domestication using gene editing. The cultivation of crops in urban environments may reduce the environmental impact of food production [37]. However, the lack of available land in cities and the need for rapid crop cycling to yield quickly and continuously mean that to date only lettuce and related 'leafy green' vegetables are cultivated in urban farms. New fruit varieties with architectures and yields suitable for urban farming have proven difficult to breed [37].

Despite all successes, gene editing approaches to the domestication of wild plants using domestication genes [11,202,207] should not be generally considered as simple one-time events in a single gene or generation. In contrast, de novo domestication literature and examples understand domestication as co-evolutionary interactions between plants and people that are complex, would require significant institutional and infrastructural investments, and can involve many disciplines. Gene editing approaches may help accelerate domestication and widespread cultivation of a new generation of soil-conserving and climate-smart crops.

## 16. Conclusions. Must I Be Cruel Only to Be Kind?

We, probably, can summarise that a revolutionary breakthrough in increasing the limit of crop yield by plant selection using targeted mutational changes in specific genes is unlikely. A replacement of inefficient photosynthetic machinery with more productive subunits for increasing the photosynthetic rate is attractive; however, there is typically a 20–30 year gap between the demonstration of innovative solutions at the experimental level and the provision of seeds to farmers [80]. Field trials involving genetically engineered plants are scarce worldwide and do not exist in Europe due to strict regulations. To date, most field trials involving genetically engineered varieties of rice, tomato, and other vegetables and crops are in Asia [214]. Thus, there seems to be little doubt that technology alone is powerless to feed the rapidly growing population.

In 1999, ecologist Peter Vitousek stated that "we are the first generation with tools to understand changes in the Earth's systems caused by human activity, and the last with the opportunity to influence the course of many of these changes" (quoted from [215]). More than ever, humanity is changing the face of the planet at an increasing rate and conducting an unprecedented "environmental breakdown". A century ago, only 15% of the Earth's surface was modified by the direct effects of human activities [216]. This proportion has now grown to 87% of the ocean and 77% of the land [217]. This has led to a global collapse of biodiversity with an average 60% decline in populations of all vertebrate species and up to 83% for freshwater species between 1970 and 2014 as measured by the Living Planet Index [218]. Human activities currently threaten approximately 1 million species with extinction, with many others already extinct [215]. Considering this rapid loss of biodiversity, the world is now facing a sixth mass extinction [219]; the first to be caused by the species *Homo sapiens* [215].

Interestingly, this problem also involves crop cultivars that suffer from modern agricultural industry trends. Multiple cultivars are replaced by single "champions" for better productiveness and logistics. Over the decades, the pressure exerted by natural and artificial selection has progressively reduced the genetic diversity of many crops, including Italian durum wheat cultivars [220,221]. Modern industrial trends dictate regional crop specialisation, which removes all but a few crop species from traditional agricultural areas. For example, in a county-level study of individual crop land cover areas in the conterminous United States of America (U.S.A.) from 1840 to 2017, Michael Crossley and colleagues found a strong and abrupt spatial concentration of most crop types in recent years. For 13 of the 18 major crops, the widespread belts that characterised agriculture in the U.S.A. early in the 20th century collapsed, with spatial concentration increasing 15-fold after 2002. The number of counties producing each crop declined by up to 97% from 1940 to 2017, and their total area declined by up to 98% despite increasing total production. Consequently, a sharp decrease in crop types within counties occurred; in 1940, 88% of counties grew > 10 crops, but this figure was only 2% in 2017. Crossley et al. showed that declining crop diversity with increasing land area is a recent phenomenon, suggesting that the corresponding environmental effects in agriculturally dominated regions have fundamentally changed [222].

The discussion above was based on the need for resource growth, taking the inevitability of human population growth as a given. In 1798, Thomas Malthus wrote in his article "An Essay on the Principle of Population" that the rate of uncontrolled population growth always advances the growth of means of subsistence such that the exponential growth of an uncontrolled population is in contrast to the arithmetic growth of subsistence resources. The critics often referred to the "green revolution" as evidence that Malthus did not consider the technological factor in the production of foodstuffs. Indeed, as described above, during the past 30 years, the crop yield per unit time and land use for crop production has increased markedly [9].

In his Nobel speech, Norman Borlaug also warned: "The green revolution has won a temporary success in man's war against hunger and deprivation; it has given man a breathing space. If fully implemented, the revolution can provide sufficient food for sustenance during the next three decades. However, the frightening power of human reproduction must also be curbed; otherwise, the success of the green revolution will be ephemeral only" [59]. Both Malthus and Borlaug drew attention to the alarming population growth, though neither could imagine the trajectory of agricultural development. Based on recent studies, Pete Smith and other scientists concluded that "Technology alone cannot provide food security in 2050. Food demand ... will need to be managed if we are to continue to prove Malthus wrong into the future" [223]. However, these authors did not speak directly about the alarmingly expanding world population.

In the analysis of 12,640 research articles over the last 50 years, the authors of [10] identified three potential levers important for human population: total food production, per capita food demand, and population size. They reported a strong and increasing focus on feeding the world through increasing food production via technology, while the focus on reducing food demand through less intensive dietary patterns has remained constant and low. Population size has declined from being the dominant lever discussed in 1969 to the least researched in 2018.

Many consider that the Earth will soon reach its limits in terms of food supplies, natural resources, and pollution, and the world population will inevitably decline or even collapse due to successive uncontrollable crises. However, these assertions rarely factor in the internal constraints that shape population dynamics [224]. There is currently no country on Earth that meets the critical needs for human well-being while staying within the environmental planetary boundaries [10]. However, changing population size and age structure may have the most profound economic, social, and geopolitical impacts in many countries [225]. These problems can be dramatically enhanced by a lack of mineral resources [226].

Tampering with human population growth is a topic loaded with delicate moral issues. Those who accept the relevant scientific evidence are often accused of being genocidal, racist, anti-poor folks, anti-religion, and generally anti-human. In fact, those who accept the scientific imperative feel that they have a moral responsibility to be concerned about the future of mankind because it is increasingly apparent that without constraints on population growth, a sixth mass extinction [219] caused by *Homo sapiens* [215] seems to be inevitable [227]. For further information, see the gloomy but sobering predictions of overly optimistic individuals and governments [228].

Undeniably, the indefinite growth of both population and consumption is impossible on a planet with finite space and resources. In fact, meeting the UN Sustainable Development Goals is already regarded as requiring a lower world population growth [229]. However, the topic remained relatively poorly documented, and many people in scientific, policy, and public arenas continued to ignore or deny that population growth is an issue until relatively recently. In 2017, over 15,000 scientists from all over the world reported in a "warning to humanity" that population growth needed to be addressed; otherwise, all efforts to reach a sustainable future would be in vain. This outlines the need to bring population growth to the forefront of international concerns and overcome the taboos surrounding this question.

In 1998, Donella Meadows wrote: "facts... show Malthus to be not dead, not wrong, maybe not right either . . . Over another few decades, we will probably put old Malthus to rest at last. It's up to us to decide whether he will rest triumphant or discredited".

We would think he is at least not discredited, and we will have to think what can be done with our growing population and remember: "I must be cruel only to be kind" (Shakespeare, Hamlet).

**Author Contributions:** Conceptualization, E.D.S.; Chapter 9, M.V.P.; all parts concerning BIOINFORMATICS and DATABASES A.V.B.; writing—original draft preparation, E.D.S.; M.V.P.; A.V.B.; writing—review and editing, E.D.S.; funding acquisition, M.V.P. All authors have read and agreed to the published version of the manuscript.

**Funding:** This work was supported by a grant from the Ministry of Science and Higher Education of the Russian Federation allocated to the Kurchatov Center for Genome Research (grant 075-15-2019-1659).

**Institutional Review Board Statement:** Not applicable.

**Informed Consent Statement:** Not applicable.

**Acknowledgments:** Many thanks to Irina Alekseenko and Tatiana Gribova for their invaluable help in preparing the manuscript.

**Conflicts of Interest:** The authors declare no conflict of interest.

## References

1. Tortell, P.D. Earth 2020: Science, society, and sustainability in the Anthropocene. *Proc. Natl. Acad. Sci. USA* **2020**, *117*, 8683–8691. [CrossRef]
2. Harte, J. *Patient Earth*; Socolow, R.H., Ed.; Holt, Rinehart, and Winston: New York, NY, USA, 1971.

3. Meadows, D.; Meadows, D.; Randers, J.; Behrens, W.W., III. *The Limits to Growth. A Report for the Club of Rome's Project on the Predicament of Mankind*; Universe Books New York: New York, NY, USA, 1972.
4. Tollefson, J. The hard truths of climate change—By the numbers. *Nature* **2019**, *573*, 324–327. [CrossRef] [PubMed]
5. NOAA's National Centers for Environmental Information (NCEI). Temperature Change and Carbon Dioxide Change. Available online: https://www.ncdc.noaa.gov/global-warming/temperature-change (accessed on 12 February 2018).
6. Hartfield, G.; Blunden, J.; Arndt, D. State of the climate in 2017. *Bull. Am. Meteorol. Soc.* **2018**, *99*, Si–S310. [CrossRef]
7. Pöschl, U. Atmospheric aerosols: Composition, transformation, climate and health effects. *Angew. Chem. Int. Ed. Engl.* **2005**, *44*, 7520–7540. [CrossRef]
8. Ramankutty, N.; Mehrabi, Z.; Waha, K.; Jarvis, L.; Kremen, C.; Herrero, M.; Rieseberg, L.H. Trends in Global Agricultural Land Use: Implications for Environmental Health and Food Security. *Annu. Rev. Plant Biol.* **2018**, *69*, 789–815. [CrossRef] [PubMed]
9. Fanzo, J.; Covic, N.; Dobermann, A.; Henson, S.; Herrero, M.; Pingali, P.; Staal, S. A research vision for food systems in the 2020s: Defying the status quo. *Glob. Food Secur.* **2020**, *26*, 100397. [CrossRef] [PubMed]
10. Tamburino, L.; Bravo, G.; Clough, Y.; Nicholas, K. From population to production: 50 years of scientific literature on how to feed the world. *Glob. Food Secur.* **2020**, *24*, 100346. [CrossRef]
11. National Academies of Sciences, Engineering, and Medicine. *Science Breakthroughs to Advance Food and Agricultural Research by 2030*; The National Academies Press: Washington, DC, USA, 2019.
12. Searchinger, T.; Waite, R.; Hanson, C.; Ranganathan, J.; Matthews, E. Creating a Sustainable Food Future: A Menu of Solutions to Feed Nearly 10 Billion People by 2050. Final Report. 2019. Available online: https://www.wri.org/research/creating-sustainable-food-future (accessed on 19 July 2019).
13. Gerten, D.; Heck, V.; Jägermeyr, J.; Bodirsky, B.; Fetzer, I.; Jalava, M.; Kummu, M.; Lucht, W.; Rockström, J.; Schaphoff, S.; et al. Feeding ten billion people is possible within four terrestrial planetary boundaries. *Nat. Sustain.* **2020**, *3*, 200–208. [CrossRef]
14. Borelli, T.; Hunter, D.; Padulosi, S.; Amaya, N.; Meldrum, G.; de Oliveira Beltrame, D.; Samarasinghe, G.; Wasike, V.W.; Güner, B.; Tan, A.; et al. Local Solutions for Sustainable Food Systems: The Contribution of Orphan Crops and Wild Edible Species. *Agronomy* **2020**, *10*, 231. [CrossRef]
15. Zafar, K.; Sedeek, K.E.M.; Rao, G.S.; Khan, M.Z.; Amin, I.; Kamel, R.; Mukhtar, Z.; Zafar, M.; Mansoor, S.; Mahfouz, M.M. Genome Editing Technologies for Rice Improvement: Progress, Prospects, and Safety Concerns. *Front. Genome Ed.* **2020**, *2*, 5. [CrossRef]
16. Zaman, Q.; Lia, C.; Chenga, H.; Hua, Q. Genome editing opens a new era of genetic improvement in polyploid crop. *Crop. J.* **2019**, *7*, 141–150. [CrossRef]
17. Piesse, M. Global Food and Water Security in 2050: Demographic Change and Increased Demand. Strategic Analysis Paper. 2020. Available online: https://www.futuredirections.org.au/wp-content/uploads/2020/02/Global-Food-and-Water-Security-in-2050-Demographic-Change-and-Increased-Demand.pdf (accessed on 4 February 2020).
18. Weber, A.P.M.; Bar-Even, A. Update: Improving the Efficiency of Photosynthetic Carbon Reactions. *Plant Physiol.* **2019**, *179*, 803–812. [CrossRef] [PubMed]
19. Fraser, E.D.G. The challenge of feeding a diverse and growing population. *Physiol. Behav.* **2020**, *221*, 112908. [CrossRef] [PubMed]
20. Rockstrom, J.; Steffen, W.; Noone, K.; Persson, A.; Chapin, F.S., 3rd; Lambin, E.F.; Lenton, T.M.; Scheffer, M.; Folke, C.; Schellnhuber, H.J.; et al. A safe operating space for humanity. *Nature* **2009**, *461*, 472–475. [CrossRef]
21. Steffen, W.; Richardson, K.; Rockstrom, J.; Cornell, S.E.; Fetzer, I.; Bennett, E.M.; Biggs, R.; Carpenter, S.R.; de Vries, W.; de Wit, C.A.; et al. Sustainability. Planetary boundaries: Guiding human development on a changing planet. *Science* **2015**, *347*, 1259855. [CrossRef]
22. Lade, S.; Steffen, W.; de Vries, W.; Carpenter, S.; Donges, J.; Gerten, D.; Hoff, H.; Newbold, T.; Richardson, K.; Rockström, J. Human impacts on planetary boundaries amplified by Earth system interactions. *Nat. Sustain.* **2020**, *3*, 119–128. [CrossRef]
23. Biermann, F.; Kim, R. Framework: A CriticalAppraisal of Approaches toDefine a "Safe OperatingSpace" for Humanity. *Annu. Rev. Environ. Resour.* **2020**, *45*, 4.1–4.25. [CrossRef]
24. Pingali, P.L. Green revolution: Impacts, limits, and the path ahead. *Proc. Natl. Acad. Sci. USA* **2012**, *109*, 12302–12308. [CrossRef]
25. Bailey-Serres, J.; Parker, J.E.; Ainsworth, E.A.; Oldroyd, G.E.D.; Schroeder, J.I. Genetic strategies for improving crop yields. *Nature* **2019**, *575*, 109–118. [CrossRef]
26. Cassman, K.G. Ecological intensification of cereal production systems: Yield potential, soil quality, and precision agriculture. *Proc. Natl. Acad. Sci. USA* **1999**, *96*, 5952–5959. [CrossRef]
27. Global Sustainable Development Report 2019: The Future Is Now—Science for Achieving Sustainable Development. 2019. Available online: https://sustainabledevelopment.un.org/globalsdreport/2019 (accessed on 11 September 2019).
28. Capstaff, N.M.; Miller, A.J. Improving the Yield and Nutritional Quality of Forage Crops. *Front. Plant Sci.* **2018**, *9*, 535. [CrossRef]
29. Richards, N.; Nielsen, B.D.; Finno, C.J. Nutritional and Non-nutritional Aspects of Forage. *Vet. Clin. Equine Pract.* **2021**, *37*, 43–61. [CrossRef]
30. Kulkarni, K.P.; Tayade, R.; Asekova, S.; Song, J.T.; Shannon, J.G.; Lee, J.D. Harnessing the Potential of Forage Legumes, Alfalfa, Soybean, and Cowpea for Sustainable Agriculture and Global Food Security. *Front. Plant Sci.* **2018**, *9*, 1314. [CrossRef]
31. Hu, H.; Scheben, A.; Edwards, D. Advances in Integrating Genomics and Bioinformatics in the Plant Breeding Pipeline. *Agriculture* **2018**, *8*, 75. [CrossRef]
32. Zhu, X.G.; Ort, D.R.; Parry, M.A.J.; von Caemmerer, S. A wish list for synthetic biology in photosynthesis research. *J. Exp. Bot.* **2020**, *71*, 2219–2225. [CrossRef]

33. Rönspies, M.; Dorn, A.; Schindele, P.; Puchta, H. CRISPR–Cas-mediated chromosome engineering for crop improvement and synthetic biology. *Nat. Plants* **2021**, *7*, 566–573. [CrossRef] [PubMed]
34. Gaudet, P.; Dessimoz, C. Gene Ontology: Pitfalls, Biases, and Remedies. *Methods Mol. Biol.* **2017**, *1446*, 189–205. [CrossRef] [PubMed]
35. Danchin, A.; Ouzounis, C.; Tokuyasu, T.; Zucker, J.D. No wisdom in the crowd: Genome annotation in the era of big data—Current status and future prospects. *Microb. Biotechnol.* **2018**, *11*, 588–605. [CrossRef]
36. Armanda, D.T.; Guinéea, J.B.; Tukkerac, A. The second green revolution: Innovative urban agriculture's contribution to food security and sustainability—A review. *Glob. Food Secur.* **2019**, *22*, 13–24. [CrossRef]
37. Kwon, C.T.; Heo, J.; Lemmon, Z.H.; Capua, Y.; Hutton, S.F.; Van Eck, J.; Park, S.J.; Lippman, Z.B. Rapid customization of Solanaceae fruit crops for urban agriculture. *Nat. Biotechnol.* **2020**, *38*, 182–188. [CrossRef]
38. Van-Erp, H. Can We Create a New Green Revolution to Avert the Coming Global Food Crisis? *Ann. Agric. Crop. Sci.* **2016**, *1*, 1001. Available online: https://austinpublishinggroup.com/agriculture-crop-sciences/fulltext/aacs-v1-id1001.php (accessed on 7 September 2016).
39. Lynch, J. Roots of the second green revolution. *Aust. Bot.* **2007**, *55*, 493–512. [CrossRef]
40. Wurschum, T.; Langer, S.M.; Longin, C.F.H.; Tucker, M.R.; Leiser, W.L. A modern Green Revolution gene for reduced height in wheat. *Plant J.* **2017**, *92*, 892–903. [CrossRef] [PubMed]
41. Spielmeyer, W.; Ellis, M.H.; Chandler, P.M. Semidwarf (sd-1), "green revolution" rice, contains a defective gibberellin 20-oxidase gene. *Proc. Natl. Acad. Sci. USA* **2002**, *99*, 9043–9048. [CrossRef] [PubMed]
42. Hedden, P. The genes of the Green Revolution. *Trends Genet.* **2003**, *19*, 5–9. [CrossRef]
43. Monna, L.; Kitazawa, N.; Yoshino, R.; Suzuki, J.; Masuda, H.; Maehara, Y.; Tanji, M.; Sato, M.; Nasu, S.; Minobe, Y. Positional cloning of rice semidwarfing gene, sd-1: Rice "green revolution gene" encodes a mutant enzyme involved in gibberellin synthesis. *DNA Res.* **2002**, *9*, 11–17. [CrossRef] [PubMed]
44. Peng, J.; Richards, D.E.; Hartley, N.M.; Murphy, G.P.; Devos, K.M.; Flintham, J.E.; Beales, J.; Fish, L.J.; Worland, A.J.; Pelica, F.; et al. 'Green revolution' genes encode mutant gibberellin response modulators. *Nature* **1999**, *400*, 256–261. [CrossRef]
45. Sasaki, A.; Ashikari, M.; Ueguchi-Tanaka, M.; Itoh, H.; Nishimura, A.; Swapan, D.; Ishiyama, K.; Saito, T.; Kobayashi, M.; Khush, G.S.; et al. Green revolution: A mutant gibberellin-synthesis gene in rice. *Nature* **2002**, *416*, 701–702. [CrossRef]
46. Wang, B.; Li, J. Understanding the Molecular Bases of Agronomic Trait Improvement in Rice. *Plant Cell* **2019**, *31*, 1416–1417. [CrossRef] [PubMed]
47. Wang, S.; Xu, S.; Chao, S.; Sun, Q.; Liu, S.; Xia, G. A Genome-Wide Association Study of Highly Heritable Agronomic Traits in Durum Wheat. *Front. Plant Sci.* **2019**, *10*, 919. [CrossRef]
48. Thomas, S.G. Novel Rht-1 dwarfing genes: Tools for wheat breeding and dissecting the function of DELLA proteins. *J. Exp. Bot.* **2017**, *68*, 354–358. [CrossRef]
49. Grant, N.P.; Mohan, A.; Sandhu, D.; Gill, K.S. Inheritance and Genetic Mapping of the Reduced Height (Rht18) Gene in Wheat. *Plants* **2018**, *7*, 58. [CrossRef] [PubMed]
50. Guo, F.; Ma, J.; Hou, L.; Shi, S.; Sun, J.; Li, G.; Zhao, C.; Xia, H.; Zhao, S.; Wang, X.; et al. Transcriptome profiling provides insights into molecular mechanism in Peanut semi-dwarf mutant. *BMC Genom.* **2020**, *21*, 211. [CrossRef]
51. Chen, X.; Xu, P.; Zhou, J.; Tao, D.; Yu, D. Mapping and breeding value evaluation of a semi-dominant semi-dwarf gene in upland rice. *Plant Divers.* **2018**, *40*, 238–244. [CrossRef] [PubMed]
52. Ferrero-Serrano, A.; Cantos, C.; Assmann, S.M. The Role of Dwarfing Traits in Historical and Modern Agriculture with a Focus on Rice. *Cold Spring Harb. Perspect. Biol.* **2019**, *11*, a034645. [CrossRef]
53. Tomlinson, L.; Yang, Y.; Emenecker, R.; Smoker, M.; Taylor, J.; Perkins, S.; Smith, J.; MacLean, D.; Olszewski, N.E.; Jones, J.D.G. Using CRISPR/Cas9 genome editing in tomato to create a gibberellin-responsive dominant dwarf DELLA allele. *Plant Biotechnol. J.* **2019**, *17*, 132–140. [CrossRef]
54. Jia, X.; Yu, L.; Tang, M.; Tian, D.; Yang, S.; Zhang, X.; Traw, M.B. Pleiotropic changes revealed by in situ recovery of the semi-dwarf gene sd1 in rice. *J. Plant Physiol.* **2020**, *248*, 153141. [CrossRef]
55. Kumar, J.; Gupta, D.S.; Gupta, S.; Dubey, S.; Gupta, P.; Kumar, S. Quantitative trait loci from identification to exploitation for crop improvement. *Plant Cell Rep.* **2017**, *36*, 1187–1213. [CrossRef]
56. Liu, F.; Wang, P.; Zhang, X.; Li, X.; Yan, X.; Fu, D.; Wu, G. The genetic and molecular basis of crop height based on a rice model. *Planta* **2018**, *247*, 1–26. [CrossRef]
57. Lieberman-Lazarovich, M.; Yahav, C.; Israeli, A.; Efroni, I. Deep Conservation of cis-Element Variants Regulating Plant Hormonal Responses. *Plant Cell* **2019**, *31*, 2559–2572. [CrossRef] [PubMed]
58. Nadolska-Orczyk, A.; Rajchel, I.K.; Orczyk, W.; Gasparis, S. Major genes determining yield-related traits in wheat and barley. *Theor. Appl. Genet.* **2017**, *130*, 1081–1098. [CrossRef]
59. Norman, B. The Green Revolution, Peace, and Humanity—Nobel Lecture. Available online: http://www.nobelprize.org/prizes/peace/1970/borlaug/lecture/ (accessed on 11 August 2021).
60. Denance, N.; Sanchez-Vallet, A.; Goffner, D.; Molina, A. Disease resistance or growth: The role of plant hormones in balancing immune responses and fitness costs. *Front. Plant Sci.* **2013**, *4*, 155. [CrossRef]
61. Nelson, R.; Wiesner-Hanks, T.; Wisser, R.; Balint-Kurti, P. Navigating complexity to breed disease-resistant crops. *Nat. Rev. Genet.* **2018**, *19*, 21–33. [CrossRef] [PubMed]

62. Pilet-Nayel, M.L.; Moury, B.; Caffier, V.; Montarry, J.; Kerlan, M.C.; Fournet, S.; Durel, C.E.; Delourme, R. Quantitative Resistance to Plant Pathogens in Pyramiding Strategies for Durable Crop Protection. *Front. Plant Sci.* **2017**, *8*, 1838. [CrossRef]
63. Wiesner-Hanks, T.; Nelson, R. Multiple Disease Resistance in Plants. *Annu. Rev. Phytopathol.* **2016**, *54*, 229–252. [CrossRef]
64. Brown, J.K.M.; Rant, J.C. Fitness costs and trade-offs of disease resistance and their consequences for breeding arable crops. *Plant Pathol.* **2013**, *62* (Suppl. 1), 83–95. [CrossRef]
65. Comont, D.; Knight, C.; Crook, L.; Hull, R.; Beffa, R.; Neve, P. Alterations in Life-History Associated with Non-target-site Herbicide Resistance in Alopecurus myosuroides. *Front. Plant Sci.* **2019**, *10*, 837. [CrossRef]
66. Ffrench-Constant, R.H.; Bass, C. Does resistance really carry a fitness cost? *Curr. Opin. Insect Sci.* **2017**, *21*, 39–46. [CrossRef] [PubMed]
67. Hu, K.; Cao, J.; Zhang, J.; Xia, F.; Ke, Y.; Zhang, H.; Xie, W.; Liu, H.; Cui, Y.; Cao, Y.; et al. Improvement of multiple agronomic traits by a disease resistance gene via cell wall reinforcement. *Nat. Plants* **2017**, *3*, 17009. [CrossRef] [PubMed]
68. Koller, T.; Brunner, S.; Herren, G.; Sanchez-Martin, J.; Hurni, S.; Keller, B. Field grown transgenic Pm3e wheat lines show powdery mildew resistance and no fitness costs associated with high transgene expression. *Transgenic Res.* **2019**, *28*, 9–20. [CrossRef] [PubMed]
69. Eom, J.S.; Chen, L.Q.; Sosso, D.; Julius, B.T.; Lin, I.W.; Qu, X.Q.; Braun, D.M.; Frommer, W.B. SWEETs, transporters for intracellular and intercellular sugar translocation. *Curr. Opin. Plant Biol.* **2015**, *25*, 53–62. [CrossRef] [PubMed]
70. Daviere, J.M.; Achard, P. Gibberellin signaling in plants. *Development* **2013**, *140*, 1147–1151. [CrossRef] [PubMed]
71. Van De Velde, K.; Ruelens, P.; Geuten, K.; Rohde, A.; Van Der Straeten, D. Exploiting DELLA Signaling in Cereals. *Trends Plant Sci.* **2017**, *22*, 880–893. [CrossRef]
72. Navarro, L.; Bari, R.; Achard, P.; Lison, P.; Nemri, A.; Harberd, N.P.; Jones, J.D. DELLAs control plant immune responses by modulating the balance of jasmonic acid and salicylic acid signaling. *Curr. Biol.* **2008**, *18*, 650–655. [CrossRef] [PubMed]
73. Beck, J.J.; Alborn, H.T.; Block, A.K.; Christensen, S.A.; Hunter, C.T.; Rering, C.C.; Seidl-Adams, I.; Stuhl, C.J.; Torto, B.; Tumlinson, J.H. Interactions Among Plants, Insects, and Microbes: Elucidation of Inter-Organismal Chemical Communications in Agricultural Ecology. *J. Agric. Food Chem.* **2018**, *66*, 6663–6674. [CrossRef]
74. Turlings, T.C.J.; Erb, M. Tritrophic Interactions Mediated by Herbivore-Induced Plant Volatiles: Mechanisms, Ecological Relevance, and Application Potential. *Annu. Rev. Entomol.* **2018**, *63*, 433–452. [CrossRef] [PubMed]
75. Karasov, T.L.; Chae, E.; Herman, J.J.; Bergelson, J. Mechanisms to Mitigate the Trade-Off between Growth and Defense. *Plant Cell* **2017**, *29*, 666–680. [CrossRef]
76. Zhang, L.; Watson, L.T. Analysis of the fitness effect of compensatory mutations. *HFSP J.* **2009**, *3*, 47–54. [CrossRef]
77. Wang, R.; Angenent, G.C.; Seymour, G.; de Maagd, R.A. Revisiting the Role of Master Regulators in Tomato Ripening. *Trends Plant Sci.* **2020**, *25*, 291–301. [CrossRef]
78. Wang, Y.; Richards, M.; Dorus, S.; Priest, B.K.; Bryson, J.J. Compensatory mutation can drive gene regulatory network evolution. *BioRxiv* **2020**. Available online: https://www.biorxiv.org/content/10.1101/2019.12.18.881276v2.full (accessed on 18 May 2018). [CrossRef]
79. Bugbee, B.; Monje, O. The limits of crop productivity: Validating theoretical estimates and determining the factors that limit crop yields in optimal environments. *Bioscience* **1992**, *42*, 494–502. [CrossRef]
80. Long, S.P.; Marshall-Colon, A.; Zhu, X.G. Meeting the global food demand of the future by engineering crop photosynthesis and yield potential. *Cell* **2015**, *161*, 56–66. [CrossRef] [PubMed]
81. Faralli, M.; Lawson, T. Natural genetic variation in photosynthesis: An untapped resource to increase crop yield potential? *Plant J.* **2020**, *101*, 518–528. [CrossRef] [PubMed]
82. Long, S.P.; Zhu, X.G.; Naidu, S.L.; Ort, D.R. Can improvement in photosynthesis increase crop yields? *Plant Cell Environ.* **2006**, *29*, 315–330. [CrossRef]
83. Moses, T. Shedding Light on the Power of Light. *Plant Physiol.* **2019**, *179*, 775–777. [CrossRef] [PubMed]
84. Shmakov, N.A.; Vasiliev, G.V.; Shatskaya, N.V.; Doroshkov, A.V.; Gordeeva, E.I.; Afonnikov, D.A.; Khlestkina, E.K. Identification of nuclear genes controlling chlorophyll synthesis in barley by RNA-seq. *BMC Plant Biol.* **2016**, *16*, 245. [CrossRef]
85. Barajas-Lopez Jde, D.; Blanco, N.E.; Strand, A. Plastid-to-nucleus communication, signals controlling the running of the plant cell. *Biochim. Biophys. Acta* **2013**, *1833*, 425–437. [CrossRef]
86. Zhu, X.G.; Long, S.P.; Ort, D.R. Improving photosynthetic efficiency for greater yield. *Annu. Rev. Plant Biol.* **2010**, *61*, 235–261. [CrossRef]
87. Dann, M.; Leister, D. Enhancing (crop) plant photosynthesis by introducing novel genetic diversity. *Philos. Trans. R. Soc. Lond. B Biol. Sci.* **2017**, *372*, 20160380. [CrossRef]
88. Leister, D. Genetic Engineering, Synthetic Biology and the Light Reactions of Photosynthesis. *Plant Physiol.* **2019**, *179*, 778–793. [CrossRef]
89. Batista-Silva, W.; da Fonseca-Pereira, P.; Martins, A.; Zsogon, A.; Nunes-Nesi, A.; Araujo, W. Engineering Improved Photosynthesis in the Era of Synthetic Biology. *Plant Comm.* **2020**, *1*, 100032. [CrossRef]
90. Simkin, A.J. Genetic Engineering for Global Food Security: Photosynthesis and Biofortification. *Plants* **2019**, *8*, 586. [CrossRef] [PubMed]
91. Simkin, A.J.; Lopez-Calcagno, P.E.; Raines, C.A. Feeding the world: Improving photosynthetic efficiency for sustainable crop production. *J. Exp. Bot.* **2019**, *70*, 1119–1140. [CrossRef] [PubMed]

92. Feng, L.; Han, Y.; Liu, G.; An, B.; Yang, J.; Yang, G.; Li, Y.; Zhu, Y. Overexpression of sedoheptulose-1,7-bisphosphatase enhances photosynthesis and growth under salt stress in transgenic rice plants. *Funct. Plant Biol.* **2007**, *34*, 822–834. [CrossRef] [PubMed]
93. Ding, F.; Wang, M.; Zhang, S.; Ai, X. Changes in SBPase activity influence photosynthetic capacity, growth, and tolerance to chilling stress in transgenic tomato plants. *Sci. Rep.* **2016**, *6*, 32741. [CrossRef]
94. Driever, S.M.; Simkin, A.J.; Alotaibi, S.; Fisk, S.J.; Madgwick, P.J.; Sparks, C.A.; Jones, H.D.; Lawson, T.; Parry, M.A.J.; Raines, C.A. Increased SBPase activity improves photosynthesis and grain yield in wheat grown in greenhouse conditions. *Philos. Trans. R. Soc. Lond. B Biol. Sci.* **2017**, *372*, 20160384. [CrossRef]
95. South, P.F.; Cavanagh, A.P.; Liu, H.W.; Ort, D.R. Synthetic glycolate metabolism pathways stimulate crop growth and productivity in the field. *Science* **2019**, *363*, eaat9077. [CrossRef] [PubMed]
96. Bito, T.; Okumura, E.; Fujishima, M.; Watanabe, F. Potential of Chlorella as a Dietary Supplement to Promote Human Health. *Nutrients* **2020**, *12*, 2524. [CrossRef]
97. Doucha, J.; Livansky, K.; Kotrbacek, V.; Zachleder, V. Production of Chlorella biomass enriched by selenium and its use in animal nutrition: A review. *Appl. Microbiol. Biotechnol.* **2009**, *83*, 1001–1008. [CrossRef]
98. Biswas, T.; Bhushan, S.; Prajapati, S.K.; Ray Chaudhuri, S. An eco-friendly strategy for dairy wastewater remediation with high lipid microalgae-bacterial biomass production. *J. Environ. Manag.* **2021**, *286*, 112196. [CrossRef]
99. Ramos-Romero, S.; Torrella, J.R.; Pages, T.; Viscor, G.; Torres, J.L. Edible Microalgae and Their Bioactive Compounds in the Prevention and Treatment of Metabolic Alterations. *Nutrients* **2021**, *13*, 563. [CrossRef] [PubMed]
100. Barros de Medeiros, V.P.; da Costa, W.K.A.; da Silva, R.T.; Pimentel, T.C.; Magnani, M. Microalgae as source of functional ingredients in new-generation foods: Challenges, technological effects, biological activity, and regulatory issues. *Crit. Rev. Food Sci. Nutr.* **2021**, 1–22. [CrossRef] [PubMed]
101. Mamaeva, A.; Namsaraev, Z.; Maltsev, Y.; Gusev, E.; Kulikovskiy, M.; Petrushkina, M.; Kuzmin, D. Simultaneous increase in cellular content and volumetric concentration of lipids in Bracteacoccus bullatus cultivated at reduced nitrogen and phosphorus concentrations. *J. Appl. Phycol.* **2018**, *30*, 2237–2246. [CrossRef]
102. Petrushkina, M.; Gusev, E.; Sorokin, B.; Zotko, N.; Mamaeva, A.; Filimonova, A.; Kuzmin, D. Fucoxanthin production by heterokont microalgae. *Algal Res.* **2017**, *24*, 387–393. [CrossRef]
103. Kreslavski, V.D.; Los, D.A.; Schmitt, F.J.; Zharmukhamedov, S.K.; Kuznetsov, V.V.; Allakhverdiev, S.I. The impact of the phytochromes on photosynthetic processes. *Biochim. Biophys. Acta Bioenerg.* **2018**, *1859*, 400–408. [CrossRef] [PubMed]
104. Sinclair, T.R.; Rufty, T.W.; Lewis, R.S. Increasing Photosynthesis: Unlikely Solution for World Food Problem. *Trends Plant Sci.* **2019**, *24*, 1032–1039. [CrossRef]
105. Tshikunde, N.M.; Mashilo, J.; Shimelis, H.; Odindo, A. Agronomic and Physiological Traits, and Associated Quantitative Trait Loci (QTL) Affecting Yield Response in Wheat (*Triticum aestivum* L.): A Review. *Front. Plant Sci.* **2019**, *10*, 1428. [CrossRef]
106. Kaiser, J. Growth spurt for height genetics. *Science* **2020**, *370*, 645. [CrossRef]
107. Manolio, T.A.; Collins, F.S.; Cox, N.J.; Goldstein, D.B.; Hindorff, L.A.; Hunter, D.J.; McCarthy, M.I.; Ramos, E.M.; Cardon, L.R.; Chakravarti, A.; et al. Finding the missing heritability of complex diseases. *Nature* **2009**, *461*, 747–753. [CrossRef]
108. Abiola, O.; Angel, J.M.; Avner, P.; Bachmanov, A.A.; Belknap, J.K.; Bennett, B.; Blankenhorn, E.P.; Blizard, D.A.; Bolivar, V.; Brockmann, G.A.; et al. The nature and identification of quantitative trait loci: A community's view. *Nat. Rev. Genet.* **2003**, *4*, 911–916. [CrossRef]
109. Ahmar, S.; Saeed, S.; Khan, M.H.U.; Ullah Khan, S.; Mora-Poblete, F.; Kamran, M.; Faheem, A.; Maqsood, A.; Rauf, M.; Saleem, S.; et al. A Revolution toward Gene-Editing Technology and Its Application to Crop Improvement. *Int. J. Mol. Sci.* **2020**, *21*, 5665. [CrossRef]
110. Cao, S.; Xu, D.; Hanif, M.; Xia, X.; He, Z. Genetic architecture underpinning yield component traits in wheat. *Theor. Appl. Genet.* **2020**, *133*, 1811–1823. [CrossRef]
111. Nordborg, M.; Weigel, D. Next-generation genetics in plants. *Nature* **2008**, *456*, 720–723. [CrossRef]
112. Birchler, J.A. Editing the Phenotype: A Revolution for Quantitative Genetics. *Cell* **2017**, *171*, 269–270. [CrossRef] [PubMed]
113. Falcon, C.M.; Kaeppler, S.M.; Spalding, E.P.; Miller, N.D.; Haase, N.; AlKhalifah, N.; Bohn, M.; Buckler, E.S.; Campbell, D.A.; Ciampitti, I.; et al. Relative utility of agronomic, phenological, and morphological traits for assessing genotype-by-environment interaction in maize inbreds. *Crop. Sci.* **2020**, *60*, 62–81. [CrossRef]
114. Mahmud, K.; Makaju, S.; Ibrahim, R.; Missaoui, A. Current Progress in Nitrogen Fixing Plants and Microbiome Research. *Plants* **2020**, *9*, 97. [CrossRef]
115. Soumare, A.; Diedhiou, A.G.; Thuita, M.; Hafidi, M.; Ouhdouch, Y.; Gopalakrishnan, S.; Kouisni, L. Exploiting Biological Nitrogen Fixation: A Route Towards a Sustainable Agriculture. *Plants* **2020**, *9*, 1011. [CrossRef] [PubMed]
116. Curatti, L.; Rubio, L.M. Challenges to develop nitrogen-fixing cereals by direct nif-gene transfer. *Plant Sci.* **2014**, *225*, 130–137. [CrossRef]
117. Bloch, S.; Ryu, M.-H.; Ozaydin, B.; Broglie, R. Harnessing atmospheric nitrogen for cereal crop production. *Curr. Opin. Biotechnol.* **2020**, *62*, 181–188. [CrossRef]
118. Xiang, N.; Guo, C.; Liu, J.; Xu, H.; Dixon, R.; Yang, J.; Wang, Y.P. Using synthetic biology to overcome barriers to stable expression of nitrogenase in eukaryotic organelles. *Proc. Natl. Acad. Sci. USA* **2020**, *117*, 16537–16545. [CrossRef]
119. Leach, J.E.; Triplett, L.R.; Argueso, C.T.; Trivedi, P. Communication in the Phytobiome. *Cell* **2017**, *169*, 587–596. [CrossRef]

120. Arif, I.; Batool, M.; Schenk, P.M. Plant Microbiome Engineering: Expected Benefits for Improved Crop Growth and Resilience. *Trends Biotechnol.* **2020**, *38*, 1385–1396. [CrossRef]
121. Kumar, A.; Dubey, A. Rhizosphere microbiome: Engineering bacterial competitiveness for enhancing crop production. *J. Adv. Res.* **2020**, *24*, 337–352. [CrossRef] [PubMed]
122. Delaux, P.M.; Schornack, S. Plant evolution driven by interactions with symbiotic and pathogenic microbes. *Science* **2021**, *371*, eaba6605. [CrossRef] [PubMed]
123. Pankievicz, V.C.S.; Irving, T.B.; Maia, L.G.S.; Ane, J.M. Are we there yet? The long walk towards the development of efficient symbiotic associations between nitrogen-fixing bacteria and non-leguminous crops. *BMC Biol.* **2019**, *17*, 99. [CrossRef] [PubMed]
124. Shelake, R.M.; Pramanik, D.; Kim, J.Y. Exploration of Plant-Microbe Interactions for Sustainable Agriculture in CRISPR Era. *Microorganisms* **2019**, *7*, 269. [CrossRef] [PubMed]
125. Koyama, H.; Kawamura, A.; Kihara, T.; Hara, T.; Takita, E.; Shibata, D. Overexpression of mitochondrial citrate synthase in Arabidopsis thaliana improved growth on a phosphorus-limited soil. *Plant Cell Physiol.* **2000**, *41*, 1030–1037. [CrossRef]
126. Yang, H.; Knapp, J.; Koirala, P.; Rajagopal, D.; Peer, W.A.; Silbart, L.K.; Murphy, A.; Gaxiola, R.A. Enhanced phosphorus nutrition in monocots and dicots over-expressing a phosphorus-responsive type I $H^+$-pyrophosphatase. *Plant Biotechnol. J.* **2007**, *5*, 735–745. [CrossRef]
127. Gevaudant, F.; Duby, G.; von Stedingk, E.; Zhao, R.; Morsomme, P.; Boutry, M. Expression of a constitutively activated plasma membrane H+-ATPase alters plant development and increases salt tolerance. *Plant Physiol.* **2007**, *144*, 1763–1776. [CrossRef]
128. Cameron, D.D.; Neal, A.L.; van Wees, S.C.; Ton, J. Mycorrhiza-induced resistance: More than the sum of its parts? *Trends Plant Sci.* **2013**, *18*, 539–545. [CrossRef]
129. Jansing, J.; Schiermeyer, A.; Schillberg, S.; Fischer, R.; Bortesi, L. Genome Editing in Agriculture: Technical and Practical Considerations. *Int. J. Mol. Sci.* **2019**, *20*, 2888. [CrossRef]
130. Van Eck, J. Applying gene editing to tailor precise genetic modifications in plants. *J. Biol. Chem.* **2020**, *295*, 13267–13276. [CrossRef]
131. Ansari, W.A.; Chandanshive, S.U.; Bhatt, V.; Nadaf, A.B.; Vats, S.; Katara, J.L.; Sonah, H.; Deshmukh, R. Genome Editing in Cereals: Approaches, Applications and Challenges. *Int. J. Mol. Sci.* **2020**, *21*, 4040. [CrossRef] [PubMed]
132. Ahmad, N.; Rahman, M.U.; Mukhtar, Z.; Zafar, Y.; Zhang, B. A critical look on CRISPR-based genome editing in plants. *J. Cell. Physiol.* **2020**, *235*, 666–682. [CrossRef]
133. Friedland, A.E.; Baral, R.; Singhal, P.; Loveluck, K.; Shen, S.; Sanchez, M.; Marco, E.; Gotta, G.M.; Maeder, M.L.; Kennedy, E.M.; et al. Characterization of Staphylococcus aureus Cas9: A smaller Cas9 for all-in-one adeno-associated virus delivery and paired nickase applications. *Genome Biol.* **2015**, *16*, 257. [CrossRef] [PubMed]
134. Muller, M.; Lee, C.M.; Gasiunas, G.; Davis, T.H.; Cradick, T.J.; Siksnys, V.; Bao, G.; Cathomen, T.; Mussolino, C. Streptococcus thermophilus CRISPR-Cas9 Systems Enable Specific Editing of the Human Genome. *Mol. Ther.* **2016**, *24*, 636–644. [CrossRef]
135. Razzaq, A.; Saleem, F.; Kanwal, M.; Mustafa, G.; Yousaf, S.; Imran Arshad, H.M.; Hameed, M.K.; Khan, M.S.; Joyia, F.A. Modern Trends in Plant Genome Editing: An Inclusive Review of the CRISPR/Cas9 Toolbox. *Int. J. Mol. Sci.* **2019**, *20*, 4045. [CrossRef] [PubMed]
136. Manghwar, H.; Lindsey, K.; Zhang, X.; Jin, S. CRISPR/Cas System: Recent Advances and Future Prospects for Genome Editing. *Trends Plant Sci.* **2019**, *24*, 1102–1125. [CrossRef]
137. Es, I.; Gavahian, M.; Marti-Quijal, F.J.; Lorenzo, J.M.; Mousavi Khaneghah, A.; Tsatsanis, C.; Kampranis, S.C.; Barba, F.J. The application of the CRISPR-Cas9 genome editing machinery in food and agricultural science: Current status, future perspectives, and associated challenges. *Biotechnol. Adv.* **2019**, *37*, 410–421. [CrossRef]
138. Chen, K.; Wang, Y.; Zhang, R.; Zhang, H.; Gao, C. CRISPR/Cas Genome Editing and Precision Plant Breeding in Agriculture. *Annu. Rev. Plant Biol.* **2019**, *70*, 667–697. [CrossRef]
139. Wang, M.; Wang, S.; Liang, Z.; Shi, W.; Gao, C.; Xia, G. From Genetic Stock to Genome Editing: Gene Exploitation in Wheat. *Trends Biotechnol.* **2018**, *36*, 160–172. [CrossRef] [PubMed]
140. Wang, W.; Pan, Q.; He, F.; Akhunova, A.; Chao, S.; Trick, H.; Akhunov, E. Transgenerational CRISPR-Cas9 Activity Facilitates Multiplex Gene Editing in Allopolyploid Wheat. *CRISPR J.* **2018**, *1*, 65–74. [CrossRef]
141. Wang, Y.; Cheng, X.; Shan, Q.; Zhang, Y.; Liu, J.; Gao, C.; Qiu, J.L. Simultaneous editing of three homoeoalleles in hexaploid bread wheat confers heritable resistance to powdery mildew. *Nat. Biotechnol.* **2014**, *32*, 947–951. [CrossRef]
142. Armario Najera, V.; Twyman, R.M.; Christou, P.; Zhu, C. Applications of multiplex genome editing in higher plants. *Curr. Opin. Biotechnol.* **2019**, *59*, 93–102. [CrossRef]
143. Hashimoto, R.; Ueta, R.; Abe, C.; Osakabe, Y.; Osakabe, K. Efficient Multiplex Genome Editing Induces Precise, and Self-Ligated Type Mutations in Tomato Plants. *Front. Plant Sci.* **2018**, *9*, 916. [CrossRef]
144. Zhou, J.; Xin, X.; He, Y.; Chen, H.; Li, Q.; Tang, X.; Zhong, Z.; Deng, K.; Zheng, X.; Akher, S.A.; et al. Multiplex QTL editing of grain-related genes improves yield in elite rice varieties. *Plant Cell Rep.* **2019**, *38*, 475–485. [CrossRef]
145. Wilson, F.M.; Harrison, K.; Armitage, A.D.; Simkin, A.J.; Harrison, R.J. CRISPR/Cas9-mediated mutagenesis of phytoene desaturase in diploid and octoploid strawberry. *Plant Methods* **2019**, *15*, 45. [CrossRef] [PubMed]
146. Li, Q.; Sapkota, M.; van der Knaap, E. Perspectives of CRISPR/Cas-mediated cis-engineering in horticulture: Unlocking the neglected potential for crop improvement. *Hortic. Res.* **2020**, *7*, 36. [CrossRef] [PubMed]
147. Rodriguez-Leal, D.; Lemmon, Z.H.; Man, J.; Bartlett, M.E.; Lippman, Z.B. Engineering Quantitative Trait Variation for Crop Improvement by Genome Editing. *Cell* **2017**, *171*, 470–480.e8. [CrossRef]

148. Xu, W.; Fu, W.; Zhu, P.; Li, Z.; Wang, C.; Wang, C.; Zhang, Y.; Zhu, S. Comprehensive Analysis of CRISPR/Cas9-Mediated Mutagenesis in Arabidopsis thaliana by Genome-wide Sequencing. *Int. J. Mol. Sci.* **2019**, *20*, 4125. [CrossRef]
149. Young, J.; Zastrow-Hayes, G.; Deschamps, S.; Svitashev, S.; Zaremba, M.; Acharya, A.; Paulraj, S.; Peterson-Burch, B.; Schwartz, C.; Djukanovic, V.; et al. CRISPR-Cas9 Editing in Maize: Systematic Evaluation of Off-target Activity and Its Relevance in Crop Improvement. *Sci. Rep.* **2019**, *9*, 6729. [CrossRef]
150. Kosicki, M.; Tomberg, K.; Bradley, A. Repair of double-strand breaks induced by CRISPR-Cas9 leads to large deletions and complex rearrangements. *Nat. Biotechnol.* **2018**, *36*, 765–771. [CrossRef] [PubMed]
151. Banakar, R.; Eggenberger, A.L.; Lee, K.; Wright, D.A.; Murugan, K.; Zarecor, S.; Lawrence-Dill, C.J.; Sashital, D.G.; Wang, K. High-frequency random DNA insertions upon co-delivery of CRISPR-Cas9 ribonucleoprotein and selectable marker plasmid in rice. *Sci. Rep.* **2019**, *9*, 19902. [CrossRef]
152. Jacobsen, T.; Ttofali, F.; Liao, C.; Manchalu, S.; Gray, B.N.; Beisel, C.L. Characterization of Cas12a nucleases reveals diverse PAM profiles between closely-related orthologs. *Nucleic Acids Res.* **2020**, *48*, 5624–5638. [CrossRef]
153. Murugan, K.; Seetharam, A.S.; Severin, A.J.; Sashital, D.G. CRISPR-Cas12a has widespread off-target and dsDNA-nicking effects. *J. Biol. Chem.* **2020**, *295*, 5538–5553. [CrossRef]
154. Graham, N.; Patil, G.B.; Bubeck, D.M.; Dobert, R.C.; Glenn, K.C.; Gutsche, A.T.; Kumar, S.; Lindbo, J.A.; Maas, L.; May, G.D.; et al. Plant Genome Editing and the Relevance of Off-Target Changes. *Plant Physiol.* **2020**, *183*, 1453–1471. [CrossRef]
155. Bharat, S.; Li, S.; Li, J.; Yan, L.; Xia, L. Base editing in plants: Current status and challenges. *Crop. J.* **2020**, *8*, 384–395. [CrossRef]
156. Kempton, H.R.; Qi, L.S. When genome editing goes off-target. *Science* **2019**, *364*, 234–236. [CrossRef]
157. Wienert, B.; Wyman, S.K.; Richardson, C.D.; Yeh, C.D.; Akcakaya, P.; Porritt, M.J.; Morlock, M.; Vu, J.T.; Kazane, K.R.; Watry, H.L.; et al. Unbiased detection of CRISPR off-targets in vivo using DISCOVER-Seq. *Science* **2019**, *364*, 286–289. [CrossRef]
158. Zuo, E.; Sun, Y.; Wei, W.; Yuan, T.; Ying, W.; Sun, H.; Yuan, L.; Steinmetz, L.M.; Li, Y.; Yang, H. Cytosine base editor generates substantial off-target single-nucleotide variants in mouse embryos. *Science* **2019**, *364*, 289–292. [CrossRef]
159. Jin, S.; Zong, Y.; Gao, Q.; Zhu, Z.; Wang, Y.; Qin, P.; Liang, C.; Wang, D.; Qiu, J.L.; Zhang, F.; et al. Cytosine, but not adenine, base editors induce genome-wide off-target mutations in rice. *Science* **2019**, *364*, 292–295. [CrossRef] [PubMed]
160. Konermann, S.; Brigham, M.D.; Trevino, A.E.; Joung, J.; Abudayyeh, O.O.; Barcena, C.; Hsu, P.D.; Habib, N.; Gootenberg, J.S.; Nishimasu, H.; et al. Genome-scale transcriptional activation by an engineered CRISPR-Cas9 complex. *Nature* **2015**, *517*, 583–588. [CrossRef]
161. Kwik, M.; Hainzl, S.; Oppelt, J.; Tichy, B.; Koller, U.; Bernardinelli, E.; Steiner, M.; Zara, G.; Nofziger, C.; Weis, S.; et al. Selective Activation of CNS and Reference PPARGC1A Promoters Is Associated with Distinct Gene Programs Relevant for Neurodegenerative Diseases. *Int. J. Mol. Sci.* **2021**, *22*, 3296. [CrossRef]
162. Weng, M.L.; Becker, C.; Hildebrandt, J.; Neumann, M.; Rutter, M.T.; Shaw, R.G.; Weigel, D.; Fenster, C.B. Fine-Grained Analysis of Spontaneous Mutation Spectrum and Frequency in Arabidopsis thaliana. *Genetics* **2019**, *211*, 703–714. [CrossRef]
163. Yang, S.; Wang, L.; Huang, J.; Zhang, X.; Yuan, Y.; Chen, J.Q.; Hurst, L.D.; Tian, D. Parent-progeny sequencing indicates higher mutation rates in heterozygotes. *Nature* **2015**, *523*, 463–467. [CrossRef] [PubMed]
164. Yang, N.; Xu, X.W.; Wang, R.R.; Peng, W.L.; Cai, L.; Song, J.M.; Li, W.; Luo, X.; Niu, L.; Wang, Y.; et al. Contributions of Zea mays subspecies mexicana haplotypes to modern maize. *Nat. Commun.* **2017**, *8*, 1874. [CrossRef]
165. Sedeek, K.E.M.; Mahas, A.; Mahfouz, M. Plant Genome Engineering for Targeted Improvement of Crop Traits. *Front. Plant Sci.* **2019**, *10*, 114. [CrossRef]
166. Kalinina, N.O.; Khromov, A.; Love, A.J.; Taliansky, M.E. CRISPR Applications in Plant Virology: Virus Resistance and Beyond. *Phytopathology* **2020**, *110*, 18–28. [CrossRef]
167. Ma, X.; Zhang, X.; Liu, H.; Li, Z. Highly efficient DNA-free plant genome editing using virally delivered CRISPR-Cas9. *Nat. Plants* **2020**, *6*, 773–779. [CrossRef]
168. Lenaghan, S.; Stewart, C. An Automated Protoplast Transformation System. In *Plant Genome Editing with CRISPR Systems. Part of the Methods in Molecular Biology Book Series (MIMB)*; Humana Press: New York, NY, USA, 2019; Volume 1917, pp. 355–363.
169. Hussain, Q.; Shi, J.; Scheben, A.; Zhan, J.; Wang, X.; Liu, G.; Yan, G.; King, G.J.; Edwards, D.; Wang, H. Genetic and signalling pathways of dry fruit size: Targets for genome editing-based crop improvement. *Plant Biotechnol. J.* **2020**, *18*, 1124–1140. [CrossRef]
170. Huang, L.; Zhang, R.; Huang, G.; Li, Y.; Melaku, G.; Zhang, S. Developing superior alleles of yield genes in rice by artificial mutagenesis using the CRISPR/Cas9 system. *Crop. J.* **2018**, *6*, 475–481. [CrossRef]
171. Miao, C.; Xiao, L.; Hua, K.; Zou, C.; Zhao, Y.; Bressan, R.A.; Zhu, J.K. Mutations in a subfamily of abscisic acid receptor genes promote rice growth and productivity. *Proc. Natl. Acad. Sci. USA* **2018**, *115*, 6058–6063. [CrossRef]
172. Adamski, N.M.; Borrill, P.; Brinton, J.; Harrington, S.A.; Marchal, C.; Bentley, A.R.; Bovill, W.D.; Cattivelli, L.; Cockram, J.; Contreras-Moreira, B.; et al. A roadmap for gene functional characterisation in crops with large genomes: Lessons from polyploid wheat. *Elife* **2020**, *9*, e55646. [CrossRef] [PubMed]
173. Igartua, E.; Contreras-Moreira, B.; Casas, A.M. TB1: From domestication gene to tool for many trades. *J. Exp. Bot.* **2020**, *71*, 4621–4624. [CrossRef]
174. Alonge, M.; Soyk, S.; Ramakrishnan, S.; Wang, X.; Goodwin, S.; Sedlazeck, F.J.; Lippman, Z.B.; Schatz, M.C. RaGOO: Fast and accurate reference-guided scaffolding of draft genomes. *Genome Biol.* **2019**, *20*, 224. [CrossRef]
175. Bayer, P.E.; Golicz, A.A.; Scheben, A.; Batley, J.; Edwards, D. Plant pan-genomes are the new reference. *Nat. Plants* **2020**, *6*, 914–920. [CrossRef]

176. Danilevicz, M.F.; Tay Fernandez, C.G.; Marsh, J.I.; Bayer, P.E.; Edwards, D. Plant pangenomics: Approaches, applications and advancements. *Curr. Opin. Plant Biol.* **2020**, *54*, 18–25. [CrossRef] [PubMed]
177. Khan, A.W.; Garg, V.; Roorkiwal, M.; Golicz, A.A.; Edwards, D.; Varshney, R.K. Super-Pangenome by Integrating the Wild Side of a Species for Accelerated Crop Improvement. *Trends Plant Sci.* **2020**, *25*, 148–158. [CrossRef]
178. Michael, T.P.; VanBuren, R. Building near-complete plant genomes. *Curr. Opin. Plant Biol.* **2020**, *54*, 26–33. [CrossRef] [PubMed]
179. Jiao, Y.; Peluso, P.; Shi, J.; Liang, T.; Stitzer, M.C.; Wang, B.; Campbell, M.S.; Stein, J.C.; Wei, X.; Chin, C.S.; et al. Improved maize reference genome with single-molecule technologies. *Nature* **2017**, *546*, 524–527. [CrossRef]
180. Mir, R.R.; Reynolds, M.; Pinto, F.; Khan, M.A.; Bhat, M.A. High-throughput phenotyping for crop improvement in the genomics era. *Plant Sci.* **2019**, *282*, 60–72. [CrossRef]
181. Alqudah, A.M.; Haile, J.K.; Alomari, D.Z.; Pozniak, C.J.; Kobiljski, B.; Borner, A. Genome-wide and SNP network analyses reveal genetic control of spikelet sterility and yield-related traits in wheat. *Sci. Rep.* **2020**, *10*, 2098. [CrossRef] [PubMed]
182. Gupta, P.K.; Kulwal, P.L.; Jaiswal, V. Association mapping in plants in the post-GWAS genomics era. *Adv. Genet.* **2019**, *104*, 75–154. [CrossRef]
183. Ashburner, M.; Ball, C.A.; Blake, J.A.; Botstein, D.; Butler, H.; Cherry, J.M.; Davis, A.P.; Dolinski, K.; Dwight, S.S.; Eppig, J.T.; et al. Gene ontology: Tool for the unification of biology. The Gene Ontology Consortium. *Nat. Genet.* **2000**, *25*, 25–29. [CrossRef] [PubMed]
184. Chris, C.S.M. Gene Ontology Data Archive. Available online: https://zenodo.org/record/3980761#.YN4uqyBR2bg (accessed on 2 July 2018).
185. Tello-Ruiz, M.K.; Naithani, S.; Gupta, P.; Olson, A.; Wei, S.; Preece, J.; Jiao, Y.; Wang, B.; Chougule, K.; Garg, P.; et al. Gramene 2021: Harnessing the power of comparative genomics and pathways for plant research. *Nucleic Acids Res.* **2020**, *49*, D1452–D1463. [CrossRef]
186. Howe, K.L.; Contreras-Moreira, B.; De Silva, N.; Maslen, G.; Akanni, W.; Allen, J.; Alvarez-Jarreta, J.; Barba, M.; Bolser, D.M.; Cambell, L.; et al. Ensembl Genomes 2020-enabling non-vertebrate genomic research. *Nucleic Acids Res.* **2020**, *48*, D689–D695. [CrossRef] [PubMed]
187. Ambrosino, L.; Colantuono, C.; Diretto, G.; Fiore, A.; Chiusano, M.L. Bioinformatics Resources for Plant Abiotic Stress Responses: State of the Art and Opportunities in the Fast Evolving-Omics Era. *Plants* **2020**, *9*, 591. [CrossRef]
188. Pan, Q.; Wei, J.; Guo, F.; Huang, S.; Gong, Y.; Liu, H.; Liu, J.; Li, L. Trait ontology analysis based on association mapping studies bridges the gap between crop genomics and Phenomics. *BMC Genom.* **2019**, *20*, 443. [CrossRef] [PubMed]
189. Wimalanathan, K.; Lawrence-Dill, C.J. Gene Ontology Meta Annotator for Plants. *BioRxiv* **2019**. [CrossRef]
190. Zhao, Y.; Wang, J.; Chen, J.; Zhang, X.; Guo, M.; Yu, G. A Literature Review of Gene Function Prediction by Modeling Gene Ontology. *Front. Genet.* **2020**, *11*, 400. [CrossRef]
191. Wimalanathan, K.; Friedberg, I.; Andorf, C.M.; Lawrence-Dill, C.J. Maize GO Annotation-Methods, Evaluation, and Review (maize-GAMER). *Plant Direct* **2018**, *2*, e00052. [CrossRef]
192. Borrill, P. Blurring the boundaries between cereal crops and model plants. *New Phytol.* **2019**, *228*, 1721–1727. [CrossRef] [PubMed]
193. Mohammadi, M.; Xaviera, A.; Beckett, T.; Beyer, S.; Chen, L.; Chikssa, H.; Cross, V.; Moreira, F.F.; French, E.; Gaire, R.; et al. Identification, deployment, and transferability of quantitative trait loci from genome-wide association studies in plants. *Curr. Plant Biol.* **2020**, *24*, 100145. [CrossRef]
194. Liu, H.J.; Yan, J. Crop genome-wide association study: A harvest of biological relevance. *Plant J.* **2019**, *97*, 8–18. [CrossRef]
195. Liu, J.; Gao, Y.; Tang, Y.; Wang, D.; Chen, X.; Yao, Y.; Guo, Y. Genome-Wide Identification, Comprehensive Gene Feature, Evolution, and Expression Analysis of Plant Metal Tolerance Proteins in Tobacco Under Heavy Metal Toxicity. *Front. Genet.* **2019**, *10*, 345. [CrossRef] [PubMed]
196. Mabhaudhi, T.; Chimonyo, V.G.P.; Hlahla, S.; Massawe, F.; Mayes, S.; Nhamo, L.; Modi, A.T. Prospects of orphan crops in climate change. *Planta* **2019**, *250*, 695–708. [CrossRef] [PubMed]
197. Gruber, K. Agrobiodiversity: The living library. *Nature* **2017**, *544*, S8–S10. [CrossRef]
198. Castaneda-Alvarez, N.P.; Khoury, C.K.; Achicanoy, H.A.; Bernau, V.; Dempewolf, H.; Eastwood, R.J.; Guarino, L.; Harker, R.H.; Jarvis, A.; Maxted, N.; et al. Global conservation priorities for crop wild relatives. *Nat. Plants* **2016**, *2*, 16022. [CrossRef]
199. Shelef, O.; Weisberg, P.J.; Provenza, F.D. The Value of Native Plants and Local Production in an Era of Global Agriculture. *Front. Plant Sci.* **2017**, *8*, 2069. [CrossRef]
200. Lemmon, Z.H.; Reem, N.T.; Dalrymple, J.; Soyk, S.; Swartwood, K.E.; Rodriguez-Leal, D.; Van Eck, J.; Lippman, Z.B. Rapid improvement of domestication traits in an orphan crop by genome editing. *Nat. Plants* **2018**, *4*, 766–770. [CrossRef]
201. Ravindran, S. QnAs with Zachary B. Lippman. *Proc. Natl. Acad. Sci. USA* **2020**, *117*, 14624–14625. [CrossRef]
202. Van Tassel, D.L.; Tesdell, O.; Schlautman, B.; Rubin, M.J.; DeHaan, L.R.; Crews, T.E.; Streit Krug, A. New Food Crop Domestication in the Age of Gene Editing: Genetic, Agronomic and Cultural Change Remain Co-evolutionarily Entangled. *Front. Plant Sci.* **2020**, *11*, 789. [CrossRef] [PubMed]
203. Fernie, A.R.; Yan, J. De Novo Domestication: An Alternative Route toward New Crops for the Future. *Mol. Plant* **2019**, *12*, 615–631. [CrossRef] [PubMed]
204. Larson, G.; Piperno, D.R.; Allaby, R.G.; Purugganan, M.D.; Andersson, L.; Arroyo-Kalin, M.; Barton, L.; Climer Vigueira, C.; Denham, T.; Dobney, K.; et al. Current perspectives and the future of domestication studies. *Proc. Natl. Acad. Sci. USA* **2014**, *111*, 6139–6146. [CrossRef] [PubMed]

205. Allaby, R. Domestication Syndrome in Plants. *Encycl. Glob. Archaeol.* **2014**, *2182*, 2184. [CrossRef]
206. Kantar, M.; Nashoba, A.; Anderson, J.; Blackman, B.; Rieseberg, L. The Genetics and Genomics of Plant Domestication. *BioScience* **2017**, *67*, 971–982. [CrossRef]
207. Iqbal, M.M.; Erskine, W.; Berger, J.D.; Nelson, M.N. Phenotypic characterisation and linkage mapping of domestication syndrome traits in yellow lupin (*Lupinus luteus* L.). *Theor. Appl. Genet.* **2020**, *133*, 2975–2987. [CrossRef]
208. Lowry, D.B.; Hoban, S.; Kelley, J.L.; Lotterhos, K.E.; Reed, L.K.; Antolin, M.F.; Storfer, A. Breaking RAD: An evaluation of the utility of restriction site-associated DNA sequencing for genome scans of adaptation. *Mol. Ecol. Resour.* **2017**, *17*, 142–152. [CrossRef]
209. Barrera-Redondo, J.; Pinero, D.; Eguiarte, L.E. Genomic, Transcriptomic and Epigenomic Tools to Study the Domestication of Plants and Animals: A Field Guide for Beginners. *Front. Genet.* **2020**, *11*, 742. [CrossRef]
210. Schreiber, M.; Stein, N.; Mascher, M. Genomic approaches for studying crop evolution. *Genome Biol.* **2018**, *19*, 140. [CrossRef]
211. Zhao, Q.; Feng, Q.; Lu, H.; Li, Y.; Wang, A.; Tian, Q.; Zhan, Q.; Lu, Y.; Zhang, L.; Huang, T.; et al. Pan-genome analysis highlights the extent of genomic variation in cultivated and wild rice. *Nat. Genet.* **2018**, *50*, 278–284. [CrossRef]
212. Boden, S.A.; Østergaard, L. How can developmental biology help feed a growing population? *Development* **2019**, *146*, dev172965. [CrossRef]
213. Trevaskis, B. Developmental Pathways Are Blueprints for Designing Successful Crops. *Front. Plant Sci.* **2018**, *9*, 745. [CrossRef]
214. Metje-Sprink, J.; Sprink, T.; Hartung, F. Genome-edited plants in the field. *Curr. Opin. Biotechnol.* **2020**, *61*, 1–6. [CrossRef] [PubMed]
215. Ganivet, E. Growth in human population and consumption both need to be addressed to reach an ecologically sustainable future. *Environ. Dev. Sustain.* **2020**, *22*, 4979–4998. [CrossRef]
216. Klein Goldewijk, K.; Beusen, A.; Van Drecht, G.; De Vos, M. The HYDE 3.1 spatially explicit database of human-induced global land-use change over the past 12,000 years. *Glob. Ecol. Biogeogr.* **2011**, *20*, 73–86. [CrossRef]
217. Watson, J.E.M.; Venter, O.; Lee, J.; Jones, K.R.; Robinson, J.G.; Possingham, H.P.; Allan, J.R. Protect the last of the wild. *Nature* **2018**, *563*, 27–30. [CrossRef]
218. The 2020 Living Planet Report. Available online: www.livingplanetindex.org/home/index (accessed on 10 September 2020).
219. Ceballos, G.; Ehrlich, P.R.; Dirzo, R. Biological annihilation via the ongoing sixth mass extinction signaled by vertebrate population losses and declines. *Proc. Natl. Acad. Sci. USA* **2017**, *114*, E6089–E6096. [CrossRef]
220. Fiore, M.C.; Mercati, F.; Spina, A.; Blangiforti, S.; Venora, G.; Dell'Acqua, M.; Lupini, A.; Preiti, G.; Monti, M.; Pe, M.E.; et al. High-Throughput Genotype, Morphology, and Quality Traits Evaluation for the Assessment of Genetic Diversity of Wheat Landraces from Sicily. *Plants* **2019**, *8*, 116. [CrossRef]
221. Taranto, F.; D'Agostino, N.; Rodriguez, M.; Pavan, S.; Minervini, A.P.; Pecchioni, N.; Papa, R.; De Vita, P. Whole Genome Scan Reveals Molecular Signatures of Divergence and Selection Related to Important Traits in Durum Wheat Germplasm. *Front. Genet.* **2020**, *11*, 217. [CrossRef]
222. Crossley, M.S.; Burke, K.D.; Schoville, S.D.; Radeloff, V.C. Recent collapse of crop belts and declining diversity of US agriculture since 1840. *Glob. Chang. Biol.* **2021**, *27*, 151–164. [CrossRef]
223. Smith, P. Malthus is still wrong: We can feed a world of 9–10 billion, but only by reducing food demand. *Proc. Nutr. Soc.* **2015**, *74*, 187–190. [CrossRef] [PubMed]
224. Leridon, H. World population outlook: Explosion or implosion? *Popul. Soc.* **2020**, *573*, 1–4. [CrossRef]
225. Vollset, S.E.; Goren, E.; Yuan, C.W.; Cao, J.; Smith, A.E.; Hsiao, T.; Bisignano, C.; Azhar, G.S.; Castro, E.; Chalek, J.; et al. Fertility, mortality, migration, and population scenarios for 195 countries and territories from 2017 to 2100: A forecasting analysis for the Global Burden of Disease Study. *Lancet* **2020**, *396*, 1285–1306. [CrossRef]
226. Gilbert, N. Environment: The disappearing nutrient. *Nature* **2009**, *461*, 716–718. [CrossRef]
227. Lidicker, W.Z., Jr. A Scientist's Warning to humanity on human population growth. *Glob. Ecol. Conserv.* **2020**, *24*, e01232. [CrossRef]
228. Taylor, K.B. The passing of western civilization. *Futures* **2020**, *122*, 102582. [CrossRef] [PubMed]
229. United Nations. *World Population Prospects 2019*; United Nations: New York, NA, USA, 2019; Volume I. Available online: https://population.un.org/wpp/ (accessed on 17 June 2019).

*plants*

MDPI

*Review*

# Advances in Genomics Approaches Shed Light on Crop Domestication

**Yang Zhao [1,2], Mengfan Feng [1,2], Dev Paudel [3], Tofazzal Islam [4], Aliya Momotaz [5], Ziliang Luo [3], Zifan Zhao [3], Ni Wei [1,2], Sicheng Li [1,2], Qing Xia [1,2], Bowen Kuang [1,2], Xiping Yang [1,2,*] and Jianping Wang [3,*]**

1 Guangxi Key Laboratory of Sugarcane Biology & National Demonstration Center for Experimental Plant Science Education, Guangxi University, Nanning 530005, China; zhaoyang@gxu.edu.cn (Y.Z.); mengfanfeng@st.gxu.edu.cn (M.F.); 1917301038@st.gxu.edu.cn (N.W.); lamina0130@126.com (S.L.); shandongxiaqing@163.com (Q.X.); kuangbowen97@163.com (B.K.)

2 State Key Laboratory for Conservation and Utilization of Subtropical Agro-Bioresources, Guangxi University, Nanning 530005, China

3 Agronomy Department, University of Florida, Gainesville, FL 32610, USA; dev.paudel@ufl.edu (D.P.); luoziliang@ufl.edu (Z.L.); zifanzhao@ufl.edu (Z.Z.)

4 Institute of Biotechnology and Genetic Engineering (IBGE), Bangabandhu Sheikh Mujibur Rahman Agricultural University, Gazipur 1706, Bangladesh; tofazzalislam@bsmrau.edu.bd

5 USDA-ARS Sugarcane Field Station, 12990 US Hwy. 441N, Canal Point, FL 33438, USA; aliya.momotaz@usda.gov

* Correspondence: xipingyang@gxu.edu.cn (X.Y.); wangjp@ufl.edu (J.W.)

**Abstract:** Crop domestication occurred ~10,000–12,000 years ago when humans shifted from a hunter–gatherer to an agrarian society. Crops were domesticated by selecting the traits in wild plant species that were suitable for human use. Research is crucial to elucidate the mechanisms and processes involved in modern crop improvement and breeding. Recent advances in genomics have revolutionized our understanding of crop domestication. In this review, we summarized cutting-edge crop domestication research by presenting its (1) methodologies, (2) current status, (3) applications, and (4) perspectives. Advanced genomics approaches have clarified crop domestication processes and mechanisms, and supported crop improvement.

**Keywords:** crop; genomics approaches; domestication; application

**Citation:** Zhao, Y.; Feng, M.; Paudel, D.; Islam, T.; Momotaz, A.; Luo, Z.; Zhao, Z.; Wei, N.; Li, S.; Xia, Q.; et al. Advances in Genomics Approaches Shed Light on Crop Domestication. *Plants* **2021**, *10*, 1571. https://doi.org/10.3390/plants10081571

Academic Editors: Ranjith Pathirana and Francesco Carimi

Received: 30 April 2021
Accepted: 6 July 2021
Published: 30 July 2021



## 1. Introduction

Crops played a major role in human cultural evolution by causing a shift from a nomadic to a sedentary society. Hence, crops are suitable as evolutionary models illuminating genetic variation and selection. Crop domestication is a major agricultural advance ensuring food security for human society. Domestication is the result of phenotypic and genetic changes mediated by breeding. It involves multigenerational selection of plant traits favoring enhanced adaptation and acclimatization to farming management practices. Approximately 12,000 years ago, most economically important crops were domesticated [1,2]. Our ancestors instinctively selected crops that were easy to harvest and those with improved yield and flavor. These simple selection strategies helped pyramid important alleles and recombinants and resulted in naturally transformed plants with beneficial traits facilitating cultivation, breeding, storage, trade, and dissemination.

Of the ~5500 food crops worldwide, 15 contribute to ~70% of the total calories consumed by humans. Rice, wheat, and maize account for >50% of the calorie demand [3]. Up to 7000 known plant species are semi-cultivated or orphan crops [4]. These natural plant resources comprise a valuable pool of genetic material that could enable future crop breeding, increase food diversity, and respond to the new challenges of global climate change and population expansion [5]. The domestication of orphan and underutilized crop plants via recently developed biotechnologies such as genome-editing and genome-enabled approaches is highly promising in crop development for smart agriculture.

Most domesticated crop species share common traits such as increased yield and seed size and decreased dormancy and seed shattering. Though crop domestication is long and slow, only a few genes are involved in it, and some of them are conserved in various species [2,6–10]. Hence, both targeted re-domestication and *de novo* wild species domestication are feasible. In these processes, targeted genes are identified, introgressed, or modified to produce new cultivars. Unlocking the potential of wild crop species domestication will improve global food security and help realize certain sustainable development goals of the United Nations such as zero poverty (No. 1) and zero hunger (No. 2). Targeted domestication, crop improvement, and mass crop cultivation are generally cost-effective approaches towards these objectives. A concerted effort under the joint leadership of the Food and Agriculture Organization (FAO), Consultative Groups of International Agricultural Research Institutions, National Agricultural Research Institutions, and various governments is required for the research, popularization, and large-scale utilization of undomesticated crops with potential.

## 2. Genomic Methods for Studying Crop Domestication

To use crop breeding knowledge, technologies, and genetic resources effectively, it is necessary to understand the mechanisms of crop domestication. Several questions should be addressed to elucidate crop domestication. Where, when, and how did crop domestication occur? Did each crop go through single or multiple domestication processes? What was the genetic architecture of crop domestication? How did selection affect domesticated species? Some of these questions are being answered through genomic evidence.

### *2.1. Population Genomics*

Population genetics and genomics have revealed that crops passed through four major stages during their evolution from wild progenitors to modern domesticated species [2]. These include (1) the onset of domestication when only one or a few wild progenitors with traits favored by humans were selected; (2) in situ propagation of selected wild progenitors to increase desirable alleles; (3) the spread and adaptation of cultivated populations to new environments; and (4) deliberate plant breeding to improve agronomic traits. During domestication, only a few individuals with traits serving human interests were selected from the wild progenitor population. Genetic drift caused by the founder effect and by selection reduced genetic diversity in domesticated crops. Genetic drift was assessed by comparing the genomes of domesticated crops and their wild relatives [11]. Advances in sequencing technologies and reduction of their costs have supported the publication of numerous high-quality studies on crop domestication using population genomics methods (Table 1). Huang et al. [12] compared the genome sequences of 446 wild and 1083 cultivated rice accessions. They found that *O. sativa japonica* originated in the middle of the Pearl River region in Southern China and was domesticated from a specific *O. rufipogon* population. There were 55 selective sweeps, and the genome signatures for selection during domestication were identified. They accounted for 5.1% of the genome regions (21.9 Mb) [12]. Hufford et al. [13] identified a few genes with strong selection in domesticated maize based on whole-genome resequencing of 75 wild, landrace, and improved maize lines. The authors also demonstrated that post-domestication diversity may have been recovered through introgression from wild relatives.

Advances in genetics, archeology, and their interdisciplinary areas have contributed to the clarification of crop domestication. Analyses of modern and ancient DNA have uncovered details about human and animal history. However, few studies have reported on the history of crop domestication, as there has been insufficient archeological evidence or DNA for genetic analysis [14]. Kistler et al. [15] sequenced 40 indigenous maize landraces and nine archeological samples from South America and compared them against 85 published maize genomes. The ancestral South American maize population was brought from its domestication center in Mexico before its domesticated traits were established. Multiple subsequent dispersal events led to maize diversity and biogeography. Scott et al. [16]

prepared whole-genome sequences of a museum specimen of Egyptian emmer wheat chaff and demonstrated that ancient Egyptian emmers already shared a common origin with modern domesticated emmer even before the crop was introduced to Egypt. The foregoing results furnished evidence for early southeastern wheat dispersal and gene flow from wild to ancient Egyptian emmer.

**Table 1.** The application of population genomics to crop domestication.

| Crop | Population Type | Population Size | Key Statistic | Discovery | Ref. |
|---|---|---|---|---|---|
| Rice | Ancestral progenitor; cultivated indica and japonica varieties | 1529 | Sequence diversity ($\pi$) population-differentiation ($F_{ST}$), cross-population extended haplotype homozygosity (XP-EHH) | Identify 55 domestication sweeps, and reveal the domesticaiton and development of cultivated rice | [12] |
| Maize | Wild, landraces and improved maize lines | 75 | $\pi$, $\rho$, $F_{ST}$, Tajima's D, normalized Fay and Wu's H, and a composite likelihood approach (XP-CLR) | Evidence of recovery of diversity after domestication, and stronger selection for domestication than improvement | [13] |
| Maize | Ancient samples, modern maizes landraces, and teosintes | 134 | Mutation load, *D*-statistics, and f3 and f4 statistic | Reveal domestication center and human-mediated spread of maize | [15] |
| Wheat | Ancient and modern domestic emmer | 64 | Haplotype structure | Uncover the history and diversity of emmar wheat | [16] |
| Cotton | Wild and domesticated cotton accessions | 352 | $\pi$, $F_{ST}$, and XP-CLR | Identify 93 domestication sweeps | [17] |

Wang et al. [17] used a genome-wide variation map for 352 wild and domesticated cotton accessions. They scanned domestication sweeps covering 74 Mb of the 'A' subgenome and 104 Mb of the 'D' subgenome and found asymmetric subgenome domestication for directional selection of long fibers. Hufford et al. [13] conducted population genomic studies and discovered that 7.6% of all maize genomic regions were under selection during domestication. However, population genomics has certain drawbacks. First, genomic signals caused by domestication or improvement might be confused because genetic diversity is reduced in both cases. Second, certain crops have undergone multiple independent domestications wherein different genomic region layers may have been selected at different times. Hence, mixtures of samples from various domestication processes could obscure signals targeted for selection. Third, introgression may bilaterally occur between wild and domesticated crops and weaken signals identified through population genomics. Fourth, certain genomic signatures identified under domestication are not directly related to any agronomic traits, and their molecular mechanisms remain unclear. Hence, they must be validated by genome-wide association studies (GWAS), quantitative trait locus (QTL) mapping, map-based cloning, and functional targeted gene analyses.

### 2.2. Genome-Wide Association Studies (GWAS)

Genome-wide association studies (GWAS) or linkage disequilibrium (LD) mapping infer the genetic basis of domestication by identifying statistically significant associations between phenotypes and genotypes or between domesticated traits and sequence variants. GWAS explores natural diversity panels comprising unrelated individuals with historical LD [18].

There has been substantial progress in mapping the QTL underlying crop domestication. Huang et al. [12] performed a GWAS for leaf sheath color and tiller angle using 446 *Oryza rufipogon* accessions (Table 2). The strongest associations occurred near the known loci *OsC1* for coloration and *PROG1* for prostrate growth. They demonstrated that the mapping resolution was threefold higher for the wild rice population than for *O. sativa*, as the former had a relatively high LD decay rate. Wang et al. [17] conducted a GWAS on fiber quality-related cotton traits using 267 accessions with two million single nucleotide polymorphisms (SNPs) and a minor allele frequency > 0.05. They identified 19 association signals, of which 16 were new discoveries. Thus, the high-density SNP set was more powerful than the previous GWAS with simple-sequence-repeat markers. Forty-three association signals were identified for seven watermelon fruit quality traits [19]. There were 208 loci significantly associated with melon fruit mass, quality, and morphological traits [20].

**Table 2.** The application of GWAS to crop domestication.

| Crop | Domestication Trait | Population Type | Population Size | Genotype Method | Model | Discovery | Ref. |
|---|---|---|---|---|---|---|---|
| Rice | Leaf sheath color and tiller angle | Ancestral progenitor *Oryza rufipogon* | 446 | Whole genome resequencing | Compressed mixed linear model | Identify assoicaitons for *OSC1* and *PROG1* | [12] |
| Cotton | Fiber quality related traits | Cotton accessions | 267 | Whole genome resequencing | Compressed mixed linear model | Identify 19 assoicaiton signals (16 were new) | [17] |
| Watermelon | Fruit quality traits | Cultivated and wild watermelon accessions | 414 | Whole genome resequencing | Linear mixed model algorithm | Identify 43 associaiton signals (35 were new) | [19] |

Unlike population genomics, GWAS directly relates genomic regions to domesticated traits and facilitates the interpretation of domestication mechanisms at molecular level. GWAS usually has a higher targeted QTL mapping resolution than QTL mapping itself, as unrelated individuals can accumulate numerous genetic recombination events since their last divergence. Moreover, GWAS requires no genetic linkage maps, and the analysis is straightforward. However, GWAS depends on the target crop diversity panel, which is usually costly to collect and maintain. It is very difficult to develop a diversity panel with minimum population structure that is powerful enough for GWAS analysis. GWAS power may also be low when rare variants are causal mutations in a study panel [11].

### 2.3. QTL Mapping

Most evolutionarily important traits are quantitative. Phenotypic variation in these traits is the result of segregations at multiple QTL, the environment, and interactions between genes and the environment [21]. A QTL is a genomic locus correlating with phenotypic trait variation in a population and may be attributed to $\geq$2 genes on the same or different chromosomes. A QTL analysis provides the genetic basis for phenotypic variation, including gene locations, numbers, and magnitudes, and their mechanisms in a biparental segregating population [22]. QTL mapping has enabled successful identification and cloning of genes underlying domestication traits. It was the first and perhaps the most widely used method for localizing the genetic basis of a trait. Several QTL analyses revealed that wild and weedy genotypes were transformed into domesticated crop species. Pourkheirandish et al. [10] performed QTL mapping on three populations developed from crosses between domesticated barley and its wild progenitor and identified and cloned *Btr1* and *Btr2*, which control grain dispersal (Table 3). They demonstrated that 1-bp and 11-bp deletions in *Btr1* and *Btr2*, respectively, made the rachis non-brittle in domesticated barley. Doust et al. [9] analyzed shattering and flowering time in a foxtail millet mapping population and found that the alleles favored during domestication had larger phenotypic effects than the genetic background or the environment. Thus, recurrent selection in breeding can substantially increase domestication-related traits. Rice seed shattering QTLs were mapped on several chromosomes with a complex genetic architecture. *OsqSH1* was identified on chromosome 1 [23], and *SH4* was localized to chromosome 4 [24]. A later study supported that *qSH1* is epistatic to *SH4* in abscission process during seed shattering. In another study on molecular cloning, the non-shattering *SH4* allele was fixed in *O. sativa* ssp. *indica* and *O. sativa* ssp. *japonica* [25]. QTL analysis and map-based cloning showed *Sh1* on chromosome 1 encoded the YABBY transcription factor and underwent three independent mutations to form non-shattering domesticated sorghum [8]. QTL mapping is a straightforward and powerful approach to identify the genes controlling crop domestication.

**Table 3.** The application of QTL to crop domestication.

| Crop | Domestication Trait | Population Type | Population Size | Marker | Discovery | Refs. |
|---|---|---|---|---|---|---|
| Rice | Seed shattering | $F_2$ | 304 | RFLP, RAPD, SNP, SSR | Localized the gene *qSH1* and gene *sh4* | [23,24] |
| Barley | Rachis non-brittle | $F_2$ | >10,000 | SNP | Localized the gene *btr1* and *btr2* | [10] |
| Foxtail millet | Shattering and flowering time | Recombinantinbred line | 182 | SNP, SSR, and sequence-tagged site markers | Two significant QTLs | [9] |

Note: SSR, simple sequence repeat; RFLP, restriction fragment length polymorphisms; RAPD, random amplified polymorphic DNA.

### 2.4. Genome Editing Using CRISPR-Cas Technology

Clustered Regularly Interspaced Short Palindromic Repeats (CRISPR) is used in gene manipulation and is revolutionary in biological research [26,27]. CRISPR was first discovered in 1987 and recognized as an adaptive immune system in archaea and bacteria [28,29]. In 2012, several research groups independently discovered that CRISPR and its associated protein (Cas) constitute a powerful genome-editing technology inducing precise DNA breaks at targeted genome locations in any living cell [26,27]. The CRISPR-Cas tool was first successfully used for plant genome editing by three independent groups in 2013 [30–32]. CRISPR-Cas has been used for programmable gene/epigenome editing and transcriptome regulation in plants [33–35]. It can also edit multiple genes (multiplexing) through simultaneous multiple guide RNA (gRNA) delivery and expression [30]. Through multiplex gene editing, CRISPR promotes basic research, accelerates plant breeding, and facilitates plant domestication and germplasm development. CRISPR-Cas can fine-tune and knock out master switches in undomesticated wild crops, enhance genomic diversity, and facilitate *de novo* domestication in one generation or a few generations [33].

The power of CRISPR-Cas toolkit in *de novo* plant domestication was demonstrated in several studies [36–39]. Zsögön et al. [39] used CRISPR-Cas to show genome editing for several domesticated genes in tomato such as *SP*, *SP5G*, *SlCLV3*, and *SlWUS* (Table 4). Furthermore, CRISPR was extended to the shuffling chromosome and used to stack multiple alleles into one tightly linked locus. CRISPR-Cas9-mediated induction of heritable chromosomal translocation was demonstrated in *Arabidopsis* [40]. Li et al. [38] applied CRISPR-Cas multiplex genome editing to four *Solanum pimpinellifolium* accessions that were salt-tolerant or highly resistant to bacterial spot disease. The genome-edited plants acquired targeted domestication traits while retaining their abiotic and biotic stress tolerance. Therefore, CRISPR-Cas multiplex editing introduces novel and retains existing plant traits and develops an ideal crop [41]. CRISPR-Cas technology could provide precise and customized modifications conducive to plant breeding. Successful application of CRISPR-Cas in tomato and wild cherry indicated that this technology could domesticate in one generation new crops resilient to environmental change [37,42,43].

The benefits of *de novo* orphan crop domestication via CRISPR-Cas gene editing include (i) high precision and accuracy, (ii) short variety development time, (iii) transgene-free product, (iv) high crop yield and nutritional value, and (v) crop resistance to biotic and abiotic stresses. However, the major challenges of *de novo* domestication via genome editing include (i) limited genomic information for wild relatives, (ii) lack of a transformation system for wild relatives, (iii) fitness cost, and (iv) limited public acceptance of genetically modified organisms (GMO).

**Table 4.** The application of CRISPR-Cas to crop domestication.

| Crop | Domestication Traits | Target Gene | Method | Discovery | Ref. |
| --- | --- | --- | --- | --- | --- |
| Rice | Panicle length, grain size, cold tolerance | *OsPIN5b*, *GS3*, *OsMYB30* | CRISPR-Cas9 system edits three genes simultaneously | Higher yield and better cold tolerance in gene-edited rice | [44] |
| Wheat | Grein length, weight and yield, TKW, Inflorescence architecture, branching and tillering | *Tagasr7-A1* (*-B1* and *–D1*), *TaDEP1*, *TaNAC2*, *TaPIN 1*, and *TaLOX2* | Transient expression of CRISPR-Cas9 in callus cells | Changes on target traits in wheat callus and regeneration of plants | [45] |
| Tomato | Fruit size, number and nutrition | *SP*, *O*, *FW2.2*, *CycB* | CRISPR-Cas9 system edits six genes simultaneously | Gene-edited tomato has at least a threefold increase in target traits | [39] |
| Cucumber | Carpel development | *CsWip1* | Optimized CRISPR/Cas9 system with *CsU6* promoter and GFP | Seven times more female flowers in gene-edited cumcumber | [46] |

Note: GFP, green fluorescent protein; TKW, thousand kernel weight.

## 3. Current Status of Research on Crop Domestication

### 3.1. Domestication Centers and Their Spread

Around 12,000 years ago, human-guided crop domestication occurred independently in the Middle East, the Fertile Crescent, China, Mesoamerica, the Andes, Near Oceania, Sub-Saharan Africa, and Eastern North America [2,47]. Of the 2500 domesticated plant species distributed in 160 families, 250 are fully domesticated [2]. Table S1 and Figure 1 show the origins and major cultivation zones of global domesticated food crops. Some of them are widely spread across several regions, whereas others are more regionally or locally important. The domestication, spread, and cultivation of food crops have demonstrated the transition from hunter–gatherer to agrarian societies.

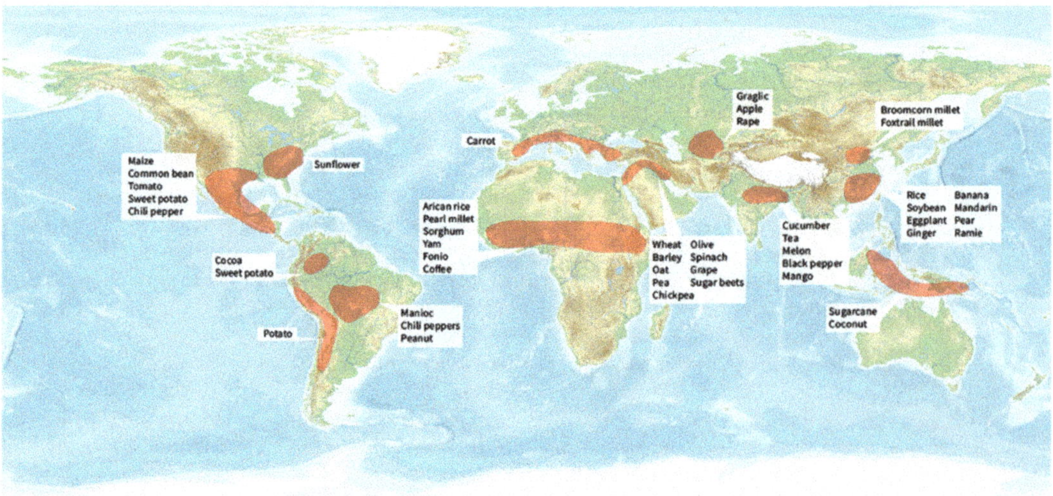


**Figure 1.** Global food crop origin and domestication. Shaded regions indicate approximate locations of centers of crop origin and domestication.

Maize is one of the most important food crops worldwide, and extensive research progress has been made on its domestication. Starch grain and phytolith evidence indicated that maize was first domesticated from wild Balsas Teosinte (*Zea mays* ssp. *parviglumis*) in Mexico ~9000 years ago [48]. After partial domestication in Mexico, maize traversed Panama, arrived in Central America ~7500 years ago, and was brought to South America ~6500 years ago. In South America, maize was fully domesticated at several independent locations ~6500–4000 years ago [15]. Two major maize movements from Mesoamerica to South America were deduced, as the Pan-American lineage shared excess ancestry with *parviglumis* compared with the strictly South American lineage. Maize was brought to the United States ~4000 years ago. In the 15th century, European colonies spread maize through the Americas and, thence, globally.

The spread of domesticated crops is slow and complex. It is affected by shifts in human society, farming practice improvements, and plant adaptability to various climates. Plants adapted to different global climates in an east to west trajectory. Archeological evidence supports that maize spread from west to east across the Amazon, which was a secondary improvement center for partially domesticated maize [15]. However, since its initial domestication in Mexico maize has spread across the Americas, including the south-to-north and lowland-to-highland directions. Gene flow from the wild relatives of maize might have improved its adaptation to various ecological niches [49].

### 3.2. Domestication Theory

The study of the inheritance of domestication genes raises a crucial question, namely, does selection act on existing variations segregating in ancestral wild populations or *de novo* mutations? Current research supports the possibility that selection acts on both variations and mutations. However, numerous domesticated traits arise from existing variations in ancestral wild populations [2]. Standing variations apparently allow rapid evolution of populations, as they lack the lag periods characteristic of *de novo* mutations [50]. Alleles selected from standing variations occur at low to moderate frequency in wild progenitors, and there are weak signatures in the genome for the selection of old mutations. Therefore, determining which mutations are affected by selection in the domestication process may help clarify the nature of selective sweeps and the rate of crop evolution.

Traits selected in domestication may distinguish crops from wild progenitors. This mechanism is known as the domestication syndrome [51]. Different crops and the same crop with multiple origins shared the same domesticated phenotypes such as loss of seed dormancy and non-shattering seeds. The existence of convergent phenotypes raises the questions as to how selection behaves in domestication and whether the same or different genes are affected by it. Molecular parallelism might explain this phenomenon. Multiple mutations in the same or different genes resulting in the same phenotype have been independently selected for domestication. Three non-shattering haplotypes at *SH1* locus were characterized in domesticated sorghum and were distributed among sorghum landraces. Thus, multiple domestications of a species may occur [8]. *SH1* was under selection for rice and maize domestication as well. However, the domestication phenotype can also be controlled by different genes. Doust et al. [52] identified novel genes controlling branching in foxtail millet. In contrast, the *teosinte brached1* ortholog had only a minor effect on this trait. Lai et al. [53] examined genome resequencing data from wild and domesticated maize and sorghum accessions and showed that the number of candidate domestication genes with parallel selection signatures was not significantly higher than that expected by chance. Certain major genes with large effects might have been repeatedly targeted by domestication selection. Alternate genes may have also produced similar phenotypes in different crop species.

Four demographic crop domestication models were proposed to elucidate the domestication process [2]. In an earlier model, a single domestication event resulted from strong selection in a small wild progenitor population and caused total reproductive isolation between the wild and domesticated species. However, archeological and genetic data suggested that genetic bottlenecks vary among crop species and introgression occurs between crops and wild relatives. Therefore, the model was modified to alternate versions wherein a single domestication event occurred with gene flow between crops, or multiple domestication events with gene flow occurred, or domestication events from interspecific hybridization occurred and were followed by clonal propagation. Crops assessed by these alternate models were reviewed by Meyer and Purugganan [2].

### 3.3. Genetic Architecture and the Molecular Basis of Genes Mediating Crop Domestication

Several QTL or GWAS studies identified genes underlying the domestication syndrome in various crops (Table S2), of which 40.5% were regulatory (transcription factors or co-regulators) and 56.0% were structural (enzymes or other proteins). Certain genes such as *Prog1* and *Prog7* control domestication traits such as prostrate rice tillers. However, the advancement of population genomics disclosed that large genomic regions such as 7.6% of the maize genome [13] and 6.9% of the cotton genome [17] were under domestication selection. This paradox could be explained by the inherent limitations of QTL or GWAS. A large proportion of the missing heritability cannot be explained by small populations containing only a few recombinants with low enough marker density to capture significant genomic regions.

Meyer and Purugganan [2] proposed that a domestication gene has a clear function associated with a domesticated trait, is under positive selection, and is fixed or almost entirely fixed at the causative mutation in all lineages under a single domestication event. The authors compiled 60 genes involved in domestication or diversification, of which 40 (66.7%) encoded transcription factors or co-regulators, 14 (23.3%) encoded enzymes, and 6 (10%) encoded transporter proteins and ubiquitin ligase. These genes can encode various traits of which some are involved in the domestication of different crops. The primary effects of causative mutations in the aforementioned genes include the creation of nonsense mutations, premature truncations, other mutations resulting in null function, *cis*-regulatory mutations, and missense mutations. Mutations with large phenotypic effects are the most common functional changes.

## 4. Domestication in Modern Crop Breeding

### 4.1. Rice

Rice is one of the most important food crops in the world, and it is also a model system to study crop domestication. Though there are tons of literature discussing rice origin and domestication, the origin and history of rice domestication remain controversial. Despite this, it is widely accepted that selections together with introgression shaped the genomes of cultivated rice [54]. With the achievements of researches on rice domestication, their applications in modern rice breeding are impressive. To meet the growth needs for food under the global climatic challenges, breeders combine genetic resources of domestication genes with those containing multiple valuable alleles to create superior cultivars [55]. In traditional crosses of diverged cultivars or germplasm, the process of selection of robust agronomic traits and removing unfavorable backgrounds could be accelerated by using molecular markers developed according to domestication genes.

Genome editing technology, which can efficiently modify target genomes predictably and precisely, is no doubt a revolutionary tool to perform molecular domestication to obtain desirable traits in laboratory [56,57]. Using this technology in rice, scientists successfully reduced seed shattering by editing *qSH1* gene [58], broke down seed dormancy by knockout *OsVP1* [59], and developed superior alleles of yield genes by editing *Gn1a* and *DEP1* genes [60]. These studies have proven the potential to improve target traits substantially in rice by editing single or a few domestication genes. Moreover, cis-regulatory elements are alternative targets for editing, which can tune gene expression levels, timing, and tissue specificity, but avoids any detrimental pleiotropic effects due to mutations in coding regions [61]. Recently, a strategy to *de novo* domesticate wild allotetraploid rice was described, and six agronomical traits were improved rapidly by genome editing of target genes [62], demonstrating the possibility to develop this polyploid wild rice to a food crop. Though still at the beginning stage, *de novo* domestication based on advanced genomics approaches shortens the process of domestication to a few years, which opens up a gate to utilize wide genetic resources in a precise way.

### 4.2. Tomato

Tomato (*Solanum lycopersicum*) originated in the Andean region, and its domestication occurred before the 15th century. Intense tomato domestication occurred in Europe in the 18th and 19th centuries [63], and tomato cultivar improvement has been ongoing since then. Wild tomato has large genetic diversity and has been extensively studied to characterize certain traits favorable for breeding [64,65]. In contrast, cultivated tomato has very low genetic diversity and has <5% of the genetic variation in their wild relatives [66]. The domestication syndrome has been studied for this crop, and several QTLs underlying growth habit and fruit size were identified [67–69].

Advances in genome editing and crop domestication enable plant geneticists to target certain genomic sites in wild plants and rapidly create improved cultivated crops. Zsögön et al. [39] edited six loci in wild tomato (*Solanum pimpinellifolium*) and generated highly productive progeny that could serve to breed improved cultivars. The six loci that were previously considered vital to tomato domestication regulated general plant growth habit (*SELF-PRUNING*) [70], fruit number (*MULTIFLORA*) [71], fruit shape (*OVATE*) [72], fruit size (*FASCIATED* and *FRUIT WEIGHT 2.2*) [73,74], and nutritional quality (*LYCOPENE BETA CYCLASE*) [75]. These genes were targeted by multiplex CRISPR-Cas9, and loss-of-function alleles were generated [39]. The $T_1$ lines were successfully edited for *SELF-PRUNING* (*SP*), *OVATE* (*O*), *FRUIT WEIGHT 2.2* (*FW2.2*), and *LYCOPENE BETA CYCLASE* (*CycB*). Compared with the wild type, the engineered plants had higher fruit numbers per truss, enhanced yield, a fourfold increase in fruit locule number, a 200% increase in fruit weight, 100% higher lycopene content, and stable β-carotene and lutein content [39]. Moreover, Brix value, fruit shape, and locule number were uniformly inherited in $T_2$ and $T_3$. Hence, the engineered traits were stable, and the wild tomato was successfully domesticated [39]. CRISPR-Cas9 was also effectively used to mutate tomato domestication gene orthologs controlling plant architecture, flower production, and fruit size in ground cherry (*Physalis pruinose*) [37], a solanaceous orphan crop.

### 4.3. Potato

Potato (*Solanum tuberosum*) is one of the most important food crops worldwide. However, cultivated potato varieties are autotetraploid and vegetatively propagated. Consequently, breeding efforts for tuber yield and quality improvement are very limited. Most potato germplasms bearing alleles controlling agronomically important traits are diploids [76]. The reinvention of inbred diploid varieties has been proposed to overcome this limitation and accelerate breeding [77]. Most diploid potato species are gametophytically self-incompatible. This trait is controlled by S-RNase genes. Recent attempts have been made to edit S-RNase genes and achieve self-compatible diploid potato varieties [78,79]. The potato genome resource and diploid potato line sequencing data identified S-RNase orthologs. CRISPR-Cas9 guided S-RNase gene knockout and successfully created self-compatible diploid potato lines that could be pollinated and generate enough seed for propagation. Thus, the domestication of wild diploid potato into an inbred crop is a novel strategy in potato genetic improvement.

### 4.4. Orphan Crops

Orphan crops are semi-cultivated species with limited regional importance such as dry bean (*Vigna* spp.) and lupin. They are often relatively less productive, not optimized for modern agriculture, and infrequently studied by the research community [80]. However, unlike several major cultivated crops, they have wide biodiversity and are adapted to poorly controlled or harsh environments. Orphan crops provide nutritional benefits and may tolerate extreme heat or cold [81]. In view of constant pressure from climate change and increased demands for food by growing populations, orphan crop domestication may become vital to food security in the future.

Ground cherry (*Physalis pruinose;* Solanaceae) is indigenous to Mexico and South America and an orphan crop distantly related to tomato [82]. However, its small fruits fall to the ground because of stem abscission, and the plant has a sprawling growth habit. Hence, its productivity is limited. Lemmon et al. [37] developed an *Agrobacterium tumefaciens*-mediated transformation system to enable editing of the domestication genes related to the aforementioned traits of ground cherry. The genomic resources were enriched by whole-genome and RNA sequencing. Orthologs of the tomato florigen repressor genes *SELFPRUNING* (*SP*) and *SELF-PRUNING 5G* (*SP5G*) in ground cherry (*Ppr-SP* and *Ppr-SP5G*) were selected for knockout in the CRISPR-Cas9 experiment. In tomato, mutations in these genes produce a compact plant architecture and flower and fruit burst. The *Ppr-SP* knockout plants were extremely compact and presented with a slight relative increase in fruit production. The *PprSP5G* knockout plants displayed a compact structure and significantly enhanced fruit production because of moderate sympodial shoot termination. The *CLAVATA* (*CLV*) ortholog (*Ppr-CLV3*) modifies fruit size by domesticating the locule number, and it was targeted with CRISPR-Cas9. The *Ppr-CLV3* mutants exhibited a 24% relative gain in fruit mass growth. In the future, other important domestication genes such as *JOINTLESS-2* (fruit abscission) will also be edited to improve the agronomic traits of ground cherry.

## 5. Crop Domestication Perspectives

From the first plant domestication at least 12,000 years ago to the present day, numerous crops have been subjected to human selection especially for the purpose of continuously increasing yield. However, domestication and modern plant breeding have steadily reduced genetic variation in crops. Consequently, modern cultivars are highly susceptible to biotic and abiotic stresses such as drought, heat, insects, and disease [83]. Wild germplasms of cultivated species have wide genetic variation and stress tolerance traits that should be exploited in modern breeding programs to develop resilient cultivars.

Domestication may involve increasing the size of certain organs such as the fruits and seeds. Therefore, genes regulating cell division, meristem size, and patterning are vital [84]. Other morphological traits such as flower number, flowering time, and nutrient composition are also important traits for selection. There is a wide range of gene functions under selection. Certain genes might have been under selection for traits never considered as targets during crop domestication. As a rule, only a few genes play major roles in domestication. Nevertheless, our understanding of domestication is constantly being reshaped by new discoveries. In the coming decade, population genomics, GWAS, and genome-editing tools will clarify genomic signatures in domestication and greatly enhance our ability to domesticate wild plant species. For example, candidate domestication genes in wild and domesticated species may be sequenced to identify selective population sweeps and functionally associate them with SNPs via GWAS. Affordable sequencing technology will enable fine mapping in orphan crops and accelerate their breeding. Regions of high divergence between cultivated and wild species may also be identified so they can be associated with domestication. There remains much to be learned about how domestication changes crop genome composition.

An important objective going forward is to determine whether this new domestication knowledge can serve as guidance for future plant breeding efforts. Current advances in plant breeding appear to indicate that we are heading in the right direction. Genome editing-tools have advanced our understanding of domestication genes and enabled us to develop new cultivars by directly incorporating domestication-related genes. For instance, the engineered ancestral progenitor of wild tomato differed in terms of fruit morphology, size, number, and nutritional value from the widely cultivated tomato [39]. Efficient crop *de novo* domestication will depend on the availability of characterized domestication genes, effective transformation methods, and open access to genome-editing technologies. And more importantly, successful *de novo* domestication should integrate genetic tools with agronomic and cultural drivers to accommodate the newly designed crops to adapt to dynamic environments and agronomic practices, and to be accepted by consumers [85].

From the aspect of genetics, rapid domestication may be realized via comparative genomics of various crop accessions mediated by next-generation sequencing and the inherent synteny between crops. This process was applied to ground cherry using tomato gene orthologs [37]. The exclusive application of domestication-related genes has helped domesticate orphan crops and develop new varieties of cultivated crops. Allele mining and gene or QTL cloning can recover superior alleles that do not pass through domestication bottlenecks [86]. Plant breeding has benefitted from the recommendation of Vavilov to collect and maintain wild crop relatives in gene banks [87]. Knowledge acquired from omics technology also complements traditional plant breeding approaches. Plant breeders with access to large datasets can develop new cultivars in a 'breeding by design' process using CRISPR-Cas genome editing [88–90]. Using the various tools available to them, breeders can enhance crop productivity, nutritive value, and biotic and abiotic stress tolerance.

**Supplementary Materials:** The following are available online at https://www.mdpi.com/article/10.3390/plants10081571/s1, Table S1: Global domesticated food crops, their origins, and major cultivation zones, Table S2: Food crop domestication genes.

**Author Contributions:** Conceptualization, X.Y. and J.W.; data curation, M.F., N.W., S.L., Q.X. and B.K.; writing—original draft preparation, Y.Z., D.P., T.I., A.M., Z.L., Z.Z. and X.Y.; writing—review and editing, X.Y. and J.W.; funding acquisition, X.Y. and J.W. All authors have read and agreed to the published version of the manuscript.

**Funding:** This work was financially supported by National Natural Science Foundation of China (31901591), Natural Science Foundation of Guangxi Province (2020AC19060, 2020JJA130036), the 'One Hundred Person' Project of Guangxi Province, Science and Technology Major Project of Guangxi Province (GK2018-266-Z01) and the USDA National Institute of Food and Agriculture Hatch project (award number: FLAAGR-005569).

**Institutional Review Board Statement:** Not applicable.

**Informed Consent Statement:** Not applicable.

**Data Availability Statement:** Not applicable.

**Conflicts of Interest:** The authors declare no conflict of interest.

## References

1. Purugganan, M.D.; Fuller, D.Q. The nature of selection during plant domestication. *Nature* **2009**, *457*, 843–848. [CrossRef]
2. Meyer, R.S.; Purugganan, M.D. Evolution of crop species: Genetics of domestication and diversification. *Nat. Rev. Genet.* **2013**, *14*, 840–852. [CrossRef]
3. Ross-Ibarra, J.; Morrell, P.L.; Gaut, B.S. Plant domestication, a unique opportunity to identify the genetic basis of adaptation. *Proc. Natl. Acad. Sci. USA* **2007**, *104*, 8641–8648. [CrossRef] [PubMed]
4. Smýkal, P.; Nelson, M.N.; Berger, J.D.; von Wettberg, E.J.B. The impact of genetic changes during crop domestication. *Agronomy* **2018**, *8*, 119. [CrossRef]
5. Lobell, D.B.; Gourdji, S.M. The influence of climate change on global crop productivity. *Plant Physiol.* **2012**, *160*, 1686–1697. [CrossRef] [PubMed]
6. Doebley, J.; Stec, A.; Hubbard, L. The evolution of apical dominance in maize. *Nature* **1997**, *386*. [CrossRef] [PubMed]
7. Tan, L.; Li, X.; Liu, F.; Sun, X.; Li, C.; Zhu, Z.; Fu, Y.; Cai, H.; Wang, X.; Xie, D.; et al. Control of a key transition from prostrate to erect growth in rice domestication. *Nat. Genet.* **2008**, *40*, 1360–1364. [CrossRef] [PubMed]
8. Lin, Z.; Li, X.; Shannon, L.M.; Yeh, C.T.; Wang, M.L.; Bai, G.; Peng, Z.; Li, J.; Trick, H.N.; Clemente, T.E. Parallel domestication of the *Shattering1* genes in cereals. *Nat. Genet.* **2012**, *44*, 720–724. [CrossRef]
9. Doust, A.N.; Lukens, L.; Olsen, K.M.; Mauro-Herrera, M.; Meyer, A.; Rogers, K. Beyond the single gene: How epistasis and gene-by-environment effects influence crop domestication. *Proc. Natl. Acad. Sci. USA* **2014**, *111*, 6178–6183. [CrossRef]
10. Pourkheirandish, M.; Hensel, G.; Kilian, B.; Senthil, N.; Chen, G.; Sameri, M.; Azhaguvel, P.; Sakuma, S.; Dhanagond, S.; Sharma, R.; et al. Evolution of the grain dispersal system in barley. *Cell* **2015**, *162*, 527–539. [CrossRef]
11. Guerra-García, A.; Piñero, D. Current approaches and methods in plant domestication studies. *Bot. Sci.* **2017**, *95*, 345. [CrossRef]

12. Huang, X.; Kurata, N.; Wei, X.; Wang, Z.X.; Wang, A.; Zhao, Q.; Zhao, Y.; Liu, K.; Lu, H.; Li, W.; et al. A map of rice genome variation reveals the origin of cultivated rice. *Nature* **2012**, *490*, 497–501. [CrossRef]
13. Hufford, M.B.; Xu, X.; van Heerwaarden, J.; Pyhajarvi, T.; Chia, J.M.; Cartwright, R.A.; Elshire, R.J.; Glaubitz, J.C.; Guill, K.E.; Kaeppler, S.M.; et al. Comparative population genomics of maize domestication and improvement. *Nat. Genet.* **2012**, *44*, 808–811. [CrossRef]
14. Schreiber, M.; Stein, N.; Mascher, M. Genomic approaches for studying crop evolution. *Genome Biol.* **2018**, *19*, 140. [CrossRef]
15. Kistler, L.; Maezumi, S.Y.; de Souza, J.; Przelomska, N.A.S.; Malaquias Costa, F.; Smith, O.; Loiselle, H.; Ramos-Madrigal, J.; Wales, N.; Ribeiro, E.R.; et al. Multiproxy evidence highlights a complex evolutionary legacy of maize in South America. *Science* **2018**, *362*, 1309–1313. [CrossRef] [PubMed]
16. Scott, M.F.; Botigue, L.R.; Brace, S.; Stevens, C.J.; Mullin, V.E.; Stevenson, A.; Thomas, M.G.; Fuller, D.Q.; Mott, R. A 3000-year-old Egyptian emmer wheat genome reveals dispersal and domestication history. *Nat. Plants* **2019**, *5*, 1120–1128. [CrossRef] [PubMed]
17. Wang, M.; Tu, L.; Lin, M.; Lin, Z.; Wang, P.; Yang, Q.; Ye, Z.; Shen, C.; Li, J.; Zhang, L.; et al. Asymmetric subgenome selection and cis-regulatory divergence during cotton domestication. *Nat. Genet.* **2017**, *49*, 579–587. [CrossRef]
18. Zhu, C.; Gore, M.; Buckler, E.S.; Yu, J. Status and prospects of association mapping in plants. *Plant Genome* **2008**, *1*, 5–20. [CrossRef]
19. Guo, S.; Zhao, S.; Sun, H.; Wang, X.; Wu, S.; Lin, T.; Ren, Y.; Gao, L.; Deng, Y.; Zhang, J.; et al. Resequencing of 414 cultivated and wild watermelon accessions identifies selection for fruit quality traits. *Nat. Genet.* **2019**, *51*, 1616–1623. [CrossRef]
20. Zhao, G.; Lian, Q.; Zhang, Z.; Fu, Q.; He, Y.; Ma, S.; Ruggieri, V.; Monforte, A.J.; Wang, P.; Julca, I.; et al. A comprehensive genome variation map of melon identifies multiple domestication events and loci influencing agronomic traits. *Nat. Genet.* **2019**, *51*, 1607–1615. [CrossRef]
21. Mackay, T.F.C. The genetic architecture of quantitative traits. *Annu. Rev. Genet.* **2001**. [CrossRef]
22. Grisel, J.E.; Crabbe, J.C. Quantitative trait loci mapping. *Alcohol. Health Res. World* **1995**, *19*, 220–227.
23. Konishi, S.; Izawa, T.; Lin, S.Y.; Ebana, K.; Yano, M. An SNP caused loss of seed shattering during rice domestication. *Science* **2006**, *312*, 1392–1396. [CrossRef]
24. Li, C.; Zhou, A.; Sang, T. Rice domestication by reducing shattering. *Science* **2006**, 1936–1939. [CrossRef]
25. Zhang, L.-B.; Zhu, Q.; Wu, Z.-Q.; Ross-Ibarra, J.; Gaut, B.S.; Ge, S.; Sang, T. Selection on grain shattering genes and rates of rice domestication. *New Phytol.* **2009**. [CrossRef]
26. Gasiunas, G.; Barrangou, R.; Horvath, P.; Siksnys, V. Cas9-crRNA ribonucleoprotein complex mediates specific DNA cleavage for adaptive immunity in bacteria. *Proc. Natl. Acad. Sci. USA* **2012**, *109*, E2579–E2586. [CrossRef] [PubMed]
27. Jinek, M.; Chylinski, K.; Fonfara, I.; Hauer, M.; Doudna, J.A.; Charpentier, E. A programmable dual-RNA-guided DNA endonuclease in adaptive bacterial immunity. *Science* **2012**. [CrossRef] [PubMed]
28. Ishino, Y.; Shinagawa, H.; Makino, K.; Amemura, M.; Nakata, A. Nucleotide sequence of the iap gene, responsible for alkaline phosphatase isozyme conversion in escherichia coli, and identification of the gene product. *J. Bacteriol.* **1987**, *169*. [CrossRef] [PubMed]
29. Makarova, K.S.; Haft, D.H.; Barrangou, R.; Brouns, S.J.J.; Charpentier, E.; Philippe Horvath, S.M.; Mojica, F.J.M.; Wolf, Y.I.; Yakunin, A.F.; van der Oost, J.; et al. Evolution and classification of the CRISPR–Cas systems. *Nat. Rev. Microbiol.* **2011**, *9*, 467–477. [CrossRef]
30. Li, J.F.; Norville, J.E.; Aach, J.; Mccormack, M.; Zhang, D.; Bush, J.; Church, G.M.; Sheen, J. Multiplex and homologous recombination-mediated genome editing in arabidopsis and nicotiana benthamiana using guide RNA and Cas9. *Nat. Biotechnol.* **2013**, *31*, 688–691. [CrossRef] [PubMed]
31. Nekrasov, V.; Staskawicz, B.; Weigel, D.; Jones, J.D.G.; Kamoun, S. Targeted mutagenesis in the model plant nicotiana benthamiana using Cas9 RNA-guided endonuclease. *Nat. Biotechnol.* **2013**, *31*. [CrossRef]
32. Shan, Q.; Wang, Y.; Li, J.; Zhang, Y.; Chen, K.; Liang, Z.; Zhang, K.; Liu, J.; Xi, J.J.; Qiu, J.L. Targeted genome modification of crop plants using a CRISPR-Cas system. *Nat. Biotechnol.* **2013**, *31*, 686. [CrossRef]
33. Haque, E.; Taniguchi, H.; Hassan, M.M.; Bhowmik, P.; Karim, M.R.; Smiech, M.; Zhao, K.; Rahman, M.; Islam, T. Application of CRISPR/Cas9 genome editing technology for the improvement of crops cultivated in tropical climates: Recent progress, prospects, and challenges. *Front. Plant Sci.* **2018**, *9*. [CrossRef]
34. Chen, K.; Wang, Y.; Zhang, R.; Zhang, H.; Gao, C. CRISPR/Cas genome editing and precision plant breeding in agriculture. *Annu. Rev. Plant Biol.* **2019**, *70*. [CrossRef]
35. Schindele, A.; Dorn, A.; Puchta, H. CRISPR/Cas brings plant biology and breeding into the fast lane. *Curr. Opin. Biotechnol.* **2019**, *61*, 7–14. [CrossRef]
36. Sedbrook, J.C.; Phippen, W.B.; Marks, M.D. New approaches to facilitate rapid domestication of a wild plant to an oilseed crop: Example pennycress (*Thlaspi arvense* L.). *Plant Sci.* **2014**, *227*, 122–132. [CrossRef]
37. Lemmon, Z.H.; Reem, N.T.; Dalrymple, J.; Soyk, S.; Swartwood, K.E.; Rodriguez-Leal, D.; Van Eck, J.; Lippman, Z.B. Rapid improvement of domestication traits in an orphan crop by genome editing. *Nat. Plants* **2018**, *4*. [CrossRef]
38. Li, T.; Yang, X.; Yu, Y.; Si, X.; Zhai, X.; Zhang, H.; Dong, W.; Gao, C.; Xu, C. Domestication of wild tomato is accelerated by genome editing. *Nat. Biotechnol.* **2018**. [CrossRef]
39. Zsögön, A.; Cermak, T.; Naves, E.R.; Notini, M.M.; Edel, K.H.; Weinl, S.; Freschi, L.; Voytas, D.F.; Kudla, J.; Peres, L.E.P. *De novo* domestication of wild tomato using genome editing. *Nat. Biotechnol.* **2018**. [CrossRef]

40. Beying, N.; Schmidt, C.; Pacher, M.; Houben, A.; Puchta, H. CRISPR–Cas9-mediated induction of heritable chromosomal translocations in *arabidopsis*. *Nat. Plants* **2020**, 1–8. [CrossRef]
41. Fernie, A.R.; Yan, J. *De Novo* domestication: An alternative route toward new crops for the future. *Mol. Plant* **2019**, *12*, 615–631. [CrossRef] [PubMed]
42. Van Eck, J. Genome editing and plant transformation of solanaceous food crops. *Curr. Opin. Biotechnol.* **2018**, *49*. [CrossRef]
43. Yamamoto, T.; Kashojiya, S.; Kamimura, S.; Kameyama, T.; Ariizumi, T.; Ezura, H.; Miura, K. Application and development of genome editing technologies to the solanaceae plants. *Plant Physiol. Biochem.* **2018**, S0981942818300718. [CrossRef]
44. Zeng, Y.; Wen, J.; Zhao, W.; Wang, Q.; Huang, W. Rational improvement of rice yield and cold tolerance by editing the three genes *OsPIN5b*, *GS3*, and *OsMYB30* with the CRISPR–Cas9 system. *Front. Plant Sci.* **2020**, *10*, 1–13. [CrossRef]
45. Zhang, Y.; Liang, Z.; Zong, Y.; Wang, Y.; Liu, J.; Chen, K.; Qiu, J.L.; Gao, C. Efficient and transgene-free genome editing in wheat through transient expression of CRISPR/Cas9 DNA or RNA. *Nat. Commun.* **2016**, *7*, 1–8. [CrossRef]
46. Hu, B.; Li, D.; Liu, X.; Qi, J.; Gao, D.; Zhao, S.; Huang, S.; Sun, J.; Yang, L. Engineering Non-transgenic gynoecious cucumber using an improved transformation protocol and optimized CRISPR/Cas9 system. *Mol. Plant* **2017**, *10*, 1575–1578. [CrossRef] [PubMed]
47. Larson, G.; Piperno, D.R.; Allaby, R.G.; Purugganan, M.D.; Andersson, L.; Arroyo-Kalin, M.; Barton, L.; Climer Vigueira, C.; Denham, T.; Dobney, K.; et al. Current perspectives and the future of domestication studies. *Proc. Natl. Acad. Sci. USA* **2014**, *111*, 6139–6146. [CrossRef]
48. Piperno, D.R.; Ranere, A.J.; Holst, I.; Iriarte, J.; Dickau, R. Starch grain and phytolith evidence for early ninth millennium B.P. maize from the Central Balsas River Valley, Mexico. *Proc. Natl. Acad. Sci. USA* **2009**, *106*, 5019–5024. [CrossRef] [PubMed]
49. Van Heerwaarden, J.; Doebley, J.; Briggs, W.H.; Glaubitz, J.C.; Goodman, M.M.; Gonzalez, J.D.J.S.; Ross-Ibarra, J. Genetic signals of origin, spread, and introgression in a large sample of maize landraces. *Proc. Natl. Acad. Sci. USA* **2011**, *108*, 1088–1092. [CrossRef] [PubMed]
50. Gepts, P. The contribution of genetic and genomic approaches to plant domestication studies. *Curr. Opin. Plant Biol.* **2014**, *18*, 51–59. [CrossRef]
51. Olsen, K.M.; Wendel, J.F. A bountiful harvest: Genomic insights into crop domestication phenotypes. *Annu. Rev. Plant Biol.* **2013**, *64*, 47–70. [CrossRef] [PubMed]
52. Doust, A.N.; Devos, K.M.; Gadberry, M.D.; Gale, M.D.; Kellogg, E.A. Genetic control of branching in foxtail millet. *Proc. Natl. Acad. Sci. USA* **2004**, *101*, 9045–9050. [CrossRef] [PubMed]
53. Lai, X.; Yan, L.; Lu, Y.; Schnable, J.C. Largely unlinked gene sets targeted by selection for domestication syndrome phenotypes in maize and sorghum. *Plant J.* **2018**, *93*, 843–855. [CrossRef] [PubMed]
54. Sang, T.; Ge, S. Understanding rice domestication and implications for cultivar improvement. *Curr. Opin. Plant Biol.* **2013**, *16*, 139–146. [CrossRef] [PubMed]
55. Sang, T.; Ge, S. Genetics and phylogenetics of rice domestication. *Curr. Opin. Genet. Dev.* **2007**, *17*, 533–538. [CrossRef] [PubMed]
56. Wang, T.; Zhang, H.; Zhu, H. CRISPR technology is revolutionizing the improvement of tomato and other fruit crops. *Hortic. Res.* **2019**, *6*, 1–13. [CrossRef]
57. Hua, K.; Zhang, J.; Botella, J.R.; Ma, C.; Kong, F.; Liu, B.; Zhu, J.K. Perspectives on the Application of Genome-Editing Technologies in Crop Breeding. *Mol. Plant* **2019**, *12*, 1047–1059. [CrossRef]
58. Sheng, X.; Sun, Z.; Wang, X.; Tan, Y.; Yu, D.; Yuan, G.; Yuan, D.; Duan, M. Improvement of the Rice "Easy-to-Shatter" Trait via CRISPR/Cas9-Mediated Mutagenesis of the *qSH1* Gene. *Front. Plant Sci.* **2020**, *11*, 1–14. [CrossRef]
59. Jung, Y.J.; Lee, H.J.; Bae, S.; Kim, J.H.; Kim, D.H.; Kim, H.K.; Nam, K.H.; Nogoy, F.M.; Cho, Y.G.; Kang, K.K. Acquisition of seed dormancy breaking in rice (*Oryza sativa* L.) via CRISPR/Cas9-targeted mutagenesis of *OsVP1* gene. *Plant Biotechnol. Rep.* **2019**, *13*, 511–520. [CrossRef]
60. Huang, L.; Zhang, R.; Huang, G.; Li, Y.; Melaku, G.; Zhang, S.; Chen, H.; Zhao, Y.; Zhang, J.; Zhang, Y.; et al. Developing superior alleles of yield genes in rice by artificial mutagenesis using the CRISPR/Cas9 system. *Crop. J.* **2018**, *6*, 475–481. [CrossRef]
61. Swinnen, G.; Goossens, A.; Pauwels, L. Lessons from Domestication: Targeting Cis-Regulatory Elements for Crop Improvement. *Trends Plant Sci.* **2016**, *21*, 506–515. [CrossRef]
62. Yu, H.; Lin, T.; Meng, X.; Du, H.; Zhang, J.; Liu, G.; Chen, M.; Jing, Y.; Kou, L.; Li, X.; et al. A route to *de novo* domestication of wild allotetraploid rice. *Cell* **2021**, *184*, 1156–1170. [CrossRef]
63. Sims, W.L. History of tomato production for industry around the world. *Acta Hortic.* **1980**, 25–26. [CrossRef]
64. Rick, C.M.; Chetelat, R.T. Utilization of related wild species for tomato improvement. *Acta Hortic.* **1995**, *412*, 21–38. [CrossRef]
65. Larry, R.; Joanne, L. Genetic resources of tomato. In *Genetic Improvement of Solanaceous Crops*; Razdan, M.K., Mattoo, A.K., Eds.; Science Publishers: Enfield, NH, USA, 2007.
66. Miller, J.C.; Tanksley, S.D. RFLP analysis of phylogenetic relationships and genetic variation in the genus *Lycopersicon*. *Theor. Appl. Genet.* **1990**, *80*, 437–448. [CrossRef] [PubMed]
67. Doganlar, S.; Frary, A.; Tanksley, S.D. The genetic basis of seed-weight variation: Tomato as a model system. *Theor. Appl. Genet.* **2000**, *100*, 1267–1273. [CrossRef]
68. Van Der Knaap, E.; Lippman, Z.B.; Tanksley, S.D. Extremely elongated tomato fruit controlled by four quantitative trait loci with epistatic interactions. *Theor. Appl. Genet.* **2002**, *104*, 241–247. [CrossRef]
69. Tanksley, S.D. The genetic, developmental, and molecular bases of fruit size and shape variation in tomato. *Plant Cell* **2004**, *16*, S181–S189. [CrossRef] [PubMed]

70. Pnueli, L.; Carmel-Goren, L.; Hareven, D.; Gutfinger, T.; Alvarez, J.; Ganal, M.; Zamir, D.; Lifschitz, E. The *SELF-PRUNING* gene of tomato regulates vegetative to reproductive switching of sympodial meristems and is the ortholog of *CEN* and *TFL1*. *Development* **1998**, *125*, 1979–1989. [CrossRef]
71. Lippman, Z.B.; Cohen, O.; Alvarez, J.P.; Abu-Abied, M.; Pekker, I.; Paran, I.; Eshed, Y.; Zamir, D. The making of a compound inflorescence in tomato and related nightshades. *PLoS Biol.* **2008**, *6*, e288. [CrossRef] [PubMed]
72. Liu, J.; Van Eck, J.; Cong, B.; Tanksley, S.D. A new class of regulatory genes underlying the cause of pear-shaped tomato fruit. *Proc. Natl. Acad. Sci. USA* **2002**, *99*. [CrossRef] [PubMed]
73. Frary, A.; Nesbitt, T.C.; Grandillo, S.; Knaap, E.; Cong, B.; Liu, J.; Meller, J.; Elber, R.; Alpert, K.B.; Tanksley, S.D. *fw2.2*: A quantitative trait locus key to the evolution of tomato fruit size. *Science* **2000**, *289*, 85–88. [CrossRef]
74. Xu, C.; Liberatore, K.L.; Macalister, C.A.; Huang, Z.; Chu, Y.H.; Jiang, K.; Brooks, C.; Ogawa-Ohnishi, M.; Xiong, G.; Pauly, M. A cascade of arabinosyltransferases controls shoot meristem size in tomato. *Nat. Genet.* **2015**, *47*, 784–792. [CrossRef]
75. Ronen, G.; Carmel-Goren, L.; Zamir, D.; Hirschberg, J. An alternative pathway to β-carotene formation in plant chromoplasts discovered by map-based cloning of Beta and old-gold color mutations in tomato. *Proc. Natl. Acad. Sci. USA* **2000**, *97*, 11102–11107. [CrossRef]
76. Spooner, D.M.; Ghislain, M.; Simon, R.; Jansky, S.H.; Gavrilenko, T. Systematics, diversity, genetics, and evolution of wild and cultivated potatoes. *Bot. Rev.* **2014**, *80*, 283–383. [CrossRef]
77. Jansky, S.H.; Charkowski, A.O.; Douches, D.S.; Gusmini, G.; Richael, C.; Bethke, P.C.; Spooner, D.M.; Novy, R.G.; De Jong, H.; De Jong, W.S. Reinventing potato as a diploid inbred line–based crop. *Crop. Sci.* **2016**, *56*, 1412. [CrossRef]
78. Enciso-Rodriguez, F.; Manrique-Carpintero, N.; Nadakuduti, S.S.; Buell, C.R.; Zarka, D.; Douches, D. Overcoming self-incompatibility in diploid potato using CRISPR-Cas9. *Front. Plant Sci.* **2019**, *10*, 376. [CrossRef]
79. Ye, M.; Peng, Z.; Tang, D.; Yang, Z.; Li, D.; Xu, Y.; Zhang, C.; Huang, S. Generation of self-compatible diploid potato by knockout of S-RNase. *Nat. Plants* **2018**, *4*, 651–654. [CrossRef]
80. Bohra, A.; Jha, U.C.; Kishor, P.B.; Pandey, S.; Singh, N.P. Genomics and molecular breeding in lesser explored pulse crops: Current trends and future opportunities. *Biotechnol. Adv.* **2014**, *32*, 1410–1428. [CrossRef] [PubMed]
81. Tadele, Z.; Bartels, D. Promoting orphan crops research and development. *Planta* **2019**, *250*. [CrossRef] [PubMed]
82. Särkinen, T.; Bohs, L.; Olmstead, R.; Knapp, S. A phylogenetic framework for evolutionary study of the nightshades (solanaceae): A dated 1000-tip tree. *BMC Evol. Biol.* **2013**, *13*, 1–15. [CrossRef]
83. Harlan, J.R.; Yeatman, C.W.; Kafton, D.; Wilkes, E.G. *Plant Genetic Resources: A Conservation Imperative*; Westview: Boulder, CO, USA, 1987.
84. Doebley, J.F.; Gaut, B.S.; Smith, B.D. The molecular genetics of crop domestication. *Cell* **2006**, *127*, 1309–1321. [CrossRef]
85. Van Tassel, D.L.; Tesdell, O.; Schlautman, B.; Rubin, M.J.; DeHaan, L.R.; Crews, T.E.; Streit Krug, A. New Food Crop Domestication in the Age of Gene Editing: Genetic, Agronomic and Cultural Change Remain Co-evolutionarily Entangled. *Front. Plant Sci.* **2020**, *11*, 1–16. [CrossRef] [PubMed]
86. Tanksley, S.D.; McCouch, S.R. Seed banks and molecular maps: Unlocking genetic potential from the wild. *Science* **1997**, *277*, 1063–1066. [CrossRef] [PubMed]
87. Vavilov, N.I. The new systematic of cultivated plants. In *The New Systematic*; Huxley, J., Ed.; Oxford University Press: London, UK, 1940; pp. 549–556.
88. Peleman, J.D.; Van Der Voort, J.R. Breeding by design. *Trends Plant Sci.* **2003**, *8*, 330–334. [CrossRef]
89. Islam, T. CRISPR-Cas technology in modifying food crops. *CAB Rev.* **2019**, *14*, 1–16. [CrossRef]
90. Islam, M.T.; Bhowmik, P.K.; Molla, K.A. *CRISPR-Cas Methods*; Springer: Berlin/Heidelberg, Germany, 2020. [CrossRef]

plants

MDPI

Article

# Analysis of Chromosome Associations during Early Meiosis in Wheat Lines Carrying Chromosome Introgressions from *Agropyron cristatum*

**Pilar Prieto [1,*], Carmen Palomino [2], Zuny Cifuentes [1] and Adoración Cabrera [2]**

[1] Plant Breeding Department, Institute for Sustainable Agriculture, Agencia Estatal Consejo Superior de Investigaciones Científicas (CSIC), Alameda del Obispo s/n, Apartado 4048, 14080 Córdoba, Spain; zmcifuentes@ias.csic.es

[2] Genetics Department, ETSIAM, Campus de Rabanales, Universidad de Córdoba, CeiA3, 14071 Córdoba, Spain; ge1pasam@uco.es (C.P.); ge1cabca@uco.es (A.C.)

* Correspondence: pilar.prieto@ias.csic.es

**Abstract:** Crested wheatgrass (*Agropyron cristatum* L. Gaertn., genome P), included in the Triticeae tribe (family Poaceae), is one of the most important grasses in temperate regions. It has been valued as a donor of important agronomic traits for wheat improvement, including tolerance to cold, drought, and high salinity, as well as resistance to leaf rust, stripe rust, and powdery mildew. For successful incorporation of beneficial alleles into wheat, it is essential that recombination between wheat and *A. cristatum* chromosomes occurs. In this work, we analysed chromosome associations during meiosis in wheat lines carrying chromosome introgressions from *A. cristatum* chromosomes 5P and 6P in the presence and absence of *Ph1* locus using fluorescence *in situ* hybridisation. The results showed that the *Ph1* locus does not affect chromosome associations between *A. cristatum* and wheat chromosomes because there were no interspecific chromosome associations; therefore, no recombination between chromosomes from wheat and *Agropyron* were observed in the absence of the *Ph1* locus. The 5P and 6P *A. cristatum* chromosomes do not have a suppressor effect on the *Ph1* locus. Wheat univalents in metaphase I suggest that *Agropyron* chromosomes might carry genes having a role in wheat homologous chromosome associations. Putative effect of the *Agropyron* genes on wheat chromosome associations does not interact with the *Ph1* locus.

**Keywords:** *Agropyron cristatum*; wheat; chromosome pairing; *Ph1*; introgression

**Citation:** Prieto, P.; Palomino, C.; Cifuentes, Z.; Cabrera, A. Analysis of Chromosome Associations during Early Meiosis in Wheat Lines Carrying Chromosome Introgressions from *Agropyron cristatum*. *Plants* **2021**, *10*, 2292. https://doi.org/10.3390/plants10112292

Academic Editors: Ranjith Pathirana and Francesco Carimi

Received: 23 September 2021
Accepted: 20 October 2021
Published: 25 October 2021



## 1. Introduction

Crested wheatgrass (*Agropyron cristatum* L. Gaertn.) is a Triticeae species (family Poaceae) and has diploid (2n = 2x = 14), tetraploid (2n = 4x = 28), and hexaploid (2n = 6x = 42) forms, all based on the basic genome P [1]. *A. cristatum* is a perennial species of economic importance that is widely cultivated in North America as an excellent source of forage, and included in the diets of beef and dairy cattle worldwide [2]. Moreover, this species shows tolerance to cold, drought, and high salinity, and is one of the most important grasses in temperate regions [3,4]. It has also been valued for stabilization of heavy metal-contaminated soils [5,6] and watershed management [7]. In a wheat breeding framework, *A. cristatum* is an appreciated donor of important agronomic traits, such as resistance to leaf rust [8], resistance to powdery mildew [9,10], resistance to stripe rust [11], high salt-tolerance [12], drought stress tolerance [13], and enhanced-grain number per spike and spike length [14]. Hybrids and fertile amphiploids between *Triticum* and *A. cristatum* have been successfully obtained [15–18] with the aim to develop wheat-*A. cristatum* introgressions [8,14,19].

Recombination between cultivated and alien chromosomes is essential for incorporating beneficial alleles into wheat from their wild relatives. In hexaploid wheat (*Triticum aestivum* L., 2n = 6x = 42, genomes AABBDD), chromosome pairing and recombination is mainly controlled by pairing the homoeologous (*Ph1b*) gene located on the long arm of chromosome 5B. This gene is responsible for the diploidization system that allows regular meiosis in polyploid wheat [20,21]. Elimination, manipulation, or 5B chromosome activity suppression have been utilized to induce homoeologous chromosome pairing in wheat, for example, the use of 5B-deficient stocks [22]. In addition, the homoeologous pairing suppressor allele, *Ph2*, located on short arm of the 3D chromosome of wheat, has also been used [23], although the recessive mutant of the homoeologous pairing suppressor *Ph1b* is probably widely used and the most effective method for the induction of homoeologous pairing thus far [24]. A deletion mutant in the *Ph1b* gene has been the determinant in the development of introgressions between alien and wheat genomes by pairing between homoeologous chromosomes [24].

Other methods that promote homoeologous pairing are based on the development of genetic crosses between wheat and *Ae. Speltoides*, which suppress the action of the *Ph1* gene due to the presence of the *Ph1* suppressor genes *Su1-Ph1* and *Su2-Ph1* [25]. Epistatic genes to *Ph1* promoting homoeologous pairing and recombination have also been reported in hybrids of wheat with *Aegilops mutica* Boiss. [26]. Individual alien chromosomes added to wheat have also been shown to enhance homoeologous pairing by suppression the activity of *Ph* genes in wheat, such as chromosome 5U of *Aegilops umbellulata* Zhuk. [27], chromosome 5E and 6E of *Elytrigia elongata* (Host) Nevski. [28], and chromosome 5Mg of *Ae. geniculata* Roth [29].

Although some information on chromosome pairing in *Agropyron* and wheat hybrids and amphiploids is available [17,30], and a possible interaction of genes from *Agropyron* with the *Ph* system in wheat has been postulated [31], little is known about the "true" pairing frequency between *A. cristatum* and wheat chromosomes. Thus, the objective of this work was to analyse chromosome associations during meiosis in wheat lines carrying chromosome introgressions from *A. cristatum* in the presence and in the absence of the *Ph1* locus in the wheat genetic background.

## 2. Results

### 2.1. Characterization of 5P and 6P *A. cristatum* Introgression Lines in Wheat in the Presence and in the Absence of the *Ph1* Locus

Addition lines for chromosomes 5P and 6P in the *Ph1* genetic background were obtained in the $F_3$ progeny from crosses between *A. cristatum* additions for chromosomes 5P and 6P in wheat CS with the CS *Ph1* mutant, respectively [32]. To detect the presence of 5P or 6P chromosomes in the presence and in the absence of the *Ph1* locus, we used specific COS molecular markers for the short and the long arms of these two chromosomes (Figure 1). Zygosity at the *Ph1* locus was predicted using *Ph1* diagnostic ABC920 SCAR marker (Figure 2).

Based on the analysis using the *Ph1* diagnostic marker, we identified homozygous plants for *Ph1* and positives for both COS108 and COS150 markers located on the short and the long arm of chromosome 5P, respectively (Figure 1a,b). Fluorescence *in situ* hybridisation (FISH) analysis demonstrated that these plants were disomic (Figure 3b) or monosomic (Figure 3c) for chromosome 5P in the *Ph1* genetic background, respectively. In addition, we identified one ditelosomic *A. cristatum* introgression line for chromosome 5PS in the *Ph1* genetic background, which only amplified the fragment for the COS108 marker located on the short arm of chromosome 5P, but not the one for the COS150 marker located on chromosome 5PL. FISH analyses demonstrated that this plant was ditelosomic for the 5PS arm (Figure 3f). The 5PS ditelosomic plant was selfed and the descendance was also evaluated using COS molecular markers. All plants obtained were ditelosomic for the 5PS chromosome arm, indicating that the 5PS arm was stably inherited to the descendance. All plants were fertile and vigorous. Similarly, positive plants for both COS440 and COS507, specific for the short and the long arm of chromosome 6P, respectively, were detected in the absence of the *Ph1* locus (Figure 1a,d) indicating that these plants carried the complete 6P chromosome in the *Ph1* mutant background. FISH analysis revealed that these plants were monosomic for chromosome 6P (Figure 3e) and no plant carrying two copies of chromosome 6P in the absence of the *Ph1* locus was obtained.

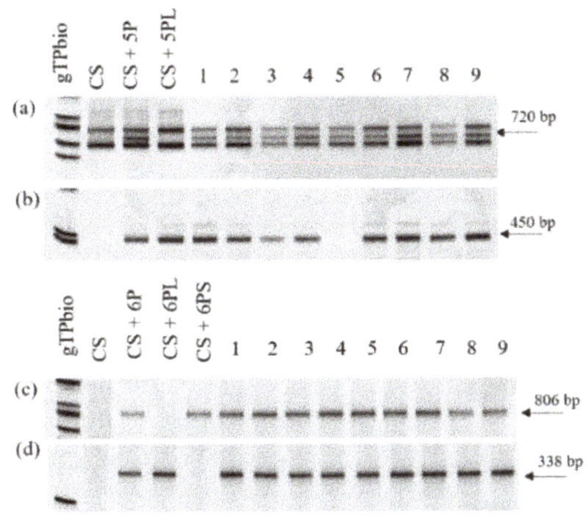


**Figure 1.** PCR amplification profiles used for the identification of chromosome 5P and 6P in the wheat addition lines. (**a**) COS108 and (**b**) COS150 mapped on the short and the long arm of chromosome 5P, respectively. (**c**) COS440 and (**d**) COS507 mapped on short and the long arm of chromosome 6P, respectively. All lines carried a complete chromosome, except for line 5 in in (**a**,**b**) which was ditelosomic for the 5PS chromosome arm.

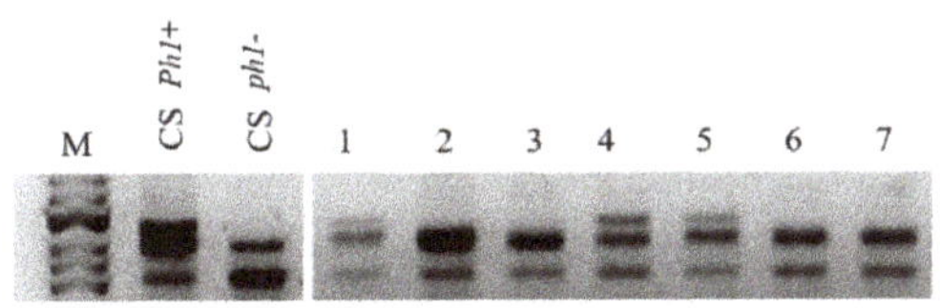


**Figure 2.** Genotypic assays for the presence of *Ph1*. The absence of *Ph1* is marked by the ABC920 SCAR marker (individuals 2, 3, 6, and 7). M: size marker; CS: *Triticum aestivum* cv Chinese Spring; *Ph1*+: wild type wheat CS; and *Ph1*−: the parental *Ph1* mutant.

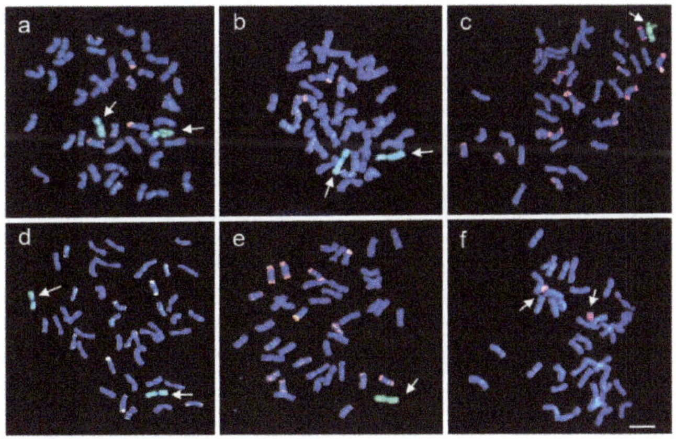


**Figure 3.** FISH to mitotic metaphase chromosome spreads. (**a**) Disomic CS+5P, *Ph1*; (**b**) Disomic CS+5P, *ph1ph1*; (**c**) Monosomic CS+5P, *ph1ph1*; (**d**) Disomic CS+6P, *Ph1*; (**e**) Monosomic CS+6P, *ph1ph1*; and (**f**) Ditelosomic CS+5P, *Ph1*. In (**a**–**e**), double FISH signals with the pAs1 probe (red) and *A. cristatum* genomic DNA (green) as probes; In (**f**), FISH signals with *A. cristatum* genomic DNA (red). Scale bar represents 10 μm.

*2.2. Early Meiosis Analysis of 5P and 6P A. cristatum Introgression Lines in Wheat in the Presence and in the Absence of the Ph1 Locus*

With the aim of assessing a putative effect of the *A. cristatum* genome on chromosome pairing and recombination during meiosis in wheat, chromosome associations were analysed in pollen mother cells (PMCs) during early meiosis by genomic *in situ* hybridisation (GISH) in plants carrying one or two copies of chromosomes 5P and 6P from *A. cristatum*, respectively, in the presence of the *Ph1* locus. Observations on chromosome associations in the presence of the *Ph1* locus were also compared to equivalent 5P and 6P addition lines in wheat in the absence of the *Ph1* locus. We unfortunately did not obtain any plant carrying two copies of chromosome 6P in the absence of the *Ph1* locus. Thus, we were only able to compare the effect of the *Ph1* locus in disomic 5P plants.

Experiments were developed in more than one hundred PMCs of each genomic combination in both early prophase I of meiosis (zygotene, pachytene) and metaphase I, and both in the presence and in the absence of the *Ph1* locus. When two copies of 5P or 6P chromosomes were present in the presence of the *Ph1* locus, both *A. cristatum* chromosomes were observed in proximity in the nucleus in early prophase (data not shown). As meiosis progressed, GISH experiments in zygotene–pachytene stages showed homologous 5P *A. cristatum* chromosomes always fully associated in pairs along the whole chromosome in all the cells analysed (Figure 4a). Similarly, homologous 6P *A. cristatum* chromosomes were also observed associated at these early meiotic stages in all the cells visualized (Figure 4b). These observations suggested that *A. cristatum* chromosomes associated correctly in pairs in the wheat background in early meiosis, and no genetic interaction between wheat and *A. cristatum* chromosomes seemed to be allowed in the presence of the *Ph1* locus at these stages. Similarly, experiments in the absence of the *Ph1* locus showed that homologous 5P *A. cristatum* chromosomes were also associated in pairs in early pachytene in all the cells analysed (Figure 4c) suggesting that chromosome interactions between wheat and *A. cristatum* chromosomes were not promoted in early meiosis in the absence of the *Ph1* locus.

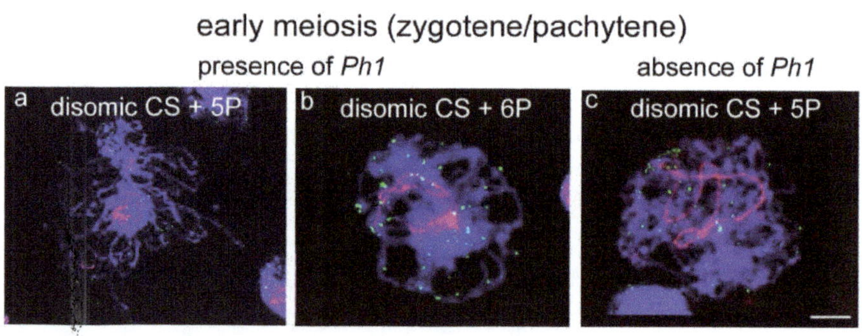


**Figure 4.** FISH to chromosome spreads in early meiosis stages (zygotene–pachytene) in wheat lines carrying chromosomes 5P or 6P from *A. cristatum*, both in the presence and in the absence of the *Ph1* locus. Simultaneous visualization of telomeres (green) and *A. cristatum* chromosomes (red). Wheat DNA was counterstained with DAPI (blue). *Agropyron cristatum* homologous chromosomes were visualized associated in pairs in all the panels. (**a**) Disomic CS+5P, *Ph1Ph1*; (**b**) Disomic CS+6P, *Ph1Ph1*; and (**c**) Disomic CS+5P *ph1ph1*. The scale bar represents 10 µm.

*2.3. Recombination between A. cristatum and Wheat Chromomomes Does Not Occur Either in the Presence or in the Absence of the Ph1 Locus*

Once we observed full chromosome associations between homologous *A. cristatum* chromosomes for both 5P and 6P during early meiosis in the wheat background either in the presence of in the absence of the *Ph1* locus, we also analysed chromosome behaviour of both 5P and 6P *A. cristatum* chromosomes during metaphase I of meiosis, although interspecific recombination events were not expected, due to the lack of interspecific chromosome pairing earlier in meiosis. Thus, wheat chromosomes were observed associated correctly in bivalents and oriented by centromeres properly at meiosis metaphase I in most of the PMCs analysed from 5P and 6P disomic additions in wheat lines carrying the *Ph1* locus (Figure 5). In addition, both homologous 5P and 6P chromosomes were also visualized correctly associated in pairs in the presence of the *Ph1* (Figure 5a,c). Moreover, in the 5P monosomic addition line, the *A. cristatum* chromosome remained always unassociated in all the cells analysed in the presence of the *Ph1* locus (Figure 5b). Our observations clearly suggested that recombination did not occur in meiosis between these *A. cristatum* and wheat chromosomes in the presence of the *Ph1* locus, even when only one copy of *A. cristatum* is present. Thus, 5P *Agropyron* chromosomes were never observed associated with wheat chromosomes by chiasmata during metaphase I in the 5P *A. cristatum* monosomic addition line, always remaining as univalent (Figure 5b) and suggesting other requirements for chromosome associations and recombination.

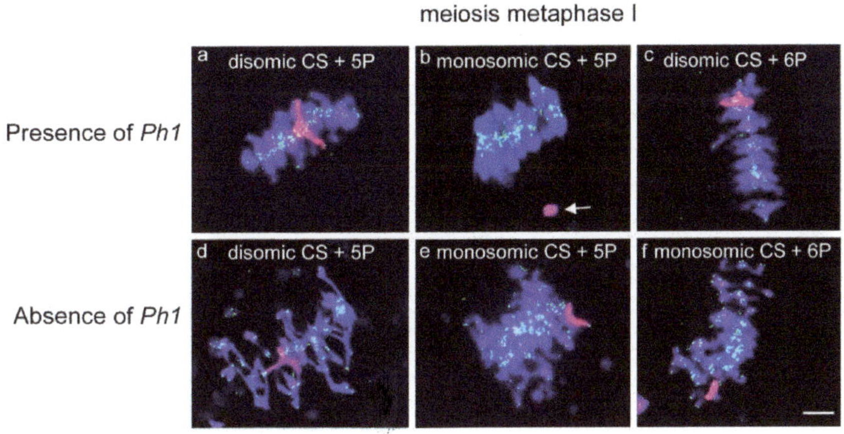


**Figure 5.** FISH analysis of chromosome associations in metaphase I in wheat lines carrying chromosomes 5P or 6P from *A. cristatum*, both in the presence and in the absence of the *Ph1* locus. Homologous 5P and 6P chromosomes (red) are visualized associated in disomic lines, independent of the presence of the *Ph1* locus. Telomeres and wheat chromosomes are visualized in green and blue, respectively. (**a**) Disomic CS+5P, *Ph1Ph1*; (**b**) Monosomic CS+5P, *Ph1Ph1*; (**c**) Disomic CS+6P, *Ph1Ph1*; (**d**) Disomic CS+5P, *ph1ph1*; (**e**) Monosomic CS+5P, *ph1ph1*; and (**f**) Monosomic CS+6P, *ph1ph1*. The scale bar is 10 µm.

We also focused on studying whether chromosome associations, and therefore overcrossing and recombination events, between *A. cristatum* and wheat chromosomes could be allowed in the absence of the *Ph1* locus. Thus, we analysed chromosome associations in PMCs in metaphase I samples from disomic and monosomic *A. cristatum* addition lines in wheat in the absence of the *Ph1* locus. As expected, homologous 5P *A. cristatum* chromosomes that associated previously in early meiosis (Figure 4c) remained associated in metaphase I in disomic addition lines in wheat in the absence of the *Ph1* (Figure 5d). In the absence of homologues, chromosomes 5P and 6P from *A. cristatum* also always remained as univalents during metaphase I in GISH experiments in 5P and 6P monosomic addition lines in the absence of the *Ph1* locus in all the cells analysed (Figure 5e,f). These results showed that interspecific chromosome associations between *A. cristatum* and wheat chromosomes are not promoted in the absence of the *Ph1* locus, suggesting that chromosome associations might also depend on other elements, such as genome homology, as *A. cristatum* and wheat species are phylogenetically distant. Furthermore, the absence of homologous chromosomes did not contribute to increasing chromosome associations between *A. cristatum* and wheat chromosomes, which did not occur either in the presence or in the absence of the *Ph1* locus. In summary, our observations suggested that the *Ph1* locus does not affect chromosome associations between *A. cristatum* and wheat chromosomes because no interspecific chromosome associations and recombination between wheat and *Agropyron* chromosomes were found in either in the presence or in the absence of the *Ph1* locus. Our results also suggest that genes carried in 5P and 6P *A. cristatum* chromosomes do not have a suppressor effect on the *Ph1* locus, as the observations were equivalent both in the presence and in the absence of the *Ph1* locus.

### *2.4. The Presence of Both 5P and 6P A. cristatum Chromosomes Affects Chromosome Associations and Recombination between Homologous Wheat Chromosomes*

Alterations in chromosome pairing between wheat chromosomes were found during GISH experiments developed in meiosis in PMCs from wheat lines carrying both the 5P and the 6P *A. cristatum* chromosomes. In fact, a high number of cells in metaphase I showed wheat chromosomes remaining as univalents both in the presence and in the absence of the *Ph1* locus when both the 5P and 6P *A. cristatum* chromosomes were present in the wheat background (Figure 6). Cells carrying one, two, or three unassociated wheat chromosomes were scored in metaphase I cells in both cases, in the presence and in the absence of the *Ph1* locus (Table 1). These results suggested that the *Agropyron* chromosomes carried genes having a role in homologous chromosome associations as correct wheat homologous pairing was disrupted in the presence of both 5P and 6P *Agropyron* chromosomes. Our observations also suggested that the putative effect of the *Agropyron* genes on homologous chromosome associations does not have any interaction with the *Ph1* locus because the observations of wheat univalents were equivalent both in the presence and in the absence of the *Ph1* locus.

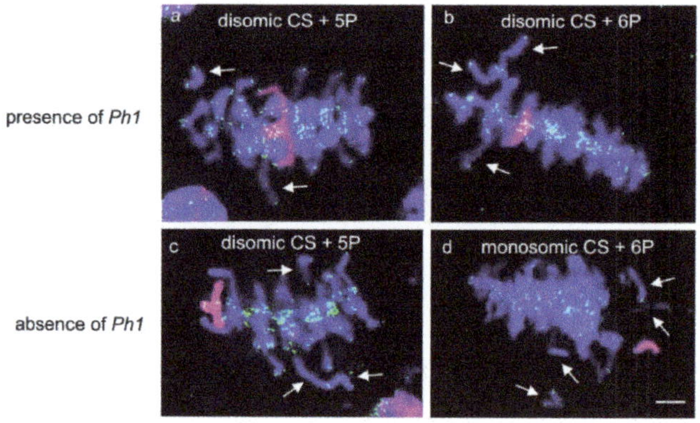


**Figure 6.** FISH analysis in metaphase I chromosome spreads from wheat lines carrying 5P or 6P *A. cristatum* chromosomes, both in the presence and in the absence of the *Ph1* locus. Representative metaphase I cells showing wheat univalents (arrowed) both in the presence and in the absence of the *Ph1* locus. Homologous 5P and 6P chromosomes (red) are visualized associated in disomic lines, independent of the presence of the *Ph1* locus. Telomeres are shown in green and wheat chromosomes in blue. (**a**) Disomic CS+5P, *Ph1Ph1;* (**b**) Disomic CS+6P, *Ph1Ph1;* (**c**) Disomic CS+5P, *ph1ph1;* (**d**) Monosomic CS+6P, *ph1ph1.* Scale bar is 10 µm.

**Table 1.** Analysis of metaphase I cells carrying unassociated (univalent) wheat chromosomes in the presence of 5P and 6P *A. cristatum* chromosomes.

| Number of Metaphase I Cells Analysed | | Number (and Percentage of Cells Carrying One, Two, or Three Wheat Chromosomes as Univalents | | | Total Number (and Percentage) of Cells Carrying Wheat Univalents |
|---|---|---|---|---|---|
| | | I | II | III | |
| Chromosome 5P | | | | | |
| Monosomic CS+5P *Ph1Ph1* | 100 | 5 (5.0%) | 11 (11.0%) | 2 (2.0%) | 18 (18.0%) |
| Disomic CS+5P *Ph1Ph1* | 132 | 14 (10.6%) | 18 (13.7%) | 3 (2.3%) | 35 (26.5%) |
| Monosomic CS+5P *ph1ph1* | 129 | 8 (6.2%) | 28 (21.7%) | 4 (3.1%) | 40 (31.0%) |
| Disomic CS+5P *ph1ph1* | 121 | 16 (13.2%) | 21 (17.3%) | 3 (2.5%) | 40 (33.0%) |
| Chromosome 6P | | | | | |
| Monosomic CS+6P *ph1ph1* | 116 | 2 (1.7%) | 15 (13.0%) | 14 (12.1%) | 31 (26.7%) |
| Disomic CS+6P *Ph1Ph1* | 117 | 10 (8.6%) | 33 (28.2%) | 3 (2.6%) | 46 (39.3%) |

## 3. Discussion

The P genome from *A. cristatum* contains a high genetic diversity, which has been demonstrated by cytological, molecular, and morphological data [33,34]. The *A. cristatum* genome has also been considered a valuable source of genes for pest resistance and abiotic stresses in the framework of wide hybridization programs to improve cereal crops [2,35]. For example, hybrids between *Triticum* and *A. cristatum* have been obtained to introduce traits from *A. cristatum* into *Triticum* [15]. Both durum and common wheat *A. cristatum* introgression lines have also been developed [8,14,18,19] and they could be a useful source of agronomic traits, such as disease resistance [9–11], abiotic stresses [12,13], thousand grain weight [14], and grain quality [36].

Understanding the genomic relationship between the donor and the recipient species is essential for successful introgression of alien chromatin into the wheat genetic background and therefore, for an effective utilization of the large *Agropyron* gene reservoir for wheat breeding purposes. However, in a plant breeding context, the development of genetic introgressions from species included in the wheat tertiary gene pool, such as *A. cristatum*, is much more difficult to achieve than from those species belonging to both primary and secondary genetic pools, because they are phylogenetically more distantly related. *A. cristatum* includes diploid and polyploid forms, all based on the basic genome P [1]. Nevertheless, genetic studies have indicated that synteny is conserved between wheat and the P genome. For example, a high transferability of COS molecular markers between *A. cristatum* and wheat has been reported [9,37]. Similar results were obtained using FISH of tandem repeats and wheat single-gene probes [38,39]. Genetic linking between *A. cristatum* and wheat genomes has also been revealed by both sequencing the transcriptome of a tetraploid *A. cristatum* [40] and genetic mapping using a wheat 660K SNP array [41]. Particularly for homologous groups 5 and 6, comparative mapping using a set of COS molecular markers showed that most of them were located on the short or long arms of wheat chromosome group 5 and 6 and were assigned to corresponding short or long arms of *A. cristatum* 5P and 6P chromosomes, respectively, indicating both homology between chromosomes 5P and 6P and wheat group 5 and 6, respectively [37]. In addition, synteny between *A. cristatum* and wheat homologous group 5 has also been demonstrated by mapping agronomically important genes, such as the grain hardness (*Ha*) and the vernalization (*VRN-1*) loci on the short and on the long arms of chromosome 5P, respectively [36,42], showing that these loci are collinear with those located on the short and the long arms, respectively, of chromosomes 5A, 5B and 5D in wheat. Together, these results indicated that a conservation of genes exists between wheat and *A. cristatum* chromosomes. However, in this work, no homoeologous chromosome pairing among wheat and *A. cristatum* has been found either in the presence or in the absence of the

*Ph1* locus. The results obtained suggest that the *Ph1* locus does not affect chromosome associations between *A. cristatum* and wheat genomes because no interspecific chromosome associations and recombination between wheat and *Agropyron* chromosomes were allowed in the absence of the *Ph1* locus. Our observations also suggested that the lack of interspecific chromosome pairing, even in the absence of the *Ph1* locus, might be due to the divergence of repetitive sequences between wheat and *A. cristatum* homoelogous group 5 and 6. In fact, specific sequences located on the distal chromosome regions (subtelomeres) might be hampering interspecific chromosome associations between these *Agropyron* and their homoeologous wheat chromosomes [43,44]. Nevertheless, translocations between wheat and *Agropyron* species have been previously achieved for transferring important traits, such as leaf and stem rust resistance genes from *A. elongatum* [45,46], wheat streak mosaic [47], and stem rust resistance genes from *A. intermedium* [48] to wheat background. Wheat *A. cristatum* Robertsonian translocations have also been obtained in the absence of the *Ph1* locus [32]. Actually, Robertsonian translocations are the most common chromosome rearrangements found between wheat and related species, both in the presence and in the absence of the *Ph1* locus [49,50], although other interstitial recombination events between wheat chromosomes and those from other relatives have been promoted in the *Ph1* mutant background [51].

Previous meiotic analyses in hybrids between wheat and the *Agropyron* species have shown high levels of chromosome pairing. For example, multivalent configurations were observed in hybrids between *T. aestivum* and tetraploid *A. cristatum* [30], and in the hybrid between *T. aestivum* and tetraploid *A. fragile*, leading to the conclusion that *A. fragile* has a genetic system that modified the *Ph1* gene activity [31]. Thus, interactions of genes from *Agropyron* with the *Ph1* system in wheat were considered the most plausible explanation for those multivalent configurations [31]. However, GISH analysis in the *T. tauschii-A. cristatum* (DDPP) amphiploid [17] revealed later that the high pairing observed was ascribable between different P chromosomes or different wheat chromosomes, but not between *Agropyron* and wheat chromosomes. The observed configurations between P chromosomes could be explained because of the segmental allopolyploid nature of tetraploid *A. cristatum*. Our results using GISH analysis support these observations as we did not find chromosome associations between *A. cristatum* and wheat chromosomes, either in the presence or in the absence of the *Ph1* locus. In addition, we observed wheat univalents that might suggest that the P genome carries genes affecting associations between wheat chromosomes themselves. Nevertheless, pairing between P and D genomes is possible and allosyndetic pairing was clearly visible in *T. tauschii-A. cristatum* associations in the DDPP amphiploid [17].

The plant material developed in this work might be very useful in studying the interactions between genes included in the P genome and those controlling chromosome associations in wheat. Previously, chromosome addition lines of *A. cristatum* in wheat in the presence of the *Ph1* locus were developed for chromosomes 1P, 2P, 3P, 4P, 5P, and 6P and the ditelosomic addition line for chromosome arms 2PS, 2PL, 4PS, 5PL, 6PS, and 6PL [15,52,53]. In fact, these available wheat-*A. cristatum* addition lines allowed the location of molecular markers and genes for several important agronomic traits on specific *A. cristatum* chromosomes and chromosome arms [8,9,36,42]. However, no ditelosomic addition or substitution lines were previously obtained for the chromosome 5PS arm. Furthermore, in this work, we have developed *A. cristatum* addition lines for the 5P and 6P chromosome in wheat background in the absence of the *Ph1* locus. The availability of CS *A. cristatum* 5PS ditelosomic lines as well as the *Agropyron* introgressions in the *Ph1* mutant background provided the opportunity to not only locate genes or markers on this chromosome arm, but also to go deeper into the knowledge of chromosome associations in the wheat background.

## 4. Materials and Methods

### 4.1. Plant Material

The plant material used in this study included common wheat (*Triticum aestivum* L.) Chinese Spring (CS), CS *ph1b* mutant [54], CS/*A. cristatum* disomic addition lines for 5P and 6P chromosomes, and ditelosomic 5PL, 6PS, and 6PL addition lines [15,52,53]. The CS/*A.cristatum* addition lines for the 5P and 6P chromosomes in the CS *ph1bph1b* genetic background were selected in descendance from the crosses between the CS/*A. cristatum* disomic addition lines for 5P and 6P chromosomes with the CS *ph1b* mutant [32].

### 4.2. DNA Marker Characterization

Genomic DNA was isolated from young frozen leaf tissue using the CTAB method [55]. Samples were stored at −20 °C until PCR amplification was carried out. The concentration of each sample was estimated using a nano-drop 1000 spectrophotometer (Thermo Scientific, Waltham, MA, USA).

Four conserved orthologous set (COS) markers named COS108, COS150, COS440, and COS580 [56] were used to identify short and long arms from 5P and 6P *A. cristatum* chromosomes, respectively, included in the addition lines. These COS markers, which have been previously transferred and determined their arm locations on *A. cristatum* chromosomes [9,37], were selected for being polymorphic between *A. cristatum* and wheat CS. In detail, each COS marker was specific for each short or long 5P and 6P chromosome arm. Primer sequences for these markers and annealing temperature (Ta) were previously given [9,37]. PCR was performed with 40 ng of template DNA in a 25 µL volume reaction mixture containing 5 µL of 1× PCR buffer 0.5 pM of each primer, 1.5 mM $MgCl_2$, 0.3 mM dNTPs, and 0.625U of Taq DNA polymerase (Promega, Madison, WI, USA). PCR conditions for the COS markers were as follows: 4 min at 94 °C, followed by 35 cycles of 45 s at 94 °C, 50 s at 58 °C, and 50 s at 72 °C. The PCR products were analysed in polyacrylamide gels (10% *w*/*v*, C: 2.67%) stained with ethidium bromide.

A PCR assay described by Wang et al. [57] was used to verify the presence of *Ph1*. Each 30 µL PCR contained 20 ng template DNA, 1× PCR buffer with $MgCl_2$, 0.25 mM dNTP, 0.17 µM primers, and 0.02U/µL of Taq DNA polymerase (Promega, Madison, WI, USA). The reaction was first denatured (94 °C/5min), and then subjected to 35 cycles of 94 °C/60 s, 51 °C/60 s, and 72 °C/60 s, followed by a final extension (72 °C/7min). The PCR products were electrophoretically separated in a 1% agarose gel and visualized by EtBr staining.

### 4.3. Somatic Cells Analyses

Chromosome spreads were prepared from root tip cells. Seeds were germinated on wet filter paper in the dark for 3 days at 4 °C, followed by a period of 24 h at 25 °C. Root tips from germinating seeds were excised and pretreated in ice water for 24 h and then fixed in a freshly prepared ethanol–acetic acid (3:1 *v*/*v*) and stored at 4 °C for at least 1 month. The plants were grown under a greenhouse held at 26 °C during the day and 18 °C during the night at a 16 h photoperiod.

The *in situ* hybridization protocol followed that described by Cabrera et al. [58]. The probe pAs1 isolated from *Aegilops tauschii* Coss. [59] was used to determine the D genome of wheat. Genomic DNA from *A. cristatum* was used as a probe to identify P genome chromosomes. The pAs1 probe was labelled with biotin-16-dUTP (corporate Roche). Total DNA of *A. cristatum* was labelled with biotin-16-dUTP (Boehringer Mannheim Biochemicals, Germany) or digoxigenin-11-dUTP (Roche Applied Science, Indianapolis, IN, USA) using nick translation. Chromosome preparations were hybridized with an *A. cristatum* genomic DNA probe or simultaneously with both pAs1 and *A. cristatum* genomic DNA probes. The final concentration of each probe was 5 ng/µL in the hybridization mix (50% formamide, 2 × SCC, 5 ng of each digoxigenin and biotin-labelled probes, 10% dextran sulfate, 0.14 µg of yeast tRNA, 0.1 µg of sonicated salmon sperm DNA, and 5 ng of glycogen. Posthybridization washes were conducted twice at 2 × SSC (5 min each) at 37 °C plus one extra wash in 1 × SSC at room temperature (RT). Biotin- and digoxigenin-labelled probes were detected with streptavidin-Cy3 conjugates (Sigma, St. Louis, MO, USA) and antidigoxigenin FITC antibodies (Roche Diagnostics, Meylan, France), respectively. The chromosomes were counterstained with 4′,6-diamidino-2-phenylindole (DAPI) and mounted in Vectashield

(Vector Laboratories, Burlingame, CA, USA). Hybridization signals were visualized using a Leica DMRB epifluorescence microscope and the images were captured with a Leica DFC 7000T camera equipped with an exposimeter spot Leica Wild MPS 52 and were processed with LEICA application suite v4.0 software (Leica, Wetzlar, Germany).

### 4.4. Meiotic Cells Analyses

GISH analysis allowed the visualization of the *A. cristatum* chromosome and their associations during meiosis in the wheat background in the presence and in the absence of the *Ph1* locus, as described previously [60]. Mature plants were used to collect spikes in meiosis, which were preserved in 100% ethanol–acetic acid 3:1 (*v*/*v*) until they were used to characterize chromosome associations. Chromosome spreads were prepared from pollen mother cells (PMCs) at meiosis. Anthers were macerated in a drop of 45% glacial acetic acid on ethanol-cleaned slides, squashed under a cover slip, and dipped in liquid nitrogen in order to fix the plant material on the slide. The cover slip was removed and the slides were air-dried and stored at 4 °C until used. Total genomic DNA from *A. cristatum* was labelled by nick translation with biotin-11-dUTP to be used as a probe for *in situ* hybridization experiments in PMCs. The conserved telomeric sequence from *A. thaliana* (AAATCCC) [61] was also labelled by nick translation with digoxigenin-11-dUTP to allow the visualization of the telomeres from wheat and *A. cristatum* chromosomes. Similar to somatic cells analysis, the final concentration of each probe was 5 ng/µL in the hybridization mix (50% formamide, 2 × SCC, 5 ng of each digoxigenin and biotin-labelled probes, 10% dextran sulfate, 0.14 µg of yeast tRNA, 0.1 µg of sonicated salmon sperm DNA, and 5 ng of glycogen). Posthybridization washes were equivalent to the ones performed for somatic cells analysis. Biotin-labelled and digoxigenin-labelled probes were detected with a streptavidin-Cy3 conjugate and antidigoxigenin-FITC, respectively. Chromosomes were counterstained with DAPI (4′,6-diamidino-2-phenylindole) and mounted in Vectashield (Vector Laboratories, Burlingame, CA, USA). Hybridization results were imagined using a Nikon Eclipse 80i epifluorescence microscope. Images were taken with a Nikon CCD camera using the Nikon 3.0 software (Nikon Instruments Europe BV, Amstelveen, The Netherlands) and processed with Photoshop 11.0.2 software for adjustment of brightness and contrast (Adobe Systems Inc., San Jose, CA, USA).

## 5. Conclusions

The development of *A. cristatum* 5P and 6P chromosomes in wheat in the *Ph1* mutant background revealed that chromosome associations between *A. cristatum* and wheat chromosomes are not performed during meiosis, even in the absence of the *Ph1* locus, although the *A. cristatum* genome might have a putative effect on homologous chromosome associations in wheat itself.

**Author Contributions:** A.C. and P.P. conceived and designed the study; Z.C. and C.P. performed the experiments; A.C. and P.P. analysed the data and wrote the paper. All authors have read and agreed to the published version of the manuscript.

**Funding:** This research was supported by grants RTI2018-093367-B100 and PID2019-103996RB-I00 from the Spanish Research Agency (Ministry of Science, Innovation and Universities).

**Institutional Review Board Statement:** Not applicable.

**Informed Consent Statement:** Not applicable.

**Acknowledgments:** The authors are very grateful to C. Calderón for her technical support. The authors deeply appreciate the comments from the independent reviewers during the revision of the manuscript.

**Conflicts of Interest:** The authors declare no conflict of interest.

## References


1. Dewey, D.R. The genomic system of classification as a guide to intergeneric hybridization with the perennial Triticeae. In *Gene Manipulation in Plant Improvement*; Gustafson, J.P., Ed.; Springer: Boston, MA, USA, 1984; pp. 209–279.
2. Asay, K.H. Breeding potentials in perennials Triticeae grasses. *Hereditas* **1992**, *116*, 167–173. [CrossRef]
3. Dewey, D.R. Salt tolerance of twenty-five strains of *Agropyron*. *Crop Sci.* **1960**, *52*, 631–635. [CrossRef]
4. Dewey, D.R. Breeding crested wheatgrass for salt tolerance. *Crop Sci.* **1962**, *2*, 403–407. [CrossRef]
5. Meng, L.; Guo, Q.; Mao, P.; Tian, X. Accumulation and tolerance characteristics of zinc in *Agropyron cristatum* plants exposed to zinc-contaminated soil. *Bull. Environ. Contam. Toxicol.* **2013**, *91*, 298–3016. [CrossRef] [PubMed]
6. Guo, Q.; Meng, L.; Mao, P.; Tian, X. An assessment of *Agropyron cristatum* tolerance to cadmium contaminated soil. *Biol. Plant.* **2014**, *58*, 174–178. [CrossRef]
7. Wang, R.R. Agropyron and *Psathyrostachys*. In *Wild Crop Relatives: Genomic and Breeding Resources, Cereals*; Kole, C., Ed.; Springer: Berlin/Heidelberg, Germany, 2011; Chapter 2; pp. 77–108.
8. Ochoa, V.; Madrid, E.; Said, M.; Rubiales, D.; Cabrera, A. Molecular and cytogenetic characterization of a common wheat-*Agropyron cristatum* chromosome translocation conferring resistance to leaf rust. *Euphytica* **2015**, *201*, 89–95. [CrossRef]
9. Copete, A.; Cabrera, A. Chromosomal location of genes for resistance to powdery mildew in *Agropyron cristatum* and mapping of conserved orthologous set molecular markers. *Euphytica* **2017**, *213*, 189. [CrossRef]
10. Li, H.; Jiang, B.; Wang, J.; Lu, Y.; Zhang, J.; Pan, C.; Yang, X.; Li, X.; Liu, W.; Li, L. Mapping of novel powdery mildew resistance gene(s) from *Agropyron cristatum* chromosome 2P. *Theor. Appl. Genet.* **2017**, *130*, 109–121. [CrossRef]
11. Zhang, Z.; Song, L.; Han, H.; Zhou, S.; Zhang, J.; Yang, X.; Li, X.; Liu, W.; Li, L. Physical localization of a locus from *Agropyron cristatum* conferring resistance to stripe rust in common wheat. *Int. J. Mol. Sci.* **2017**, *18*, 2403. [CrossRef]
12. McGuire, P.E.; Dvorák, J. High salt-tolerance potential in wheatgrasses. *Crop Sci.* **1981**, *21*, 702–705. [CrossRef]
13. Bayat, H.; Nemati, H.; Tehranifar, A.; Gazanchian, A. Screening different crested wheatgrass (*Agropyron cristatum* (L.) Gaertner.) accessions for drought stress tolerance. *Arch. Agron. Soil Sci.* **2016**, *62*, 769–780. [CrossRef]
14. Zhang, J.; Zhang, J.; Liu, W.; Han, H.; Lu, Y.; Yang, X.; Li, X.; Li, L. Introgression of *Agropyron cristatum* 6P chromosome segment into common wheat for enhanced thousand-grain weight and spike length. *Theor. Appl. Genet.* **2015**, *128*, 1827–1837. [CrossRef]
15. Chen, Q.; Jahier, J.; Cauderon, Y. Production and cytogenetic analysis of BC1, BC2, and BC3 progenies of an intergeneric hybrid between *Triticum aestivum* (L.) Thell. and tetraploid *Agropyron cristatum* (L.) Gaertn. *Theor. Appl. Genet.* **1992**, *84*, 698–703. [CrossRef] [PubMed]
16. Martín, A.; Rubiales, D.; Cabrera, A. Meiotic pairing in a trigeneric hybrid *Triticum tauschii-Agropyron cristatum-Hordeum chilense*. *Hereditas* **1998**, *129*, 113–118. [CrossRef]
17. Martín, A.; Cabrera, A.; Esteban, E.; Hernández, P.; Ramírez, M.C.; Rubiales, D. A fertile amphiploid between diploid wheat (*Triticum tauschii*) and crested wheatgrass (*Agropyron cristatum*). *Genome* **1999**, *42*, 519–524. [CrossRef]
18. Soliman, M.H.; Rubiales, D.; Cabrera, A. A Fertile Amphiploid between durum wheat (*Triticum turgidum*) and the Agroticum amphiploid (*Agropyron cristatum* x *T. tauschii*). *Hereditas* **2001**, *135*, 183–186. [CrossRef] [PubMed]
19. Soliman, M.H.; Cabrera, A.; Sillero, J.C.; Rubiales, D. Genomic constitution and expression of disease resistance in *Agropyron cristatum* x durum wheat derivatives. *Breed. Sci.* **2007**, *57*, 17–21. [CrossRef]
20. Riley, R.; Chapman, V. Genetic control of the cytologically diploid behaviour of hexaploid wheat. *Nature* **1958**, *182*, 713–715. [CrossRef]
21. Sears, E.R. Genetic control of chromosome pairing in wheat. *Ann. Rev. Genet.* **1976**, *10*, 31–51. [CrossRef]
22. Sears, E.R. Transfer of alien genetic material to wheat. In *Wheat Science-Today and Tomorrow*; Evans, L.T., Peacock, W.J., Eds.; Cambridge University Press: Cambridge, UK, 1981; pp. 75–89.
23. Mello-Sampayo, T. Genetic regulation of meitic chromosome pairing by chromosome 3D of *Triticum aestivum*. *Nat. New Biol.* **1971**, *230*, 22–23. [CrossRef]
24. Qui, L.; Friebe, B.; Zhang, P.; Gill, B.S. Homoeologous recombination, chromosome engineering and crop improvement. *Chromosome Res.* **2007**, *15*, 3–19. [CrossRef]
25. Dvorak, J.; Deal, K.R.; Luo, M.C. Discovery and mapping of wheat *Ph1* suppressors. *Genetics* **2006**, *174*, 17–27. [CrossRef]
26. Dover, G.A.; Riley, R. Variation at two loci affecting homoeologous meiotic pairing in *Triticum aestivum* × *Aegilops mutica* hybrids. *Nat. New Biol.* **1972**, *235*, 61–62. [CrossRef]
27. Riley, R.; Chapman, V.; Miller, T. The determination of meiotic chromosome pairing. In Proceedings of the 4th International Wheat Genetics Symposium, University of Missouri, Colombia, MO, USA, 6–11 August 1973; Sears, E.R., Sears, L.M.S., Eds.; pp. 731–738.
28. Dvorak, J. Chromosomal distribution of genes in diploid *Elytrigia elongata* that promote or suppress pairing of wheat homoeologous chromosomes. *Genome* **1987**, *29*, 34–40. [CrossRef]
29. Liu, W.; Rouse, M.; Friebe, B.; Jin, Y.; Gill, B.S.; Pumphrey, M.O. Discovery and molecular mapping of a new gene conferring resistance to stem rust, Sr53, derived from *Aegilops geniculata* and characterization of spontaneous translocation stocks with reduced alien chromatin. *Chromosom. Res.* **2011**, *19*, 669–682. [CrossRef]
30. Jahuar, P.P. Chromosome pairing in hybrids between hexaploid bread wheat and tetraploid crested wheatgrass (*Agropyron cristatum*). *Hereditas* **1992**, *116*, 107–109. [CrossRef]
31. Ahmad, F.; Comeau, A. A new intergeneric hybrid between *Triticum aestivum* L., and *Agropyron fragile* (Roth) Candary: Variation in *A.fragile* for suppression of the wheat *Ph*-locus activity. *Plant Breed.* **1991**, *106*, 275–283. [CrossRef]

32. Copete-Parada, A.; Palomino, C.; Cabrera, A. Development and characterization of wheat-*Agropyron cristatum* introgression lines induced by gametocidal genes and *Ph1* mutant. *Agronomy* **2021**, *11*, 277. [CrossRef]
33. Che, Y.H.; Li, H.J.; Yang, Y.P.; Yang, X.M.; Li, X.Q.; Li, L.H. On the use of SSR markers for the genetic characterization of the *Agropyron cristatum* (L.) Gaertn. in northern China. *Genet. Resour. Crop Evol.* **2008**, *55*, 389–396. [CrossRef]
34. Copete, A.; Moreno, R.; Cabrera, A. Characterization of a world collection of *Agropyron cristatum* accessions. *Genet. Resour. Crop Evol.* **2018**, *65*, 1455–1469. [CrossRef]
35. Dong, Y.S.; Zhou, R.H.; Xu, S.J.; Li, L.H.; Cauderon, Y.; Wang, R.R.-C. Desirable characteristics in perennial Triticeae collected in China for wheat improvement. *Hereditas* **1992**, *116*, 175–178. [CrossRef]
36. Cabrera, A.; Castellano, L.; Recio, R.; Alvarez, J.B. Chromosomal location and molecular characterization of three grain hardness genes in *Agropyron cristatum*. *Euphytica* **2019**, *215*, 165. [CrossRef]
37. Said, M.; Copete, A.; Gaál, E.; Molnár, I.; Cabrera, A.; Doležel, J.; Vrána, J. Uncovering macrosyntenic relationships between tetraploid *Agropyron cristatum* and bread wheat genomes using COS markers. *Theor. Appl. Genet.* **2019**, *132*, 2881–2898. [CrossRef] [PubMed]
38. Linc, G.; Gaál, E.; Molnár, I.; Icsó, D.; Badaeva, E.; Molnár-Láng, M. Molecular cytogenetic (FISH) and genome analysis of diploid wheatgrasses and their phylogenetic relationship. *PLoS ONE* **2017**, *12*, e0173623. [CrossRef]
39. Said, M.; Hřibová, E.; Danilova, T.V.; Karafiátová, M.; Čížková, J.; Friebe, B.; Doležel, J.; Gill, B.S.; Vrána, J. The *Agropyron cristatum* karyotype, chromosome structure and cross-genome homeology as revealed by fluorescence *in situ* hybridization with tandem repeats and wheat single-gene probes. *Theor. Appl. Genet.* **2018**, *131*, 2213–2227. [CrossRef] [PubMed]
40. Zhou, S.; Yan, B.; Li, F.; Zhang, J.; Zhang, J.; Ma, H.; Liu, W.; Lu, Y.; Yang, X.; Liu, X.; et al. RNA-Seq analysis provides first insights into the phylogenetic relationships and interespecific variation between *Agropyron cristatum* and wheat. *Front. Plant Sci.* **2017**, *8*, 1644. [CrossRef]
41. Zhou, S.; Zhang, J.; Che, Y.; Liu, W.; Lu, Y.; Yang, X.; Li, X.; Jia, J.; Liu, X.; Li, L. Construction of *Agropyron* Gaertn. Genetic linkage maps using a wheat 660 K SNP array reveals a homoeologous relationship with the wheat genome. *Plant Biotechnol. J.* **2018**, *16*, 818–827. [CrossRef]
42. Cabrera, A.; Copete-Parada, A.; Madrid, E. Cloning and characterization of a putative orthologue of the wheat vernalization (*VRN1*) gene in perennial wheatgrass (*Agropyron cristatum*). *Plant Breed.* **2020**, *139*, 1290–1298. [CrossRef]
43. Calderón, M.D.C.; Rey, M.D.; Cabrera, A.; Prieto, P. The subtelomeric region is important for chromosome recognition and pairing during meiosis. *Sci. Rep.* **2014**, *4*, 6488. [CrossRef]
44. Aguilar, M.; Prieto, P. Sequence analysis of wheat subtelomeres reveals a high polymorphism among homoeologous chromosomes. *Plant Genome* **2020**, *13*, e20065. [CrossRef]
45. Sears, E.R. *Agropyron-wheat* transfers induced by homoeologous pairing. In Proceedings of the 4th International Wheat Genetic Symposium, Columbia, MO, USA, 6–11 August 1973; Sears, E.R., Sears, L.M.S., Eds.; pp. 191–199.
46. Sears, E.R. Analysis of *wheat-Agropyron* recombinant chromosomes. In Proceedings of the 8th Eucarpia Congress, Madrid, Spain, 23–25 May 1977; pp. 63–72.
47. Wang, R.C.; Liang, G.H.; Heyne, E.G. Effectiveness of *ph* gene in inducing homoeologous chromosome pairing *Agrotricum*. *Theor. Appl. Genet.* **1977**, *51*, 139–142. [CrossRef]
48. Friebe, B.; Zeller, F.J.; Mukai, Y.; Forster, B.P.; Bartos, P.; Mclntosh, R.A. Characterization of rest-resistant *wheat-Agropyron intermedium* derivatives by C-banding, *in situ* hybridization and isozyme analysis. *Theor. Appl. Genet.* **1992**, *83*, 775–782. [CrossRef]
49. Prieto, P.; Ramírez, C.; Cabrera, A.; Ballesteros, J.; Martín, A. Development and cytogenetic characterisation of a double goat grass-barley chromosome substitution in tritordeum. *Euphytica* **2006**, *147*, 337–342. [CrossRef]
50. Rey, M.D.; Calderón, M.C.; Rodrigo, M.J.; Zacarías, L.; Alós, E.; Prieto, P. Novel bread wheat lines enriched in carotenoids carrying *Hordeum chilense* chromosome arms in the *ph1b* background. *PLoS ONE* **2015**, *10*, e0134598. [CrossRef]
51. Rey, M.D.; Calderón, M.C.; Prieto, P. The use of the *ph1b* mutant to induce recombination between the chromosomes of wheat and barley. *Front. Plant Sci.* **2015**, *6*, 160. [CrossRef]
52. Chen, Q.; Lu, Y.L.; Jahier, J.; Bernard, M. Identification of wheat-*Agropyron cristatum* monosomic addition lines by RFLP analysis using a set of assigned wheat DNA probes. *Theor. Appl. Genet.* **1994**, *89*, 70–75. [CrossRef] [PubMed]
53. Jubault, M.; Tanguy, A.M.; Abélard, P.; Coriton, O.; Dusautoir, J.C.; Jahier, J. Attempts to induce homoeologous pairing between wheat and *Agropyron cristatum* genomes. *Genome* **2006**, *49*, 190–193. [CrossRef]
54. Sears, E.R. An induced mutant with homoeologous pairing in common wheat. *Can. J. Genet. Cytol.* **1977**, *19*, 585–593. [CrossRef]
55. Murray, M.G.; Thompson, W.F. Rapid isolation of high molecular weight plant DNA. *Nucleic Acids Res.* **1980**, *8*, 4321–4326. [CrossRef] [PubMed]
56. Quraishi, U.M.; Abrouk, M.; Bolot, S.; Pont, C.; Throude, M.; Guilhot, N.; Confolent, C.; Bortolini, F.; Praud, S.; Murigneux, A.; et al. Genomics in cereals: From genome-wide conserved orthologous set (COS) sequences to candidate genes for trait dissection. *Funct. Integr. Genom.* **2009**, *9*, 473–484. [CrossRef]
57. Wang, X.; Lai, J.; Liu, G.; Chen, F. Development of a Scar marker for the *Ph1* locus in common wheat and its application. *Crop Sci.* **2002**, *42*, 1365–1368. [CrossRef]
58. Cabrera, A.; Martín, A.; Barro, F. In-situ comparative mapping (ISCM) of *Glu-1* loci in *Triticum* and *Hordeum*. *Chromosome Res.* **2002**, *10*, 49–54. [CrossRef] [PubMed]

59. Rayburn, A.L.; Gill, B.S. Isolation of a D-genome specific repeated DNA sequence from *Aegilops squarrosa*. *Plant Mol. Biol. Rep.* **1986**, *4*, 102–109. [CrossRef]
60. Calderón, M.C.; Rey, M.D.; Martín, A.; Prieto, P. Homoeologous chromosomes from two *Hordeum* species can recognize and associate during meiosis in wheat in the presence of the *Ph1* locus. *Front. Plant Sci.* **2018**, *9*, 585. [CrossRef] [PubMed]
61. Richards, E.J.; Ausubel, F.M. Isolation of a higher eukaryotic telomere from *Arabidopsis thaliana*. *Cell* **1988**, *53*, 127–136. [CrossRef]

*plants*

MDPI

*Article*

# Powdery Mildew Resistance Genes in Single-Plant Progenies Derived from Accessions of a Winter Barley Core Collection

**Antonín Dreiseitl [1,*] and Zdeněk Nesvadba [2]**

1 Department of Integrated Plant Protection, Agrotest Fyto Ltd., Havlíčkova 2787, CZ-767 01 Kroměříž, Czech Republic
2 Gene Bank, Crop Research Institute, Drnovská 507, CZ-161 06 Praha 6-Ruzyně, Czech Republic; nesvadba@vurv.cz
* Correspondence: dreiseitl@vukrom.cz; Tel.: +420-577-317-139

**Abstract:** The main problems of crop gene banks comprise heterogeneity of accessions, resulting from mechanical admixtures or out-crossing during their multiplication, and especially the mislabeling of accessions. These discrepancies can adversely affect the results of many expensive research and breeding projects that are based on the use of gene bank resources. To tackle these problems, 860 single-plant progenies (SPPs) of 172 accessions of the Czech winter barley core collection were grown and tested with a set of 53 isolates representing the global virulence/avirulence diversity of powdery mildew. Seventy-one resistance phenotypes encompassed the diversity of known specific resistances and their combinations. Based on testing groups of five SPPs, 94 accessions had one phenotype found in all five SPPs (homogeneous accessions), whereas in 78 accessions (45.3%) more than one phenotype was identified (heterogeneous accessions). In three varieties, specific resistances against the whole set of isolates were detected, but due to high adaptability of the pathogen, they are not recommended for breeding resistant cultivars. Selected SPPs were integrated in the gene bank and are now a reliable source of genotypically pure seed with defined powdery mildew resistance genes that can be used by breeders and researchers. The results obtained can be used to verify authenticity of accession genotype and pedigree, particularly for older varieties for which no other original criteria are available.

**Keywords:** *Blumeria graminis* f. sp. *hordei*; gene bank; *Hordeum vulgare*; isolates of the pathogen; infection response arrays; resistance gene postulation; winter barley core collection

**Citation:** Dreiseitl, A.; Nesvadba, Z. Powdery Mildew Resistance Genes in Single-Plant Progenies Derived from Accessions of a Winter Barley Core Collection. *Plants* **2021**, *10*, 1988. https://doi.org/10.3390/plants10101988

Academic Editors: Ranjith Pathirana and Francesco Carimi

Received: 28 July 2021
Accepted: 22 September 2021
Published: 23 September 2021



## 1. Introduction

Barley (*Hordeum vulgare* L.) is an important cereal crop used mainly as human food, feed for domestic animals and for malt products. Powdery mildew, caused by the fungus *Blumeria graminis* (D.C.) Golovin ex Speer, f. sp. *hordei* Em. Marchal (*Bgh*), is a worldwide disease that can cause frequent epidemics on barley particularly in Central [1] and Northwest Europe [2,3]. To reduce the disease and its effect on yield and quality, genetic resistance is an efficient and environmentally acceptable way [4,5] and has played an essential part in disease management for about seven decades [6].

Breeding barley resistant to powdery mildew began in Europe in the interwar period [7]. The first resistance genes used in commercial varieties included *Mlg* in spring and *Mlh* and *Mlra* in winter cultivars. The first cultivar with a gene located at the Mla locus was Vogelsanger Gold (*Mla6*) derived from wild barley (*Hordeum vulgare* subsp. *spontaneum*) and many subsequent cultivars were bred with other new resistances, mainly alleles of Mla locus [6,8]. Later, it was found that even European landraces carry some specific resistance genes completely ineffective in the field [9].

*Bgh* is a highly adaptable pathogen, and its Central European population is extremely diverse because it reflects a great diversity of resistances used in breeding programmes [10]. More than 70 resistance genes [9] have been described and most of them are present in cultivars either singly or in various combinations, whereas cultivars without a resistance gene make up only a small proportion.

Recently different resistance genes at Mla locus and *mlo* gene were molecularly isolated [11]. Isolates selected from the global population of the pathogen can reveal the detailed genotypic diversity of the host and this knowledge was used to characterize cultivars including verification of their homogeneity and authenticity of their genotype and pedigree [12,13]. This is particularly needed for older accessions for which there is little to no genomic information available, like molecular marker profiles generated on the original seed stocks. Hence, molecular technology alone cannot be used for confirming their identity without additional testing. If new methods are used for characterizing accessions of old varieties without confirmation their seed authenticity and purity, it can lead to a series of wrong results, loss of the original cultivar identity and decreased genetic diversity in gene banks.

This contribution describes the identification of the powdery mildew resistance genes in single-plant progenies (SPPs) of accessions of the core collection (CC) of the Czech winter barley gene bank. Results from a similar project have already been reported [8], but more than 85% of the accessions showed heterogeneity in their resistance and identification of resistances was limited. Therefore, the aims of this study were i) to grow a technically-manageable group of homogeneous samples (lines) in order to record the most frequent genotypes present in each accession; ii) to identify resistance genes in each of them, and iii) to evaluate their authenticity. Together, this aims at providing breeders and researchers with a large set of single-line barley germplasm with well-defined powdery mildew resistance genes.

## 2. Results

All 860 SPPs derived from 172 gene bank accessions of winter barley were characterized by homogeneous infection response arrays (IRAs) confirming the genotypic uniformity of these single-line samples. From the tests 71 IRAs were recorded that represented the phenotypes of specific resistance genes and their combinations (Table 1). Based on testing groups of five SPPs, 94 accessions had one identical phenotype found in all SPPs (homogeneous accessions) whereas, 78 accessions (45.3%) were heterogeneous in which two, three, four and five genotypes were found in 46, 19, 10 and 3 accessions, respectively. Based on the name of each accession that was tested and IRAs, 298 variants were found. The resistance genes of all 860 SPPs are presented in Table S1.

**Table 1.** Infection response arrays produced by 11 *Blumeria graminis* f. sp. *hordei* isolates on 71 barley genotypes and their powdery mildew resistance genes.

| *Ml* Gene(s) | Race I | J-462 | EA30 | PF512 | C-132 | 3-33 | 65 | GH | 54 | Z-6 | E-6 |
|---|---|---|---|---|---|---|---|---|---|---|---|
| none | 4 [1] | 4 | 4 | 4 | 4 | 4 | 4 | 4 | 4 | 4 | 4 |
| *a3* | 0 | 4 | 0 | 0 | 4 | 0 | 0 | 0 | 0 | 4 | 0 |
| *a6* | 0 | 4 | 2 | 4 | 4 | 0 | 0 | 0 | 0 | 4 | 0 |
| *a6, aLo* | 0 | 0 | 2 | 4 | 4 | 0 | 0 | 0 | 0 | 4 | 0 |
| *a6, Dt6, g, h* | 0 | 4 | 0 | 4 | 1–2 | 0 | 0 | 0 | 0 | 2 | 0 |
| *a6, h* | 0 | 4 | 2 | 4 | 1–2 | 0 | 0 | 0 | 0 | 4 | 0 |
| *a6, h, Lu, ra, Ru2* | 0 | 4 | 0–1 | 2–3 | 1–2 | 0 | 0 | 0 | 0 | 1–2 | 0 |
| *a6, h, ra* | 0 | 4 | 0–1 | 4 | 1–2 | 0 | 0 | 0 | 0 | 4 | 0 |
| *a6, IM9* | 0 | 0 | 0 | 4 | 0 | 0 | 0 | 0 | 0 | 0 | 0 |
| *a6, ra* | 0 | 4 | 0–1 | 4 | 4 | 0 | 0 | 0 | 0 | 4 | 0 |
| *a7* | 0 | 0 | 1–2 | 4 | 4 | 0 | 0 | 1–2 | 1–2 | 4 | 4 |
| *a7, h* | 0 | 0 | 1–2 | 4 | 1–2 | 0 | 0 | 1–2 | 1–2 | 4 | 4 |
| *a8* | 0 | 4 | 4 | 4 | 4 | 4 | 4 | 4 | 4 | 4 | 4 |
| *a8, Dr2* | 0 | 4 | 4 | 4 | 4 | 4 | 4 | 2–3 | 4 | 4 | 4 |
| *a8, Dr2, ra* | 0 | 4 | 0–1 | 4 | 4 | 4 | 0–1 | 2–3 | 4 | 4 | 4 |
| *a8, Dr2, ra, VIR* | 0 | 4 | 0–1 | 1 | 4 | 4 | 0–1 | 2–3 | 4 | 4 | 4 |

**Table 1.** *Cont.*

| *Ml* Gene(s) | Race I | J-462 | EA30 | PF512 | C-132 | 3-33 | 65 | GH | 54 | Z-6 | E-6 |
|---|---|---|---|---|---|---|---|---|---|---|---|
| *a8, h* | 0 | 4 | 4 | 4 | 1–2 | 1–2 | 4 | 1–2 | 1–2 | 4 | 4 |
| *a8, h, Ln, ra* | 0 | 4 | 0–1 | 0–1 | 1–2 | 1–2 | 0–1 | 1–2 | 1–2 | 0–1 | 0–1 |
| *a8, h, Lu, Ru2* | 0 | 4 | 4 | 2–3 | 1–2 | 1–2 | 1–2 | 1–2 | 1–2 | 1–2 | 4 |
| *a8, h, ra* | 0 | 4 | 0–1 | 4 | 1–2 | 1–2 | 0–1 | 1–2 | 1–2 | 4 | 4 |
| *a8, h, ra, Ru2* | 0 | 4 | 0–1 | 2–3 | 1–2 | 1–2 | 0–1 | 1–2 | 1–2 | 2–3 | 4 |
| *a8, He2* | 0 | 4 | 4 | 4 | 4 | 4 | 4 | 4 | 2–3 | 4 | 4 |
| *a8, Lu, ra* | 0 | 4 | 0–1 | 4 | 4 | 4 | 0–1 | 1–2 | 4 | 1–2 | 4 |
| *a8, ra* | 0 | 4 | 0–1 | 4 | 4 | 4 | 0–1 | 4 | 4 | 4 | 4 |
| *a8, Ru2* | 0 | 4 | 4 | 2–3 | 4 | 4 | 2–3 | 2–3 | 2–3 | 2–3 | 4 |
| *a8, VIR* | 0 | 4 | 4 | 1 | 4 | 4 | 4 | 4 | 4 | 4 | 4 |
| *a12* | 1 | 4 | 4 | 4 | 4 | 1 | 1 | 1 | 1 | 4 | 4 |
| *a12, aLo, g, Lu* | 0 | 0 | 0 | 4 | 4 | 0 | 0 | 1 | 1 | 1–2 | 4 |
| *a12, g* | 0 | 4 | 0 | 4 | 4 | 0 | 0 | 1 | 1 | 4 | 4 |
| *a13* | 0 | 0 | 0 | 0 | 4 | 0 | 0 | 0 | 0 | 0 | 4 |
| *aLo* | 0 | 0 | 4 | 4 | 4 | 4 | 4 | 4 | 4 | 4 | 4 |
| *aLo, Dr2* | 0 | 0 | 4 | 4 | 4 | 4 | 4 | 2 | 4 | 4 | 4 |
| *aLo, Dr2, ra* | 0 | 0 | 0–1 | 4 | 4 | 4 | 0–1 | 2 | 4 | 4 | 4 |
| *aLo, g* | 0 | 0 | 0 | 4 | 4 | 0 | 0 | 4 | 4 | 4 | 4 |
| *aLo, h* | 0 | 0 | 4 | 4 | 1–2 | 1–2 | 4 | 1–2 | 1–2 | 4 | 4 |
| *aLo, h, Lu, ra* | 0 | 0 | 0–1 | 4 | 1–2 | 1–2 | 0–1 | 1–2 | 1–2 | 1–2 | 4 |
| *aLo, h, Lu, ra, Ru2* | 0 | 0 | 0–1 | 2–3 | 1–2 | 1–2 | 0–1 | 1–2 | 1–2 | 1–2 | 4 |
| *aLo, h, ra* | 0 | 0 | 0–1 | 4 | 1–2 | 1–2 | 0–1 | 1–2 | 1–2 | 4 | 4 |
| *aLo, Lu* | 0 | 0 | 4 | 4 | 4 | 4 | 1–2 | 1–2 | 4 | 1–2 | 4 |
| *aLo, Lu, ra* | 0 | 0 | 0–1 | 4 | 4 | 4 | 0–1 | 1–2 | 4 | 1–2 | 4 |
| *aLo, Lu, ra, Ru2* | 0 | 0 | 0–1 | 2–3 | 4 | 4 | 0–1 | 1–2 | 2–3 | 1–2 | 4 |
| *aLo, Lu, Ru2* | 0 | 0 | 4 | 2–3 | 4 | 4 | 1–2 | 1–2 | 2–3 | 1–2 | 4 |
| *aLo, ra* | 0 | 0 | 0–1 | 4 | 4 | 4 | 0–1 | 4 | 4 | 4 | 4 |
| *aLo, ra, Ru2* | 0 | 0 | 0–1 | 2–3 | 4 | 4 | 0–1 | 4 | 2–3 | 2–3 | 4 |
| *aLo, Ru2* | 0 | 0 | 4 | 2–3 | 4 | 4 | 2–3 | 2–3 | 2–3 | 2–3 | 4 |
| *aLo, VIR* | 0 | 0 | 4 | 1 | 4 | 4 | 4 | 4 | 4 | 4 | 4 |
| *at, h* | 2 | 4 | 2 | 2 | 1–2 | 2 | 2 | 1–2 | 1–2 | 1–2 | 2 |
| *Ch* | 2 | 4 | 4 | 4 | 4 | 4 | 4 | 4 | 4 | 4 | 4 |
| *Ch, Dr2* | 2 | 4 | 4 | 4 | 4 | 4 | 4 | 2 | 4 | 4 | 4 |
| *Ch, Dr2, Lu, Ru2* | 2 | 4 | 4 | 2–3 | 4 | 4 | 1–2 | 2 | 2–3 | 1–2 | 4 |
| *Ch, Dr2, ra* | 2 | 4 | 0–1 | 4 | 4 | 4 | 0–1 | 2 | 4 | 4 | 4 |
| *Ch, Dr2, ra, VIR* | 2 | 4 | 0–1 | 1 | 4 | 4 | 0–1 | 2 | 4 | 4 | 4 |
| *Ch, h, ra* | 2 | 4 | 0–1 | 4 | 1–2 | 1–2 | 0–1 | 1–2 | 1–2 | 4 | 4 |
| *Ch, Lu, ra* | 2 | 4 | 0–1 | 4 | 4 | 4 | 0–1 | 1–2 | 4 | 1–2 | 4 |
| *Ch, ra* | 2 | 4 | 0–1 | 4 | 4 | 4 | 0–1 | 4 | 4 | 4 | 4 |
| *Ch, ra, VIR* | 2 | 4 | 0–1 | 1 | 4 | 4 | 0–1 | 4 | 4 | 4 | 4 |
| *Ch, Ru2* | 2 | 4 | 4 | 2–3 | 4 | 4 | 2–3 | 2–3 | 2–3 | 2–3 | 4 |
| *Dr2, ra* | 4 | 4 | 0–1 | 4 | 4 | 4 | 0–1 | 2 | 4 | 4 | 4 |
| *Dr2, ra, VIR* | 4 | 4 | 0–1 | 1 | 4 | 4 | 0–1 | 2 | 4 | 4 | 4 |
| *g* | 0 | 4 | 0 | 4 | 4 | 0 | 0 | 4 | 4 | 4 | 4 |
| *g, Ln* | 0 | 4 | 0 | 0–1 | 4 | 0 | 0 | 4 | 0–1 | 0–1 | 0–1 |
| *h* | 4 | 4 | 4 | 4 | 1–2 | 1–2 | 4 | 1–2 | 1–2 | 4 | 4 |
| *h, Lu, ra* | 4 | 4 | 0–1 | 4 | 1–2 | 1–2 | 0–1 | 1–2 | 1–2 | 1–2 | 4 |
| *h, ra* | 4 | 4 | 0–1 | 4 | 1–2 | 1–2 | 0–1 | 1–2 | 1–2 | 4 | 4 |
| *IM9, St* | 0 | 0 | 0 | 0 | 0 | 0 | 0 | 0 | 0 | 0 | 4 |
| *La, ra* | 4 | 4 | 0–1 | 4 | 4 | 2–3 | 0–1 | 2–3 | 4 | 4 | 4 |
| *Lu, Ru2* | 4 | 4 | 4 | 2–3 | 4 | 4 | 1–2 | 1–2 | 2–3 | 1–2 | 4 |
| *ra* | 4 | 4 | 0–1 | 4 | 4 | 4 | 0–1 | 4 | 4 | 4 | 4 |
| *Ru2* | 4 | 4 | 4 | 2–3 | 4 | 4 | 2–3 | 2–3 | 2–3 | 2–3 | 4 |
| *VIR* | 4 | 4 | 4 | 1 | 3 | 4 | 4 | 3 | 3 | 4 | 4 |
| *Wo* | (2) [2] | (3) | (3) | (3) | (3) | (3) | (3) | (3) | (3) | (3) | (3) |

[1] Phenotypes (infection responses) of host-pathogen interactions evaluated according to Torp et al. [14], where 0 = resistant and 4 = susceptible. [2] Parentheses indicate smaller number of colonies.

Twenty-three known *Ml* resistance genes were identified in 761 SPPs (Table 2), the most frequent being *aLo*, *ra*, *a8*, *h* and *Ch* in 237, 220, 154, 139 and 112 SPPs, respectively. Some genes, especially *a8*, *Ch*, Dr2 and *ra*, may also be present in other varieties because their phenotype is masked by commonly occurring resistance genes such as *g*, *a6*, *a7*, *a12* or *a13*. Conversely, the least frequent were *He2* found in two SPPs, *Ln* in six SPPs of two accessions, *a3* in seven SPPs of three accessions, and *at*, *Dt6*, *St*, and *Wo* each present in all five SPPs of Local (Merkez-Kaza), Duet, Traminer and Wong, respectively.

**Table 2.** Number of specific resistance genes found in 860 single-plant progenies derived from 172 winter barley gene bank accessions.

| *Ml* Genes | Number | *Ml* Genes | Number |
|---|---|---|---|
| *a3* | 7 | *IM9* | 10 |
| *a6* | 72 | *La* | 9 |
| *a7* | 25 | *Ln* | 6 |
| *a8* | 154 | *Lu* | 92 |
| *a12* | 14 | *ra* | 220 |
| *a13* | 21 | *Ru2* | 73 |
| *aLo* | 237 | *St* | 5 |
| *at* | 5 | *VIR* | 23 |
| *Ch* | 112 | *Wo* | 5 |
| *Dr2* | 89 | Sum | 1363 |
| *Dt6* | 5 | | |
| *g* | 38 | Effective (*e*) | 20 |
| *h* | 139 | Unknown (*u*) | 28 |
| *He2* | 2 | *none* | 51 |

In total, 1363 known genes were recorded in 761 SPPs with identified resistances equivalent to an average of 1.79 specific genes per SPP. In 525 of them (69.0%) genes were located at the Mla locus. IRAs of 28 SPPs did not correspond with the IRAs of the reported resistances and were, therefore, designated as unknown (*u*), 20 SPPs relating mostly to Bonita, Marconee and Mc Nair 601 were resistant to all isolates (*e*) and conversely in 51 SPPs no resistance gene was found when in 10 varieties (Bankuti 14, Bordia, Dagestanskij-Samuricum 293, Krakovski, Krasnodarskij 2929, Krusevacki, Nakaizumi Zairai, Opolski 152, Stupicky dvourady and Zalarinec) susceptible SPPs with no detected resistance gene (designated 'none' in Table S1) predominated.

## 3. Discussion

The resistance of 19 of the 172 accessions is presented in the catalogue of European varieties [6], but only two of which (Frolic and Perga) had identical resistance to those reported here and only *Mla7* was found in Marinka whereas *Mla7* and *Mlg* are listed in the catalogue for this cultivar.

In nine accessions (Breustedts Atlas, Breustedts Schladener I, Carsten Zweizeilige, Dana, Eckendorfer Glatta, Engelens Dea, Fimbull II, Hauters Wintergerste and Strengs Dura), there were more resistance genes in addition to those published and reflects the high resolution of the large set of carefully selected isolates used herein. An example of the refinement of earlier results is *MlaLo*, the most frequent gene found here, including the first five out of nine accessions mentioned above. This gene was not known until recently. Formerly [14,15], RT0 was found in some varieties after testing with the Japanese isolate Race I [16], which is typical for *Mla8* often present in spring varieties [17]. The same response was subsequently detected and is typical for the newly-discovered gene *MlaLo* [18], which is allelic to *Mla8* [19] and, in contrast to *Mla8*, is characteristic of winter barley.

In the previous paper [8], approximately 50 plants grown from seeds of an accession stored in the gene bank were used for resistance tests and heterogeneity was found in 147 out of 172 accessions of the core collection. In this paper, only five SPPs were tested from each accession, among which different genotypes in 78 accessions (45.3%) were found. In 77 of them, heterogeneity had also been uncovered previously, while in Alterna, heterogeneity was revealed for the first time and was caused by the presence of two genotypes with similar IRAs conditioned by the genes *Mla8* and *Ml(Ch)*.

The number of heterogeneous accessions in the collection is high, but not all heterogeneity could be detected for several reasons [8], including testing only a limited number of SPPs. Heterogeneity of gene bank accessions can have several causes relating to breeding methods or collecting landraces, out-crossing and mechanical admixtures. Cv. Will is an example of heterogeneity resulting from a breeding method without selecting for mildew resistance in which two genotypes (*Mla7* and *Mla7, Mlh*) were recorded. *Mla7*, which was present in all five SPPs confirms that they originated from an identical crossing programme and that the presence or absence of *Mlh* has resulted in the existence of a second line. A similar and previously mentioned case could be Marinka which possesses *Mla7* and *Mlg* according to the catalogue [6], whereas, a line possessing only *Mla7* was identified herein once again [8].

The best example of the absence of resistance gene selection in a population derived from an identical origin is Ragusa 34 ¬-40. Among five SPPs, five genotypes contained five *Ml* genes in different combinations (*a8; a8, h; a8, h, ra; Ch, h, ra* and *Ch, Dr2, ra*). Since such a number of genes can result in more different genotypes, many SPPs of this accession should be tested to uncover all existing combinations. Conversely Ventitre is an example of an accession that is composed of genotypes that could not have an identical source. In two SPPs there was a genotype with three *Ml* genes (*aLo, ra, Ru2*), whereas, in three other lines only one different gene (*Mla8*) was found. These two genotypes have no resistances in common and the accession is a mixture of different varieties.

Beside detection of heterogeneity the results enable to uncover mislabelled accessions. In seven accessions as well as in two other varieties published elsewhere, the resistance genes differed from those listed in the catalogue (Table 3). For example, although *Ml(Bw)* has been recorded in Angela [20], in our tests Angela has three different *Ml* genes (*h, (Lu)* and *ra*) and clearly demonstrates the difference between "our" accession and that used previously. In the catalogue, Borwina has *Mla6* although at the time of its registration in the Czech Republic (1983) this variety had a resistance phenotype that differed from all other winter varieties. Therefore, it was considered a new gene and designated according to this variety *Ml(Bw)* [21]. This resistance was soon discovered in other Central European winter varieties [22], and subsequently in many Chinese barleys [23]. Later it was found to be identical to the *Ml(Ru2)* present in one (P15) of the near-isogenic lines of the spring variety Pallas [24,25]. *Ml(Ru2)* has so far only been found in winter varieties (except P15) and in our study in 73 SPPs. Thus, in addition to the *Ml* genes *aLo, ra* and *h, Ml(Ru2)* can be included among the typical resistances of winter barley. Two genotypes were found in the group of Borwina SPPs, both carry *Ml(Ru2)*, but differ in having five other resistance genes.

Old varieties are predominant in CCs and have undergone several propagation cycles, each of which might result in possible errors. Mislabeling and contamination of accessions could occur mainly at the time when the rules for working in the nascent gene banks were not sufficiently specified and technical equipment was not able to maintain seed with a high degree of purity. Therefore, the differences found in the resistance of nine of the varieties (Table 3) can be explained by the probable mislabeling of the accessions. However, an identical commercial name used for different varieties cannot be ruled out as with two spring varieties Opal [6,22], Freya in which the spring type has *Mlg* [26] and winter type has *Mla6* (this contribution) and Zenit in which both spring and winter types have *Mla13* [8,22].

In this report three varieties with resistance genes effective against all the pathogen isolates used were found. In other tests many varieties with a similar type of resistance to different sets of pathogen isolates have been identified. This indicates a lack or less intensive directional selection of the pathogen due to the absence or low proportion of barley varieties with appropriate resistances rather than the effectiveness and especially the durability of these resistances. Many of varieties with reported effective resistances, including Bonita studied here, were used in a set of 95 differential varieties in the current (2021) population study, and although only 72 isolates were obtained, new and rare virulences to several these resistances were detected [27].

**Table 3.** Powdery mildew resistance genes found in identically designated winter barley accessions (discrepancies probably resulted from mislabeling of varieties).

| Accession | *Ml* Resistance Gene(s) | |
|---|---|---|
| | **Present Study** | **Previous Studies** |
| Angela | *h, Lu, ra* | *Bw* (=*Ru2*), *ra* [1] |
| Borwina | *a8, h, ra, Ru2* | *a6* [2] |
| Borwina | *aLo, Lu, Ru2* | |
| Capri | *aLo, Lu, Ru2* | *g* [2] |
| Erfa | *aLo, Lu* | *h, u* [2] |
| Jutta | a8 | *a12, g* [2] |
| Leon | *aLo, Dr2, ra* | *none* [2] |
| Leon | *a8* | |
| Leon | *none* | |
| Nelly | *a13* | *a7, Ab* [1] |
| Pamina | *aLo, Lu* | *a9, g* [2] |
| Pamina | *a6, ra* | |
| Vogelsanger Gold | *a8* | *a6, h, ra* [2] |

[1] Anonymous [20]. [2] Brown and Jørgensen [6].

The rapid adaptation of the pathogen excludes the successful use of specific resistance genes even though they appear to be fully effective, because specific resistance genes are quickly overcome and for this reason their use in breeding barley for resistance against powdery mildew is no longer recommended [9,28]. In winter barley, it is advisable to focus on the accumulation of minor genes that are predominantly non-specific [29–31]. An alternative strategy could be to use the resistance of *Hordeum bulbosum*, the only representative of the secondary genepool of cultivated barley [32], although the resistance of the three so far derived genes [33–35] to pathogen adaptation has not been sufficiently tested.

## 4. Materials and Methods

The following parts, especially 4.2. and 4.3. are similar to those previously described [8].

### 4.1. Plant Material and Pathogen Isolates

A set of individually sown plants from each of 172 accessions of the CC of the Czech gene bank of winter barley originating from 35 countries were grown in rows in the field and five single-plants of each accession were harvested and investigated. For resistance tests, 53 selected reference isolates of *Bgh* were used, which had been collected in 11 countries in all nonpolar continents over a period of 63 years (1953–2016) and represents the global virulence/avirulence diversity of the pathogen [36].

### 4.2. Testing Procedure

About 20 seeds of an ear of each SPP was sown in a pot (80 mm diameter) and grown in a mildew-proof greenhouse under natural daylight. The primary leaves were excised when the second leaves were emerging, and segments 20 mm long were cut from the middle part of healthy fully-expanded leaves. One leaf segment of each SPP was placed on the surface of water agar in a 150 mm Petri dish and each dish was separately inoculated with the pathogen isolates in a concentration of ca. 10 conidia $mm^{-2}$.

### 4.3. Evaluation

Seven days after inoculation, infection responses (IR = phenotype of SPP × isolate interaction) on the middle part of the adaxial side of leaf segments were scored on a scale

0–4, where 0 = no mycelium and sporulation, and 4 = strong mycelial growth and sporulation (Figure 1) [14]; IRs 3, 3–4 and 4 were considered susceptible. Each SPP was tested with a minimum of one replication but most SPPs were included in two or more replications. A set of 53 IRs provided an IRA for each SPP. Based on the gene-for-gene model [37] the resistance genes in SPPs were postulated by comparing the IRAs with previously determined IRAs of standard barley genotypes possessing known resistance genes.

**Figure 1.** Infection responses (IRs) produced by a *Blumeria graminis* f. sp. *hordei* isolate on 30 barley genotypes each represented with a triplet of leaf segments seven days after inoculation; three IRs are shown here, IR 0—full resistance (green leaf segments), IR 4—full susceptibility (leaf segments with colonies of white conidia of the pathogen) and IR 2—moderate resistance (colonies of grey mycelium surrounded with chlorotic/necrotic spots) on a triplet in the middle of the last long row. This figure is a demonstration picture from other tests.

## 5. Conclusions

- The main problems of gene banks include mislabeling of accessions, heterogeneity resulting from mechanical admixtures or out-crossing during their multiplication and low germination. All these problems can adversely affect the results of research and breeding projects that are based on the use of gene bank resources. Therefore, the highest priority of plant gene banks curators must be to provide breeders and researchers with authentic seed of original genotypes.
- From all 172 accessions of the given CC, 860 homogeneous lines (SPPs) were created and their genetic basis of resistance to powdery mildew was studied.
- More than one genotype was identified among SPPs of 78 accessions (=45.3% heterogeneous accessions).
- Only 21 accessions (12.2%) were found to have data previously published on their resistance and the resistance genes identified here often differed.
- Selected lines (SPPs) of the CC accessions were multiplied in the field and stored in the gene bank, and are freely available for the use of breeders and researchers [38].
- Seeds of many accessions were requested from other gene banks and from each of them SPPs were also grown. These are studied in a similar way to assess their homogeneity and authenticity. Accessions whose authenticity is questionable will be replaced with genuine ones.
- Rules for replacing accessions with questionable identity and using genotypes derived from heterogeneous accessions require an international agreement.

**Supplementary Materials:** The following are available online at https://www.mdpi.com/article/10.3390/plants10101988/s1, Table S1. Eight hundred and sixty single-plant progenies of 172 winter barley gene bank accessions, their country of origin and postulated *Ml* resistance genes against powdery mildew.

**Author Contributions:** Z.N. provided barley resources (SPPs) and unpublished supporting information about them and corrected details of the contribution. A.D. did all other scientifically important activities including the original idea, methodology, experimental planning, scoring and evaluation of resistance tests and writing of the draft. All authors have read and agreed to the published version of the manuscript.

**Funding:** This research was funded by the Ministry of Agriculture of the Czech Republic, institutional support nos. MZE-RO1118 (AD) and MZE-RO0418 (ZN).

**Institutional Review Board Statement:** Not applicable.

**Informed Consent Statement:** Not applicable.

**Data Availability Statement:** All relevant data are involved in the article and supplementary Table S1.

**Acknowledgments:** Great effort during reproduction of all needed SPPs of the given accessions precisely completed by stuff of the gene bank and the excellent technical assistance of Dagmar Krejčířová for conducting the resistance tests is highly appreciated.

**Conflicts of Interest:** Authors declare no conflict of interest.

## References


1. Dreiseitl, A. Differences in powdery mildew epidemics in spring and winter barley based on 30-year variety trials. *Ann. Appl. Biol.* **2011**, *159*, 49–57. [CrossRef]
2. Jørgensen, J.H. Erysiphe graminis, powdery mildew of cereals and grasses. In *Advances in Plant Pathology*; Academic Press: London, UK, 1988; Volume 6, pp. 137–157.
3. Jensen, H.P.; Christensen, E.; Jørgensen, J.H. Powdery mildew resistance genes in 127 Northwest European spring barley varieties. *Plant Breed.* **1992**, *108*, 210–228. [CrossRef]
4. Wolfe, M.S. The genetics of barley mildew. *Rev. Plant Pathol.* **1972**, *51*, 507–522.
5. Jørgensen, J.H. Genetics of powdery mildew resistance in barley. *Crit. Rev. Plant Sci.* **1994**, *13*, 97–119. [CrossRef]
6. Brown, J.K.M.; Jørgensen, J.H. A catalogue of mildew resistance genes in European barley varieties. In Proceedings of the Integrated Control of Cereal Mildews: Virulence and Their Change, Proceedings of the Second European Workshop on Integrated Control of Cereal Mildews, Roskilde, Denmark, 23–25 January 1990; Jørgensen, J.H., Ed.; Risø National Laboratory: Roskilde, Denmark, 1991; pp. 263–286.
7. Honecker, L. Beiträge zum Mehltauproblem bei der Gerste mit besonderer Berücksichtigung der züchterischen Seite. *Pflanzenbau* **1931**, *8*, 78–84.
8. Dreiseitl, A. Genotype heterogeneity in accessions of a winter barley core collection assessed on postulated specific powdery mildew resistance genes. *Agronomy* **2021**, *11*, 513. [CrossRef]
9. Dreiseitl, A. Specific resistance of barley to powdery mildew, its use and beyond. A concise critical review. *Genes* **2020**, *11*, 971. [CrossRef]
10. Dreiseitl, A. Great pathotype diversity and reduced virulence complexity in a Central European population of *Blumeria graminis* f. sp. *hordei* in 2015–2017. *Eur. J. Plant Pathol.* **2019**, *53*, 801–811. [CrossRef]
11. Sanchez-Martin, J.; Keller, B. NLR immune receptors and diverse types of non-NLR proteins control race-specific resistance in Triticeae. *Curr. Opin. Plant Biol.* **2021**, *62*, 102053. [CrossRef]
12. Dreiseitl, A. Heterogeneity of powdery mildew resistance revealed in accessions of the ICARDA wild barley collection. *Front. Plant Sci.* **2017**, *8*, 202. [CrossRef]
13. Dreiseitl, A.; Zavřelová, M. Identification of barley powdery mildew resistances in gene bank accessions and the use of gene diversity for verifying seed purity and authenticity. *PLoS ONE* **2018**, *13*, e0208719. [CrossRef] [PubMed]
14. Torp, J.; Jensen, H.P.; Jørgensen, J.H. *Powdery Mildew Resistance Genes in 106 Northwest European Spring Barley Cultivars. Year-Book, 1978*; Royal Veterinary and Agricultural University: Copenhagen, Denmark, 1978; pp. 75–102.
15. Dreiseitl, A. Powdery mildew resistance in winter barley cultivars. *Plant Breed.* **2007**, *126*, 268–273. [CrossRef]
16. Hiura, U.; Heta, H. Studies on the disease resistance in barley. III. Further studies on the physiologic races of *Erysiphe graminis hordei* in Japan. *Ber. Ohara Inst. Landwirtsch. Biol.* **1955**, *10*, 135–156.
17. Jørgensen, J.H.; Jensen, H.P. Powdery mildew resistance gene Ml-a8 (Reg1h8) in Northwest European spring barley varieties. *Barley Genet. Newsl.* **1983**, *13*, 51–52.
18. Dreiseitl, A. Dissimilarity of barley powdery mildew resistances Heils Hanna and Lomerit. *Czech J. Genet. Plant Breed.* **2011**, *47*, 95–100. [CrossRef]

19. Dreiseitl, A. A novel way to identify specific powdery mildew resistance genes in hybrid barley cultivars. *Sci. Rep.* **2020**, *10*, 18930. [CrossRef] [PubMed]
20. *Beschreibende Sortenliste Getreide, Mais Ölfrüchte Leguminosen Hackfrüchte*; Bundessortenamt: Hannover, Germany, 2003; pp. 43–44.
21. Dreiseitl, A. Analysis of breeding Czechoslovak barley varieties for resistance to fungal diseases particularly powdery mildew. *Pol'nohospodárstvo* **1993**, *39*, 467–475.
22. Dreiseitl, A.; Jørgensen, J.H. Powdery mildew resistance in Czech and Slovak barley cultivars. *Plant Breed.* **2000**, *119*, 203–209. [CrossRef]
23. Dreiseitl, A.; Yang, J. Powdery mildew resistance in a collection of Chinese barley varieties. *Genet. Resour. Crop Evol.* **2007**, *54*, 259–266. [CrossRef]
24. Kølster, P.; Munk, L.; Stølen, O.; Løhde, J. Near-isogenic barley lines with genes for resistance to powdery mildew. *Crop Sci.* **1986**, *26*, 903–907. [CrossRef]
25. Dreiseitl, A. Identity of barley powdery mildew resistances Bw and Ru2. *Czech J. Genet. Plant Breed.* **2012**, *48*, 185–188. [CrossRef]
26. Nover, I.; Lehmann, C.O. Resistenzeigenschaften in Gersten- und Weizen-sortiment Gatersleben 14. Prüfung von Sommergersten auf ihr Verhalten gegen Mehltau (*Erysiphe graminis* f. sp. DC. *hordei* Marchal). *Kulturpflanze* **1972**, *19*, 283–298. [CrossRef]
27. Dreiseitl, A. Rare Virulences in a Central European Population of *Blumeria graminis* f. sp. *hordei*. Unpublished work. 2021.
28. McDonald, B.A.; Linde, C. Pathogen population genetics, evolutionary potential, and durable resistance. *Annu. Rev. Phytopathol.* **2002**, *40*, 349–379. [CrossRef] [PubMed]
29. Aghnoum, R.; Marcel, T.C.; Johrde, A.; Pecchioni, N.; Schweizer, P.; Niks, R.E. Basal host resistance of barley to powdery mildew: Connecting quantitative trait loci and candidate genes. *Mol. Plant Microbe Interact.* **2010**, *23*, 91–102. [CrossRef] [PubMed]
30. Spies, A.; Korzun, V.; Bayles, R.; Rajaraman, J.; Himmelbach, A.; Hedley, P.E.; Schweizer, P. Allele mining in barley genetic resources reveals genes of race-non-specific powdery mildew resistance. *Front. Plant Sci.* **2012**, *2*, 113. [CrossRef] [PubMed]
31. Novakazi, F.; Krusell, L.; Jensen, J.D.; Orabi, J.; Jahoor, A.; Bengtsson, T. You had me at "MAGIC"!: Four barley MAGIC populations reveal novel resistance QTL for powdery mildew. *Genes* **2020**, *11*, 1512. [CrossRef]
32. Von Bothmer, R.; Sato, K.; Komatsuda, T.; Yasuda, S.; Fischbeck, G. *Diversity in Barley (Hordeum vulgare)*; von Bothmer, R., van Hintum, T., Knüpffer, H., Sato, K., Eds.; Elsevier Science B.V.: Amsterdam, The Netherlands, 2003; Chapter 2; pp. 9–27.
33. Xu, J.; Kasha, K.J. Transfer of a dominant gene for powdery mildew resistance and DNA from *Hordeum bulbosum* into cultivated barley (*Hordeum vulgare*). *Theor. Appl. Genet.* **1992**, *84*, 771–777. [CrossRef]
34. Pickering, R.A.; Hill, A.M.; Michel, M.; Timmerman-Vaughan, G.M. The transfer of a powdery mildew resistance gene from *Hordeum bulbosum* L. to barley (*H. vulgare* L.) chromosome 2 (2I). *Theor. Appl. Genet.* **1995**, *91*, 1288–1292. [CrossRef]
35. Hoseinzadeh, P.; Ruge-Wehling, B.; Schweizer, P.; Stein, N.; Pidon, H. High resolution mapping of a *Hordeum bulbosum*-derived powdery mildew resistance locus in barley using distinct homologous introgression lines. *Front. Plant Sci.* **2020**, *11*, 225. [CrossRef]
36. Dreiseitl, A. A novel resistance against powdery mildew found in winter barley cultivars. *Plant Breed.* **2019**, *138*, 840–845. [CrossRef]
37. Flor, H.H. Host-parasite interaction in flax rust—Its genetics and other implications. *Phytopathology* **1955**, *45*, 680–685.
38. Available online: https://www.gzr.cz/?lang=en (accessed on 15 July 2021).

*plants*

MDPI

*Article*

# From Traditional Breeding to Genome Editing for Boosting Productivity of the Ancient Grain Tef [*Eragrostis tef* (Zucc.) Trotter]

**Muhammad Numan [1], Abdul Latif Khan [2], Sajjad Asaf [2], Mohammad Salehin [1], Getu Beyene [3], Zerihun Tadele [4] and Ayalew Ligaba-Osena [1,*]**

1 Laboratory of Molecular Biology and Biotechnology, Department of Biology, University of North Carolina at Greensboro, Greensboro, NC 27412, USA; m_numan@uncg.edu (M.N.); m_salehin@uncg.edu (M.S.)

2 Natural and Medical Sciences Research Center, Biotechnology and OMICs Laboratory, University of Nizwa, Nizwa 616, Oman; abdullatif@unizwa.edu.om (A.L.K.); sajjadasaf@unizwa.edu.om (S.A.)

3 Donald Danforth Plant Science Center, St. Louis, MO 63132, USA; GDuguma@danforthcenter.org

4 Institute of Plant Sciences, University of Bern, Altenbergrain 21, CH-3013 Bern, Switzerland; zerihun.tadele@ips.unibe.ch

* Correspondence: alosena@uncg.edu

**Abstract:** Tef (*Eragrostis tef* (Zucc.) Trotter) is a staple food crop for 70% of the Ethiopian population and is currently cultivated in several countries for grain and forage production. It is one of the most nutritious grains, and is also more resilient to marginal soil and climate conditions than major cereals such as maize, wheat and rice. However, tef is an extremely low-yielding crop, mainly due to lodging, which is when stalks fall on the ground irreversibly, and prolonged drought during the growing season. Climate change is triggering several biotic and abiotic stresses which are expected to cause severe food shortages in the foreseeable future. This has necessitated an alternative and robust approach in order to improve resilience to diverse types of stresses and increase crop yields. Traditional breeding has been extensively implemented to develop crop varieties with traits of interest, although the technique has several limitations. Currently, genome editing technologies are receiving increased interest among plant biologists as a means of improving key agronomic traits. In this review, the potential application of clustered regularly interspaced short palindromic repeats (CRISPR) and CRISPR-associated proteins (CRISPR-Cas) technology in improving stress resilience in tef is discussed. Several putative abiotic stress-resilient genes of the related monocot plant species have been discussed and proposed as target genes for editing in tef through the CRISPR-Cas system. This is expected to improve stress resilience and boost productivity, thereby ensuring food and nutrition security in the region where it is needed the most.

**Keywords:** CRSIPR-Cas; drought tolerance; *Eragrostis tef*; genome editing; stress resilience

**Citation:** Numan, M.; Khan, A.L.; Asaf, S.; Salehin, M.; Beyene, G.; Tadele, Z.; Ligaba-Osena, A. From Traditional Breeding to Genome Editing for Boosting Productivity of the Ancient Grain Tef [*Eragrostis tef* (Zucc.) Trotter]. *Plants* **2021**, *10*, 628. https://doi.org/10.3390/plants10040628

Academic Editor: Francesco Carimi

Received: 8 February 2021
Accepted: 22 March 2021
Published: 25 March 2021



## 1. Introduction

The world population is increasing at an alarming rate, demanding an increase in food production. The Green Revolution of the 1960s has led to a substantial increase in major cereal production, but that is unlikely to meet the urgent demand for higher food production [1] under the current climate scenario. To meet world food demands, the production of major crops alone is insufficient, as they are less suited to extreme climate and low-input conditions [2]. There is an increasing interest in underutilized crops such as tef (*Eragrostis tef* (Zucc.) Trotter); millets, including proso millet (*Panicum miliaceum* Mill.) and finger millet (*Eleusine coracana* Gaertn.); and quinoa (*Chenopodium quinoa* Willd.), which are more versatile due to their resilience to marginal growing conditions, and outstanding nutritional values. Despite its valuable traits, the grain yield of tef is very low. In 2018, the average yield of tef in Ethiopia was only 1.7 ton ha$^{-1}$ as compared to maize (4 ton ha$^{-1}$) and wheat (2.7 ton ha$^{-1}$) [3]. Tef is a cereal crop originating in the Horn of Africa, which is widely cultivated in Ethiopia and Eritrea. In Ethiopia, tef is a staple food for about 70% of the population. The crop is annually cultivated on 2.9 million hectares of land, producing about 4.5 million tons of grain [4]. Tef is tolerant to marginal soil and unfavorable climate conditions, which makes it a potential crop for arid and semiarid areas as well as poorly drained soils [5]. Tef is also one of the most nutrient-dense crops, containing high amounts of macro- and micro-nutrients (primarily calcium and iron), amino acids and vitamins [6]. Tef cultivation in Ethiopia and around the globe has increased in recent years due to its many health-related benefits. Since the absence of gluten epitopes has been confirmed in tef by antibody assays [7], it has been recommended as an alternative diet for people suffering from celiac disease, the immune reaction to consuming gluten containing foods such as wheat (*Triticum aestivum* L.), barley (*Hordeum vulgare* L.) and rye (*Secale cereal* L.), which affects 0.6–1.0 percent of the population globally [8,9]. In addition to the extensive use of tef grain for human consumption, the straw of tef is more nutritious and palatable as a livestock feed compared to the straw from cereals such as barley and wheat [10]. Moreover, tef straw is used as construction material because it serves as an organic binder for mud used for plastering walls for local houses [11]. Various agronomic traits, such as panicle architecture, tilling, grain size and plant height, have been targets for the improvement of tef yield. Grain yield is a highly complex trait which has several components, including seed weight, form and size of panicles, florets per panicle and number of fertile tillers [12,13]. Other important traits that determine grain yield include shoot biomass, panicle weight and the number of tillers in a plant [14]. Furthermore, certain agronomic traits such as shattering proneness, lodging tolerance, dry matter yield, leaf area, and plant height directly or indirectly influence grain yield in crops [15,16].

The main factors causing yield loss in tef include susceptibility to lodging, weed competition, drought, small grain size and soil acidity [5]. Although tef shows several agronomic and nutritionally desirable traits, it is under tremendous pressure due to harsh environmental stress conditions [5]. The crop is relatively resistant to diseases and insect pests as compared to other cereal crops. Among abiotic stressors, tef yield is significantly reduced by drought and soil acidity. Weed competition has broad about a range of effects on the yield of tef in Ethiopia [17]. Many direct and indirect strategies of weed control are employed by farmers [18]. Hand weeding and frequent tillage are the two commonly used methods of weed control in tef production. Furthermore, weeds can be controlled by herbicide application with proper management of spray times and frequency. However, the herbicides must be specific to broad-leaved weeds to avoid damaging tef plants. Taken together, hand weeding, the use of herbicides and resistant tef varieties are viable alternatives in order to overcome yield loss due to weeds. With proper weed control methods, improved tef varieties such as *Kora* and *Quncho* have been shown to produce higher yields [19].

Drought is a major abiotic stress which has significant effects on crop yield in most African countries. Water scarcity has resulted in a fragile ecosystem in Africa's arid and semiarid regions. In sub-Saharan Africa, about 1.1 billion people live in drier environments; however, this number is expected to double by 2050, and is expected to reach 2.5 billion people [20]. Drought stress after planting [21] and during the flowering and grain filling stage has serious effects on crop yields, and up to 60% of yield loss has been reported in pearl millet at these stages [21,22]. In tef, drought has been reported to cause about 40% yield loss [23].

The other major cause of low productivity in tef is lodging, which is the displacement of the stalks from the vertical position due to wind and rain [24]. Lodging occurs frequently before grain maturity, significantly reducing the grain yield [25]. Tef is primarily susceptible to stem lodging [26,27]. Panicle length is also associated with lodging tolerance [25]. Semi-dwarf varieties of tef are lodging-tolerant and produce higher yields than tall varieties [28]. Lodging limits the use of inputs such as N-fertilizers, exacerbating the susceptibility of the plant to lodging [29].

To overcome the effects of the constrains mentioned above and to improve the tef productivity, it is important to develop resistant and high-yield verities. There are several approaches to increasing crop productivity as well as stress tolerance in crops. Among these strategies, genome editing techniques have recently received increased attention. Previous studies have suggested that the productivity of many cereal crops such as maize [30,31], rice [32–35], wheat [30,36] and other monocots [37,38] have been improved using the clustered regularly interspaced short palindromic repeats (CRISPR) system. In rice (*Oryza sativa* L.), CRISPR-associated proteins (CRISPR-Cas) systems have been used to improve tolerance to drought [39], cold [40] and salt stress [41,42], ultimately boosting productivity [39]. In wheat, two efficient and simple CRISPR-Cas methods have been developed to improve productivity and stress resilience [43–45]. The CRISPR-Cas technology used in these monocots is expected to be transferred to tef. Therefore, the aim of this review is to highlight the potential of CRISPR-Cas-mediated gene-editing in trait improvement in tef.

## 2. Mechanisms of Tolerance to Lodging and Environmental Constraints in Tef

### 2.1. Lodging Tolerance

Lodging is the process by which cereal shoots are displaced from an upright position to a horizontal position [46]. Lodging is considered a complex phenomenon, influenced by several factors, such as diseases, agronomic practice, crop history, soil type, landscape, geography, rain and wind [47]. Stem lodging is the bending or breaking of stem internodes (lower culm internodes), whereas root lodging is the failure of the root to maintain its integrity in the soil [48]. The application of fertilizers aggravates lodging, and hence the yield potential of tef. Lodging stress can be reduced by controlling/decreasing plant height. However, reducing plant height by inhibiting plant growth regulators or introducing dwarfing genes could lead to crop yield reductions [47]; hence, researchers have suggested targeting traits other than plant height to reduce yield loss due to lodging. A recent study by Merchuk-Ovnat, et al. [49] suggested that early lodging is likely caused by a rapid increase in inflorescence weight [49]. This group also observed variations among the tested tef population in terms of lodging time and strength, with some populations possessing the strength to hold the inflorescence in the grain filling season up to a certain point before they were bent to the ground. Due to its weak stem, tef has high chance of succumbing to lodging due to rain or wind [50]. Modification of the stem's chemical composition, such as its cellulose, lignin, structural carbohydrate and silica composition, is expected to increase lodging-, disease-, and pest-resistance [51]. Silicon (Si) is a beneficial plant nutrient that has been shown to increase tolerance to lodging, diseases and pests, as well as to abiotic stresses such as drought, salinity, heavy metal stresses, and extreme temperature in various crops, ultimately leading to increased grain yield [52–56]. We recently performed greenhouse experiments to study whether tef benefits from Si application. Our findings revealed that Si improves grain and biomass yield, stress resilience, and regulates the expression of Si-transporter genes in tef [57]. However, conclusive evidence showing the mechanism of silicon-induced stress resilience is lacking [58].

Although lodging is the main cause of low yield in tef [59], both physiological and molecular aspects are understudied, and biotechnological, molecular and breeding techniques [47] are not well developed to prevent lodging. A partnership formed by the 'Tef Improvement Project' has recently developed semi-dwarf and lodging-tolerant tef varieties, which are currently being disseminated in farmer's fields in Ethiopia [60].

Lodging tolerance has been shown to be improved by modulating the biosynthesis of plant growth regulators (PGRs). For example, the inhibition of gibberellic acid (GA) has been shown to reduce plant height [46,61] and decrease lodging susceptibility. Shorter internodes are associated with reduced plant height [62]. During the Green Revolution of the 1960s and 1970s, inhibition or alteration of GA in rice and wheat was mainly targeted for developing semi-dwarf varieties, which ultimately boosted the yield of these crops [63]. In tef, mutation in the *α-Tubulin* gene is associated with agronomically important traits such as semi-dwarfism and lodging tolerance [59]. Blösch, et al. [25] have reported that panicle angle contributes to lodging tolerance in tef. Jifar, et al. [28] also identified some lodging tolerance genotypes (*RIL-91*, *RIL-244* and *RIL-11*).

Genes associated with dwarfism in plants have been widely studied [64–69]. The two prominent genes of the 1960s Green Revolution were the semi-dwarf (*SD1*) gene in rice [66,70,71] and reduced height-1 (*RHT-B1b* and *RHT-D1b*) in wheat [72]. SD1 belongs to the gibberellin biosynthetic pathway, whereas *RHT* is a GA response regulator and is a *DELLA* protein family gene. *DELLA* proteins are important components of the signal transduction pathway of GA, encoded by the wild-type allele of *RHT-B1b* and *RHT-D1b* [73]. In the Green Revolution wheat varieties, introduction of a stop codon in the N-terminus of the two reduced height-1 (*RHT-B1* and *RHT-D1*) loci was responsible for the semi-dwarf and lodging tolerance traits [72]. In rice, the enzyme gibberellin 20-oxidase (*GA20*) encoded by the *SD1* gene is responsible for the biosynthesis of GA [65,74]. A frame shift mutation due to a 383-bp deletion in the *sd1* allele has been shown to greatly reduce the level of *GA20* oxidase [66]. Mutation of the *sd1* and *RHT* homologs in tef could potentially lead to lodging tolerance and significantly improve grain yield. Similarly, genetic loci (*DW1*, *DW2*, *DW3* and *DW4*) that control plant height across several environmental conditions have been identified in sorghum. Recently, scientists have transferred these mutations into a single sorghum line and managed to release a semi-dwarf commercial variety that contains mutations in three loci (*DW1*, *DW2* and *DW4*) [75,76]. This suggests that these mutations could also be introduced into tef to develop semi-dwarf varieties with improved stress tolerance and enhanced grain yield.

### 2.2. Drought Tolerance

Understanding the degree of stress tolerance in crop plants is important in devising alternative strategies for improving yield and quality. Drought is one of the most important abiotic stresses affecting plant growth and development. Plants have developed various mechanisms of drought tolerance [77,78]. The mechanisms that have been reported in tef include modifications of stomatal conductance, osmotic adjustment, development of a deep rooting system and maintenance of cell membrane stability [79,80]. Development of a deep root system and osmotic adjustment are major drought stress tolerance mechanisms in many crops, including tef [79]. The association of plant height, root depth and thickness to drought stress tolerance was previously reported in tef [79]. Recently, crosstalk between plant height and drought tolerance was reported from a study on tef and other small cereals where semi-dwarf plants were found to be drought-tolerant [81]. Osmotic adjustment is also known to enable tef leaves to maintain leaf turgor pressure (LTP) [79,82] under extreme drought conditions by retrieving and absorbing water even from dry soils. Modification of root growth parameters in response to water scarcity is another strategy used to mitigate drought stress [83,84]. For example, the increase in root length of cowpea, peanut and soybean plants when exposed to drought enabled them to absorb deep soil water [84]. Similarly, developing deep-rooted tef plants with an extensive and broad root system is a desirable trait to withstand drought stress [79].

### 2.3. Weed Competition and Herbicide Tolerance

Weed competition is another important plant trait in areas of low-input integrated weed management systems [85]. The competitive ability of crops has been divided into two broad categories; the first category is the crop's ability to reduce competitor fitness, whereas the second is the crop's ability to resist yield losses and withstand its neighbor's competitive impact [86]. Different terms have been used for these aspects in the literature, such as "tolerance ability" and "suppressive ability" [87,88].

In Ethiopia, smallholder farmers have adopted some cultural methods to mitigate the impact of weed competition. Hand weeding and frequent tillage are common practices used to control weeds in tef production [17]. Herbicides are not widely used, mainly due to economic reasons and shortages of supplies. An alternative strategy in weed management is the use of cultivars with competitive ability due to their sustainability [88,89]. However, information on tef varieties with high weed competitive ability is limited as compared to other cereals such as oats (*Avena sativa* L.), barley (*Hordeum vulgare* L.) and wheat (*Triticum aestivum* L.) [86]. Tef varieties can be improved using genetic modification tools such as the CRISPR system to improve weed tolerance and enhance productivity. Potential genes for weed resistance and yield improvement can be overexpressed in tef or engineered through the CRISPR-Cas system to minimize the impact of weed competition.

Herbicide-resistant varieties have been developed in crops such as soybean by targeting key genes in amino acid synthesis or other functions. Among these genes, acetolactate synthase (ALS) is involved in the synthesis of branched-chain amino acids such as isoleucine, leucine and valine [90]. ALS is the target site for five non-competitive inhibitor families—sulfonylaminocarbonyltriazolinones, pyrimidinylthiobenzoates, triazolopyrimidines, imidazolinones and sulfonylureas [91]. Plants engineered in the *ALS* gene are resistant to non-selective herbicides, whereas all non-engineered plants, including weeds, are sensitive to the non-selective herbicides. A similar principle was implemented to develop glyphosate-resistant plants in which the *EPSPS* (*5-enolpyruvylshikimate-3-phosphate synthase*) gene was targeted. The *EPSPS* gene is involved in the shikimate cycle [92]. Overexpression or knockout of the above-mentioned genes might contribute towards developing tef plants with resistance to non-selective herbicides.

### *2.4. Panicle Architecture*

Panicle architecture and grain size are important yield traits in cereal crops such as rice, wheat and barley [93–95]. There is a direct relationship between agronomic traits such as panicle number, number of spikelets in panicle, spikelet filling percentage, grain size and number and crop yield [96]. For example, in rice, higher grain yield in a hybrid variety is associated with the number of spikelets in a panicle [96,97]. In some crops, genes that control panicle number and grain size have been identified and modified to increase yield [98–100]. For example, *OsSPL14* (squamosa promoter binding protein-like 14) gene and microRNA "*OsmiR397*" promoted panicle branching and increased grain size in rice, which ultimately lead to high grain yield [99,101]. In tef, homologs of these genes remain to be identified and characterized to determine their role in increasing grain size and to improve yield.

## **3. Status of Tef Improvement**

### *3.1. Traditional Breeding: Past and Current Status of Tef Improvement*

Scientific research on tef started in Ethiopia in 1950s [102]. Early breeding work focused on germplasm enhancement through collection, characterization, evaluation and conservation, as well as genetic improvement in which pure lines were selected from already existing germplasm [11,103] (Figure 1). Since flower opening characteristics were revealed in tef in 1974, [104], hybridization was used as a means of tef improvement. Molecular approaches in tef including marker development, genetic linkage maps, genetic and molecular diversity analysis were initiated during 1995–1998 [11]. Further progress was made during 1998–2003, including the initiation of interspecific hybridization, in vitro culture and mutagenesis in order to improve disease and lodging resistance. Over the last two decades, there has been progress in the area of tef genetic architecture and genomics research [105,106] (Figure 1). From a total of 42 improved tef varieties released by the National Research Program in Ethiopia, 18 were developed using the hybridization technique [107].

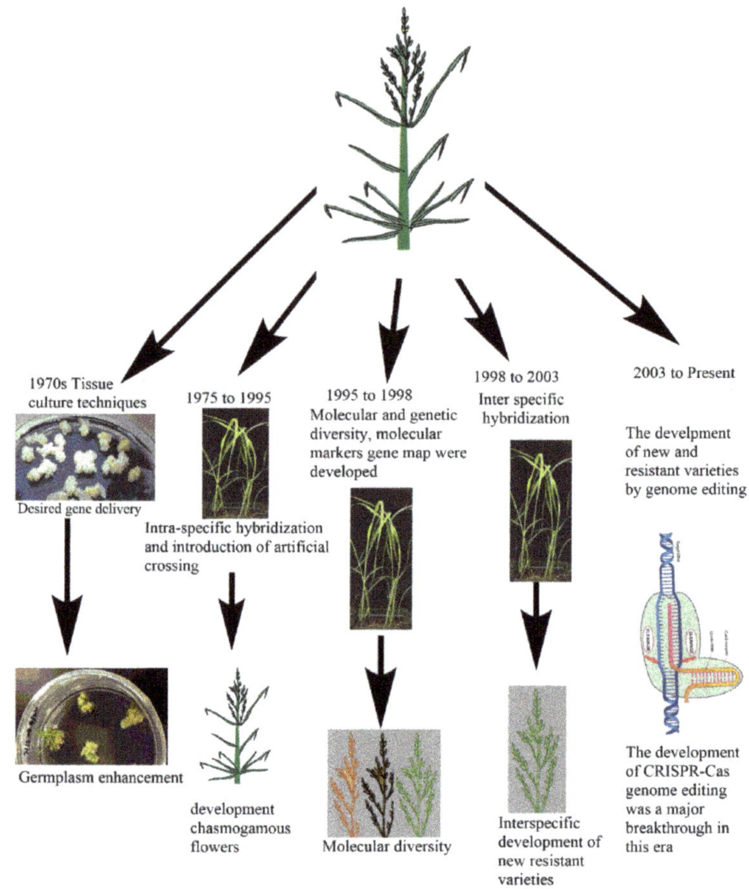


**Figure 1.** Improvement of tef varieties over the last 50 years. The improvement of tef started back in 1970s with tissue culture techniques, followed by hybridization, the study of molecular diversity, molecular marker analysis, the development of resistant varieties by interspecific hybridization and mutation and the recently emerged clustered regularly interspaced short palindromic repeats (CRISPR)-associated proteins (CRISPR-Cas) genome editing technique. Note: (The pictures used in this figure were either taken in the author's labs or drawn using ChemBioDraw software).

### *3.2. Molecular Marker Development*

The application of molecular markers in tef improvement was initiated during 1995–1998 [11]. Molecular markers near target genes are utilized for marker-assisted selection (MAS) or marker-assisted breeding (MAB) [108]. They enable the effective use of alleles during the selection of phenotypes. The most commonly used markers are microsatellites (simple sequence repeats; SSRs), amplified fragment length polymorphism (AFLPs) and single nucleotide polymorphisms (SNPs) [108]. During the selection of molecular markers, some important factors are considered, such as the quality and quantity of required DNA, procedures for marker assays, the level of polymorphism and the cost of the marker [109]. In tef, the SSRs and expressed sequence tag (EST), restriction fragment length polymorphisms (RFLPs) and random amplified polymorphic DNA (RAPD) have been developed [110,111]. Through SSR analysis, Abraha, et al. [112] identified and improved some important traits in tef, including grain yield, days to maturity, panicle length and plant height. Similarly, variability in tef accessions was identified using AFLP markers, which can be used in seed multiplication and breeding programs [113]. Application of these markers could play a great role in environmental stress tolerance in tef for improved productivity. Targeting induced local lesions in genomes (TILLING) is another genetic method used to identify small deletions or single base pair changes (mutation detection) in specific target genes [114]. In tef, targeting induced local lesions in genomes (TILLING) was used for targeting and improving valuable agronomic traits such as drought tolerance, seed size and dwarfism [115].

## 4. Potential of Genome Editing Technologies for Tef Improvement

Genome editing is one of the most recently developed technologies that has great potential to improve abiotic stress tolerance and boost productivity in tef. In a given genome, DNA can be replaced, inserted or deleted at an endogenous loci through a robust genetic engineering technique using sequence-specific nucleases (SSNs) [116]. SSNs such as CRISPR and CRISPR-associated protein 9 (CRISPR-Cas9) [117–120], transcriptional activator-like effector nuclease (TALEN) [121–123] and zinc finger nuclease (ZFN) [124,125] have been implicated in rapid genome editing in recent years. In addition to these, plant scientists use other techniques such as base editing, prime editing [126] and CRSIPR-Cpf1 [127]. Recently, CRISPR-Cpf1 has successfully used the prime genome editing in wheat Lin, et al. [128] and rice Lin, et al. [128], Li, et al. [129] genomes. These genome editing tools have been used in model plants, but with advances in genome editing, these procedures are now customized for wide variety of plant species and are usually specific to genotype [130] (Figure 2). However, to adopt advanced genetic engineering technologies in tef, there must be a well-established transformation and regeneration system, which is currently underdeveloped or non-existent for underutilized crops including tef. Recent advances in transgenic technologies have revealed promising tools for enhancing transformation and regeneration of transgenic lines. For example, overexpression of the maize embryogenic regulator genes baby boom (*Bbm*) and Wuschel2 (*Wus2*) has been shown to produce high transformation frequencies in numerous previously non-transformable monocot species, including maize inbred lines, sorghum (*Sorghum bicolor* (L.) Moench), sugarcane (*Saccharum officinarum* L.) and indica rice (*Oryza sativa* ssp. *indica*) [131]. More recently, Debernardi et al. [132] reported that expression of a fusion protein combining wheat growth-regulating factor 4 (GRF4) and its cofactor GRF-interacting factor 1 (GIF1) has been shown to substantially increase the efficiency and speed of regeneration in wheat, triticale and rice and increase the number of transformable wheat genotypes. These approaches have great potential for genetic improvement of tef and other recalcitrant economically important crops.

Since its first application as a plant genome editing technique [120,133,134], CRISPR-Cas has been widely applied in crop improvement programs [135,136]. Major crops that have benefited from the CRISPR-Cas technique include rice [32–35], maize [30,31], wheat [30,36] and other monocots [38]. In rice (*Oryza sativa*), the CRISPR-Cas system has been used to enhance drought [39], cold [40] and salt [41,42] tolerance, and to boost productivity [39]. Recently, in wheat, which is one of the plant species that is considered recalcitrant to genetic transformation via the *Agrobacterium* method, two efficient and simple CRISPR-Cas methods were developed [43–45]. Taken together, CRISPR-Cas technology has been widely implemented in both monocots and dicots, and has great potential to be implemented in tef improvement so that the performance of the crop against diverse environmental stresses will be enhanced, with the ultimate goal of boosting productivity.

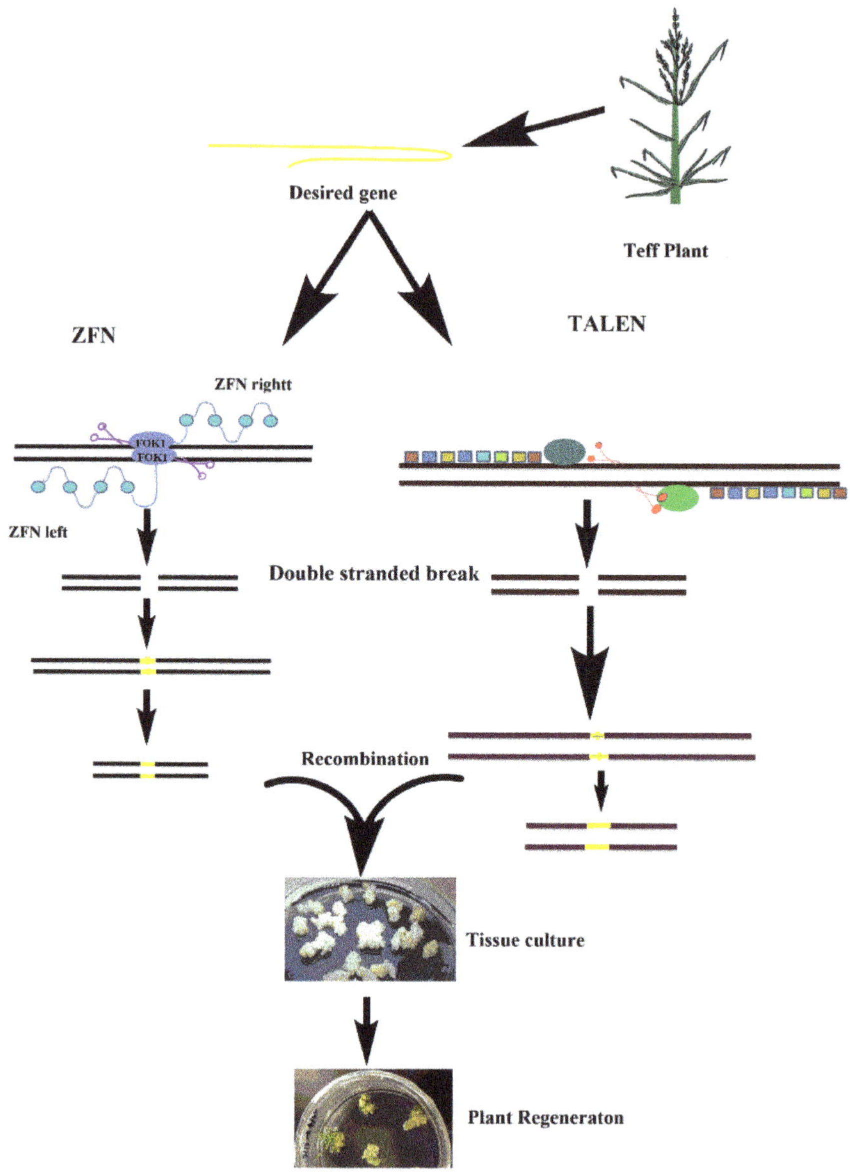


**Figure 2.** A schematic view of genome editing by zinc finger nuclease (ZFN) and transcriptional activator-like effector nuclease (TALEN) in tef. A desired gene is selected from tef and integrated with ZFN and TALEN and then transferred to a cell through a vector, which will then introduce a break into the double-stranded DNA and integrate the gene of interest into the host genome. Transformed cells are used to regenerate to whole plants. (Note: the pictures used in this figure were either taken in the author's labs or drawn using ChemBioDraw software).

*Candidate Tef Genes for CRISPR-Cas Technology*

The CRISPR-Cas system has proven efficient because it uses a single guide RNA through pairing of DNA targeting [137,138]. Targeting of DNA is essential for genome editing across all organisms [139]. In order to edit any plant gene using the CRISPR-Cas system, it is not necessary to integrate into the genome. For example, a guide RNA and Cas can be expressed transiently in the protoplast to edit a plant genome, and the protoplast can be regenerated into whole plant. Cas is a class II CRISPR system which is used in various organisms as a gene editing tool [138,140]. The basic mechanism involved in CRISPR-Cas editing is transformation to cells, followed by its integration with the host genome, and expression, where it cuts the specific locus of interest on the chromosome. The genome cleavage requires the Cas system, together with a single guided RNA (sgRNA): fusion of trans-activating (tracr RNA) and CRISPR RNAs (crRNA), followed by the recognition of the desired DNA sequence and protospacer-adjacent motifs (PAMs) (Figure 3) [138].

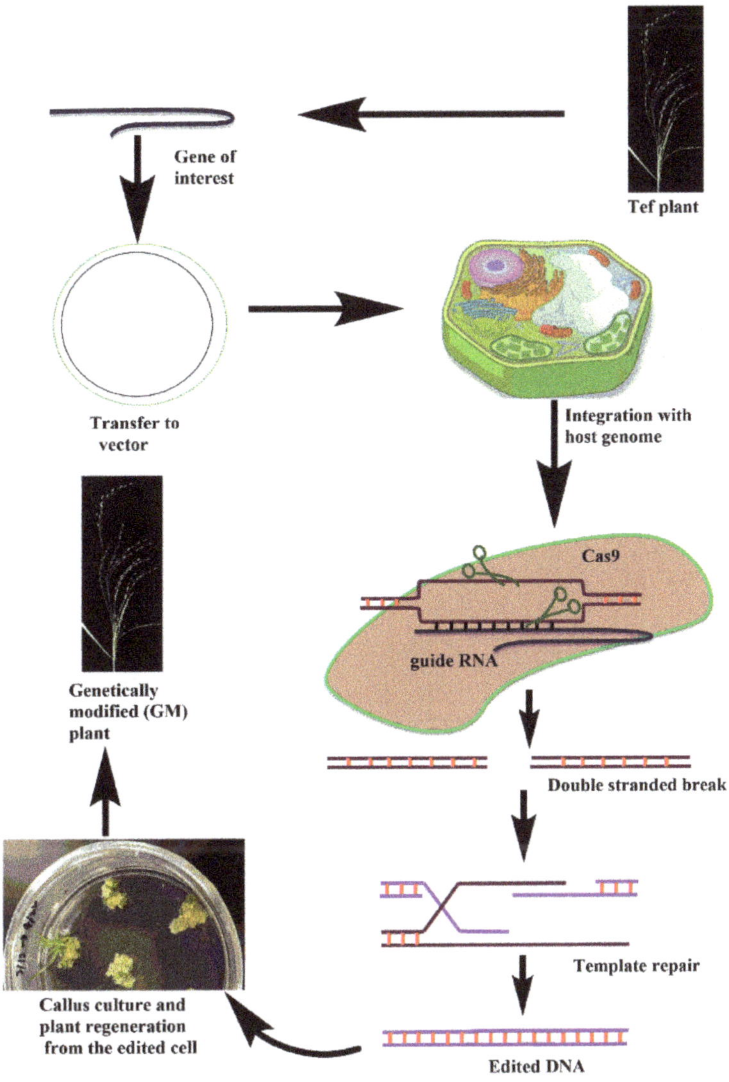


**Figure 3.** Illustration of the CRISPR-Cas system for tef genome editing. The gene of interest is transferred into a binary vector, which will be transferred into the target tissue (e.g., embryogenic calli) via *Agrobacterium* transformation, where the CRISPR-Cas protein machinery binds and breaks the double-stranded DNA of the gene of interest. CRISPR-edited lines will be regenerated from rthe callus. (Note: the pictures used in this figure were either taken in the author's labs or drawn using ChemBioDraw software).

To utilize CRISPR-Cas technology in tef improvement, identification of target genes that regulate agronomically important traits is crucial. In this review, we explored the draft genome sequence of tef [141] to identify genes that are possible targets for improved yield and abiotic stress tolerance. We reviewed the literature for genes which are negative regulators of abiotic stress tolerance, and those that regulate plant height and yield attributes in monocots, including rice, maize, wheat and finger millet, which is closely related to tef. We then searched for homologs in tef (Table 1) from the Ensembl plant database using *CoGeBlast*-comparative genomics databases [142]. The tef homologs were aligned with those in other monocots using the *Mega X clustlaw* alignment tool [143,144]. After alignment, a phylogenetic tree was constructed using the *Mega X maximum likelihood* tool [144] (Figure 4). It can be observed from Figure 4 that the tef homologs showed maximum bootstrap values with those of the other monocots.

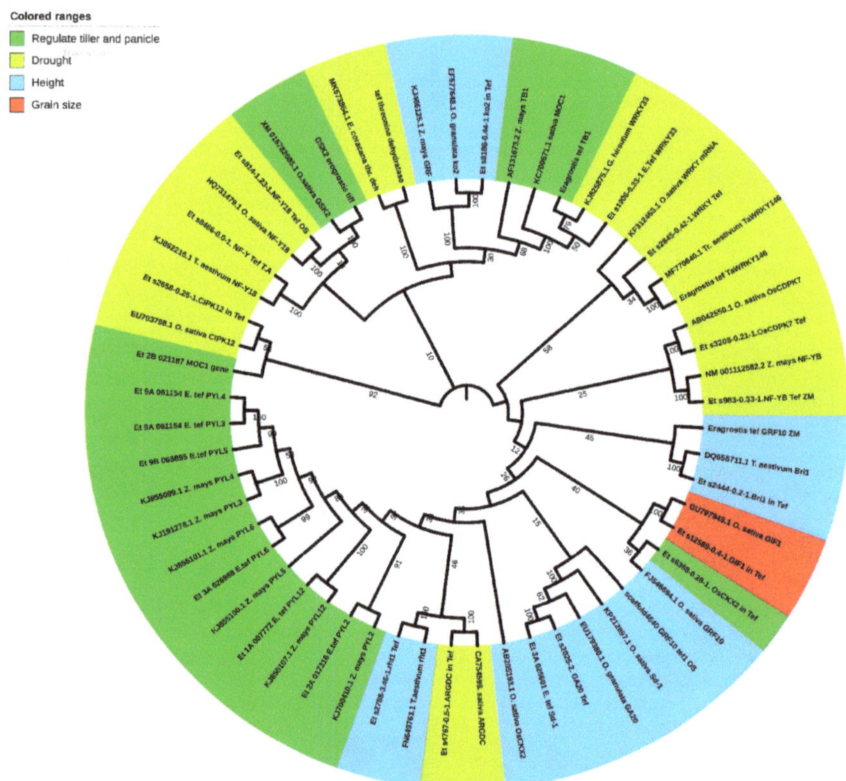


**Figure 4.** Phylogenetic tree of stress-resistant genes in tef and related monocots. The tree was constructed by using specific gene sequences downloaded from NCBI and Ensembl Plants. Bootstrap values (1000 pseudoreplicates) are shown on the nodes of the branches.

Tef is tolerant to poor soil conditions including waterlogging and drought [145]. However, tef yield is reduced by lodging, terminal drought and diseases. Therefore, tef is expected to benefit from CRISPR-Cas genome editing technology. The draft genome sequence of tef has been released [141]. Two complete homologous chromosomes with syntenic gene pairs have been reported in the tef genome due to its allotetraploid genome. The subgenomes are small (~300 Mb), with a low number of transposable elements (TE) and a high density of genes as compared to other polyploid grasses [141]. One of the major obstacles for the targeted breeding of tef is the presence of genes in two genomes (AA and BB: tef is allotetraploid, with 2n = 4x = 40 chromosomes). Gene redundancy poses a difficulty in mutagenesis for developing lodging-resistant and semi-dwarf varieties [146]. This obstacle can be overcome by techniques such as targeted genome engineering and marker assisted selection. In a plant genome, the majority of genes have variable expression patterns; therefore, the two sub-genomes are more likely to affect agronomic traits with different frequencies [141,147]. To utilize CRISPR-Cas technology in tef improvement, the identification of target genes that regulate agronomically important traits is crucial.

**Table 1.** Summary of genes involved in key agronomic traits of selected crops. Homologs of these genes in tef were downloaded from the genomic database to identify potential candidate genes for CRISPR-Cas-mediated gene editing in tef.

| Gene | Plant Name | Accession Number | Reference |
|---|---|---|---|
| **Plant Height** | | | |
| $KO_2$ | *Oryza sativa Japonica* | AY660664 | [148] |
| GA regulatory factor-like (GRF) mRNA | *Zea mays* | KJ466125 | [149] |
| growth-regulating factor 10 (GRF10) | *Oryza sativa Indica* | FJ546694 | [150] |
| GA20-oxidase (GA20ox2) | *Oryza granulata* | EU179380 | [151] |
| BRI1 | *Triticum aestivum* | DQ655711 | [152] |
| Sd-1 (used in green revl) | *Oryza sativa* | KP212897.1 | [70] |
| RHT1 | *Triticum aestivum* | FN649763 | [153] |
| **Number of Tillers and Panicle Branches** | | | |
| OsCKX2 | *Oryza sativa* | AB205193.1 | [154] |
| teosinte branched1 (tb1) | *switchgrass* | AF131673.2 | [155] |
| GSK2 | *Oryza sativa* | XM_015782085 | [156] |
| PYL2 | *Oryza sativa* | KJ700410.1 | [157] |
| PYL3, | *Oryza sativa* | KJ191278.1 | |
| PYL4, | *Oryza sativa* | KJ855099.1 | |
| PYL5, | *Oryza sativa* | KJ855100.1 | |
| PYL6 | *Oryza sativa* | KJ855101.1 | |
| PYL12 | *Oryza sativa* | KJ855107.1 | |
| monoculm1 MOC1 | *Oryza sativa Japonica* | KC700671.1 | [158] |
| **Grain Size** | | | |
| GIF1A | *Oryza sativa* | GU797949 | [159] |
| **Drought Tolerance** | | | |
| GhWRKY33 | *Gossypium hirsutum* | KJ825875.1 | [160] |
| WRKY mRNA | *Triticum aestivum* | KT865879 | [161] |
| threonine dehydratase mRNA | *Eleusine coracana* | MK573864 | [162] |
| OsCDPK7 | *Oryza sativa Japonica* | AB042550 | [163] |
| TaWRKY146 | *Triticum aestivum* | MF770640.1 | [164] |
| NF-Y18 | *Oryza sativa Japonica* | HQ731479 | [165] |
| Arginine decarboxylase (ADC) | *Oryza sativa Japonica* | CA754598.1 | [166] |
| CIPK12 | *Oryza sativa Japonica* | EU703798 | [166] |
| NF-YB | *Zea mays* | NM_001112582 | [167] |

## 5. Constraints and Solutions Related to CRISPR-Cas Genome Editing

The stable transfer of the transgene into the target site using CRISPR-Cas during the transformation process may cause the insertion of plasmid DNA or unwanted genes, which makes it a genetically modified (GM) crop. This limits the use of CRISPR-Cas system for sustainable agriculture and biotechnology because in some countries the use of GMOs is either tightly regulated or totally prohibited [168]. Although genetic segregation is the process by which the foreign DNA can be removed, this is not applicable to some clonally propagated plants. Moreover, in some countries, CRISPR-Cas products are still not acceptable because foreign DNA materials are used in the process, although these foreign materials are removed at the end [168]. In plants, DNA-free genome editing has been conducted using two approaches; these are pre-assembled ribonucleoproteins

(RNPs) [169,170] and the delivery of a combination of guide RNA and mRNA-encoding Cas [43]. However, CRISPR-Cas RNA transient expression efficiency is low, suggesting a need for additional optimization. Following this approach, the addition of a protectant for stabilizing RNA could prove to be a promising strategy [171].

Another major drawback of the CRISPR-Cas system is its non-specificity. In this case, Cas cleaves DNA at non-target sites that are not complementary of single guide RNA [172]. This drawback impedes CRISPR-Cas potential applications, particularly when genome alteration needs to be precise, as in the case of gene therapy. Off-target sites may not change plant breeding as much as the chemical and physical alterations used in traditional breeding procedures, which generate many alterations in plants [173]. These off-target alterations can be removed by performing backcrossing to the original plant. However, this takes several generations of investigation, and the improvement of the process will be slow.

In plants, the specificity of the CRISPR-Cas system has been estimated by deliberate non-target investigation [174]. For RNPs, non-target alterations were hardly recognized by thorough sequencing, indicating that RNPs enhance the specificity of the editing system [172]. However, no study has been reported on Cas specificity in plants. Several impartial strategies which include Digenome-seq, high-throughput genome-wide translocation sequencing (HTGTS), genome-wide unbiased identification of double stranded breaks (DSBs) enabled by sequencing (GUIDE-seq) and breaks labeling, enrichment on streptavidin, and sequencing (BLESS) have been used to investigate non-specific changes in human cells [175–178], and these strategies need to be administered in plants to evaluate Cas specificity at the genome level. The need for improving its specificity is a major challenge for CRISPR-Cas genome editing, which requires attention. Various strategies have been established for improving specificity [179], including high-fidelity Cas variants and the Cas paired nickase strategy [180–182].

## 6. Conclusions

Climate change and global warming are expected to trigger major abiotic stresses, which are expected to reduce crop yields and ultimately lead to food shortages in the foreseeable future. Since agricultural crops fulfill most of the world's food supply, it should be the topmost priority of plant biologists to take concrete measures to cope with climate change and the forecasted food shortages. Climate change and global warming are manifested by abiotic stress factors that could reduce crop productivity. The goal of this review was to provide an insight on the potential of advanced tools such as CRISPR-Cas for use by plant biologists in order to improve stress resilience, modify plant architecture and improve productivity. Application of this cutting-edge technology in underutilized/orphan crops such as tef will provide several benefits. It is expected to improve food security in the Horn of Africa, a region which is very vulnerable to the negative impact of climate change, and which has been experiencing frequent food insecurity and adding to the global refugee crisis. It will also enhance the acceptance of tef as a healthy and nutritious grain, which will play a role in reducing micronutrient deficiency.

**Author Contributions:** M.N. and A.L.-O. conceived the review; M.N. wrote the draft of the manuscript; A.L.-O., M.S., Z.T., G.B., A.L.K. and S.A. contributed and edited the manuscript. All authors have read and agreed to the published version of the manuscript.

**Funding:** This manuscript was support by the University of North Carolina at Greensboro (Grant # 133504 to A.L.-O.).

**Institutional Review Board Statement:** Not applicable.

**Informed Consent Statement:** Not applicable.

**Data Availability Statement:** Not applicable.

**Conflicts of Interest:** All the authors declare no conflict of interest.

## References


1. Cheng, A. Shaping a sustainable food future by rediscovering long-forgotten ancient grains. *Plant Sci.* **2018**, *269*, 136–142. [CrossRef]
2. Ray, D.K.; Gerber, J.S.; MacDonald, G.K.; West, P.C. Climate variation explains a third of global crop yield variability. *Nat. Commun.* **2015**, *6*, 1–9. [CrossRef]
3. Cochrane, L.; Bekele, Y.W. Average crop yield (2001–2017) in Ethiopia: Trends at national, regional and zonal levels. *Data Brief* **2018**, *16*, 1025. [CrossRef] [PubMed]
4. CSA. *Agricultural Sample Survey 2015/2016*; CSA: Addis Ababa, Ethiopia, 2016.
5. Tadele, Z.; Assefa, K. Increasing food production in Africa by boosting the productivity of understudied crops. *Agronomy* **2012**, *2*, 240–283. [CrossRef]
6. El-Alfy, T.S.; Ezzat, S.M.; Sleem, A.A. Chemical and biological study of the seeds of *Eragrostis tef* (Zucc.) Trotter. *Nat. Prod. Res.* **2012**, *26*, 619–629. [CrossRef]
7. Spaenij-Dekking, L.; Kooy-Winkelaar, Y.; Koning, F. The Ethiopian cereal tef in celiac disease. *N. Engl. J. Med.* **2005**, *353*, 1748–1749. [CrossRef]
8. Saturni, L.; Ferretti, G.; Bacchetti, T. The gluten-free diet: Safety and nutritional quality. *Nutrients* **2010**, *2*, 16–34. [CrossRef]
9. Gujral, N.; Freeman, H.J.; Thomson, A.B. Celiac disease: Prevalence, diagnosis, pathogenesis and treatment. *World J. Gastroenterol. WJG* **2012**, *18*, 6036. [CrossRef] [PubMed]
10. Yami, A. *17. Tef Straw: A Valuable Feed Resource to Improve Animal Production and Productivity*; Tef Improvement: Bern, Switzerland, 2013; Volume 233.
11. Assefa, K.; Yu, J.K.; Zeid, M.; Belay, G.; Tefera, H.; Sorrells, M. Breeding tef [*Eragrostis tef* (Zucc.) Trotter]: Conventional and molecular approaches. *Plant Breed.* **2011**, *130*, 1–9. [CrossRef]
12. Fang, X.; Li, Y.; Nie, J.; Wang, C.; Huang, K.; Zhang, Y.; Zhang, Y.; She, H.; Liu, X.; Ruan, R. Effects of nitrogen fertilizer and planting density on the leaf photosynthetic characteristics, agronomic traits and grain yield in common buckwheat (*Fagopyrum esculentum* M.). *Field Crops Res.* **2018**, *219*, 160–168. [CrossRef]
13. Zhao, H.; Mo, Z.; Lin, Q.; Pan, S.; Duan, M.; Tian, H.; Wang, S.; Tang, X. Relationships between grain yield and agronomic traits of rice in southern China. *Chil. J. Agric. Res.* **2020**, *80*, 72–79. [CrossRef]
14. Belete, T.; Mathewos, T.; Daba, G. Correlation of yield and yield related traits of Tef (*Eragrostis tef* (Zucc.) Trotter) varieties in Ethiopia. *J. Genet. Environ. Resour. Conserv.* **2020**, *8*, 35–39.
15. Tilahun, Z.M. Effect of row spacing and nitrogen fertilizer levels on yield and yield components of rice varieties. *World Sci. News* **2019**, *116*, 180–193.
16. Arefaine, A.; Adhanom, D.; Tekeste, N. Response of Teff (*Eragrostis tef* (Zucc.) Trotter) to Seeding Rate and Methods of Sowing on Yield and Yield Attributes in a Subhumid Environment, Northern Ethiopia. *Int. J. Agron.* **2020**, *2020*, 1–7. [CrossRef]
17. Haftamu, G.; Mitiku, H.; Yamoah, C. Tillage frequency, soil compaction and N-fertilizer rate effects on yield of teff (*Eragrostis tef* (Zucc.) Trotter) in central zone of Tigray, Northern Ethiopia. *Momona Ethiop. J. Sci.* **2009**, *1*, 82–94.
18. Menalled, F.D. Sustainable Agriculture and Integrated Weed Management. In *Weed Control: Sustainability, Hazards, and Risks in Cropping Systems Worldwide*; CRC Press: Boca Raton, FL, USA, 2018; Volume 3.
19. Abrha, B.; Tsegay, A.; Gebrehiwot, K. Economic analysis of tef (*Eragrostis tef* (zucc.) trotter) yield in response to soil water, weed and fertilizer managements in the northern highlands of Ethiopia. *J. Drylands* **2017**, *2*, 675–682.
20. Rockström, J.; Falkenmark, M. Agriculture: Increase water harvesting in Africa. *Nat. News* **2015**, *519*, 283. [CrossRef]
21. Matsuura, A.; Tsuji, W.; An, P.; Inanaga, S.; Murata, K. Effect of pre-and post-heading water deficit on growth and grain yield of four millets. *Plant Prod. Sci.* **2012**, *15*, 323–331. [CrossRef]
22. Winkel, T.; Renno, J.-F.; Payne, W. Effect of the timing of water deficit on growth, phenology and yield of pearl millet (*Pennisetum glaucum* (L.) R. Br.) grown in Sahelian conditions. *J. Exp. Bot.* **1997**, *48*, 1001–1009. [CrossRef]
23. Abraha, M.T.; Hussein, S.; Laing, M.; Assefa, K. Genetic management of drought in tef: Current status and future research directions. *Glob. J. Crop Soil Sci. Plant Breed.* **2015**, *3*, 156–161.
24. Assefa, K.; Chanyalew, S.; Tadele, Z. Tef, *Eragrostis tef* (Zucc.) Trotter. *Millets and Sorghum: Biology and Genetic Improvement*. 2017, pp. 226–266. Available online: https://www.researchgate.net/publication/312353091_Tef_Eragrostis_tef_Zucc_Trotter_Biology_and_Genetic_Improvement (accessed on 19 March 2021).
25. Blösch, R.; Plaza-Wüthrich, S.; de Reuille, P.B.; Weichert, A.; Routier-Kierzkowska, A.-L.; Cannarozzi, G.; Robinson, S.; Tadele, Z. Panicle angle is an important factor in tef lodging tolerance. *Front. Plant Sci.* **2020**, *11*, 1–12. [CrossRef]
26. Van Delden, S.; Vos, J.; Ennos, A.; Stomph, T. Analysing lodging of the panicle bearing cereal teff (*Eragrostis tef*). *New Phytol.* **2010**, *186*, 696–707. [CrossRef] [PubMed]
27. Ketema, S. *Tef (Eragrostis Tef) Breeding, Genetic Resources, Agronomy, Utilization and Role in Ethiopian Agriculture*; Institute of Agricultural Research: Zaria, Nigeria, 1993.
28. Jifar, H.; Tesfaye, K.; Assefa, K.; Chanyalew, S.; Tadele, Z. Semi-dwarf tef lines for high seed yield and lodging tolerance in Central Ethiopia. *Afr. Crop Sci. J.* **2017**, *25*, 419–439. [CrossRef]
29. Paff, K.; Asseng, S. A review of tef physiology for developing a tef crop model. *Eur. J. Agron.* **2018**, *94*, 54–66. [CrossRef]
30. Debbarma, J.; Sarki, Y.N.; Saikia, B.; Boruah, H.P.D.; Singha, D.L.; Chikkaputtaiah, C. Ethylene response factor (ERF) family proteins in abiotic stresses and CRISPR–Cas9 genome editing of ERFs for multiple abiotic stress tolerance in crop plants: A review. *Mol. Biotechnol.* **2019**, *61*, 153–172. [CrossRef] [PubMed]

31. Tiwari, Y.K.; Yadav, S.K. High temperature stress tolerance in maize (*Zea mays* L.): Physiological and molecular mechanisms. *J. Plant Biol.* **2019**, *62*, 93–102. [CrossRef]
32. Zhou, H.; Liu, B.; Weeks, D.P.; Spalding, M.H.; Yang, B. Large chromosomal deletions and heritable small genetic changes induced by CRISPR/Cas9 in rice. *Nucleic Acids Res.* **2014**, *42*, 10903–10914. [CrossRef] [PubMed]
33. Jiang, W.; Zhou, H.; Bi, H.; Fromm, M.; Yang, B.; Weeks, D.P. Demonstration of CRISPR/Cas9/sgRNA-mediated targeted gene modification in Arabidopsis, tobacco, sorghum and rice. *Nucleic Acids Res.* **2013**, *41*, e188. [CrossRef]
34. Saika, H.; Mori, A.; Endo, M.; Toki, S. Targeted deletion of rice retrotransposon Tos17 via CRISPR/Cas9. *Plant Cell Rep.* **2019**, *38*, 455–458. [CrossRef]
35. Lee, K.; Eggenberger, A.L.; Banakar, R.; McCaw, M.E.; Zhu, H.; Main, M.; Kang, M.; Gelvin, S.B.; Wang, K. CRISPR/Cas9-mediated targeted T-DNA integration in rice. *Plant Mol. Biol.* **2019**, *99*, 317–328. [CrossRef]
36. Dayani, S.; Sabzalian, M.R.; Mazaheri-Tirani, M. CRISPR/Cas9 Genome Editing in Bread Wheat (*Triticum aestivum* L.) Genetic Improvement. In *Advances in Plant Breeding Strategies: Cereals*; Springer: Berlin/Heidelberg, Germany, 2019; pp. 453–469.
37. Jaganathan, D.; Ramasamy, K.; Sellamuthu, G.; Jayabalan, S.; Venkataraman, G. CRISPR for crop improvement: An update review. *Front. Plant Sci.* **2018**, *9*, 985. [CrossRef] [PubMed]
38. Basso, M.F.; Ferreira, P.C.G.; Kobayashi, A.K.; Harmon, F.G.; Nepomuceno, A.L.; Molinari, H.B.C.; Grossi-de-Sa, M.F. Micro RNA s and new biotechnological tools for its modulation and improving stress tolerance in plants. *Plant Biotechnol. J.* **2019**, *17*, 1482–1500. [CrossRef] [PubMed]
39. Shi, J.; Gao, H.; Wang, H.; Lafitte, H.R.; Archibald, R.L.; Yang, M.; Hakimi, S.M.; Mo, H.; Habben, J.E. ARGOS 8 variants generated by CRISPR-Cas9 improve maize grain yield under field drought stress conditions. *Plant Biotechnol. J.* **2017**, *15*, 207–216. [CrossRef] [PubMed]
40. Shen, C.; Que, Z.; Xia, Y.; Tang, N.; Li, D.; He, R.; Cao, M. Knock out of the annexin gene OsAnn3 via CRISPR/Cas9-mediated genome editing decreased cold tolerance in rice. *J. Plant Biol.* **2017**, *60*, 539–547. [CrossRef]
41. Zhang, A.; Liu, Y.; Wang, F.; Li, T.; Chen, Z.; Kong, D.; Bi, J.; Zhang, F.; Luo, X.; Wang, J. Enhanced rice salinity tolerance via CRISPR/Cas9-targeted mutagenesis of the OsRR22 gene. *Mol. Breed.* **2019**, *39*, 47. [CrossRef] [PubMed]
42. Sadanandom, A.; Srivastava, A.K.; Zhang, C. Targeted mutagenesis of the SUMO protease, Overly Tolerant to Salt1 in rice through CRISPR/Cas9-mediated genome editing reveals a major role of this SUMO protease in salt tolerance. *BioRxiv* **2019**. [CrossRef]
43. Zhang, Y.; Liang, Z.; Zong, Y.; Wang, Y.; Liu, J.; Chen, K.; Qiu, J.-L.; Gao, C. Efficient and transgene-free genome editing in wheat through transient expression of CRISPR/Cas9 DNA or RNA. *Nat. Commun.* **2016**, *7*, 1–8. [CrossRef]
44. Wang, W.; Akhunova, A.; Chao, S.; Akhunov, E. Optimizing multiplex CRISPR/Cas9-based genome editing for wheat. *BioRxiv* **2016**. [CrossRef]
45. Howells, R.M.; Craze, M.; Bowden, S.; Wallington, E.J. Efficient generation of stable, heritable gene edits in wheat using CRISPR/Cas9. *BMC Plant Biol.* **2018**, *18*, 215. [CrossRef] [PubMed]
46. Berry, P.; Sterling, M.; Spink, J.; Baker, C.; Sylvester-Bradley, R.; Mooney, S.; Tams, A.; Ennos, A. Understanding and reducing lodging in cereals. *Adv. Agron.* **2004**, *84*, 215–269.
47. Dagnaw, H. Influence of Nitrogen Fertilizer Rates and Varieties on Grain Yield, Grain Nutrition and Injera Sensory Quality of Tef [*Eragrostis tef* (Zucc.) Trotter] Varieties. Master's Thesis, Adis Ababa University, Addis Ababa, Ethiopia, 2018.
48. Sterling, M.; Baker, C.; Berry, P.; Wade, A. An experimental investigation of the lodging of wheat. *Agric. For. Meteorol.* **2003**, *119*, 149–165. [CrossRef]
49. Merchuk-Ovnat, L.; Bimro, J.; Yaakov, N.; Kutsher, Y.; Amir-Segev, O.; Reuveni, M. In-Depth Field Characterization of Teff [*Eragrostis tef* (Zucc.) Trotter] Variation: From Agronomic to Sensory Traits. *Agronomy* **2020**, *10*, 1107. [CrossRef]
50. Assefa, K.; Cannarozzi, G.; Girma, D.; Kamies, R.; Chanyalew, S.; Plaza-Wüthrich, S.; Blösch, R.; Rindisbacher, A.; Rafudeen, S.; Tadele, Z. Genetic diversity in tef [*Eragrostis tef* (Zucc.) Trotter]. *Front. Plant Sci.* **2015**, *6*, 177. [CrossRef] [PubMed]
51. Wu, W.; Ma, B.L. A new method for assessing plant lodging and the impact of management options on lodging in canola crop production. *Sci. Rep.* **2016**, *6*, 1–17. [CrossRef] [PubMed]
52. Mariani, L.; Ferrante, A. Agronomic management for enhancing plant tolerance to abiotic stresses—Drought, salinity, hypoxia, and lodging. *Horticulturae* **2017**, *3*, 52. [CrossRef]
53. Epstein, E. The anomaly of silicon in plant biology. *Proc. Natl. Acad. Sci. USA* **1994**, *91*, 11–17. [CrossRef]
54. Ma, J.F. Role of silicon in enhancing the resistance of plants to biotic and abiotic stresses. *Soil Sci. Plant Nutr.* **2004**, *50*, 11–18. [CrossRef]
55. De Carvalho, D.D.; Costa, F.T.; Duran, N.; Haun, M. Cytotoxic activity of violacein in human colon cancer cells. *Toxicol. In Vitro* **2006**, *20*, 1514–1521. [CrossRef]
56. Liang, Y.; Nikolic, M.; Bélanger, R.; Gong, H.; Song, A. *Silicon in Agriculture*; Springer: Dordrecht, The Netherlands, 2015; Volume 10, pp. 978–994.
57. Ligaba-Osena, A.; Guo, W.; Choi, S.C.; Limmer, M.A.; Seyfferth, A.L.; Hankoua, B.B. Silicon enhances biomass and grain yield in an ancient crop tef [*Eragrostis tef* (Zucc.) Trotter]. *Front. Plant Sci.* **2020**, *11*, 608503. [CrossRef]
58. Deshmukh, R.K.; Ma, J.F.; Bélanger, R.R. Role of silicon in plants. *Front. Plant Sci.* **2017**, *8*, 1858. [CrossRef]
59. Jöst, M.; Esfeld, K.; Burian, A.; Cannarozzi, G.; Chanyalew, S.; Kuhlemeier, C.; Assefa, K.; Tadele, Z. Semi-dwarfism and lodging tolerance in tef (*Eragrostis tef*) is linked to a mutation in the α-Tubulin 1 gene. *J. Exp. Bot.* **2015**, *66*, 933–944. [CrossRef]
60. Cannarozzi, G.; Chanyalew, S.; Assefa, K.; Bekele, A.; Blösch, R.; Weichert, A.; Klauser, D.; Plaza-Wüthrich, S.; Esfeld, K.; Jöst, M. Technology generation to dissemination: Lessons learned from the tef improvement project. *Euphytica* **2018**, *214*, 1–20. [CrossRef]

61. Rademacher, W. Growth retardants: Effects on gibberellin biosynthesis and other metabolic pathways. *Annu. Rev. Plant Biol.* **2000**, *51*, 501–531. [CrossRef] [PubMed]
62. Sanvicente, P.; Lazarevitch, S.; Blouet, A.; Guckert, A. Morphological and anatomical modifications in winter barley culm after late plant growth regulator treatment. *Eur. J. Agron.* **1999**, *11*, 45–51. [CrossRef]
63. Hedden, P. The genes of the Green Revolution. *Trends Genet.* **2003**, *19*, 5–9. [CrossRef]
64. Itoh, H.; Ueguchi-Tanaka, M.; Sentoku, N.; Kitano, H.; Matsuoka, M.; Kobayashi, M. Cloning and functional analysis of two gibberellin 3β-hydroxylase genes that are differently expressed during the growth of rice. *Proc. Natl. Acad. Sci. USA* **2001**, *98*, 8909–8914. [CrossRef] [PubMed]
65. Monna, L.; Kitazawa, N.; Yoshino, R.; Suzuki, J.; Masuda, H.; Maehara, Y.; Tanji, M.; Sato, M.; Nasu, S.; Minobe, Y. Positional cloning of rice semidwarfing gene, sd-1: Rice "green revolution gene" encodes a mutant enzyme involved in gibberellin synthesis. *DNA Res.* **2002**, *9*, 11–17. [CrossRef] [PubMed]
66. Sasaki, A.; Ashikari, M.; Ueguchi-Tanaka, M.; Itoh, H.; Nishimura, A.; Swapan, D.; Ishiyama, K.; Saito, T.; Kobayashi, M.; Khush, G.S. Green revolution: A mutant gibberellin-synthesis gene in rice. *Nature* **2002**, *416*, 701. [CrossRef]
67. Hong, Z.; Ueguchi-Tanaka, M.; Umemura, K.; Uozu, S.; Fujioka, S.; Takatsuto, S.; Yoshida, S.; Ashikari, M.; Kitano, H.; Matsuoka, M. A rice brassinosteroid-deficient mutant, ebisu dwarf (d2), is caused by a loss of function of a new member of cytochrome P450. *Plant Cell* **2003**, *15*, 2900–2910. [CrossRef] [PubMed]
68. Multani, D.S.; Briggs, S.P.; Chamberlin, M.A.; Blakeslee, J.J.; Murphy, A.S.; Johal, G.S. Loss of an MDR transporter in compact stalks of maize br2 and sorghum dw3 mutants. *Science* **2003**, *302*, 81–84. [CrossRef]
69. Asano, K.; Hirano, K.; Ueguchi-Tanaka, M.; Angeles-Shim, R.B.; Komura, T.; Satoh, H.; Kitano, H.; Matsuoka, M.; Ashikari, M. Isolation and characterization of dominant dwarf mutants, Slr1-d, in rice. *Mol. Genet. Genom.* **2009**, *281*, 223–231. [CrossRef]
70. Spielmeyer, W.; Ellis, M.H.; Chandler, P.M. Semidwarf (sd-1),"green revolution" rice, contains a defective gibberellin 20-oxidase gene. *Proc. Natl. Acad. Sci. USA* **2002**, *99*, 9043–9048. [CrossRef] [PubMed]
71. Muangprom, A.; Thomas, S.G.; Sun, T.-p.; Osborn, T.C. A novel dwarfing mutation in a green revolution gene from *Brassica rapa*. *Plant Physiol.* **2005**, *137*, 931–938. [CrossRef]
72. Peng, J.; Richards, D.E.; Hartley, N.M.; Murphy, G.P.; Devos, K.M.; Flintham, J.E.; Beales, J.; Fish, L.J.; Worland, A.J.; Pelica, F. 'Green revolution'genes encode mutant gibberellin response modulators. *Nature* **1999**, *400*, 256. [CrossRef] [PubMed]
73. Wang, Y.; Chen, L.; Du, Y.; Yang, Z.; Condon, A.G.; Hu, Y.-G. Genetic effect of dwarfing gene Rht13 compared with Rht-D1b on plant height and some agronomic traits in common wheat (*Triticum aestivum* L.). *Field Crops Res.* **2014**, *162*, 39–47. [CrossRef]
74. Itoh, H.; Tatsumi, T.; Sakamoto, T.; Otomo, K.; Toyomasu, T.; Kitano, H.; Ashikari, M.; Ichihara, S.; Matsuoka, M. A rice semi-dwarf gene, Tan-Ginbozu (D35), encodes the gibberellin biosynthesis enzyme, ent-kaurene oxidase. *Plant Mol. Biol.* **2004**, *54*, 533–547. [CrossRef]
75. Hilley, J.L.; Weers, B.D.; Truong, S.K.; McCormick, R.F.; Mattison, A.J.; McKinley, B.A.; Morishige, D.T.; Mullet, J.E. Sorghum Dw2 encodes a protein kinase regulator of stem internode length. *Sci. Rep.* **2017**, *7*, 1–13. [CrossRef] [PubMed]
76. Yamaguchi, M.; Fujimoto, H.; Hirano, K.; Araki-Nakamura, S.; Ohmae-Shinohara, K.; Fujii, A.; Tsunashima, M.; Song, X.J.; Ito, Y.; Nagae, R. Sorghum Dw1, an agronomically important gene for lodging resistance, encodes a novel protein involved in cell proliferation. *Sci. Rep.* **2016**, *6*, 1–11. [CrossRef] [PubMed]
77. Mickelbart, M.V.; Hasegawa, P.M.; Bailey-Serres, J. Genetic mechanisms of abiotic stress tolerance that translate to crop yield stability. *Nat. Rev. Genet.* **2015**, *16*, 237–251. [CrossRef] [PubMed]
78. Duque, A.S.; de Almeida, A.M.; da Silva, A.B.; da Silva, J.M.; Farinha, A.P.; Santos, D.; Fevereiro, P.; de Sousa Araújo, S. Abiotic stress responses in plants: Unraveling the complexity of genes and networks to survive. In *Abiotic Stress-Plant Responses and Applications in Agriculture*; IntechOpen: London, UK, 2013; pp. 49–101. [CrossRef]
79. Ayele, M.; Blum, A.; Nguyen, H.T. Diversity for osmotic adjustment and root depth in tef [*Eragrostis tef* (Zucc.) Trotter]. *Euphytica* **2001**, *121*, 237–249. [CrossRef]
80. Araya, A.; Stroosnijder, L.; Girmay, G.; Keesstra, S. Crop coefficient, yield response to water stress and water productivity of teff (*Eragrostis tef* (Zucc.). *Agric. Water Manag.* **2011**, *98*, 775–783. [CrossRef]
81. Plaza-Wüthrich, S.; Blösch, R.; Rindisbacher, A.; Cannarozzi, G.; Tadele, Z. Gibberellin deficiency confers both lodging and drought tolerance in small cereals. *Front. Plant Sci.* **2016**, *7*, 643. [CrossRef]
82. Kusaka, M.; Lalusin, A.G.; Fujimura, T. The maintenance of growth and turgor in pearl millet (*Pennisetum glaucum* [L.] Leeke) cultivars with different root structures and osmo-regulation under drought stress. *Plant Sci.* **2005**, *168*, 1–14. [CrossRef]
83. Farooq, M.; Wahid, A.; Kobayashi, N.; Fujita, D.; Basra, S. Plant drought stress: Effects, mechanisms and management. In *Sustainable Agriculture*; Springer: Berlin/Heidelberg, Germany, 2009; pp. 153–188.
84. Merrill, S.D.; Tanaka, D.L.; Hanson, J.D. Root length growth of eight crop species in Haplustoll soils. *Soil Sci. Soc. Am. J.* **2002**, *66*, 913–923. [CrossRef]
85. Renton, M.; Chauhan, B.S. Modelling crop-weed competition: Why, what, how and what lies ahead? *Crop Prot.* **2017**, *95*, 101–108. [CrossRef]
86. Andrew, I.; Storkey, J.; Sparkes, D. A review of the potential for competitive cereal cultivars as a tool in integrated weed management. *Weed Res.* **2015**, *55*, 239–248. [CrossRef]
87. Hansen, P.K.; Kristensen, K.; Willas, J. A weed suppressive index for spring barley (*Hordeum vulgare*) varieties. *Weed Res.* **2008**, *48*, 225–236. [CrossRef]
88. Gebrehiwot, H.G.; Aune, J.B.; Netland, J.; Eklo, O.M.; Torp, T.; Brandsæter, L.O. Weed-Competitive Ability of Teff (*Eragrostis tef* (Zucc.) Trotter) Varieties. *Agronomy* **2020**, *10*, 108. [CrossRef]

89. Laizer, H.C.; Chacha, M.N.; Ndakidemi, P.A. Farmers' knowledge, perceptions and practices in managing weeds and insect pests of common bean in Northern Tanzania. *Sustainability* **2019**, *11*, 4076. [CrossRef]
90. Dezfulian, M.H.; Foreman, C.; Jalili, E.; Pal, M.; Dhaliwal, R.K.; Roberto, D.K.A.; Imre, K.M.; Kohalmi, S.E.; Crosby, W.L. Acetolactate synthase regulatory subunits play divergent and overlapping roles in branched-chain amino acid synthesis and Arabidopsis development. *BMC Plant Biol.* **2017**, *17*, 71. [CrossRef]
91. Singh, S.; Kumar, V.; Dhanjal, D.S.; Singh, J. Herbicides and Plant Growth Regulators: Current Developments and Future Challenges. In *Natural Bioactive Products in Sustainable Agriculture*; Singh, J., Yadav, A.N., Eds.; Springer: Singapore, 2020. [CrossRef]
92. Küpper, A.; Borgato, E.A.; Patterson, E.L.; Netto, A.G.; Nicolai, M.; Carvalho, S.J.P.d.; Nissen, S.J.; Gaines, T.A.; Christoffoleti, P.J. Multiple Resistance to Glyphosate and Acetolactate Synthase Inhibitors in Palmer Amaranth (*Amaranthus palmeri*) Identified in Brazil. *Weed Sci.* **2017**, *65*, 317–326. [CrossRef]
93. Xue, W.; Xing, Y.; Weng, X.; Zhao, Y.; Tang, W.; Wang, L.; Zhou, H.; Yu, S.; Xu, C.; Li, X. Natural variation in Ghd7 is an important regulator of heading date and yield potential in rice. *Nat. Genet.* **2008**, *40*, 761–767. [CrossRef]
94. Guillen-Portal, F.R.; Stougaard, R.N.; Xue, Q.; Eskridge, K.M. Compensatory mechanisms associated with the effect of spring wheat seed size on wild oat competition. *Crop Sci.* **2006**, *46*, 935–945. [CrossRef]
95. Gupta, P.K.; Rustgi, S.; Kumar, N. Genetic and molecular basis of grain size and grain number and its relevance to grain productivity in higher plants. *Genome* **2006**, *49*, 565–571. [CrossRef]
96. Huang, M.; Zou, Y.-B.; Jiang, P.; Bing, X.; Md, I.; Ao, H.-J. Relationship between grain yield and yield components in super hybrid rice. *Agric. Sci. China* **2011**, *10*, 1537–1544. [CrossRef]
97. Yao, F.; Huang, J.; Cui, K.; Nie, L.; Xiang, J.; Liu, X.; Wu, W.; Chen, M.; Peng, S. Agronomic performance of high-yielding rice variety grown under alternate wetting and drying irrigation. *Field Crops Res.* **2012**, *126*, 16–22. [CrossRef]
98. Terao, T.; Nagata, K.; Morino, K.; Hirose, T. A gene controlling the number of primary rachis branches also controls the vascular bundle formation and hence is responsible to increase the harvest index and grain yield in rice. *Theor. Appl. Genet.* **2010**, *120*, 875–893. [CrossRef]
99. Zhang, Y.-C.; Yu, Y.; Wang, C.-Y.; Li, Z.-Y.; Liu, Q.; Xu, J.; Liao, J.-Y.; Wang, X.-J.; Qu, L.-H.; Chen, F. Overexpression of microRNA OsmiR397 improves rice yield by increasing grain size and promoting panicle branching. *Nat. Biotechnol.* **2013**, *31*, 848–852. [CrossRef] [PubMed]
100. Huang, X.; Qian, Q.; Liu, Z.; Sun, H.; He, S.; Luo, D.; Xia, G.; Chu, C.; Li, J.; Fu, X. Natural variation at the DEP1 locus enhances grain yield in rice. *Nat. Genet.* **2009**, *41*, 494–497. [CrossRef]
101. Miura, K.; Ikeda, M.; Matsubara, A.; Song, X.-J.; Ito, M.; Asano, K.; Matsuoka, M.; Kitano, H.; Ashikari, M. OsSPL14 promotes panicle branching and higher grain productivity in rice. *Nat. Genet.* **2010**, *42*, 545–549. [CrossRef]
102. Assefa, K.; Chanyalew, S.; Metaferia, G. *Conventional and Molecular Tef Breeding*; Tef Improvement: Bern, Switzerland, 2011; Volume 33.
103. Tadele, Z.; Ferede Haile, B.; Abreha, E.; Assefa, K.; Chanyalew, S.; Mekbib, F. *Morpho-Physiologic, Genotype X Environment Interaction and In Vitro Evaluation for Drought Tolerance in Tef Eragrostis tef (Zucc.) Trotter, Ethiopia*; Haramaya University: Haromaya, Ethiopia, 2018.
104. Berehe, T. Breakthrough in tef breeding technique. *FAO Inf. Bull. Cerealimprovement Prod. Near East Proj.* **1975**, *3*, 11–23.
105. Belay, G.; Tefera, H.; Getachew, A.; Assefa, K.; Metaferia, G. Highly client-oriented breeding with farmer participation in the Ethiopian cereal tef [*Eragrostis tef* (Zucc.) Trotter]. *Afr. J. Agric. Res.* **2008**, *3*, 22–28.
106. Belay, G.; Tefera, H.; Tadesse, B.; Metaferia, G.; Jarra, D.; Tadesse, T. Participatory variety selection in the Ethiopian cereal tef (*Eragrostis tef*). *Exp. Agric.* **2006**, *42*, 91–101. [CrossRef]
107. Chanyalew, S.; Assefa, K.; Tadele, Z. Tef [*Eragrostis tef* (Zucc.) Trotter] Breeding. In *Advances in Plant Breeding Strategies: Cereals*; Springer: Berlin/Heidelberg, Germany, 2019; pp. 373–403.
108. Ibitoye, D.; Akin-Idowu, P. Marker-assisted-selection (MAS): A fast track to increase genetic gain in horticultural crop breeding. *Afr. J. Biotechnol.* **2010**, *9*, 8889–8895.
109. Jiang, G.-L. Molecular markers and marker-assisted breeding in plants. *Plant Breed. Lab. Fields* **2013**, 45–83. [CrossRef]
110. Yu, J.-K.; Sun, Q.; Rota, M.L.; Edwards, H.; Tefera, H.; Sorrells, M.E. Expressed sequence tag analysis in tef (*Eragrostis tef* (Zucc.) Trotter). *Genome* **2006**, *49*, 365–372. [CrossRef]
111. Bai, G.; Ayele, M.; Tefera, H.; Nguyen, H.T. Amplified fragment length polymorphism analysis of tef [*Eragrostis tef* (Zucc.) Trotter]. *Crop Sci.* **1999**, *39*, 819–824. [CrossRef]
112. Abraha, M.T.; Shimelis, H.; Laing, M.; Assefa, K.; Amelework, B. Assessment of the genetic relationship of tef (*Eragrostis tef*) genotypes using SSR markers. *S. Afr. J. Bot.* **2016**, *105*, 106–110. [CrossRef]
113. Ayele, M.; Nguyen, H. Evaluation of amplified fragment length polymorphism markers in tef, *Eragrostis tef* (Zucc.) Trotter, and related species. *Plant Breed.* **2000**, *119*, 403–409. [CrossRef]
114. Kashtwari, M.; Wani, A.A.; Rather, R.N. TILLING: An alternative path for crop improvement. *J. Crop Improv.* **2019**, *33*, 83–109. [CrossRef]
115. Tadele, Z. Orphan crops: Their importance and the urgency of improvement. *Planta* **2019**, *250*, 677–694. [CrossRef]
116. Voytas, D.F. Plant Genome Engineering with Sequence-Specific Nucleases. *Annu. Rev. Plant Biol.* **2013**, *64*, 327–350. [CrossRef]
117. Jinek, M.; Chylinski, K.; Fonfara, I.; Hauer, M.; Doudna, J.A.; Charpentier, E. A programmable dual-RNA–guided DNA endonuclease in adaptive bacterial immunity. *Science* **2012**, *337*, 816–821. [CrossRef]
118. Cong, L.; Ran, F.A.; Cox, D.; Lin, S.; Barretto, R.; Habib, N.; Hsu, P.D.; Wu, X.; Jiang, W.; Marraffini, L.A. Multiplex genome engineering using CRISPR/Cas systems. *Science* **2013**, *339*, 819–823. [CrossRef]

119. Feng, Z.; Zhang, B.; Ding, W.; Liu, X.; Yang, D.-L.; Wei, P.; Cao, F.; Zhu, S.; Zhang, F.; Mao, Y. Efficient genome editing in plants using a CRISPR/Cas system. *Cell Res.* **2013**, *23*, 1229. [CrossRef]
120. Shan, Q.; Wang, Y.; Li, J.; Zhang, Y.; Chen, K.; Liang, Z.; Zhang, K.; Liu, J.; Xi, J.J.; Qiu, J.-L. Targeted genome modification of crop plants using a CRISPR-Cas system. *Nat. Biotechnol.* **2013**, *31*, 686. [CrossRef] [PubMed]
121. Christian, M.; Cermak, T.; Doyle, E.L.; Schmidt, C.; Zhang, F.; Hummel, A.; Bogdanove, A.J.; Voytas, D.F. Targeting DNA double-strand breaks with TAL effector nucleases. *Genetics* **2010**, *186*, 757–761. [CrossRef] [PubMed]
122. Zhang, Y.; Zhang, F.; Li, X.; Baller, J.A.; Qi, Y.; Starker, C.G.; Bogdanove, A.J.; Voytas, D.F. Transcription activator-like effector nucleases enable efficient plant genome engineering. *Plant Physiol.* **2013**, *161*, 20–27. [CrossRef] [PubMed]
123. Shan, Q.; Wang, Y.; Chen, K.; Liang, Z.; Li, J.; Zhang, Y.; Zhang, K.; Liu, J.; Voytas, D.F.; Zheng, X. Rapid and efficient gene modification in rice and *Brachypodium* using TALENs. *Mol. Plant* **2013**, *6*, 1365–1368. [CrossRef] [PubMed]
124. Zhang, F.; Maeder, M.L.; Unger-Wallace, E.; Hoshaw, J.P.; Reyon, D.; Christian, M.; Li, X.; Pierick, C.J.; Dobbs, D.; Peterson, T. High frequency targeted mutagenesis in *Arabidopsis thaliana* using zinc finger nucleases. *Proc. Natl. Acad. Sci. USA* **2010**, *107*, 12028–12033. [CrossRef]
125. Sander, J.D.; Dahlborg, E.J.; Goodwin, M.J.; Cade, L.; Zhang, F.; Cifuentes, D.; Curtin, S.J.; Blackburn, J.S.; Thibodeau-Beganny, S.; Qi, Y. Selection-free zinc-finger-nuclease engineering by context-dependent assembly (CoDA). *Nat. Methods* **2011**, *8*, 67. [CrossRef] [PubMed]
126. Jiang, Z.; Hong, X.; Zhang, S.; Yao, R.; Xiao, Y. CRISPR base editing and prime editing: DSB and template-free editing systems for bacteria and plants. *Synth. Syst. Biotechnol.* **2020**, *5*, 277–292.
127. Alok, A.; Sandhya, D.; Jogam, P.; Rodrigues, V.; Bhati, K.K.; Sharma, H.; Kumar, J. The rise of the CRISPR/Cpf1 system for efficient genome editing in plants. *Front. Plant Sci.* **2020**, *11*, 264. [CrossRef] [PubMed]
128. Lin, Q.; Zong, Y.; Xue, C.; Wang, S.; Jin, S.; Zhu, Z.; Wang, Y.; Anzalone, A.V.; Raguram, A.; Doman, J.L. Prime genome editing in rice and wheat. *Nat. Biotechnol.* **2020**, *38*, 582–585. [CrossRef]
129. Li, S.; Zhang, X.; Wang, W.; Guo, X.; Wu, Z.; Du, W.; Zhao, Y.; Xia, L. Expanding the scope of CRISPR/Cpf1-mediated genome editing in rice. *Mol. Plant* **2018**, *11*, 995–998. [CrossRef]
130. Mohanta, T.K.; Bashir, T.; Hashem, A.; Abd_Allah, E.F.; Bae, H. Genome editing tools in plants. *Genes* **2017**, *8*, 399. [CrossRef] [PubMed]
131. Lowe, K.; Wu, E.; Wang, N.; Hoerster, G.; Hastings, C.; Cho, M.-J.; Scelonge, C.; Lenderts, B.; Chamberlin, M.; Cushatt, J. Morphogenic regulators Baby boom and Wuschel improve monocot transformation. *Plant Cell* **2016**, *28*, 1998–2015. [CrossRef] [PubMed]
132. Debernardi, J.M.; Tricoli, D.M.; Ercoli, M.F.; Hayta, S.; Ronald, P.; Palatnik, J.F.; Dubcovsky, J. A GRF–GIF chimeric protein improves the regeneration efficiency of transgenic plants. *Nat. Biotechnol.* **2020**, *38*, 1274–1279. [CrossRef]
133. Li, J.-F.; Norville, J.E.; Aach, J.; McCormack, M.; Zhang, D.; Bush, J.; Church, G.M.; Sheen, J. Multiplex and homologous recombination–mediated genome editing in Arabidopsis and *Nicotiana benthamiana* using guide RNA and Cas9. *Nat. Biotechnol.* **2013**, *31*, 688. [CrossRef]
134. Nekrasov, V.; Staskawicz, B.; Weigel, D.; Jones, J.D.; Kamoun, S. Targeted mutagenesis in the model plant *Nicotiana benthamiana* using Cas9 RNA-guided endonuclease. *Nat. Biotechnol.* **2013**, *31*, 691. [CrossRef] [PubMed]
135. Paul, J.W.; Qi, Y. CRISPR/Cas9 for plant genome editing: Accomplishments, problems and prospects. *Plant Cell Rep.* **2016**, *35*, 1417–1427. [CrossRef]
136. Demirci, Y.; Zhang, B.; Unver, T. CRISPR/Cas9: An RNA-guided highly precise synthetic tool for plant genome editing. *J. Cell. Physiol.* **2018**, *233*, 1844–1859. [CrossRef] [PubMed]
137. Liang, G.; Zhang, H.; Lou, D.; Yu, D. Selection of highly efficient sgRNAs for CRISPR/Cas9-based plant genome editing. *Sci. Rep.* **2016**, *6*, 1–8. [CrossRef]
138. Ran, F.A.; Hsu, P.D.; Wright, J.; Agarwala, V.; Scott, D.A.; Zhang, F. Genome engineering using the CRISPR-Cas9 system. *Nat. Protoc.* **2013**, *8*, 2281. [CrossRef]
139. Samanta, M.K.; Dey, A.; Gayen, S. CRISPR/Cas9: An advanced tool for editing plant genomes. *Transgenic Res.* **2016**, *25*, 561–573. [CrossRef]
140. Fu, Y.; Sander, J.D.; Reyon, D.; Cascio, V.M.; Joung, J.K. Improving CRISPR-Cas nuclease specificity using truncated guide RNAs. *Nat. Biotechnol.* **2014**, *32*, 279. [CrossRef] [PubMed]
141. VanBuren, R.; Wai, C.M.; Wang, X.; Pardo, J.; Yocca, A.E.; Wang, H.; Chaluvadi, S.R.; Han, G.; Bryant, D.; Edger, P.P. Exceptional subgenome stability and functional divergence in the allotetraploid Ethiopian cereal teff. *Nat. Commun.* **2020**, *11*, 1–11. [CrossRef] [PubMed]
142. Joyce, B.; Baltzell, A.; Bomhoff, M.; Lyons, E. Comparative Genomics Using CoGe, Hook, Line, and Sinker. In *Bioinformatics in Aquaculture: Principles and Methods*; John Wiley & Sons: Hoboken, NJ, USA, 2017. [CrossRef]
143. Kumar, S.; Tamura, K.; Nei, M. MEGA3: Integrated software for molecular evolutionary genetics analysis and sequence alignment. *Brief. Bioinform.* **2004**, *5*, 150–163. [CrossRef]
144. Kumar, S.; Stecher, G.; Li, M.; Knyaz, C.; Tamura, K. MEGA X: Molecular evolutionary genetics analysis across computing platforms. *Mol. Biol. Evol.* **2018**, *35*, 1547–1549. [CrossRef]
145. Ketema, S. *Tef-Eragrostis tef (Zucc.)*; Bioversity International: Rome, Italy, 1997; Volume 12.
146. Zhu, Q.; Smith, S.M.; Ayele, M.; Yang, L.; Jogi, A.; Chaluvadi, S.R.; Bennetzen, J.L. High-throughput discovery of mutations in tef semi-dwarfing genes by next-generation sequencing analysis. *Genetics* **2012**, *192*, 819–829. [CrossRef]
147. Miller, D. Teff Grass: A new alternative. In Proceedings of the 2009 California Alfalfa & Forage Symposium and Western Seed Conference, Reno, NV, USA, 2–4 December 2009; UC Cooperative Extension, Plant Sciences Department, University of California: Davis, CA, USA, 2009.

148. Shi, L.; Wei, X.; Adedze, Y.; Sheng, Z.; Tang, S.; Hu, P.; Wang, J. Characterization and gene cloning of the rice (*Oryza sativa* L.) dwarf and narrow-leaf mutant dnl3. *Genet. Mol. Res.* **2016**, *15*, 1–12. [CrossRef]
149. Zhao, Z.; Xue, Y.; Yang, H.; Li, H.; Sun, G.; Zhao, X.; Ding, D.; Tang, J. Genome-Wide Identification of miRNAs and Their Targets Involved in the Developing Internodes under Maize Ears by Responding to Hormone Signaling. *PLoS ONE* **2016**, *11*, e0164026. [CrossRef] [PubMed]
150. Zhou, M.; Gu, L.; Li, P.; Song, X.; Wei, L.; CHen, Z.; Cao, X. Degradome sequencing reveals endogenous small RNA targets in rice (*Oryza sativa* L. ssp. *indica*). *Front. Biol.* **2010**, *5*, 67–90. [CrossRef]
151. Yang, Y.-H.; Zhang, F.-M.; Ge, S. Evolutionary rate patterns of the Gibberellin pathway genes. *BMC Evol. Biol.* **2009**, *9*, 206. [CrossRef] [PubMed]
152. Singla, B.; Khurana, J.P.; Khurana, P. Characterization of three somatic embryogenesis receptor kinase genes from wheat, *Triticum aestivum*. *Plant Cell Rep.* **2008**, *27*, 833–843. [CrossRef]
153. Pinthus, M.; Levy, A. The relationship between the Rht 1 and Rht 2 dwarfing genes and grain weight in *Triticum aestivum* L. spring wheat. *Theor. Appl. Genet.* **1983**, *66*, 153–157. [CrossRef]
154. Gouda, G.; Gupta, M.K.; Donde, R.; Mohapatra, T.; Vadde, R.; Behera, L. Marker-assisted selection for grain number and yield-related traits of rice (*Oryza sativa* L.). *Physiol. Mol. Biol. Plants Int. J. Funct. Plant Biol.* **2020**, *26*, 885. [CrossRef] [PubMed]
155. Liu, Y.; Wang, W.; Yang, B.; Currey, C.; Fei, S.-Z. Functional analysis of the teosinte branched 1 gene in the tetraploid switchgrass (*Panicum virgatum* L.) by CRISPR/Cas9-directed mutagenesis. *BioRxiv* **2020**, 1–33. [CrossRef] [PubMed]
156. Xiao, Y.; Zhang, G.; Liu, D.; Niu, M.; Tong, H.; Chu, C. GSK2 stabilizes OFP3 to suppress brassinosteroid responses in rice. *Plant J.* **2020**, *102*, 1187–1201. [CrossRef]
157. Miao, C.; Xiao, L.; Hua, K.; Zou, C.; Zhao, Y.; Bressan, R.A.; Zhu, J.-K. Mutations in a subfamily of abscisic acid receptor genes promote rice growth and productivity. *Proc. Natl. Acad. Sci. USA* **2018**, *115*, 6058–6063. [CrossRef]
158. Xu, C.; Wang, Y.; Yu, Y.; Duan, J.; Liao, Z.; Xiong, G.; Meng, X.; Liu, G.; Qian, Q.; Li, J. Degradation of MONOCULM 1 by APC/C TAD1 regulates rice tillering. *Nat. Commun.* **2012**, *3*, 1–9. [CrossRef] [PubMed]
159. Bhatia, D.; Joshi, S.; Das, A.; Vikal, Y.; Sahi, G.K.; Neelam, K.; Kaur, K.; Singh, K. Introgression of yield component traits in rice (*Oryza sativa* ssp. *indica*) through interspecific hybridization. *Crop Sci.* **2017**, *57*, 1557–1573. [CrossRef]
160. Wang, N.-N.; Xu, S.-W.; Sun, Y.-L.; Liu, D.; Zhou, L.; Li, Y.; Li, X.-B. The cotton WRKY transcription factor (GhWRKY33) reduces transgenic Arabidopsis resistance to drought stress. *Sci. Rep.* **2019**, *9*, 1–13. [CrossRef] [PubMed]
161. Satapathy, L.; Kumar, D.; Kumar, M.; Mukhopadhyay, K. Functional and DNA–protein binding studies of WRKY transcription factors and their expression analysis in response to biotic and abiotic stress in wheat (*Triticum aestivum* L.). *3 Biotech* **2018**, *8*, 40. [CrossRef]
162. Hittalmani, S.; Mahesh, H.; Shirke, M.D.; Biradar, H.; Uday, G.; Aruna, Y.; Lohithaswa, H.; Mohanrao, A. Genome and transcriptome sequence of finger millet (*Eleusine coracana* (L.) Gaertn.) provides insights into drought tolerance and nutraceutical properties. *BMC Genom.* **2017**, *18*, 465. [CrossRef]
163. Wang, Y.; Tong, X.; Qiu, J.; Li, Z.; Zhao, J.; Hou, Y.; Tang, L.; Zhang, J. A phosphoproteomic landscape of rice (*Oryza sativa*) tissues. *Physiol. Plant.* **2017**, *160*, 458–475. [CrossRef] [PubMed]
164. Ma, J.; Gao, X.; Liu, Q.; Shao, Y.; Zhang, D.; Jiang, L.; Li, C. Overexpression of TaWRKY146 increases drought tolerance through inducing stomatal closure in Arabidopsis thaliana. *Front. Plant Sci.* **2017**, *8*, 2036. [CrossRef] [PubMed]
165. Soltani Najafabadi, M. Improving rice (*Oryza sativa* L.) drought tolerance by suppressing a NF-YA transcription factor. *Iran. J. Biotechnol.* **2012**, *10*, 40–48.
166. Capell, T.; Escobar, C.; Liu, H.; Burtin, D.; Lepri, O.; Christou, P. Over-expression of the oat arginine decarboxylase cDNA in transgenic rice (*Oryza sativa* L.) affects normal development patterns in vitro and results in putrescine accumulation in transgenic plants. *Theor. Appl. Genet.* **1998**, *97*, 246–254. [CrossRef]
167. Li, X.-Y.; Mantovani, R.; Hooft van Huijsduijnen, R.; Andre, I.; Benoist, C.; Mathis, D. Evolutionary variation of the CCAAT-binding transcription factor NF-Y. *Nucleic Acids Res.* **1992**, *20*, 1087–1091. [CrossRef] [PubMed]
168. Sprink, T.; Eriksson, D.; Schiemann, J.; Hartung, F. Regulatory hurdles for genome editing: Process-vs. product-based approaches in different regulatory contexts. *Plant Cell Rep.* **2016**, *35*, 1493–1506. [CrossRef]
169. Liang, Z.; Chen, K.; Li, T.; Zhang, Y.; Wang, Y.; Zhao, Q.; Liu, J.; Zhang, H.; Liu, C.; Ran, Y. Efficient DNA-free genome editing of bread wheat using CRISPR/Cas9 ribonucleoprotein complexes. *Nat. Commun.* **2017**, *8*, 14261. [CrossRef] [PubMed]
170. Woo, J.W.; Kim, J.; Kwon, S.I.; Corvalán, C.; Cho, S.W.; Kim, H.; Kim, S.-G.; Kim, S.-T.; Choe, S.; Kim, J.-S. DNA-free genome editing in plants with preassembled CRISPR-Cas9 ribonucleoproteins. *Nat. Biotechnol.* **2015**, *33*, 1162. [CrossRef] [PubMed]
171. Latorre, A.; Latorre, A.; Somoza, Á. Modified RNAs in CRISPR/Cas9: An old trick works again. *Angew. Chem. Int. Ed.* **2016**, *55*, 3548–3550. [CrossRef] [PubMed]
172. Gerashchenkov, G.; Rozhnova, N.; Kuluev, B.; Kiryanova, O.Y.; Gumerova, G.; Knyazev, A.; Vershinina, Z.; Mikhailova, E.; Chemeris, D.; Matniyazov, R. Design of Guide RNA for CRISPR/Cas Plant Genome Editing. *Mol. Biol.* **2020**, *54*, 24–42. [CrossRef]
173. Salvi, S.; Druka, A.; Milner, S.G.; Gruszka, D. Induced genetic variation, TILLING and NGS-based cloning. In *Biotechnological Approaches to Barley Improvement*; Springer: Berlin/Heidelberg, Germany, 2014; pp. 287–310.
174. Zhang, D.; Li, Z.; Li, J.-F. Targeted gene manipulation in plants using the CRISPR/Cas technology. *J. Genet. Genom.* **2016**, *43*, 251–262. [CrossRef]
175. Crosetto, N.; Mitra, A.; Silva, M.J.; Bienko, M.; Dojer, N.; Wang, Q.; Karaca, E.; Chiarle, R.; Skrzypczak, M.; Ginalski, K. Nucleotide-resolution DNA double-strand break mapping by next-generation sequencing. *Nat. Methods* **2013**, *10*, 361. [CrossRef]

176. Frock, R.L.; Hu, J.; Meyers, R.M.; Ho, Y.-J.; Kii, E.; Alt, F.W. Genome-wide detection of DNA double-stranded breaks induced by engineered nucleases. *Nat. Biotechnol.* **2015**, *33*, 179. [CrossRef]
177. Kim, D.; Bae, S.; Park, J.; Kim, E.; Kim, S.; Yu, H.R.; Hwang, J.; Kim, J.-I.; Kim, J.-S. Digenome-seq: Genome-wide profiling of CRISPR-Cas9 off-target effects in human cells. *Nat. Methods* **2015**, *12*, 237. [CrossRef]
178. Tsai, S.Q.; Zheng, Z.; Nguyen, N.T.; Liebers, M.; Topkar, V.V.; Thapar, V.; Wyvekens, N.; Khayter, C.; Iafrate, A.J.; Le, L.P. GUIDE-seq enables genome-wide profiling of off-target cleavage by CRISPR-Cas nucleases. *Nat. Biotechnol.* **2015**, *33*, 187. [CrossRef]
179. Hsu, P.D.; Lander, E.S.; Zhang, F. Development and applications of CRISPR-Cas9 for genome engineering. *Cell* **2014**, *157*, 1262–1278. [CrossRef] [PubMed]
180. Fauser, F.; Schiml, S.; Puchta, H. Both CRISPR/C as-based nucleases and nickases can be used efficiently for genome engineering in A rabidopsis thaliana. *Plant J.* **2014**, *79*, 348–359. [CrossRef] [PubMed]
181. Slaymaker, I.M.; Gao, L.; Zetsche, B.; Scott, D.A.; Yan, W.X.; Zhang, F. Rationally engineered Cas9 nucleases with improved specificity. *Science* **2016**, *351*, 84–88. [CrossRef] [PubMed]
182. Kleinstiver, B.P.; Pattanayak, V.; Prew, M.S.; Tsai, S.Q.; Nguyen, N.T.; Zheng, Z.; Joung, J.K. High-fidelity CRISPR–Cas9 nucleases with no detectable genome-wide off-target effects. *Nature* **2016**, *529*, 490. [CrossRef]

plants

MDPI

Article

# Agromorphological and Physiological Performance of Ethiopian Common Bean (*Phaseolus vulgaris* L.) Genotypes under Different Agroecological Conditions

**Shiferaw Girsil Tigist [1,2], Julia Sibiya [2], Assefa Amelework [3,*] and Gemechu Keneni [4]**

1 Ethiopian Institute of Agricultural Research, Melkassa Agricultural Research Centre, Adama P.O. Box 430, Ethiopia; tshiferaw2006@gmail.com
2 African Center for Crop Improvement, University of KwaZulu-Natal, Private Bag X01, Scottsville, Pietermaritzburg 3209, South Africa; sibiyaj@ukzn.ac.za
3 Agricultural Research Council, Vegetable, Industrial and Medicinal Plant, Private Bag X293, Pretoria 0001, South Africa
4 Ethiopian Institute of Agricultural Research, Holeta Agricultural Research Centre, Addis Ababa P.O. Box 2003, Ethiopia; gemechukeneni@yahoo.com
* Correspondence: assefaa@arc.agric.za

**Abstract:** The objectives of this study were to assess the agronomic performance of common bean genotypes, previously selected for their response to infestation, by Mexican bean weevil and to identify promising lines that can be used as parents in a downstream breeding program. Field experiments were conducted using 144 genotypes under three different agro-ecologies in an unbalanced incomplete block design with three replications. Data on 15 agro-morphological traits were collected, and multivariate methods were used to examine the patterns of variation among the genotypes. The genotypes revealed a high level of phenotypic diversity for all agronomic traits. Six principal components, which contributed 84% of the total variation among the genotypes, were identified. The 15 agro-morphological traits classified the genotypes into three distinct major clusters and sub-clusters. The clustering patterns of the genotypes were according to the seed size, whereby the small and medium beans were distinctly separated from the large-seeded beans. The study established the existence of considerable genetic variations among common bean genotypes. Unique genotypes, such as Nasir, Awash Melka, and RAZ-36 from Cluster I, RAZ-2, RAZ-11, and RAZ-42 from Cluster II, and SER-125, SCR-15, MAZ-200, MAZ-203, and RAZ-120 from Cluster III, were selected based on their distinct agronomic performance. The selected genotypes could be useful for the common bean breeding program.

**Keywords:** agro-morphological traits; cluster analysis; common bean; genetic diversity; principal component analysis

**Citation:** Tigist, S.G.; Sibiya, J.; Amelework, A.; Keneni, G. Agromorphological and Physiological Performance of Ethiopian Common Bean (*Phaseolus vulgaris* L.) Genotypes under Different Agroecological Conditions. *Plants* **2023**, *12*, 2342. https://doi.org/10.3390/plants12122342

Academic Editors: Ranjith Pathirana and Francesco Carimi

Received: 20 April 2023
Revised: 7 June 2023
Accepted: 9 June 2023
Published: 16 June 2023



## 1. Introduction

The common bean (*Phaseolus vulgaris* L.) is the third most important source of calories, after maize and cassava, and the second most important source of dietary protein and minerals in the human diet [1]. In Africa, the major common bean-producing countries include Burundi, DR Congo, Ethiopia, Kenya, Rwanda, Tanzania, and Uganda, indicating that East Africa is the most suitable bean production region on the continent [1–4]. In Ethiopia, most of the traditional foods, especially during the fasting seasons, are prepared from pulse crops, such as chickpeas, field pea, faba beans, and lentils. However, recently there has been a growing interest in common beans, particularly among low-income farmers, since the prices of other highland pulses are rising [5,6].

The major common bean production areas are Oromiya, the Southern Nations, Nationalities, and Peoples (SNNP), and the Amahara regions. These regions cover about 98% (51% Oromiya, 27% SNNPR, and 20% Amhara) of the common bean production in the country [7]. Although farmers from different parts of the country grow different types of beans, the most predominant types being white and red small beans [3]. Both white and red small beans are produced in the Oromiya region and account for 61% and 44% of bean production in the country, respectively. In the Oromiya region, only two zones (East Shewa and West Arsi) cover 76% of the total bean production of the region [7]. In East Shewa, white beans (34%) are the most dominant types and are mainly grown for export, while in West Arsi, farmers only produce red beans [7,8].

The genetic improvement strategy of the National Common Bean Research Program in Ethiopia is focused mainly on consumer preferences and resistance to biotic and abiotic stresses. More than 55 improved common bean varieties have been released and adopted by farmers [9]. Despite the success of developing acceptable common bean genotypes, harnessing the genetic potential of the crop by delivering varieties with high yield and related quality traits is still hindered by the narrow genetic base used in the breeding program [10]. The National Bean Breeding Program relies mostly on exotic germplasms sourced from the Center for Tropical Agriculture (CIAT) and national breeding programs in neigbhoring countries.

Since the common bean's introduction in the 16th century, farmers have been preserving and discovering important genotypes that are adapted to their local environments and needs, which has led to the evolution of morphologically diverse landraces [11,12]. Landraces have been used as a source of desirable genes in breeding for biotic and abiotic stresses [13]. A number of researchers have reported on the wide genetic diversity in the Ethiopian common bean genotypes for a number of important traits [3,14,15]. However, the potential of the local landraces as sources of breeding material is not yet well-known and exploited. The objective of the present study, therefore, was to assess the performance of common bean genotypes for yield and yield components across different agro-ecologies and to select promising parents for breeding.

## 2. Results

### 2.1. Agronomic Performance

The analyses of variance for each location revealed a highly significant variability ($p < 0.001$) among the genotypes for all the traits studied. In addition, the performance of the genotypes was highly influenced by the prevailing environment. Thus, a combined analysis of variance was conducted over three locations, which showed highly significant genotypes by environment interactions for all the traits (Table 1). The mean squares partitioned for genotype, environment, and genotype by environment interaction indicated that environment (location) effects were more important for the variability recorded in all the traits except for the pod per plant and hundred seed weight. The percentage contribution of the genotypic effect ranged from 0.9% for days to 50% flowering to 68% for pod per plant, while the environmental effects ranged from 99% to 13% for the abovementioned traits. For pod per plant, the genotype main effect (68%) and genotype by environment interaction (19%) had more influence than the environment main effect. However, for the grain yield per plant, both the genotype (38%) and the environmental effects (42%) were important for the expression of the traits (Table 1). The coefficient of determination ($R^2$) estimated for all the traits ranged from 0.82 for plant height to 0.99 for hundred seed weight.

The minimum, maximum, mean, and standard deviation, as well as the coefficient of variation values of the 15 agro-morphological traits recorded at the three locations, are presented in Table 2. The range for DTF was recorded from 37 to 52 days, with a mean of 44 days. The PH, LA, and TCC ranged from 32–56 cm, 0.80–5.80 $m^2$, and 35–57 $\mu mol/m^2$, respectively. The difference in DTM of late and early maturing genotypes was 25 days, with a mean value of 94 days, while GFP ranged from 37 to 59 days. The number of PPP and SPP ranged from 13–51 and 3–6, respectively. The genotypes revealed a high variation in seed size, ranging from small (12 g) to large (59 g). The minimum AGBM was 28 gm, and the maximum was 53 gm per plant, with the mean being 37 gm per plant. The GY showed a wide variation, with the values ranging from 15 to 42 gm per plant, and the mean yield was 26 gm per plant. In addition, the range of HI, BPR, and EGR were 40–92, 30–56, and 29–82, respectively. The coefficient of variation recorded in the traits studied ranged from 2.3% to 12.9%.

**Table 1.** Combined analysis of variance of 15 agro-morphological traits recorded on 144 common bean genotypes at three locations.

| Traits | Mean Square | | | | | | | |
|---|---|---|---|---|---|---|---|---|
| | Replication (DF = 2) | Block (DF = 11) | iBlock (DF = 11) | Genotype (G) (DF = 121) | Environment (E) (DF = 2) | G × E Interaction (GEI) (DF =286) | Error (DF = 862) | $R^2$ |
| Days to 50% flowering (DTF) | 1.87 | 71.38 | 44.09 | 49.00 ** | 5174.08 ** | 8.99 ** | 2.41 | 0.93 |
| Plant height (PH) | 219.97 | 153.7 | 116.38 | 204.82 ** | 19,842.36 ** | 197.00 ** | 32.57 | 0.82 |
| Total chlorophyll content (TCC) | 227.23 | 386.6 | 106.05 | 111.18 ** | 2050.41 ** | 65.64 ** | 4.78 | 0.91 |
| Leaf area (LA) | 0.59 | 2.02 | 1.47 | 1.50 ** | 224.60 ** | 1.02 ** | 0.04 | 0.97 |
| Days to 90% maturity (DTM) | 50.43 | 539.4 | 124.7 | 182.58 ** | 19,915.01 ** | 54.92 ** | 4.55 | 0.96 |
| Grain filling period (GFP) | 33.71 | 351.6 | 89.16 | 137.80 ** | 6043.10 ** | 51.46 ** | 6.41 | 0.9 |
| Pods per plant (PPP) | 55.55 | 971.8 | 682.06 | 540.76 ** | 105.54 ** | 148.81 ** | 7.41 | 0.95 |
| Seeds per pod (SPP) | 0.77 | 14.49 | 8.91 | 4.24 ** | 9.17 ** | 1.80 ** | 0.21 | 0.88 |
| Hundred seed weight (HSW) | 12.83 | 2571 | 536.61 | 675.19 ** | 589.14 ** | 24.16 ** | 2.1 | 0.99 |
| Aboveground biomass (AGBM) | 30.2 | 300.9 | 88.8 | 153.36 ** | 4539.12 ** | 82.36 ** | 4.14 | 0.94 |
| Grain yield (g/plants) (GY) | 24.28 | 568.1 | 281.35 | 219.75 ** | 240.55 ** | 111.65 ** | 4.3 | 0.95 |
| Harvest index (HI) | 15.62 | 1620 | 1042.57 | 783.12 ** | 12,005.66 ** | 708.42 ** | 23.46 | 0.95 |
| Grain production efficiency (GPE) | 60.61 | 915.4 | 558.21 | 436.05 ** | 1359.47 ** | 184.89 ** | 10.54 | 0.93 |
| Biomass production rate (%) (BPR) | 17.38 | 652.6 | 75.51 | 164.86 ** | 1352.50 ** | 102.99 ** | 5.15 | 0.93 |
| Economic growth rate (%) (EGR) | 37.79 | 2672 | 832.04 | 851.69 ** | 2826.06 ** | 483.27 ** | 24.76 | 0.93 |

** Significant at $p < 0.001$.

**Table 2.** Summary statistics on 15 agro-morphological traits evaluated on 144 common bean genotypes at three locations.

| Trait | Min | Max | Mean ± SE | SD | CV% |
|---|---|---|---|---|---|
| Days to 50% flowering (Days) | 37 | 51.9 | 44.17 ± 0.12 | 1.55 | 3.51 |
| Plant height (cm) | 31.7 | 56.2 | 44.20 ± 0.30 | 5.71 | 12.91 |
| Total chlorophyll content (μmol/$m^2$) | 35 | 57.1 | 45.01 ± 0.17 | 2.19 | 4.86 |
| Leaf area ($m^2$/plant) | 0.8 | 5.8 | 3.00 ± 0.03 | 0.19 | 6.39 |
| Days to 90% maturity (Days) | 79.1 | 103.8 | 94.41 ± 0.23 | 2.13 | 2.26 |
| Grain filling period (Days) | 37.4 | 58.6 | 50.25 ± 0.18 | 2.53 | 5.04 |
| Pods per plant (No) | 13.1 | 50.9 | 27.70 ± 0.28 | 2.72 | 9.83 |
| Seeds per pod (No) | 2.1 | 6.1 | 4.21 ± 1.06 | 0.46 | 10.84 |
| Hundred seed weight (gm) | 12.1 | 58.8 | 24.84 ± 0.27 | 1.45 | 5.84 |
| Above ground biomass (gm/plants) | 27.7 | 53.2 | 36.74 ± 0.19 | 2.04 | 5.54 |
| Grain yield (gm/plant) | 14.6 | 41.8 | 25.64 ± 0.21 | 2.07 | 8.09 |
| Harvest index | 39.76 | 92.1 | 69.99 ± 0.47 | 4.84 | 6.92 |
| Grain production efficiency (gm/plants) | 14 | 54.5 | 29.52 ± 0.28 | 3.25 | 11 |
| Biomass production rate (%) | 30.3 | 56.2 | 39.18 ± 0.20 | 2.27 | 5.79 |
| Economic growth rate (%) | 28.5 | 82 | 51.48 ± 0.43 | 4.98 | 9.67 |

### *2.2. Principal Component Analysis (PCA)*

The PCA grouped the 15 phenotypic traits into 15 components, which accounted for the entire (100%) variability among the studied genotypes. However, the principal components, with an eigenvalue of less than one, were eliminated. The first six principal components (PCs), accounting for 83.7% of the variability observed among the studied common bean genotypes, were maintained. Table 3 presents the eigenvectors and values, the percentage of total variance, and the total cumulative variance for the 15 phenotypic traits used in this study. The first principal component (PC1) explained 29.7% of the total phenotypic variation among the 144 common bean genotypes was mainly due to the additive effects of the GY, GPE, EGR, AGBM, and BPR. The second PC, which accounted for 19.4% of the total variation, was well associated with phenological traits, such as DTF, DTM, and the GFP. Likewise, the third PC, which accounted for about 11% of the total variance of the genotypes, was due to the discriminatory effect of the HSW and PPP. The variation in PH, DTF, and SPP constituted a large part of the total variation explained by the fourth PC. The fifth and sixth PCs accounted for 7.7% and 6.7% of the total variation, chiefly due to the contrast between the TCC and SPP, and HI and LA, respectively.

**Table 3.** Principal component (PC) analysis of various agro-morphological traits estimated at three locations.

| Trait | PC1 | PC2 | PC3 | PC4 | PC5 | PC6 |
|---|---|---|---|---|---|---|
| AGBM | 0.41 | 0.00 | −0.03 | 0.17 | −0.22 | 0.19 |
| BPR | 0.38 | −0.23 | 0.06 | 0.15 | −0.24 | 0.15 |
| DTF | −0.12 | 0.31 | 0.02 | 0.51 | −0.26 | −0.30 |
| DTM | 0.02 | 0.55 | −0.21 | 0.06 | 0.10 | 0.05 |
| EGR | 0.42 | −0.14 | 0.13 | 0.11 | −0.04 | −0.16 |
| GFP | 0.10 | 0.46 | −0.25 | −0.23 | 0.26 | 0.23 |
| GPE | 0.43 | 0.12 | −0.08 | −0.18 | 0.19 | 0.08 |
| GY | 0.45 | 0.04 | 0.02 | 0.01 | 0.05 | −0.06 |
| HI | 0.19 | 0.11 | 0.05 | −0.27 | 0.22 | −0.64 |
| HSW | 0.10 | −0.21 | −0.63 | −0.02 | 0.04 | 0.05 |
| LA | 0.09 | 0.27 | 0.16 | 0.07 | −0.22 | 0.44 |
| PH | 0.09 | 0.13 | −0.38 | 0.50 | −0.04 | −0.27 |
| PPP | 0.10 | 0.26 | 0.52 | 0.21 | 0.31 | 0.01 |
| TCC | 0.09 | −0.27 | 0.08 | 0.22 | 0.48 | −0.02 |
| SPP | 0.11 | 0.16 | 0.08 | −0.42 | −0.53 | −0.28 |
| Eigenvalue | 4.45 | 2.91 | 1.68 | 1.36 | 1.15 | 1.00 |
| % total variation | 29.67 | 19.37 | 11.23 | 9.09 | 7.65 | 6.66 |
| % cumulative variation | 29.67 | 49.04 | 60.27 | 69.36 | 77.01 | 83.67 |

AGBM = aboveground biomass (gm/plant); BPR = biomass production rate; DTF = days to 50% flowering (days); DTM = days to 90% maturity (days); EGR = economic growth rate (%); GFP = grain filling period (days); GPE = grain production efficiency (gm/plant); GY = grain yield (gm/plant); HI = harvest index; HSW = hundred seed weight (gm/100 seed); LA = leaf area ($m^2$/plant); PH = plant height (cm); PPP = pods per plant (No); TCC = total chlorophyll content (μmol/$m^2$); SPP = seeds per pod (No).

To select genotypes with the best performance, the contribution of each trait was determined by the PCA. It was found that yield had a significant effect on the phenotypic variation among the 144 genotypes. Hence, the top ten best genotypes were selected from both small and medium market classes based on grain yield performance. The mean performance of the top ten high-yielding genotypes from both small and medium-seeded genotypes is presented in Table 4.

**Table 4.** Mean performance of the top ten selected common bean genotypes for seed color and yield and yield-related traits.

| Genotype | SC | DTF | PH | TCC | LA | DTM | PPP | SPP | HSW | AGBM | GY | HI | GPE | BPR | EGR | GFP |
|---|---|---|---|---|---|---|---|---|---|---|---|---|---|---|---|---|
| **Top ten small-seeded genotypes** | | | | | | | | | | | | | | | | |
| Nasir | Red | 41.6 | 53.3 | 48.0 | 2.5 | 95.4 | 36.0 | 5.1 | 24.6 | 46.1 | 41.8 | 63.5 | 54.5 | 48.3 | 77.5 | 53.9 |
| SER-125 | Red | 41.8 | 41.7 | 47.1 | 2.6 | 90.7 | 26.0 | 3.5 | 25.6 | 39.7 | 36.5 | 77.3 | 42.6 | 44.3 | 75.9 | 48.9 |
| Awash Melka | White | 46.6 | 52.8 | 49.1 | 3.0 | 93.9 | 33.7 | 4.5 | 21.7 | 46.6 | 34.6 | 63.5 | 34.0 | 52.6 | 76.9 | 47.3 |
| RAZ-36 | White | 42.7 | 45.0 | 53.4 | 3.2 | 96.2 | 46.3 | 3.1 | 18.1 | 45.5 | 33.1 | 66.9 | 41.3 | 47.1 | 63.3 | 53.6 |
| 241757 | Red | 47.0 | 47.2 | 41.5 | 2.9 | 95.7 | 29.8 | 4.4 | 22.7 | 43.7 | 32.9 | 76.3 | 34.0 | 45.6 | 68.7 | 48.7 |
| 230526 | Red | 42.9 | 41.1 | 40.4 | 3.4 | 96.6 | 27.0 | 5.0 | 23.6 | 37.0 | 32.2 | 86.9 | 40.3 | 38.3 | 59.8 | 53.7 |
| RAZ-44 | White | 42.8 | 48.3 | 50.1 | 2.9 | 96.2 | 31.2 | 4.1 | 18.1 | 42.1 | 31.4 | 82.5 | 39.2 | 43.8 | 60.7 | 53.4 |
| 241734 | Red | 43.4 | 46.1 | 45.6 | 4.0 | 101.1 | 30.0 | 4.6 | 22.1 | 44.4 | 31.3 | 72.1 | 41.6 | 44.0 | 54.5 | 57.7 |
| 214665 | Red | 43.1 | 44.4 | 43.7 | 3.4 | 99.3 | 27.4 | 5.4 | 22.8 | 41.1 | 30.1 | 74.6 | 39.5 | 41.4 | 53.7 | 56.2 |
| NC-51 | Red | 42.1 | 41.1 | 42.9 | 2.6 | 95.1 | 26.6 | 3.8 | 24.1 | 38.2 | 29.2 | 74.9 | 37.1 | 40.0 | 54.9 | 53.0 |
| **Top ten medium-seeded genotypes** | | | | | | | | | | | | | | | | |
| 207935 | Carioca | 44.9 | 51.1 | 49.6 | 3.2 | 95.6 | 24.2 | 5.7 | 29.4 | 53.2 | 41.2 | 80.7 | 46.8 | 56.2 | 82.0 | 50.7 |
| SCR-11 | Red | 42.0 | 45.0 | 49.9 | 2.7 | 92.3 | 25.4 | 3.9 | 29.2 | 44.9 | 36.9 | 56.6 | 44.2 | 48.8 | 74.3 | 50.3 |
| RAZ-40 | White | 41.4 | 37.8 | 49.5 | 3.1 | 89.6 | 20.3 | 3.7 | 36.7 | 35.8 | 32.6 | 62.4 | 32.2 | 40.6 | 60.7 | 48.1 |
| NC-28 | Cream | 40.9 | 45.0 | 47.3 | 3.1 | 99.4 | 32.0 | 3.1 | 28.9 | 42.3 | 31.8 | 75.8 | 45.0 | 42.6 | 55.0 | 58.6 |
| 211302 | Brown | 39.8 | 38.3 | 47.8 | 2.8 | 89.0 | 21.6 | 4.2 | 36.5 | 42.3 | 31.7 | 77.7 | 39.0 | 47.2 | 66.1 | 49.2 |
| SCR-15 | Red | 43.3 | 38.9 | 47.6 | 2.8 | 94.0 | 27.1 | 3.7 | 38.3 | 41.5 | 31.3 | 89.0 | 36.5 | 43.8 | 62.1 | 50.7 |
| SCR-26 | Red | 43.6 | 49.4 | 47.2 | 3.0 | 92.6 | 23.9 | 4.2 | 27.7 | 42.9 | 29.2 | 67.5 | 31.8 | 46.1 | 57.8 | 49.0 |
| 228077 | Red | 42.9 | 43.3 | 37.5 | 3.4 | 100.7 | 26.3 | 5.7 | 25.9 | 38.4 | 28.4 | 75.8 | 39.3 | 38.1 | 48.8 | 57.8 |
| KK25/MAIAWA/19 | Red | 43.6 | 47.2 | 42.0 | 2.8 | 95.4 | 20.8 | 5.6 | 36.9 | 33.1 | 28.2 | 77.3 | 33.7 | 34.9 | 54.7 | 51.9 |
| RAZ-120 | White | 45.7 | 45.0 | 50.3 | 2.8 | 90.7 | 28.6 | 3.7 | 26.4 | 38.3 | 27.8 | 75.1 | 27.4 | 42.6 | 63.1 | 45.0 |

SC = seed color; DTF = days to 50% flowering (days); PH = plant height (cm); TCC = total chlorophyll content (μmol/$m^2$); LA = leaf area ($m^2$/plant); DTM = days to 90% maturity (days); PPP = pods per plant (No); SPP = seeds per pod (No); HSW = hundred seed weight (gm/100 seed); AGBM = aboveground biomass (gm/plant); GY = grain yield (gm/plant); HI = harvest index; GPE = grain production efficiency (gm/plant); BPR = biomass production rate (%); EGR = economic growth rate (%), GFP = grain filling period (days).

Genotypes, such as Nasir, SER-125, Awash Melka, RAZ-36, 241757, 230526, RAZ-44, 241734, 214665, and NC-51, were selected from the small-seed market class, and they had a grain yield ranging from 29.2 to 41.8 g/plant. The top ten selected high-yielding genotypes from the medium market class included 207935, SCR-11, RAZ-40, NC-28, 211302, SCR-15, SCR-26, 228077, KK25/NAGAGA/19, and RAZ-120. These genotypes produced a grain yield ranging from 27.8 to 41.2 g/plant. There was no single genotype that showed consistent superiority for all the traits among the selected genotypes. However, the improved small-seeded variety, Nasir, exhibited the highest GY and GPE while genotype 207935 showed the highest AGBM, BPR, and EGR of all the tested genotypes. Based on the field performance of the 144 genotypes, 45% of the selected genotypes were landraces (241757, 230526, 241734, 214665, 207935, 211302, NC-51, NC-28, and 228077), 25% were resistant lines (RAZ-36, RAZ-44, RAZ-40, KK25/NAGAGA/19, and RAZ-120), 25% were released varieties (Nasir, SER-125, SCR-15, SCR-26, and Awash Melka) and 5% were advanced breeding lines (SCR-11).

### 2.3. *Correlations of Yield and Its Components*

The correlation among the 15 agro-morphological traits is presented in Table 5. Grain yield was highly significantly and positively ($p < 0.001$) correlated with AGBM, HI, GPE, BPR, and EGR. Similarly, GFP and PPP were highly significant ($p < 0.01$), and SPP and HSW had a significant ($p < 0.05$) correlation with GY. Biomass production rate was found to be negatively and highly significantly ($p < 0.001$) correlated with DTF, DTM, and GFP but highly ($p < 0.001$) positively correlated with AGBM and GPE. The total chlorophyll content, on the other hand, revealed a negative and significant association with DTF, DTM, GFP, and SPP. Similarly, HSW had a negative and significant correlation with LA, DTF, PPP, and SPP and a positive and significant association with PH. The days to 50% flowering had a significant negative association with HSW, GPE, and BPR. The relationship between PPP and SPP with HSW was also significant but negative.

**Table 5.** Correlation analysis among 15 agro-morphological traits in 144 common bean genotypes recorded at three locations.

| Trait | DTF | PH | TCC | LA | DTM | GFP | PPP | SPP | HSW | AGBM | HI | GPE | BPR | EGR | GY |
|---|---|---|---|---|---|---|---|---|---|---|---|---|---|---|---|
| DTF | 1.00 | | | | | | | | | | | | | | |
| PH | 0.31 *** | 1.00 | | | | | | | | | | | | | |
| TCC | −0.26 ** | 0.05 | 1.00 | | | | | | | | | | | | |
| LA | 0.14 | 0.03 | −0.12 | 1.00 | | | | | | | | | | | |
| DTM | 0.52 *** | 0.29 *** | −0.33 *** | 0.31 *** | 1.00 | | | | | | | | | | |
| GFP | 0.03 | 0.16 | −0.23 ** | 0.28 *** | 0.87 *** | 1.00 | | | | | | | | | |
| PPP | 0.21 * | −0.04 | 0.04 | 0.24 ** | 0.27 *** | 0.20 * | 1.00 | | | | | | | | |
| SPP | −0.02 | −0.04 | −0.24 ** | 0.14 | 0.14 | 0.17 | −0.04 | 1.00 | | | | | | | |
| HSW | −0.24 ** | 0.25 ** | 0.11 | −0.22 ** | −0.10 | 0.01 | −0.61 *** | −0.18 * | 1.00 | | | | | | |
| AGBM | −0.09 | 0.22 ** | 0.15 | 0.24 ** | 0.09 | 0.15 | 0.10 | 0.22 ** | 0.18 * | 1.00 | | | | | |
| HI | −0.07 | 0.06 | 0.08 | 0.07 | 0.12 | 0.18 * | 0.13 | 0.26 ** | −0.01 | 0.14 | 1.00 | | | | |
| GPE | −0.34 *** | 0.12 | 0.05 | 0.15 | 0.26 ** | 0.50 *** | 0.23 ** | 0.21 * | 0.20 * | 0.69 *** | 0.40 *** | 1.00 | | | |
| BPR | −0.30 *** | 0.11 | 0.27 ** | 0.09 | −0.34 *** | −0.22 ** | −0.01 | 0.12 | 0.19 * | 0.88 *** | 0.08 | 0.55 *** | 1.00 | | |
| EGR | −0.16 | 0.09 | 0.19 * | 0.01 | −0.22 ** | −0.16 | 0.21 * | 0.12 | 0.15 | 0.70 *** | 0.31 *** | 0.72 *** | 0.77 *** | 1.00 | |
| GY | −0.16 | 0.15 | 0.09 | 0.12 | 0.12 | 0.23 ** | 0.26 ** | 0.18 * | 0.18 * | 0.75 *** | 0.39 *** | 0.92 *** | 0.68 *** | 0.92 *** | 1.00 |

PH = plant height (cm); LA = leaf area ($m^2$/plant); TCC = total chlorophyll content ($\mu mol/m^2$); DTF = days to 50% flowering (days); DTM = days to 90% maturity (days); PPP = pods per plant (No); SPP = seeds per pod (No); HSW = hundred seed weight (gm/100 seed); AGBM = aboveground biomass (gm/plant); GY = grain yield (gm/plant); HI = harvest index; GPE = grain production efficiency (gm/plant); BPR = biomass production rate (%); EGR = economic growth rate (%), GFP = grain filling period (days). *** = significant ($p < 0.001$); ** = significant ($p < 0.01$); * = significant ($p < 0.05$).

### 2.4. *Cluster Analysis*

The relationship among the 144 common bean genotypes was revealed by using the neighbor-joining algorithm using the unweighted pair group method (UPGMA). The cluster analysis on the mean of 15 phenotypic traits clearly classified the 144 genotypes into three major clusters and seven sub-clusters (Figure 1). The first cluster (Cluster I) was composed of 36 (25%) of the genotypes and was dominated by small-seeded beans. This cluster was further divided into two sub-clusters (sub-Cluster Ia and Ib), with 18 genotypes each. With regard to genotype status, Cluster I consisted of 26 landraces, two resistant lines, and five varieties. The second cluster (Cluster II) consisted of the largest number, mainly small-seeded genotypes (49%). This cluster was further sub-divided into three sub-clusters, with 26, 22, and 23 genotypes, respectively. Cluster III consisted mainly of large and medium-seeded genotypes. This cluster was comprised of 37 genotypes, which were further sub-divided into two sub-Clusters, with 20 and 17 genotypes, respectively. Of the 16 resistant lines, 50% were in Cluster III, together with large-seeded released varieties.

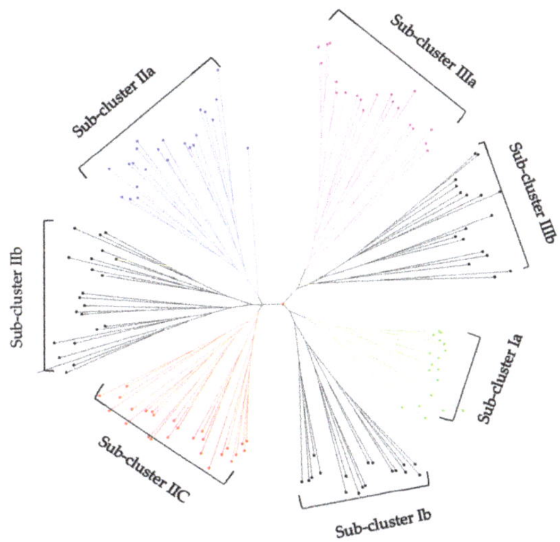


**Figure 1.** Dendrogram generated, based on hierarchical cluster analysis using UPGMA cluster algorithm, based on morphological data of 144 common bean genotypes (Supplementary Table S1).

## 2.5. *Performances of Genotypes in Different Clusters*

Table 6 summarizes the cluster means of the 15 phenotypic traits for the three main clusters and seven sub-clusters. The mean performance of the clusters showed the presence of considerable phenotypic variation among genotypes within each cluster. Genotypes in Cluster I revealed the highest mean values for all the traits except for PH, HSW, and TCC. Genotypes in Cluster III had the highest mean values for PH, HSW, and TCC.

Sub-cluster Ia contained genotypes that had a large LA and a large number of SPP. Genotypes grouped in sub-Cluster Ib were characterized by tall plants with a large number of PPP, as well as the highest AGBM, GY, and EGR. Although sub-Clusters Ia and Ib consisted of genotypes with small-seed sizes, genotypes in sub-Cluster Ib were much smaller than those in sub-Cluster Ia. Genotypes in sub-Clusters IIa and IIb were relatively early maturing, with a short GFP. However, sub-Cluster IIc consisted of genotypes that were late maturing and took long to fully fill the grain. In general, the genotypes clustered in sub-Clusters IIa and IIc were low-performing genotypes that had an extended period of vegetative growth and the highest total chlorophyll content.

Out of the two sub-Clusters under Cluster III, sub-Cluster IIIb included the best-performing genotypes in traits, such as GY, HI, GPE, BPR, and EGR. These genotypes also had a high TCC, a short flowering time, and were of medium seed size. Sub-Cluster IIIa, on the other hand, consisted of tall genotypes with large seed sizes. The genetic distance averaged for all the genotypes in each cluster revealed that the genotypes in each respective cluster were diverse. The smallest mean genetic distance was observed among genotypes clustered in Cluster I sub-Cluster Ib, while the highest genetic distance was found among genotypes grouped in Cluster III sub-Cluster IIIa. Generally, cluster analysis allows the selection of unique and genetically complementary genotypes for breeding and conservation. Genotypes Nasir, Awash Melka, and RAZ-36 from Cluster I, RAZ-2, RAZ-11, and RAZ-42 from Cluster II, and SER125, SCR-15, MAZ-200, MAZ-203, and RAZ-120 from

Cluster III were selected as potential parental genotypes. The selected genotypes have unique attributes, including grain yield, earliness, and seed color, shape, and size.

**Table 6.** The cluster means of 15 agro-morphological traits for the common bean genotypes, based on data recorded at three locations.

| Trait | Cluster Means | | | | | | |
|---|---|---|---|---|---|---|---|
| | C-I (*n* = 36) | | C-II (*n* = 71) | | | C-III (*n* = 37) | |
| | SC-Ia (*n* = 18) | SC-Ib (*n* = 18) | SC-IIa (*n* = 26) | SC-IIb (*n* = 22) | SC-IIc (*n* = 23) | SC-IIIa (*n* = 20) | SC-IIIb (*n* = 17) |
| PH | 43.4 | 47.2 | 43.0 | 40.8 | 44.2 | 47.1 | 44.9 |
| LA | 3.41 | 3.13 | 3.15 | 2.82 | 3.01 | 2.91 | 2.87 |
| TCC | 44.3 | 45.7 | 43.3 | 45.9 | 44.2 | 44.7 | 48.0 |
| DTF | 43.2 | 45.4 | 44.3 | 43.9 | 45.6 | 44.1 | 42.2 |
| DTM | 95.7 | 96.2 | 94.6 | 87.4 | 99.1 | 94.7 | 93.1 |
| GFP | 52.4 | 50.9 | 50.4 | 43.6 | 53.6 | 50.6 | 50.9 |
| PPP | 27.0 | 37.8 | 25.5 | 26.0 | 34.3 | 17.9 | 26.1 |
| SPP | 4.8 | 4.3 | 4.4 | 4.0 | 4.2 | 3.6 | 4.2 |
| HSW | 23.2 | 19.7 | 21.4 | 21.1 | 16.9 | 42.2 | 32.4 |
| AGBM | 38.8 | 41.3 | 36.2 | 33.6 | 32.8 | 36.1 | 40.8 |
| GY | 28.2 | 32.2 | 22.8 | 21.9 | 22.0 | 25.3 | 30.7 |
| HI | 75.8 | 75.6 | 64.6 | 65.9 | 68.7 | 67.2 | 76.6 |
| GPE | 34.5 | 36.3 | 26.1 | 21.8 | 25.9 | 29.5 | 37.1 |
| BPR | 40.5 | 43.2 | 38.7 | 38.7 | 33.3 | 38.2 | 44.0 |
| EGR | 53.9 | 64.2 | 45.6 | 50.8 | 41.6 | 49.6 | 61.0 |
| Genetic distance | 0.45 | 0.53 | 0.51 | 0.59 | 0.56 | 0.60 | 0.58 |

PH = plant height (cm); LA = leaf area ($m^2$/plant); TCC = total chlorophyll content ($\mu$mol/$m^2$); DTF = days to 50% flowering (days); DTM = days to 90% maturity (days); PPP = pods per plant (No); SPP = seeds per pod (No); HSW = hundred seed weight (gm/100 seed); AGBM = aboveground biomass (gm/plant); GY = grain yield (gm/plant); HI = harvest index; GPE = grain production efficiency (gm/plant); BPR = biomass production rate (%); EGR = economic growth rate (%), GFP = grain filling period (days); C = cluster; SC = sub cluster; *n* = number.

## 3. Discussion

### 3.1. Agronomic Performance

The present study examined the genetic variability and agronomic performance of 144 selected common bean genotypes for 15 yield and yield-related traits in three locations. The highly significant genotype mean squares for all the characters demonstrated that the genotypes exhibited a wide genetic variability for yield and yield-related traits. The observed highly significant environmental main effects suggested that the three locations were diverse in terms of weather- and location-related factors, such as temperature, rainfall, relative humidity, wind, altitude, soil physical and chemical properties. The three test locations represented three different agro-ecologies, with Melkassa representing the dryland agro-ecology, Arsi Negele representing the highly productive highland agro-ecology and Alem Tena representing the middling agro-ecology. Ceccarelli et al. [16] indicated that the genotype and environment components are recognized as the primary sources of variability in agronomic and genetic studies. Similarly, the highly significant genotype by environmental interaction indicated that genotypic performance is highly variable across different environments. Ceccarelli [17] also indicated that the expression of morphological and physiological plant characteristics associated with yield in optimal and stress conditions is different. Therefore, the discrimination and characterization of genotype adaptation across environments are crucial for optimizing the deployment of genetic resources.

In this study, the means and ranges of phenological traits and yield-related traits, such as the number of PPP, the number of SPP, HSW, and SW, revealed a wide range of genetic variation. A high phenotypic variation for these traits in the common bean was also reported by different authors [10,18–24]. The high phenotypic variation observed in this study may be attributed to the genetic variations among the genotypes and the environmental variations in the tested locations. In this study, more than 75% of the genotypes were landraces, suggesting that there was ample genetic variability among the landraces that can be exploited in future common bean improvement programs. This was also confirmed by other researchers that the Ethiopian common bean landraces were represented by high phenotypic diversity [3,10,24]. Similarly, the common bean grown in different parts of the world revealed a significant variation in yield and yield-related traits [18,20,25–29].

Principal component analysis (PCA)A Principal Component Analysis (PCA) was conducted to measure the relative contribution of each trait with regard to the total variation in the studied common bean genotypes. The first six components, with an eigenvalue of $\geq$1 explained 84% of the total variation were identified. However, about 50% of the phenotypic variation was explained by the first two components. Similar results were reported for agro-morphological traits in the common bean by several researchers [10,21,27,30,31]. In the present study, about 30% of the phenotypic variations observed were due to the variation in GY and AGBM. However, phenological traits also contributed significantly to discriminating the genotypes. The significant discriminatory effect of DTF was also reported by Burle et al. [21] and Fisseha [10]. Likewise, about 11% of the variations detected among the tested genotypes were due to the variation in seed weight. In previous studies, this trait was reported as the most important trait used to differentiate the two common bean gene pools [32]. However, the contribution of the trait in this study was relatively low compared to other previously reported results [10,21]. This could be due to the fact that most of the genotypes were selected from small (74%) and medium (15%) seed sizes, as reported by De Lima et al. [27].

The top 20 common bean genotypes were selected as potential parents for breeding programs, based on PCA1 values, which constitute the additive effect of GY, GPE EGR, and AGBM. The principal component analysis showed that grain yield had the most significant role in discriminating the 144 genotypes. The selection of the top genotypes was conducted according to the common bean market preferences in the major common bean-producing regions in Ethiopia, where the Mesoamerican beans (small-seeded) have more market demand than the Andean (large-seeded) genotypes. Based on their agronomic performance, the selected genotypes were composed of nine landraces, five resistant lines, three varieties, and three advanced breeding lines. As can be expected, the released varieties in the selected small-seeded group topped the rank in grain yield. The majority (45%) of the selected genotypes were landraces, suggesting that landraces can be used as a good source of valuable genes for future common bean breeding programs in Ethiopia [33]. Although the local landraces were found to be better adapted, genetically diverse, and agronomically suitable, the National Bean Breeding Program has been entirely dependent on the exotic germplasm. The SCR lines (SCR-11 and SCR-15) were the two top selected genotypes from the medium-sized red bean group. These lines are red beans that were developed for drought-prone areas carrying drought tolerance and with recessive genes for resistance to bean common mosaic virus [34]. The lines with Zabrotes-resistance genes, such as RAZ-36, RAZ-40, RAZ-44, and RAZ-120, and the Malawian resistance variety (KK25/MAIAWA/19), were found to be agronomically suitable.

### 3.2. Correlations of Yield and Its Components

Yield is a complex trait and is the outcome of the interaction of a number of genes and traits. Moreover, the expression of the traits is highly influenced by environmental factors, such as temperature, moisture, and light. It is also well known that the overall yield performance of genotypes is determined by the interaction of the traits rather than the expression of individual traits [16]. Blum [35] also indicated that yield per se is not under direct genetic control but under the control of the integrated effects of a multitude of physiological and biochemical processes. Hence, an understanding of the association between yield and yield-related traits is very crucial in order to exploit the genetic variability through selection. In the present study, grain yield had a significant positive association with the GFP, the number of PPP, HSW, AGBM, and HI. A selection based on these traits can be used as an indirect selection criterion for the improvement of grain yield in the common bean. Several researchers have also reported the positive significant correlation of grain yield with the above-mentioned traits [10,24,25,28,36,37]. Different authors [10,34,37,38] have also reported a strong positive correlation between HSW and GY. Some reports, on the other hand, have indicated a strong negative correlation between GY and HSW [24,28,39,40]. The variation in the sets of traits and the strength of the association might be a result of the variations in the environmental conditions and the genotypes used.

### 3.3. Cluster Analysis

The hierarchical cluster analysis conducted on the means of 15 agro-morphological traits resulted in three distinct major clusters and seven sub-clusters. For the traits under consideration, the within-cluster variation was found to be the lowest, while the between-cluster variation was the highest [41,42]. The mean performance of the genotypes grouped under the different clusters and sub-clusters showed considerable phenotypic variation. The clustering patterns were according to the seed size, where small and medium-seeded genotypes were clustered in Cluster I and II, while all the large-seeded genotypes were grouped in Cluster III. Several authors, such as Singh et al. [43], Burle et al. [21], Madakbaş and Ergin [44], and Boros et al. [20], support the present result. Based on hundred-seed weight, genotypes with HSW < 25 g are categorized as small-seeded, HSW $\geq$ 25–41 g as medium- seeded and HSW > 41 g as large-seeded. The clustering of genotypes, based on their seed size (gene pools), was clearly observed in the molecular genetic diversity analysis using SNP markers [45]. The clustering of landraces across all clusters indicated that Ethiopian landrace collections had a wide genetic variation for yield and yield-related traits. In addition, a large number (82%) of the genotypes was found to have a small to medium seed size, suggesting that the Ethiopian common bean genotypes are predominantly from the Mesoamerican gene pool, as supported by Asfaw et al. [3].

## 4. Materials and Methods

### 4.1. Description of the Study Site

The study was conducted at three on-station trial sites in the Oromiya region of central Ethiopia. The sites were Melkassa (8°24′52.04″ N, 39°19′41.22″ E), Alem Tena (8°17′32.29″ N, 38°56′48.77″ E), and Arsi Negele (7°22′30.29″ N, 38°40′17.78″ E), which are located at an altitude of 1550, 1611, and 1960 meters above sea level (m.a.s.l.), respectively. The climatic data of Melkassa and Alem Tena were collected from Melkassa and Debrie Zeit Agricultural Research Centers, respectively. However, the weather station at Arsi Negele was not functional, and the weather data is not included in this study. The weather data on rainfall and temperature for the two sites are presented in Figure 2. The soil types of Melkassa and Alem Tena are sandy loamy, while the soil is clay in Arsi Negele.

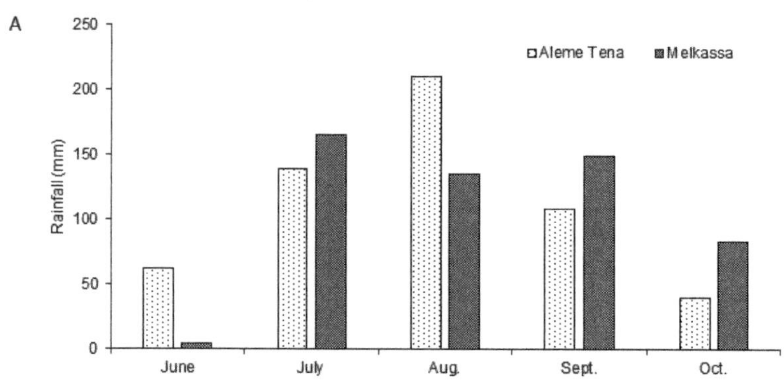


**Figure 2.** *Cont.*

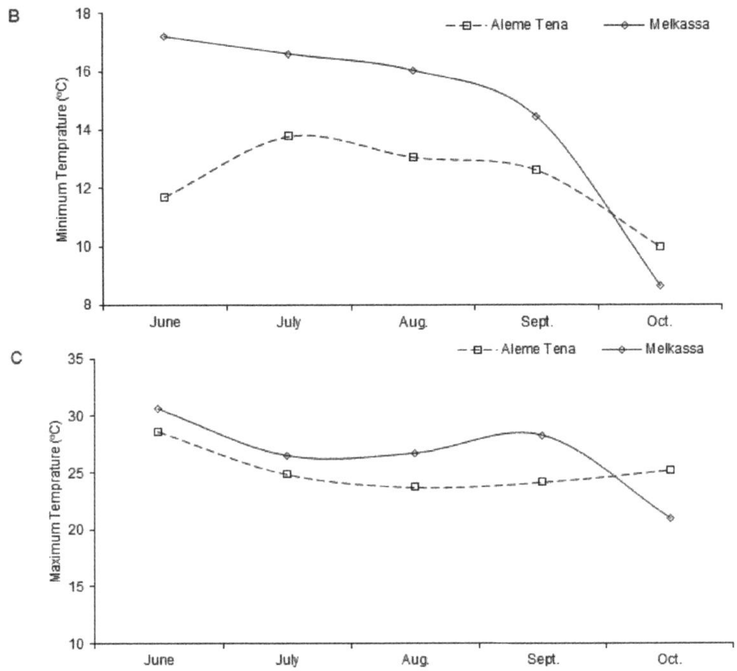


**Figure 2.** Climate data (**A**) rainfall (in mm), (**B**) minimum and (**C**) maximum temperatures (in °C) of Melkassa and Alem Tena sites during the growing.

### 4.2. Experimental Material and Experimental Design

A total of 144 common bean genotypes were selected on the basis of the prior screening of the genotypes for their response to bruchid infestation under laboratory conditions. The genotypes comprised 109 landraces, 16 released varieties, and 19 pre-release breeding lines. The 109 common bean landraces were collected from different regions of Ethiopia, and of the 19 pre-released genotypes, 16 were resistant to the Mexican bean weevil. The genotypes were grown during the off-season under irrigation for seed increase and to offset any differences in seed age and the effects of the prior growing environments [46]. The 144 genotypes were planted in a 12 × 12 alpha lattice design with three replications. The common bean genotypes were planted in 3 m long three rows, an inter-row spacing of 1 m, and an intra-row spacing of 40 cm. Weeds were controlled by frequent hand-weeding throughout the experimental period. Di-ammonium phosphate (DAP) fertilizer was applied during planting at a rate of 100 kg/ha [47], and other agronomic practices were carried out according to the cultivation practices recommended for each site.

### 4.3. Data Collection

In this study, a total of 15 phenological and agronomic traits were evaluated based on the IBPGR [48] common bean descriptors. For the agronomic traits, five randomly selected plants were sampled for data collection, while the phenological traits, such as days to 50% flowering (DTF) and days to 90% maturity (DTM), were recorded on a whole plot basis. Data on the following agronomic traits were collected: Plant height (PH), pods per plant (PPP), seeds per pod (SPP), hundred seed weight (HSW), the aboveground biomass (AGBM), and grain yield (GY).

In addition, the harvest index (HI) was measured as a proportion of grain yield to the aboveground biomass, and the grain-filling period (GFP) was calculated by subtracting the number of days to 90% maturity from the days to 50% flowering. Grain production efficiency (GPE) was calculated as a proportion of the grain-filling period to the duration of the vegetative period, and biomass production rate (BPR) was estimated by dividing the aboveground biomass weight by the days to 90% physiological maturity. Economic growth rate (EGR) was calculated as a proportion of grain yield to the grain seed fill period. Other physiological parameters, such as leaf area (LA) measured by a leaf area meter (LICOR model LI-3000) and total chlorophyll content (TCC) measured by a non-destructive, hand-held chlorophyll meter (SPAD-502 Chlorophyll Meter), were also included.

### 4.4. Data Analysis

Data were subjected to the analysis of the unbalanced incomplete block design procedure using GenStat Version 19 [49]. The homogeneity of variances among the three locations was examined by using Bartlett's test for each of the studied agro-morphological traits. Bartlett's test showed that all the traits had an equal error variance. All the agro-morphological traits were checked successively for normality using GenStat, and all the traits showed a normal distribution. The three locations were treated as environments, and a combined analysis of variance over the environments was done to estimate the variance component. Genotypes and environments were considered as fixed effects and replications, and blocks as random effects, and a combined analysis over environments was estimated from the linear additive model, which is expressed as:

$$Y_{ijklm} = \mu + r_i + b_j + \varphi_k + G_l + E_m + GE_{lm} + \varepsilon_{ijklm}$$

where $\mu$ = the overall mean, $r_i$ = the effect due to $i^{th}$ replication, $b_j$ = the effect due to the $j^{th}$ block within the $i^{th}$ replication, $\varphi_k$ = the effect due to the $k^{th}$ incomplete block within the $j^{th}$ block, $G_l$ = genotypic effect of the $l^{th}$ genotype, $E_m$ = environmental effect of the $m^{th}$ environment, $GE_{lm}$ = the interaction effect of the $l^{th}$ genotype and the $m^{th}$ environment.

The data were also subjected to the Principal Component Analysis (PCA) procedure using Genstat Version 19. For multivariate analysis, the data were standardized to a mean of zero, and a variance of unity was made to avoid the differences in scales used for recording data on the different characters [50]. The top ten highest-yielding genotypes were selected based on the traits that had the highest contribution to the first principal component, i.e., grain yield.

The correlation coefficients between characters were estimated based on the following formula:

$$r = {}^{\mathrm{Cov_{xy}}}\Big/_{\mathrm{sqrt}\left[\sigma_x^2 + \sigma_y^2\right]}$$

where $\mathrm{Cov_{xy}}$ = co-variance of traits x and y, ${\sigma_x}^2$ = variance of x and ${\sigma_y}^2$ = variance of y.

A hierarchical cluster analysis was performed to examine the grouping patterns of the genotypes based on their dissimilarity matrix with respect to the corresponding means of all the fifteen characters. The dissimilarity matrix was calculated using the Dice similarity index [51], and the cluster analysis was done by using the unweighted pair group method, the arithmetic mean (UPGMA), using DARwin 6.0 software [52]. A dendrogram was then generated on the dissimilarity matrix, and a bootstrap analysis was performed for node construction using 10,000 bootstrap values. The group means for all 15 agro-morphological traits were calculated and compared. Promising parental genotypes were selected.

## 5. Conclusions

The study identified a considerably wide genetic diversity among the 144 common bean genotypes for all the 15 phenotypic traits studied. Traits such as the GY, HSW, and AGBM were found to be the most important traits in differentiating germplasm into different clusters. It was also found that the Ethiopian common bean landraces showed a wide range of variation for all 15 of the agro-morphological traits studied, which suggests these germplasms can be used as valuable sources of genes in the National Common Bean Improvement programs. Genetically unique genotypes, such as Nasir, Awash Melka, and RAZ-36 from cluster I RAZ-2, RAZ-11, and RAZ-42 from Cluster II and SER-125, SCR-15,

MAZ-200, MAZ-203, and RAZ-120 from Cluster III, were identified as suitable parental genotypes. Released varieties, Nasir and Awash Melka, are the top high-yielding varieties that have been adopted in most of the bean growing areas. SER-125 and SCR-15, on the other hand, is a recently released variety that possesses most of the farmers' preferred traits. The selected genotypes could be useful for the common bean-breeding program.

**Supplementary Materials:** The following supporting information can be downloaded at: https://www.mdpi.com/article/10.3390/plants12122342/s1, Table S1: Lists of genotype in each Cluster and Sub-cluster.

**Author Contributions:** S.G.T. conceptualization, implementation, data analysis and writing of manuscript. J.S., Supervision and writing manuscript. A.A., data analysis and writing of manuscript. G.K., Supervision and writing of manuscript. All authors have read and agreed to the published version of the manuscript.

**Funding:** Revolution for Africa (AGRA) and African Centre for Crop Improvement (ACCI), University of KwaZulu-Natal.

**Data Availability Statement:** Data is contained within the article.

**Conflicts of Interest:** The authors declare no conflict of interest.

## References

1. Broughton, W.J.; Hernandez, G.; Blair, M.; Beebe, S.; Gepts, P.; Vanderleyden, J. Beans (*Phaseolus* spp.)–Model food legumes. *Plant Soil* **2003**, *252*, 55–128. [CrossRef]
2. Buruchara, R.; Chirwa, R.; Sperling, L.; Mukankusi, C.; Rubyogo, J.C.; Muthoni, R.; Abang, M.M. Development and delivery of bean varieties in Africa: The Pan-Africa Bean Research Alliance (PABRA) model. *Afr. Crop Sci. J.* **2011**, *19*, 227–245.
3. Asfaw, A.; Blair, M.W.; Almekinders, C. Genetic diversity and population structure of common bean (*Phaseolus vulgaris* L.) landraces from the East African highlands. *Theor. Appl. Genet.* **2009**, *120*, 1–12. [CrossRef] [PubMed]
4. Hillocks, R.J.; Madata, C.S.; Chirwa, R.; Minja, E.M.; Msolla, S. Phaseolus bean improvement in Tanzania, 1959–2005. *Euphytica* **2006**, *150*, 215–231. [CrossRef]
5. Karanja, D.; Endire, S.G.; Ruraduma, C.; Kimani, P.M.; Kweka, S.O.; Louis, B. *Value-Added Bean Technologies for Enhancing Foodsecurity, Nutrition, Income and Resilience to Cope with Climate Change and Variability Challenges in Eastern Africa*; International Livestock Research Institute (ILRI): Nairobi, Kenya, 2011.
6. Ferris, S.; Kaganzi, E. *Evaluating Marketing Opportunities for Haricot Beans in Ethiopia. IPMS (Improving Productivity and Market success) of Ethiopian Farmers Project Working Paper 7*; ILRI (International Livestock Research Institute): Nairobi, Kenya, 2008.
7. CSA. *Report on Area and Production of Crops: Agricultural Sample Survey on Private Peasant Holdings of 2014/2015 Meher Season*; Central Statistic Authority: Addis Ababa, Ethiopia, 2015.
8. Legesse, D.; Kumssa, G.; Assefa, T.; Taha, M.; Gobena, J.; Alema, T.; Abebe, A.; Mohhamed, Y.; Terefe, H. Production and Marketing of White Pea Beans in the Rift Valley, Ethiopia. *Natl. Bean Res. Program Ethiop. Inst. Agric. Res.* **2006**, *2*, 88–92.
9. Assefa, T.; Rubyogo, J.C.; Sperling, L.; Amsalu, B.; Deressa, A.; Reda, F.R.; Kirkby, R.; Buruchara, R. Creating partnerships for enhanced impact: Bean variety delivery in Ethiopia. *J. Crop Sci. Soci.* **2006**, *12*, 1–17.
10. Fisseha, Z. Genetic Diversity and Population Structure of Common Bean (*Phaseolus vulgaris* L.) Germplasm from Ethiopia. Ph.D. Thesis, Addis Ababa University, Addis Ababa, Ethiopia, 2015.
11. Sperling, L. The effect of the civil war on Rwanda's bean seed systems and unusual bean diversity. *Biodiv. Conser.* **2001**, *10*, 989–1009. [CrossRef]
12. Wortmann, C.S.; Kirkby, R.A.; Eledu, C.A.; Allen, D.J. *Atlas of Common Bean (Phaseolus vulgaris L.) production in Africa*; CIAT-Pan-African Bean Research Alliance: Kampala, Uganda, 1998.
13. Shashidhar, H.; Kanbar, A.; Toorchi, M.; Raveendra, G.M.; Kundur, P.; Vimarsha, H.S.; Soman, R.; Kumar, N.G.; Bekele, B.D.; Bhavani, P. Breeding for drought resistance using whole plant architecture: Conventional and molecular approach. In *Plant Breeding from Laboratories to Fields*; Andersen, S.B., Ed.; Tech Open Access Publisher: Vienna, Austria, 2013; pp. 151–166.
14. Fisseha, Z.; Tesfaye, K.; Dagne, K.; Blair, M.W.; Harvey, J.; Kyallo, M.; Gepts, P. Genetic diversity and population structure of common bean (*Phaseolus vulgaris* L.) germplasm of Ethiopia as revealed by microsatellite markers. *Afr. J. Biotechnol.* **2016**, *15*, 2824–2847.
15. Dagnew, K.; Haileselassie, T.; Feyissa, T. Genetic diversity study of common bean (*Phaseolus vulgaris* L.) germplasm from Ethiopia using inter simple sequence repeat (ISSR) markers. *Afr. J. Biotech.* **2014**, *13*, 3638–3649.
16. Ceccarelli, S.; Acevedo, E.; Grando, S. Breeding for yield stability in unpredictable environments: Single traits, interaction between traits, and architecture of genotypes. *Euphytica* **1991**, *56*, 169–185. [CrossRef]
17. Ceccarelli, S. Specific adaptation and breeding for marginal conditions. *Euphytica* **1994**, *77*, 205–219. [CrossRef]

18. Teame, G.; Ephrem, S.; Getachew, B. Performance evaluation of common bean (*Phaseolus vulgaris* L.) varieties in Raya Valley, Northern Ethiopia. *Afr. J. Plant Sci.* **2017**, *11*, 4B6E27E62320. [CrossRef]
19. Scarano, D.; Rubio, F.; José, J.; Rao, R.; Corrado, G. Scientia Horticulturae Morphological and genetic diversity among and within common bean (*Phaseolus vulgaris* L.) landraces from the Campania region. *Sci. Hortic.* **2014**, *180*, 72–78. [CrossRef]
20. Boros, L.; Wawer, A.; Borucka, K. Morphological, phonological and agronomical characterization of variability among common bean (*Phaseolus vulgaris* L.) local population from the national center for plant genetic resources: Polish genebank. *J. Hort. Res.* **2014**, *22*, 123–130.
21. Burle, M.L.; Fonseca, J.R.; del Peloso, M.J.; Melo, L.C.; Temple, S.R.; Gepts, P. Integrating phenotypic evaluations with a molecular diversity assessment of a Brazilian collection of common bean landraces. *Crop Sci.* **2011**, *51*, 2668–2680. [CrossRef]
22. Stoilova, T.; Pereira, G.; De Sousa, M.T.; Carnide, V. Diversity in common bean landraces (*Phaseolus vulgaris* L.) from Bulgaria and Portugal. *J. Cent. Eur. Agric.* **2005**, *6*, 443–448.
23. Stoilova, T.; Pereira, G.; De Sousa, M.T. Morphological characterization of a small common bean (*Phaseolus vulgaris* L.) collection under different environments. *J. Cent. Eur. Agric.* **2013**, *14*, 854–864. [CrossRef]
24. Negash, K. Studies on Genetic Divergence in Common Bean (*Phaseolus vulgaris* L.) Introductions of Ethiopia. Ph.D. Thesis, Addis Ababa University, Addis Ababa, Ethiopia, 2006.
25. Prakash, J.; Ram, R.B.; Meena, M.L. Genetic variation and characters interrelationship studies for quantitative and qualitative traits in french bean (*Phaseolus vulgaris* L.) under Lucknow conditions. *Legume Res.* **2015**, *38*, 425. [CrossRef]
26. Awan, F.K.; Khurshid, M.Y.; Afzal, O.; Ahmed, M.; Chaudhry, A.N. Agro-morphological evaluation of some exotic common bean (*Phaseolus vulgaris* L.) genotypes under rainfed conditions of Islamabad, Pakistan. *Pakistan J. Bot.* **2014**, *46*, 259–264.
27. De Lima, M.S.; Carneiro, J.E.d.S.; Carneiro, P.C.S.; Pereira, C.S.; Vieira, R.F.; Cecon, P.R. Characterization of genetic variability among common bean genotypes by morphological descriptors. *Crop Breed. Appl. Biotechnol.* **2012**, *12*, 76–84. [CrossRef]
28. Duran, L.A.; Blair, M.W.; Giraldo, M.C.; Macchiavelli, R.; Prophete, E.; Nin, J.C.; Beaver, J.S. Morphological and molecular characterization of common bean landraces and cultivars from the Caribbean. *Crop Sci.* **2005**, *45*, 1320–1328. [CrossRef]
29. Oscar, J.G.; Blair, M.W.; Frankow-Lindberg, B.E.; Gullberg, U. Molecular and phenotypic diversity of common bean landraces from Nicaragua. *Crop Sci.* **2004**, *44*, 1412–1418.
30. Okii, D.; Tukamuhabwa, P.; Odong, T.; Namayanja, A.; Mukabaranga, J.; Paparu, P.; Gepts, P.; Sciences, E. Morphological diversity of tropical common bean germplasm. *Afr. Crop Sci. J.* **2014**, *22*, 59–67.
31. Marzooghian, A.; Moghaddam, M.; Valizadeh, M.; Kooshki, M.H. Genetic diversity of common bean genotypes as revealed by seed storage proteins and some agronomic traits. *Plant Breed. Seed Sci.* **2013**, *67*, 125–137. [CrossRef]
32. Singh, S.P.; Gutiérrez, J.; Molina, A.; Urrea, C.; Gepts, P. Genetic diversity in cultivated common bean: II. Marker-based analysis of morphological and agronomic traits. *Crop Sci.* **1991**, *31*, 23–29. [CrossRef]
33. Darkwa, K.; Ambachew, D.; Mohammed, H.; Asfaw, A.; Blair, M.W. Evaluation of common bean (*Phaseolus vulgaris* L.) genotypes for drought stress adaptation in Ethiopia. *Crop J.* **2016**, *4*, 367–376. [CrossRef]
34. Negahi, A.; Bihamat, R.; Mohammed, I.; Negahi, Z.; Mohammed, A. Evaluation of genetic variation of some agronomical and morphological traits in Iranian and exotic common bean (*Phaseolus vulgaris* L.). *Agric. Comm.* **2014**, *2*, 22–26.
35. Blum, A. *Plant Breeding for Stress Environments*; CRC Press Inc.: Boca Raton, FL, USA, 1988.
36. Karasu, A.; Oz, M. A Study on coefficient analysis and association between agronomical characters in drey bean (*Phaseolus vulgaris* L.). *Bulgarian J. Agricul. Sci.* **2010**, *16*, 203–211.
37. Mohammad, A.; Tajik, M.; Ebadi, A.G. The study of relationship between differnt traits in common bean (*Phaseolus vulgaris* L.) with multivariate statstical methods. *Am. J. Agric. Environ. Sci.* **2008**, *3*, 806–809.
38. Roy, S.K.; Karim, M.A.; Islam, A.K.M.; Bari, M.N.; Mian, M.A.K.; Tetsushi, H. Relationship between yield and its component characters of bush bean (*Phaseolus vulgaris* L.). *S. Pac. Stud.* **2006**, *27*, 2–12.
39. Ahmed, S. Correlation and path analysis for agro-morphological traits in rajmash beans under Baramulla-Kashmir region. *Afr. J. Agric. Res.* **2013**, *8*, 2027–2032.
40. Kumar, A.; Singh, A.; Singh, P.; Singh, S.B.; Singh, V. Relationship and path analysis for green pod yield and its contributing characters over environments in French bean (*Phaseolus vulgaris* L.). *Legume Res.* **2009**, *32*, 270–273.
41. Hair, J.F.; Anderson, R.E.; Tatham, R.L.; Black, W.C. *Multivariate Data Analysis*; Macmillan Publishing Company: New York, NY, USA, 1995.
42. Singh, R.K.; Chaudhary, B.D. *Biometrical Methods in Quantitative Genetic Analysis*; Kalyani Publishers: New Delhi, India, 1985.
43. Singh, S.; Gepts, P.; Debouck, D. Races of common bean (*Phaseolus vulgaris*, Fabaceae). *Econ. Bot.* **1991**, *45*, 379–396. [CrossRef]
44. Madakbaş, S.Y.; Ergin, N. Morphological and phe-nological characterization of Turkish bean (*Phaseolus vulgaris* L.) genotypes and their present variation states. *Afr. J. Agric. Res.* **2011**, *6*, 6155–6166.
45. Girsil, T.S. Genetic Studies on Host Plant Resistance to Mexican Bean Weevil (*Zabrotes subfasciatus* Boheman) in Ethiopian Common Bean (*Phaseolus vulgaris* L.) Germplasms. Ph.D. Thesis, African Center for Crop Improvement School of Agricultural, Earth and Environmental Sciences College of Agriculture, Engineering and Science University of KwaZulu-Natal, Pietermaritzburg, South Africa, 2017.
46. Liao, M.; Hocking, P.J.; Dong, B.; Delhaize, E.; Richardson, A.E.; Ryan, P.R. Variation in early phosphorus-uptake efficiency among wheat genotypes grown on two contrasting Australian soils Australian. *J. Agricul. Res.* **2008**, *59*, 57–166. [CrossRef]

47. Assefa, T.; Beebe, S.E.; Rao, I.M.; Cuasquer, J.B.; Duque, M.C.; Rivera, M.; Battisti, A.; Lucchin, M. Pod harvest index as a selection criterion to improve drought resistance in white pea bean. *Field Crop Res.* **2013**, *148*, 24–33. [CrossRef]
48. IBPGR. *Descriptors for Phaseolus vulgaris*; IBPGR, International Plant Genetic Resources Institute: Rome, Italy, 1982.
49. Payne, R.W.; Murray, D.A.; Harding, S.A.; Baird, D.B.; Soutar, D.M. *GenStat for Windows Introduction*, 12th ed.; VSN International: Hemel Hempstead, UK, 2017.
50. Manly, B.F.J. *Multivariate Statistical Methods: A Primer*; Chapman and Hall: London, UK, 1986.
51. Shena, Z.; Zhang, K.; Ma, L.; Duan, J.; Ao, Y. Analysis of the genetic relationships and diversity among 11 populations of *Xanthoceras sorbifolia* using phenotypic and microsatellite marker data. *Electro. J. Biotech.* **2017**, *26*, 33–39. [CrossRef]
52. Perrier, X.; Jacquemoud-Collet, J. DARwin Software: Dissimilarity Analysis and Representation for Windows. 2006. Available online: http://www.darwin.cirad.fr/darwin (accessed on 12 May 2021).

*plants*

MDPI

*Article*

# The Metabolic Profile of Young, Watered Chickpea Plants Can Be Used as a Biomarker to Predict Seed Number under Terminal Drought

**Sarah J. Purdy [1,*], David Fuentes [2], Purushothaman Ramamoorthy [3], Christopher Nunn [4], Brent N. Kaiser [5] and Andrew Merchant [6]**

1 New South Wales Department of Primary Industries, 4 Marsden Park Road, Calala, NSW 2340, Australia
2 Charles Perkins Centre, Sydney Mass Spectrometry, The University of Sydney, John Hopkins Drive, Sydney, NSW 2000, Australia
3 Plant Breeding Institute, Sydney Institute of Agriculture, School of Life and Environmental Sciences, The University of Sydney, 12656 Newell Hwy, Narrabri, NSW 2390, Australia
4 CSIRO Agriculture and Food, Australian Cotton Research Institute, 21888 Kamilaroi Hwy, Narrabri, NSW 2390, Australia
5 Sydney Institute of Agriculture, The University of Sydney, 380 Werombi Road, Sydney, NSW 2006, Australia
6 The School of Life, Earth and Environmental Science, The University of Sydney, 380 Werombi Road, Sydney, NSW 2006, Australia
* Correspondence: sarah.purdy@dpi.nsw.gov.au

**Abstract:** Chickpea is the second-most-cultivated legume globally, with India and Australia being the two largest producers. In both of these locations, the crop is sown on residual summer soil moisture and left to grow on progressively depleting water content, finally maturing under terminal drought conditions. The metabolic profile of plants is commonly, correlatively associated with performance or stress responses, e.g., the accumulation of osmoprotective metabolites during cold stress. In animals and humans, metabolites are also prognostically used to predict the likelihood of an event (usually a disease) before it occurs, e.g., blood cholesterol and heart disease. We sought to discover metabolic biomarkers in chickpea that could be used to predict grain yield traits under terminal drought, from the leaf tissue of young, watered, healthy plants. The metabolic profile (GC-MS and enzyme assays) of field-grown chickpea leaves was analysed over two growing seasons, and then predictive modelling was applied to associate the most strongly correlated metabolites with the final seed number plant$^{-1}$. Pinitol (negatively), sucrose (negatively) and GABA (positively) were significantly correlated with seed number in both years of study. The feature selection algorithm of the model selected a larger range of metabolites including carbohydrates, sugar alcohols and GABA. The correlation between the predicted seed number and actual seed number was $R^2$ adj = 0.62, demonstrating that the metabolic profile could be used to predict a complex trait with a high degree of accuracy. A previously unknown association between D-pinitol and hundred-kernel weight was also discovered and may provide a single metabolic marker with which to predict large seeded chickpea varieties from new crosses. The use of metabolic biomarkers could be used by breeders to identify superior-performing genotypes before maturity is reached.

**Keywords:** carbohydrates; D-pinitol; kernel weight; metabolomics; phenotype; predictive modelling; yield

**Citation:** Purdy, S.J.; Fuentes, D.; Ramamoorthy, P.; Nunn, C.; Kaiser, B.N.; Merchant, A. The Metabolic Profile of Young, Watered Chickpea Plants Can Be Used as a Biomarker to Predict Seed Number under Terminal Drought. *Plants* **2023**, *12*, 2172. https://doi.org/10.3390/plants12112172

Academic Editors: Ranjith Pathirana and Francesco Carimi

Received: 9 March 2023
Revised: 16 May 2023
Accepted: 18 May 2023
Published: 30 May 2023



## 1. Introduction

New crop varieties with improved tolerance to climatic challenges are urgently required to enhance food security. The current processes to breed new varieties are often slow, as molecular markers are not available for many traits, so each new candidate variety must be grown to maturity before being assessed.

Metabolic biomarkers are routinely used in medicine to diagnose a condition based on its association with a specific metabolite. Examples include human chorionic gonadotropin (hCG) in urine and pregnancy, and serum creatinine in chronic kidney disease [1]. In addition to these association-based methods, medical techniques also prognostically utilise metabolic biomarkers, i.e., to predict a future outcome. An example is the prediction of cardiovascular disease, where the likelihood of a patient suffering a cardiovascular event (heart attack or stroke) is calculated using a risk prediction algorithm such as QRISK2 [2,3]. These online calculators used by general practitioners take measures of blood chemistry, along with blood pressure and physical, lifestyle and socio-economic factors, to predict the likelihood of a cardiac event within a specified time period (usually 10 years).

In plant research, metabolite profiling has frequently been used to associate the abundance of a particular metabolite with stress responses, such as increases in proline or jasmonic acid during chilling or herbivory, respectively [4,5], but more rarely has a predictive approach taken.

Metabolite profiling approaches have been used in plant studies to associate large-scale re-programming of the metabolome during treatments such as dark-induced senescence [6], cold acclimation [7] and drought response [8]. Many metabolites showed strong diurnal variation or fluctuation in abundance due to natural phenomena such as the variation in daylength or temperature [9–11], which has led to them being considered too variable to reliably be used to predict a trait. Indeed, the quantification of metabolites was described as being a "snapshot of a specific moment in time" [12] because of its dependence on development and the environment. However, despite the well-documented effect of the environment, it was demonstrated in a HPLC-QTOF MS study of 2475 mass peaks (noting that some metabolites result in more than one peak) that 75% were heritable and that ~800 had a heritability ($H^2$) of $\geq$0.7 [13].

Metabolite profiling in combination with statistical methods was successfully used to predict biomass in Arabidopsis in a number of studies [14–18]. By combining a negative correlation with starch and a positive correlation with enzyme activities, approximately a third of the variation in the biomass of an Arabidopsis inbred family could be accounted for [16]. The statistical methods applied to generate the biomass predictive models started with pairwise associations between single metabolites and biomass using rank correlation, but it was found that the predictive power of this method was low [14,18]. Both the previous authors then used a multivariate approach to combine groups of metabolites into a model; in the case of Meyer et al. [14], this was a canonical correlation analysis (CCA), while Sulpice et al. [18] used a partial least square regression (PLS) to identify combinations of key predictors that most strongly correlated with the trait. Just two metabolites, starch and, to a lesser extent, protein, were found to be the key predictive metabolites in the study by Sulpice et al. (2009) [18]. Whilst all of these studies were performed under tightly controlled laboratory conditions for the model species, Arabidopsis, they demonstrated that the metabolic profile of plants was strongly related to phenotype, such that it could be used to predict it.

In crop species, metabolic markers for drought tolerance were identified in rice samples taken before a drought event and those taken during drought [19]. The strongest correlation was from the marker, gluconic acid, under control (non-drought-stressed) conditions with a correlation of 0.72 with shoot dry weight under drought. In maize, a positive correlation between control levels of myo-inisitol and grain yield under drought was identified [20]. These two examples are evidence that predictive markers of drought tolerance can be identified. Prognostic biomarkers for chip quality were also identified in potato, where the abundance of glucose and fructose was found to positively correlate with discolouration during frying (low chip quality). When either of these hexoses were used as markers to predict chip quality in new crosses, the correlation (RS) between predicted and measured quality was 0.67 [21]. Predictive models for yield were also developed from the biomarker profile for the bioenergy grass Miscanthus [22]. Importantly, this study was conducted under field conditions and over two years, demonstrating that a metabolic profile that correlated with performance was consistent over multiple years. However, this study also identified that many markers were not consistent, so it was important to identify only those that were robust and repeatable [22].

In studies comparing molecular versus metabolic markers, molecular markers tended to have stronger predictive power, but only slightly. In a comparison between marker types in maize, metabolic markers correlated 0.6–0.8 with the trait, whereas molecular markers were slightly higher: 0.72–0.81 [23]. However, as the authors pointed out, this meant that the 130 metabolites that were studied were almost as effective as the 38,000 molecular markers (SNPs) [23].

Chickpea is the second-most-cultivated legume in the world [24]. India and Australia were the largest global producers of chickpea, producing 9 and 2 M tonnes, respectively in 2017 [25]. In both these countries, chickpea is grown on receding soil moisture and often experiences terminal drought stress during its late reproductive growth stage. As a predominantly indeterminate species, chickpea plants continue to produce vegetative and floral biomass while water availability remains adequate [26]. The extended flowering period, combined with progressively receding water availability, results in greater reproductive inhibition as the season progresses, as drought negatively impacts pollen viability and the ability of plants to accumulate biomass [27,28]. It was estimated that up to 50% of annual global yields of chickpea are lost due to drought [29]. Yield, particularly under the influence of drought, is a complex trait that exhibits strong G $\times$ E interactions. There are no molecular markers available for this trait, so the process for yield improvement is slow. We hypothesised that metabolic biomarkers could be identified from young, healthy plants that, under field-grown conditions over two growing seasons, were (a) consistently correlated with yield traits under terminal drought and (b) strongly correlated enough to produce predictive models. Such models could be used by breeders to identify genotypes that perform well under adverse conditions, without having to expose them to the relevant stimuli (e.g., drought).

## 2. Results

### 2.1. Phenotype

Seed weight (g plant$^{-1}$) was significantly different between years in the replicated genotypes and between rainfed and irrigated treatments in both replicated and non-replicated genotypes (Table 1, Tables S1 and S2). The hundred kernel weight (HKW) did not significantly change between years, and no significant differences were observed between treatments (Table 1). Significant differences that reflected the change in seed weight (g plant$^{-1}$) were observed from seed number plant$^{-1}$, which was different between years for replicated genotypes and between treatments for replicated and non-replicated genotypes (Table 1).

**Table 1.** Yield parameters. Significant differences (Student's *t*-test) beneath the values indicate significant differences between years for the replicated genotypes (** $p$ = <0.01). Significant differences next to the values for replicated and unique genotypes show significant differences between the rainfed and irrigated plants (** $p$ = <0.01). For replicated genotypes n = 23, and for unique genotype n = 13.

| | Seed Weight (g Plant$^{-1}$) | | | HKW (g) | | Seed Number (Plant$^{-1}$) | | |
|---|---|---|---|---|---|---|---|---|
| | Rainfed | Irrigated | | Rainfed | Irrigated | Rainfed | Irrigated | |
| Year 1 replicated genotypes | 4.85 | 8.26 | ** | 21.42 | 21.41 | 23.20 | 38.18 | ** |
| Year 2 replicated genotypes | 7.16 | 12.74 | ** | 21.11 | 22.4 | 34.68 | 58.36 | ** |
| Significantly different between years: | ** | ** | | | | ** | ** | |
| Year 1 non-replicated genotypes | 5.15 | 8.49 | ** | 31.59 | 32.27 | 20.55 | 30.37 | ** |
| Year 2 non-replicated genotypes | 7.63 | 12.73 | ** | 15.48 | 17.18 | 53.04 | 77.10 | ** |

### 2.2. Metabolomic Profiling

Nine metabolites were profiled by gas chromatography mass spectrometry (GC-MS) from leaf tissue from the rainfed plants (prior to withholding irrigation), in both years (Table 2 and Table S3). The most abundant metabolite in most genotypes in year 1 was malic acid, but in year 2 the sugar alcohol, D-pinitol, was in greater abundance overall than malic acid (Table 2). All the tested metabolites were significantly different in concentration between year 1 and year 2: the organic acids, such as lactic, malonic, malic and citrate, were lower in year 2, whereas the carbohydrates, including sugar alcohols, such as chiro-inositol and D-pinitol, were all in greater abundance in year 2. Exceptions were the observations regarding acids, succinate and γ-aminobutyric acid (GABA), which were also in greater abundance in year 2 (Table 2).

**Table 2.** GCMS metabolic profile of the leaf tissue. All values are in μg mL. n = 4.

| | Lactic Acid | Malonic Acid | Succinate | Malic Acid | GABA | Citric Acid | D-Pinitol | Chiro-Inositol | Myo-Inositol |
|---|---|---|---|---|---|---|---|---|---|
| Year 1 replicated averages | 0.52 | 0.88 | 0.25 | 8.53 | 0.51 | 3.05 | 3.85 | 0.57 | 0.91 |
| Year 2 replicated averages | 0.32 | 0.70 | 0.97 | 6.43 | 1.00 | 1.50 | 6.97 | 0.72 | 1.12 |
| Significantly different between years: | ** | ** | ** | ** | ** | ** | ** | ** | ** |
| Year 1 non-replicated averages | 0.54 | 0.88 | 0.25 | 8.40 | 0.54 | 2.93 | 4.17 | 0.57 | 0.89 |
| Year 2 non-replicated averages | 0.30 | 0.69 | 0.96 | 6.51 | 1.04 | 1.51 | 5.87 | 0.75 | 1.07 |

GABA—γ-aminobutyric acid; Significant differences between years for the replicated genotypes are shown using a Student's two-tailed *t*-test, ** $p = < 0.01$.

Carbohydrates were analysed by enzyme assays. The most-abundant carbohydrate in both years was starch, averaging 72–76 mg $g^{-1}$ dry weight (DW) (Table 3 and Table S4). Glucose and sucrose were in significantly greater abundance in year 2 in the replicated genotypes, but no significant differences between years were observed for either fructose or starch (Table 3).

**Table 3.** Enzyme assay profile of the leaf tissue. NSC = sum of non-structural carbohydrates. All values are in mg $g^{-1}$ DW. n = 4. Significant differences between years for the replicated genotypes are shown using a Student's two-tailed *t*-test, ** $p = < 0.01$.

| | D-Glucose | D-Fructose | Sucrose | Starch | NSC |
|---|---|---|---|---|---|
| Year 1 replicated averages | 8.06 | 8.84 | 21.97 | 76.32 | 115.2 |
| Year 2 replicated averages | 11.22 | 7.84 | 66.93 | 72.87 | 158.9 |
| Significantly different between years: | ** | | ** | | ** |
| Year 1 non-replicated averages | 9.45 | 8.71 | 21.59 | 66.46 | 106.2 |
| Year 2 non-replicated averages | 10.79 | 7.60 | 60.51 | 72.42 | 151.3 |

### 2.3. Pearson's Correlations

Pearson's correlations were carried out to identify relationships between metabolites and seed number plant$^{-1}$, and HKW in year 1 and year 2 (Table 4). Six out of 13 metabolites showed a significant correlation with seed number in at least one year, which were malic acid, GABA, D-pinitol, D-glucose, sucrose and starch (Table 4). Only two metabolites, sucrose and D-pinitol, showed a significant correlation with the trait in both years ($p = <0.05$), but GABA was significantly correlated in year 2, and $p = 0.06$ in year 1, so a consistent trend was observed (Table 4).

**Table 4.** Pearson's correlation between seed number plant$^{-1}$ and hundred-kernel weight (HKW), and the metabolites. Significant correlations (* $p$ = < 0.05, ** $p$ = < 0.01, and *** $p$ = < 0.001) are coloured red and purple, and correlations where $p$ = < 0.1 are coloured pink. GABA = γ-aminobutyric acid.

| | Seed Number Plant$^{-1}$ | | Hundred-Kernel Weight | |
|---|---|---|---|---|
| | Year 1 | Year 2 | Year 1 | Year 2 |
| | *Rs* | *Rs* | *Rs* | *Rs* |
| Lactose | 0.08 | 0.12 | −0.08 | 0.07 |
| Malonic acid | 0.02 | 0.01 | −0.04 | 0.21 |
| Succinate | 0.16 | 0.03 | −0.02 | 0.02 |
| Malic acid | 0.48 *** | −0.10 | −0.28 | −0.09 |
| GABA | 0.31 | 0.34 * | −0.1 | −0.16 |
| Citrate | 0.09 | −0.17 | −0.2 | −0.11 |
| D_pinitol | −0.58 *** | −0.52 *** | 0.62 *** | 0.55 *** |
| Chiro_inositol | −0.04 | 0.08 | 0.11 | −0.15 |
| Myo_inositol | −0.07 | −0.09 | −0.06 | 0.17 |
| D-glucose | −0.53 *** | −0.11 | 0.6 *** | 0.18 |
| D-fructose | −0.04 | 0.04 | 0.18 | −0.02 |
| Sucrose | −0.39 * | −0.43 ** | 0.29 | 0.32 * |
| Starch | 0.31 | −0.15 | −0.53 | 0.11 |

Fewer significant correlations were observed for HKW and the metabolites; indeed, only D-pinitol was significantly positively correlated in both years of testing (Table 4).

D-Pinitol

As the strongest Pearson's correlation in both years was between HKW and D-pinitol, the relationship was further investigated. Both the Kabuli and Desi types were included in the year 1 trial, so the concentration in the two types was determined (Table S1 and Figure 1). The concentration of D-pinitol was significantly higher in the Kabuli types, with an average of 4.8 mg g$^{-1}$ DW compared to 3.8 mg g$^{-1}$ DW in the Desi types (Figure 1). The average HKW for Desi and Kabuli was 20.9 g and 40.4 g, respectively (Supplementary Table S2). Only Desi genotypes were included in the year 2 trial, but these were comprised of types of different origin, specifically, Australian breeding' lines, Australian varieties, Indian varieties and the ICRISAT reference set (Figure 1). The concentration of D-pinitol was observed to be the highest in the Indian and Australian varieties, which were both significantly higher than those sourced from the ICRISAT reference set, while the Australian breeders' lines were in between the two (Figure 1). The HKW values corresponded to this trend, being 19.7 g, 19 g, 14.2 g and 25 g for the four sources, respectively (Table S2).

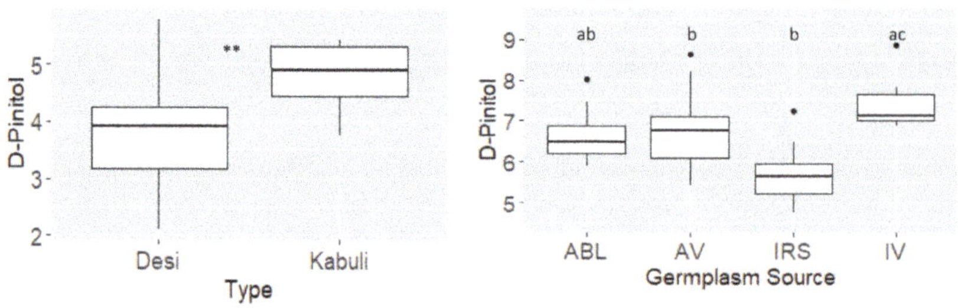


**Figure 1.** Relationships between D-pinitol and seed size. (Left) Concentration of D-pinitol (mg g$^{-1}$ DW) in Desi (n = 29) and Kabuli (n = 7) varieties of chickpea in year 1. (Right) Concentration of D-pinitol and hundred-kernel weight (HKW) in genotypes sourced from Australian breeding lines (ABL) (n = 8), Australian varieties (AV) (n = 13), Indian varieties (IV) (n = 6) and the ICRISAT reference set (IRS) (n = 9). ** $p$ = < 0.01 (Student's $t$-test), and letters above the bars show significant differences (Tukey's HSD Test, $p$ = < 0.05).

### 2.4. Predictive Modelling

As HKW was observed to be a largely fixed trait, not changing between years and treatments, we decided to focus the predictive modelling on seed number plant$^{-1}$, as this was the yield parameter that showed the strongest change in response to terminal drought (Table 2). Significant correlations between seed number and multiple metabolites were observed, so a multi-variate linear regression modelling approach was employed to develop the model. The step akaike information criteria (AIC) feature selection procedure reduced the number of variables to seven, which included sucrose, D-pinitol and GABA that were identified in the Spearman's rank correlations and also chiro-inositol, fructose, starch and the total non-structural carbohydrate (NSC) abundance (Table 5).

**Table 5.** Coefficients of significantly correlated variables.

| Coefficients | | | | | |
|---|---|---|---|---|---|
| | Estimate | Std. Error | t Value | Pr | |
| (Intercept) | −4.81 | 13.75 | −0.35 | 0.73 | |
| GABA | 46.27 | 8.13 | 5.69 | 0.00 | *** |
| D-pinitol | −4.64 | 1.35 | −3.44 | 0.00 | ** |
| Chiro-inositol | 44.03 | 18.26 | 2.41 | 0.02 | * |
| D-fructose | 1.74 | 0.61 | 2.85 | 0.01 | ** |
| Sucrose | 1.37 | 0.38 | 3.57 | 0.00 | *** |
| Starch | 1.32 | 0.37 | 3.60 | 0.00 | *** |
| NSC | −1.29 | 0.35 | −3.71 | 0.00 | *** |

GABA = γ-aminobutyric acid; NSC = non-structural carbohydrates; *** $p = < 0.001$, ** $p = < 0.01$, * $p = < 0.05$.

These selected variables were then used to train the model using a leave-one-out cross-validation (LOOCV) approach (see Materials and Methods). A strong significant correlation was observed between the predicted and actual seed number plant$^{-1}$ with $R^2$—adjusted = 0.623, demonstrating that 62% of the variation in final seed number, under terminal drought (rainfed) conditions, could be explained by seven metabolites measured from healthy, young leaves early in the growing season (Figure 2).

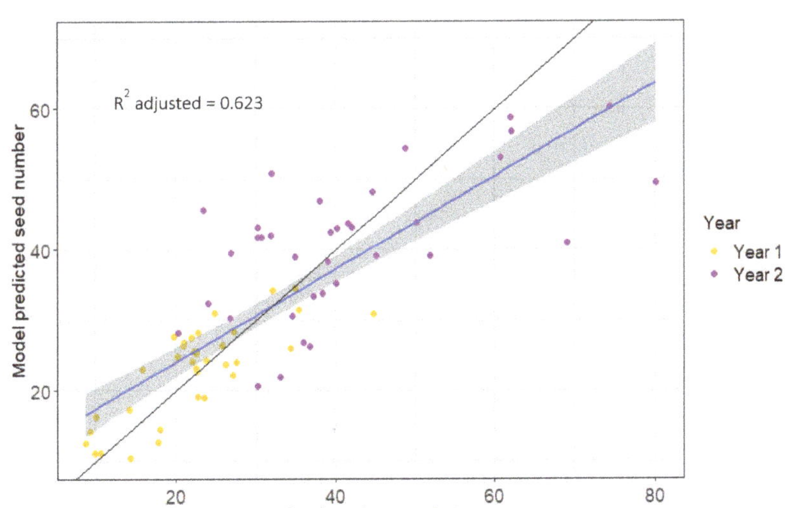


**Figure 2.** Model output showing actual seed number versus predicted seed number. Black line shows a 1:1 relationship and blue line shows linear regression between actual versus predicted seed number, grey shaded area shows confidence interval (0.95).

## 3. Discussion

We observed no significant differences in HKW between years or treatments. Complementary to our findings, was an observation that HKW did not significantly differ between chickpea genotypes under drought or watered conditions in a glasshouse trial, whereas seed number did show significant differences and also a strong correlation with yield under drought [30]. The authors concluded that HKW did not change in response to drought because, once a seed enters the phase of rapid dry weight accumulation, it has priority for assimilates over seeds in the early stage of development [30]. Hundred-kernel weight was previously shown to be the most heritable yield trait in both chickpea and broadbean, which is evidence that it is less susceptible to environmental influences [31,32]. This shows replicability in our finding that seed number, rather than size, is a major determining factor in yield under terminal drought.

Significant differences in the metabolic profile between the two years were observed, highlighting the dynamic nature of metabolites. Despite this, the correlation analyses showed that several metabolites were consistently correlated with yield traits in both years. They were sucrose, GABA (seed number) and D-pinitol (seed number and HKW). These three metabolites were all reported to increase during water stress in multiple species including Arabidopsis [33,34], rice [35], sesame [36], soybean, ricebean and other tropical legumes [37–40]. The metabolic changes that occur during stress events were linked to the ability of particular genotypes to survive or succumb. For example, a drought-tolerant variety of sorghum accumulated greater amounts of sugars and sugar alcohols during drought stress than a susceptible cultivar [41]. Similarly, in soybean, the drought tolerance of a wild accession was attributed to its capacity to accumulate a greater abundance of osmoprotective compounds during drought compared to a more susceptible line [40]. These examples link adjustments in the metabolic profile during drought (including the accumulation of carbohydrates, GABA and sugar alcohols) with improved yield performance. However, in our study, the metabolites were profiled from young, healthy, watered plants before the drought stress was imposed and a correlation between these metabolites and yield under drought was still observed. This could suggest that, to some extent, the higher yielding (high seed number) genotypes observed in our study showed a level of pre-adaptation to drought conditions. A similar observation was made in sesame, where it was reported that drought-tolerant genotypes had a higher concentration of GABA even under well-watered conditions [36]. As chickpea has been bred to complete its lifecycle under terminal drought, it appears that the metabolic adaptations that facilitate performance under these conditions have been selected and are still observable under well-watered conditions.

The positive correlation between D-pinitol and HKW and the negative relationship with seed number are results of an existing negative relationship between these two yield traits, as previously reported, which the authors attributed to parallel demands for photosynthates and nutrients [30,42]. The stronger of the two relationships was between HKW and D-pinitol.

D-pinitol is a free cyclitol that is found throughout the genus Leguminosae [39]. In mammalian systems, D-pinitol is regarded as a bioactive compound because it possesses insulin-like properties and can lower blood glucose in patients with type 2 diabetes [43,44]. In plants, the predominant association of D-pinitol is as a compatible solute, with accumulation frequently observed during abiotic stress [37–39]. Accumulation was observed to increase as photosynthesis declined due to drought stress, providing evidence that carbon is diverted away from the primary metabolism and into D-pinitol [39]. In transgenic tobacco that overexpressed a myo-inositol O-methyl transferase gene, *IMT1*, which catalyses the first step in the biosynthesis of the cyclic sugar alcohol D-pinitol, large quantities of ononitol accumulated [45,46]. When the transgenic plants were exposed to drought or salt treatments, they were able to retain photosynthetic performance relative to controls [46]. Therefore, D-pinitol plays a role in protecting remobilisation to the seed during filling under terminal drought conditions. The majority of scientific publications regarding D-pinitol refer to its role in stress protection (usually drought or salt) [39,46–49]. However, our results point to a more central role for D-pinitol in seed size, as consistent relationships were found, with large seeded Kabuli types having a higher concentration than smaller seeded Desi types, and Desi varieties that had presumably been selected for seed size amongst other attributes, had more than non-varieties. Of interest is the observation that ciceritol, an α-d-digalactoside of D-pinitol, accounts for 36–43% of the total sugars in chickpea seeds [50,51]. It is, therefore, possible that the increased abundance of leaf D-pinitol in larger-seeded varieties provides more of the pre-cursor material for remobilisation to the seed later in development. It would be interesting to experiment with the exogenous feeding of either D-pinitol or its pre-cursor, myo-inositol, to observe whether corresponding changes in seed size or seed ciceritol are observed.

Our results also suggest that D-pinitol could be used to identify new crosses that can produce larger seeds even before flowering occurs. For, example, if large kernel size varieties were the main aim, a cross could be made between a larger seed size parent and a parent with smaller seeds but another desirable trait (e.g., disease tolerance). The resultant progeny could be screened for leaf D-pinitol concentration before flowering occurred, and only the highest-accumulating lines could be taken forward. This would save time and money by avoiding growing plants to maturity that do not show the desirable kernel trait.

By combining a core set of markers, we were able to develop a model that could predict the number of mature seeds under drought conditions to a high degree of accuracy ($R^2$ adj = 0.62). Selecting plants for abiotic stress tolerance, e.g., drought, flooding, frost and heat, is very challenging because, for large-scale breeding programs, crosses need to be screened outdoors, which is dependent upon the relevant climatic conditions occurring in a given year. The ability to predict genotypic performance under abiotic stress from non-stressed plants is of huge benefit. While, ideally, molecular markers would be more reliable for trait prediction than metabolites because they are not subject to environmental perturbations, for many crops and particularly for complex traits, they are simply not available.

Our study shows that the leaf metabolic profile of well-watered, young plants can be measured 80 days before harvest to identify, with a high degree of accuracy, which genotypes are more likely to produce higher seed numbers under terminal drought conditions. Given the close correlation between yield (g/area) and seed number plant$^{-1}$, it is very likely that yield could also be the focus of our model. D-pinitol concentrations in the leaf are strongly and consistently associated with seed size, and this could be used as a means of early selection. The second year of our study and the year after (2018 and 2019, respectively) were the driest on record for eastern Australia. In 2021, the net value of the national welfare lost to this drought event was AUD 53 billion [52]. Extreme weather conditions, including drought, are predicted to increase in frequency and severity as part of our changing climate. Metabolite-assisted breeding offers a means to accelerate the selection of superior crosses that continue to produce viable yields under extreme climatic conditions.

## 4. Materials and Methods

### 4.1. Field Trial

The field site was located at the University of Sydney's IA Watson Grains Research Centre, Narrabri, NSW Australia (30°16′31.7″ S, 149°48′10.7″ E). The field trial used in 2017 (year 1) was previously described [53]. The trials were sown on 5 and 7 June in year 1 (2017) and 2018 (year 2), respectively.

The field sites were 0.6 ha in total, which was divided in half, into an irrigated and rainfed treatment in an incomplete block design. For this study, the metabolite data and associated yield parameters were only collected from the rainfed side of the field. Therefore, this represents a fully randomised block design. Thirty-six genotypes were grown each year, with four replicates of each plot in each treatment (rainfed or irrigated). Plots were initially 1.6 × 6 m, which were then cut back to 4 m before podding commenced. Each plot and the perimeter of the whole trial was surrounded by a double-row of buffer plots. Seeds were planted using a five-row mechanical planter, and the row spacing was set to 0.32 m.

Seeds were pre-treated with fungicide and treated with granulated inoculant (Nodulator®, Group-N Granular Legume Inoculant, BASF Australia Limited, Southbank, VIC, Australia) at a rate of 3.2 kg $ha^{-1}$, and Granulock Z Extra fertiliser (Granulock®, Incitec Pivot Limited, Port Lincoln, SA, Australia) at a rate of 50 kg $ha^{-1}$ was applied at the time of sowing [53].

The irrigated treatment received 25 mm irrigation (total 100 mm) approximately every two weeks from mid-August in year 1 and from May in year 2, which was homogeneously applied to the field using a lateral move irrigator. Supplementary irrigation was supplied only to the irrigated treated plots at three timepoints in year 1, but in year 2 the residual soil moisture at the start of the season was so low following the previous dry year that supplementary irrigation was supplied to both treatments until anthesis and, thereafter, only to the irrigated treatment in year 2. A total of 92 mm of irrigation was supplied during the experiment in year 1. In year 2, a total of 70 mm was applied prior to planting (in two applications), and a further 190 mm was applied to the irrigated treatment and 110 mm to the rainfed treatment over the course of the experiment. The biomarker harvests took place before drought treatment (withdrawal of irrigation) was imposed. Therefore, the "treatments" had been equally watered at the time of the biomarker leaf harvest.

### 4.2. Plant Material

Forty-nine genotypes were tested over the two years, with thirty-six included each year, and twenty-three lines being tested in both years (Table S1). In year 1, both Desi and Kabuli types were included, but in year 2 only Desi types were cultivated. Genotypes were selected from current cultivars bred for the northern NSW region: older Australian varieties and lines sourced from ICRISAT including Indian varieties (denoted by the "ICCV" prefix) and lines from the ICRISAT reference set (denoted by the "ICC" prefix). The ICC and ICCV selections were based on pre-breeding observations and publications reporting interesting rooting/biomass/morphology and/or drought response [54,55].

### 4.3. Yield Harvest

At maturity (around day after sowing (DAS) 160), a 50 × 50 cm quadrat was placed around an area of the plot, and all plants within it were counted and then cut at the base. The plants were placed in paper bags, dried to a constant weight and then threshed to remove the seeds. Cleaned seed was weighed, and then both values were divided by the number of plants to give seed yield g $plant^{-1}$. Hundred-kernel weight (HKW) was automated using a seed counter (Contador, Pfeuffer, Kitzingen, Germany). The average number of seeds per plant was calculated as (seed yield g $plant^{-1}$/HKW) × 100). All plots with both treatments (rainfed and irrigated) were harvested, but only data from the rainfed plots (from which the biomarkers were harvested) were used for the model development. Machine-harvested plot yields are not included in this study because diverse genotypes were used, and the combine harvester more effectively harvested taller, larger-seeded genotypes than those with smaller seeds and stature.

### 4.4. Biomarker Harvest Protocol

Biomarker harvests took place at DAS 74 and DAS 80 in years 1 and 2, respectively. This timepoint was selected because it was the earliest that an entire stem could be harvested from each plot that would yield 20 mg dry weight of leaf material. This harvest point was when the earliest-flowering genotypes had their first emerged petals. All genotypes were at Biologische Bundesanstalt, Bundessortenamt und Chemische Industrie (BBCH) Scale 55–59. Samples were harvested from the rainfed side of the field (before the irrigation applications were ceased and drought effects took effect). Harvests were carried out on clear days between 12:00–2 pm to control for diurnal effects. A single stem that was representative of canopy height was selected from each plot, cut at the base with scissors and placed in a Whirl-Pak sample bag (Whirl-Pak, Filtration Group, https://www.whirl-pak.com/ (accessed on 15 May 2023)). Samples were frozen in liquid nitrogen, stored at −80 °C and then freeze-dried (Virtis FreezeMobile, Gardiner, MT, USA). Samples were split into leaf and stem tissues, transferred to 2 mL microcentrifuge tubes and ball-milled to a fine powder (Geno/Grinder 2010, Spex SamplePrep, Metuchen, NJ, USA). In this manuscript, only data from the leaf samples are shown.

### 4.5. Metabolite Extraction

Soluble sugars and starch were enzymatically analysed, as previously described [56,57], and GC-MS protocols were as previously described [58,59]. Metabolite extraction: Approximately 20 mg (actual weight recorded) of each freeze-dried, ball-milled plant tissue sample was weighed into 2 mL screw cap micro centrifuge tubes. Metabolites were extracted four times with 1 mL of 80% ($v/v$) ethanol, and the resulting supernatants were pooled; two extractions were at 80 °C for 20 min and 10 min, respectively, and the remaining two were at room temperature. A 0.5 mL aliquot of soluble metabolite extract and the remaining pellet containing the insoluble fraction (including starch) were dried down in a heat block at 50 °C until all the solvent had evaporated. The dried-down residue from the soluble fraction was then resuspended in 0.5 mL of distilled water. Samples were stored at −20 °C for analysis.

### 4.6. Soluble Sugar Analysis

Soluble sugars of samples extracted in the previous step were enzymatically quantified using a Megazyme protocol (Megazyme Sucrose, D-glucose and D-fructose Assay Procedure, K-SUFRG 04/18, Megazyme International, Co Wicklow, Ireland) by the stepwise addition of hexokinase, phosphoglucose isomerase and β-fructosidase [60]. Samples were photometrically quantified (Benchmark Plus, BioRad, Hercules, CA, USA) by measuring the change in wavelength at 340 nm for 20 min after the addition of each enzyme. Sucrose, glucose and fructose were then quantified from standard curves included on each 96-well plate.

### 4.7. Starch Quantification

Starch was quantified using a modified Megazyme protocol (Megazyme Total Starch Assay Procedure, AOAC method 996.11, Megazyme International, Co Wicklow, Ireland). Briefly, the dried pellet was resuspended in 0.4 mL of 0.2 M KOH, vortexed vigorously and heated to 90 °C in a water bath for 15 min to facilitate gelatinisation of the starch. A total of 1.28 mL of 0.15 M NaOAc (pH 3.8) was added to each tube (to neutralise the sample) before the addition of 20 µL α-amylase and 20 µL amyloglucosidase (Megazyme International, Co Wicklow Ireland). After incubation at 50 °C for 30 min and centrifugation for 5 min, a 0.02 mL aliquot was combined with 0.6 mL of GOPOD reagent (Megazyme International, Co Wicklow, Ireland). A total of 0.2 mL of this reaction was photometrically assayed (Benchmark Plus, BioRad, Hercules, CA, USA) on a 96-well microplate at 510 nm against a water-only blank. Starch was quantified from known standard curves on the same plate. Each sample and standard were tested in duplicate. Each plate contained a control sample of known concentration for both soluble sugars and starch analysis.

### 4.8. Gas Chromatography Mass Spectrometry

For the carbohydrates, sugar alcohols and organic acid analyses, gas chromatography (GC) techniques used by Merchant et al. (2006) [58] were followed accordingly. First, 50 µL of dried extract were suspended in 450 µL anhydrous pyridine, to which a solution of 1:10 ratio mixture of trimethylchloroacetamide (TMCS) and bis-trimethylsilyl-trifluroacetamide (BSTFA) was added for derivatisation. Samples were incubated for 35 min at 75 °C and analysed by GC-MS within 24 h. The analysis was carried out on an Agilent 6890 Gas Chromatograph with QQQ 7000 Mass selective detector (Agilent Technologies, Santa Clara, CA, USA). Samples were injected in a split splitless injector at 300 °C with a 20:1 split injection onto a HP-5 column (30 m, 0.25 mm ID, 0.25 µm film thickness) with helium carrier gas at a constant flow of 1 mL/min. The temperature program had an initial oven temperature set of 60 °C for 2 min, ramping to 220 °C at 10 °C $min^{-1}$ for

5 min and then to 300 °C at 10 °C min$^{-1}$ for 5 min. GC-MS results were identified based on retention times relative to standards and extracted ions. Peak areas were integrated, and their relative quantities were calculated by Mass Hunter software (version B.07.01, Agilent Technologies) and used for peak integration.

### 4.9. Statistics and Modelling: Linear Modelling and Feature Selection

All statistical tests, modelling and feature selection were carried out in R [61]. Student's *t*-tests were two-sided, assuming unequal variances ($p$ = <0.05).

A multivariate linear regression model was constructed to analyse the relationship between seed number plant$^{-1}$ in rainfed (terminal drought)-treated plants and 14 metabolites analysed from well-watered conditions early in the growing season. The modelling and model evaluation and trait prediction (below) were conducted in R using the Caret (Classification and Regression Training) package [61,62]. In order to simplify the model by reducing the number of variables, the Step Akaike Information Criteria (AIC) [63,64] was applied. This maximum-likelihood estimation (MLE) feature selection technique tests whether the AIC value is increased or decreased with the step-wise addition of each explanatory variable (metabolite), with a lower value being the desired outcome. Both a forwards and backwards approach were tested, and the backwards method was found to produce the highest adjusted $R^2$ value. The backwards elimination method sequentially removes variables that do not show a significant ($p$ = <0.05) relationship to the trait, leaving only the minimum significantly correlated set [65,66]. A backwards approach is preferable if there is a high likelihood of collinearity amongst variables [67], which is often the case with metabolites, e.g., Ceusters et al. [68].

### 4.10. Model Evaluation and Trait Prediction

The leave-one-out cross-validation (LOOCV) approach was utilised to train the model [65]. In this approach, the model is repeatedly re-fitted using a different training and test set each time. With each iteration, a single test value (genotype) is omitted from the training set, and the mean square error (MSE) of the predicted versus actual value for that genotype is calculated. The process is repeated until all values have been used as the test value (n = 72). The test MSE is the average of all the calculated MSE's. A linear regression between the predicted and actual values was then plotted, and the adjusted $R^2$ and $p$ values were determined.

**Supplementary Materials:** The following supporting information can be downloaded at https://www.mdpi.com/article/10.3390/plants12112172/s1. Table S1: Genotypes tested. Table S2: Yield trait values. Table S3: GCMS metabolic profile of the leaf tissue. Table S4: Enzyme assay profile of the leaf tissue.

**Author Contributions:** Conceptualisation, S.J.P. and A.M.; methodology, S.J.P., D.F. and C.N.; validation, S.J.P. and C.N.; formal analysis, S.J.P., P.R. and D.F.; investigation, S.J.P. and P.R.; resources, A.M.; writing—original draft preparation, S.J.P., P.R. and A.M.; writing—review and editing, A.M., S.J.P., B.N.K. and P.R.; supervision, A.M.; funding acquisition, B.N.K. and A.M. All authors have read and agreed to the published version of the manuscript.

**Funding:** This study was funded by the Australian Research Council Industrial Transformation Research Hub for Legumes for Sustainable Agriculture (IH140100013).

**Data Availability Statement:** All data are included in this manuscript or the supporting documents.

**Acknowledgments:** We wish to acknowledge the technical support provided by Sabina Yasmin.

**Conflicts of Interest:** The authors declare no conflict of interest.

## References

1. Bidin, M.Z.; Shah, A.M.; Stanslas, J.; Seong, C.L.T. Blood and urine biomarkers in chronic kidney disease: An update. *Clin. Chim. Acta* **2019**, *495*, 239–250. [CrossRef] [PubMed]

2. Collins, G.S.; Altman, D.G. Predicting the 10 year risk of cardiovascular disease in the United Kingdom: Independent and external validation of an updated version of QRISK2. *BMJ* **2012**, *344*, e4181. [CrossRef] [PubMed]
3. Hippisley-Cox, J.; Coupland, C.; Vinogradova, Y.; Robson, J.; Minhas, R.; Sheikh, A.; Brindle, P. Predicting cardiovascular risk in England and Wales: Prospective derivation and validation of QRISK2. *BMJ* **2008**, *336*, 1475–1482. [CrossRef]
4. Wanner, L.A.; Junttila, O. Cold-induced freezing tolerance in Arabidopsis. *Plant Physiol.* **1999**, *120*, 391–399. [CrossRef] [PubMed]
5. Wang, L.; Wu, J.Q. The Essential Role of Jasmonic Acid in Plant-Herbivore Interactions—Using the Wild Tobacco Nicotiana attenuata as a Model. *J. Genet. Genom.* **2013**, *40*, 597–606. [CrossRef] [PubMed]
6. Sequera-Mutiozabal, M.I.; Erban, A.; Kopka, J.; Atanasov, K.E.; Bastida, J.; Fotopoulos, V.; Alcazar, R.; Tiburcio, A.F. Global Metabolic Profiling of Arabidopsis Polyamine Oxidase 4 (AtPAO4) Loss-of-Function Mutants Exhibiting Delayed Dark-Induced Senescence. *Front. Plant Sci.* **2016**, *7*, 173. [CrossRef]
7. Hannah, M.A.; Wiese, D.; Freund, S.; Fiehn, O.; Heyer, A.G.; Hincha, D.K. Natural genetic variation of freezing tolerance in arabidopsis. *Plant Physiol.* **2006**, *142*, 98–112. [CrossRef]
8. Nam, K.H.; Kim, D.Y.; Kim, H.J.; Pack, I.S.; Kim, H.J.; Chung, Y.S.; Kim, S.Y.; Kim, C.G. Global metabolite profiling based on GC-MS and LC-MS/MS analyses in ABF3-overexpressing soybean with enhanced drought tolerance. *Appl. Biol. Chem.* **2019**, *62*, 15. [CrossRef]
9. Hoffman, D.E.; Jonsson, P.; Bylesjo, M.; Trygg, J.; Antti, H.; Eriksson, M.E.; Moritz, T. Changes in diurnal patterns within the Populus transcriptome and metabolome in response to photoperiod variation. *Plant. Cell. Environ.* **2010**, *33*, 1298–1313. [CrossRef]
10. Janska, A.; Marsik, P.; Zelenkova, S.; Ovesna, J. Cold stress and acclimation—What is important for metabolic adjustment? *Plant. Biol.* **2010**, *12*, 395–405. [CrossRef]
11. Guy, C.; Kaplan, F.; Kopka, J.; Selbig, J.; Hincha, D.K. Metabolomics of temperature stress. *Physiol. Plant.* **2008**, *132*, 220–235. [CrossRef]
12. Riedelsheimer, C.; Lisec, J.; Czedik-Eysenberg, A.; Sulpice, R.; Flis, A.; Grieder, C.; Altmann, T.; Stitt, M.; Willmitzer, L.; Melchinger, A.E. Genome-wide association mapping of leaf metabolic profiles for dissecting complex traits in maize. *Proc. Natl. Acad. Sci. USA* **2012**, *109*, 8872–8877. [CrossRef] [PubMed]
13. Keurentjes, J.J.; Fu, J.; de Vos, C.H.; Lommen, A.; Hall, R.D.; Bino, R.J.; van der Plas, L.H.; Jansen, R.C.; Vreugdenhil, D.; Koornneef, M. The genetics of plant metabolism. *Nat. Genet.* **2006**, *38*, 842–849. [CrossRef] [PubMed]
14. Meyer, R.C.; Steinfath, M.; Lisec, J.; Becher, M.; Witucka-Wall, H.; Torjek, O.; Fiehn, O.; Eckardt, A.; Willmitzer, L.; Selbig, J.; et al. The metabolic signature related to high plant growth rate in Arabidopsis thaliana. *Proc. Natl. Acad. Sci. USA* **2007**, *104*, 4759–4764. [CrossRef] [PubMed]
15. Steinfath, M.; Gartner, T.; Lisec, J.; Meyer, R.C.; Altmann, T.; Willmitzer, L.; Selbig, J. Prediction of hybrid biomass in Arabidopsis thaliana by selected parental SNP and metabolic markers. *Theor. Appl. Genet.* **2010**, *120*, 239–247. [CrossRef]
16. Sulpice, R.; Trenkamp, S.; Steinfath, M.; Usadel, B.; Gibon, Y.; Witucka-Wall, H.; Pyl, E.T.; Tschoep, H.; Steinhauser, M.C.; Guenther, M.; et al. Network Analysis of Enzyme Activities and Metabolite Levels and Their Relationship to Biomass in a Large Panel of Arabidopsis Accessions. *Plant Cell.* **2010**, *22*, 2872–2893. [CrossRef]
17. Sulpice, R.; Nikoloski, Z.; Tschoep, H.; Antonio, C.; Kleessen, S.; Larhlimi, A.; Selbig, J.; Ishihara, H.; Gibon, Y.; Fernie, A.R.; et al. Impact of the Carbon and Nitrogen Supply on Relationships and Connectivity between Metabolism and Biomass in a Broad Panel of Arabidopsis Accessions(1[W][OA]). *Plant. Physiol.* **2013**, *162*, 347–363. [CrossRef]
18. Sulpice, R.; Pyl, E.T.; Ishihara, H.; Trenkamp, S.; Steinfath, M.; Witucka-Wall, H.; Gibon, Y.; Usadel, B.; Poree, F.; Piques, M.C.; et al. Starch as a major integrator in the regulation of plant growth. *Proc. Natl. Acad. Sci. USA* **2009**, *106*, 10348–10353. [CrossRef]
19. Degenkolbe, T.; Do, P.T.; Kopka, J.; Zuther, E.; Hincha, D.K.; Kohl, K.I. Identification of drought tolerance markers in a diverse population of rice cultivars by expression and metabolite profiling. *PLoS ONE* **2013**, *8*, e63637. [CrossRef]
20. Obata, T.; Witt, S.; Lisec, J.; Palacios-Rojas, N.; Florez-Sarasa, I.; Yousfi, S.; Araus, J.L.; Cairns, J.E.; Fernie, A.R. Metabolite Profiles of Maize Leaves in Drought, Heat, and Combined Stress Field Trials Reveal the Relationship between Metabolism and Grain Yield. *Plant. Physiol.* **2015**, *169*, 2665–2683. [CrossRef] [PubMed]
21. Steinfath, M.; Strehmel, N.; Peters, R.; Schauer, N.; Groth, D.; Hummel, J.; Steup, M.; Selbig, J.; Kopka, J.; Geigenberger, P.; et al. Discovering plant metabolic biomarkers for phenotype prediction using an untargeted approach. *Plant. Biotechnol. J.* **2010**, *8*, 900–911. [CrossRef] [PubMed]
22. Maddison, A.L.; Camargo-Rodriguez, A.; Scott, I.M.; Jones, C.M.; Elias, D.M.O.; Hawkins, S.; Massey, A.; Clifton-Brown, J.; McNamara, N.P.; Donnison, I.S.; et al. Predicting future biomass yield in Miscanthus using the carbohydrate metabolic profile as a biomarker. *Glob. Change Biol. Bioenergy* **2017**, *9*, 1264–1278. [CrossRef] [PubMed]
23. Riedelsheimer, C.; Czedik-Eysenberg, A.; Grieder, C.; Lisec, J.; Technow, F.; Sulpice, R.; Altmann, T.; Stitt, M.; Willmitzer, L.; Melchinger, A.E. Genomic and metabolic prediction of complex heterotic traits in hybrid maize. *Nat. Genet.* **2012**, *44*, 217–220. [CrossRef]
24. Berger, J.D.; Turner, N.C.; Siddique, K.H.M.; Knights, E.J.; Brinsmead, R.B.; Mock, I.; Edmondson, C.; Khan, T.N. Genotype by environment studies across Australia reveal the importance of phenology for chickpea (*Cicer arietinum* L.) improvement. *Aust. J. Agric. Res.* **2004**, *55*, 1071–1084. [CrossRef]
25. FAO. FAOSTAT: Crops. Available online: http://www.fao.org/faostat/en/#data/QC (accessed on 15 May 2023).

26. Ambika; Hegde, V.; Nimmy, M.S.; Bharadwaj, C.; Tripathi, S.; Singh, R.K.; Kumar, R. Unraveling genetics of semi-determinacy and identification of markers for indeterminate stem growth habit in chickpea (*Cicer arietinum* L.). *Sci. Rep.* **2021**, *11*, 21837. [CrossRef]
27. Pang, J.Y.; Turner, N.C.; Du, Y.L.; Colmer, T.D.; Siddique, K.H.M. Pattern of Water Use and Seed Yield under Terminal Drought in Chickpea Genotypes. *Front. Plant Sci.* **2017**, *8*, 1375. [CrossRef]
28. Leport, L.; Turner, N.C.; French, R.J.; Barr, M.D.; Duda, R.; Davies, S.L.; Tennant, D.; Siddique, K.H.M. Physiological responses of chickpea genotypes to terminal drought in a Mediterranean-type environment. *Eur. J. Agron.* **1999**, *11*, 279–291. [CrossRef]
29. Sabaghpour, S.H.; Malhotra, R.S.; Sarparast, R.; Safikhani, M.; Alizadeh, S.H.; Jahangeri, A.; Khalaf, G. Registration of 'Arman'—A Kabuli chickpea cultivar. *Crop. Sci.* **2006**, *46*, 2704–2705. [CrossRef]
30. Pushpavalli, R.; Zaman-Allah, M.; Turner, N.C.; Baddam, R.; Rao, M.V.; Vadez, V. Higher flower and seed number leads to higher yield under water stress conditions imposed during reproduction in chickpea. *Funct. Plant Biol.* **2015**, *42*, 162–174. [CrossRef]
31. Tuba Bicer, B.T.; Sakar, D. Heritability and gene effects for yield and yield components in chickpea. *Hereditas* **2008**, *145*, 220–224. [CrossRef] [PubMed]
32. Toker, C. Estimates of broad-sense heritability for seed yield and yield criteria in faba bean (*Vicia faba* L.). *Hereditas* **2004**, *140*, 222–225. [CrossRef] [PubMed]
33. Fàbregas, N.; Fernie, A.R. The metabolic response to drought. *J. Exp. Bot.* **2019**, *70*, 1077–1085. [CrossRef] [PubMed]
34. Lozano-Elena, F.; Fabregas, N.; Coleto-Alcudia, V.; Cano-Delgado, A.I. Analysis of metabolic dynamics during drought stress in Arabidopsis plants. *Sci. Data* **2022**, *9*, 90. [CrossRef] [PubMed]
35. Mathan, J.; Singh, A.; Ranjan, A. Sucrose transport and metabolism control carbon partitioning between stem and grain in rice. *J. Exp. Bot.* **2021**, *72*, 4355–4372. [CrossRef] [PubMed]
36. You, J.; Zhang, Y.; Liu, A.; Li, D.; Wang, X.; Dossa, K.; Zhou, R.; Yu, J.; Zhang, Y.; Wang, L.; et al. Transcriptomic and metabolomic profiling of drought-tolerant and susceptible sesame genotypes in response to drought stress. *BMC Plant. Biol.* **2019**, *19*, 267. [CrossRef]
37. Ford, C.W. Accumulation of Low-Molecular Weight Solutes in Water-Stressed Tropical Legumes. *Phytochemistry* **1984**, *23*, 1007–1015. [CrossRef]
38. Wanek, W.; Richter, A. Biosynthesis and accumulation of D-ononitol in *Vigna umbellata* in response to drought stress. *Physiol. Plant* **1997**, *101*, 416–424. [CrossRef]
39. Dumschott, K.; Dechorgnat, J.; Merchant, A. Water Deficit Elicits a Transcriptional Response of Genes Governing d-pinitol Biosynthesis in Soybean (Glycine max). *Int. J. Mol. Sci.* **2019**, *20*, 2411. [CrossRef]
40. Wang, X.X.; Guo, R.; Li, M.X.; Liu, Y.; Zhao, M.L.; Fu, H.; Liu, X.Y.; Wang, S.Y.; Shi, L.X. Metabolomics reveals the drought-tolerance mechanism in wild soybean (Glycine soja). *Acta Physiol. Plant* **2019**, *41*, 161. [CrossRef]
41. Kumar, M.; Kumar Patel, M.; Kumar, N.; Bajpai, A.B.; Siddique, K.H.M. Metabolomics and Molecular Approaches Reveal Drought Stress Tolerance in Plants. *Int. J. Mol. Sci.* **2021**, *22*, 9108. [CrossRef]
42. Gowda, C.L.L.; Upadhyaya, H.D.; Dronavalli, N.; Singh, S. Identification of Large-Seeded High-Yielding Stable Kabuli Chickpea Germplasm Lines for Use in Crop Improvement. *Crop. Sci.* **2011**, *51*, 198–209. [CrossRef]
43. Gao, Y.F.; Zhang, M.N.; Wu, T.C.; Xu, M.Y.; Cai, H.N.; Zhang, Z.S. Effects of D-Pinitol on Insulin Resistance through the PI3K/Akt Signaling Pathway in Type 2 Diabetes Mellitus Rats. *J. Agric. Food Chem.* **2015**, *63*, 6019–6026. [CrossRef] [PubMed]
44. Kim, M.J.; Yoo, K.H.; Kim, J.H.; Seo, Y.T.; Ha, B.W.; Kho, J.H.; Shin, Y.G.; Chung, C.H. Effect of pinitol on glucose metabolism and adipocytokines in uncontrolled type 2 diabetes. *Diabetes Res. Clin. Pract.* **2007**, *77*, S247–S251. [CrossRef]
45. Vernon, D.M.; Bohnert, H.J. A Novel Methyl Transferase Induced by Osmotic-Stress in the Facultative Halophyte Mesembryanthemum-Crystallinum. *Embo J.* **1992**, *11*, 2077–2085. [CrossRef] [PubMed]
46. Sheveleva, E.; Chmara, W.; Bohnert, H.J.; Jensen, R.G. Increased salt and drought tolerance by D-ononitol production in transgenic *Nicotiana tabacum* L. *Plant Physiol.* **1997**, *115*, 1211–1219. [CrossRef]
47. Sengupta, S.; Patra, B.; Ray, S.; Majumder, A.L. Inositol methyl tranferase from a halophytic wild rice, *Porteresia coarctata* Roxb. (Tateoka): Regulation of pinitol synthesis under abiotic stress. *Plant Cell. Environ.* **2008**, *31*, 1442–1459. [CrossRef] [PubMed]
48. Ahn, C.H.; Hossain, M.A.; Lee, E.; Kanth, B.K.; Park, P.B. Increased salt and drought tolerance by D-pinitol production in transgenic *Arabidopsis thaliana*. *Biochem. Biophys. Res. Commun.* **2018**, *504*, 315–320. [CrossRef]
49. Streeter, J.G.; Lohnes, D.G.; Fioritto, R.J. Patterns of pinitol accumulation in soybean plants and relationships to drought tolerance. *Plant Cell. Environ.* **2001**, *24*, 429–438. [CrossRef]
50. Sánchez-Mata, M.C.; Peñuela-Teruel, M.J.; Cámara-Hurtado, M.; Díez-Marqués, C.; Torija-Isasa, M.E. Determination of Mono-, Di-, and Oligosaccharides in Legumes by High-Performance Liquid Chromatography Using an Amino-Bonded Silica Column. *J. Agric. Food Chem.* **1998**, *46*, 3648–3652. [CrossRef]
51. Aguilera, Y.; Martin-Cabrejas, M.A.; Benitez, V.; Molla, E.; Lopez-Andreu, F.J.; Esteban, R.M. Changes in carbohydrate fraction during dehydration process of common legumes. *J. Food Compos. Anal.* **2009**, *22*, 678–683. [CrossRef]
52. Wittwer, G.; Waschik, R. Estimating the economic impacts of the 2017–2019 drought and 2019–2020 bushfires on regional NSW and the rest of Australia. *Aust. J. Agric. Resour. Econ.* **2021**, *65*, 918–936. [CrossRef]
53. Huang, J.; Purushothaman, R.; McBratney, A.; Bramley, H. Soil Water Extraction Monitored Per Plot Across a Field Experiment Using Repeated Electromagnetic Induction Surveys. *Soil. Syst.* **2018**, *2*, 11. [CrossRef]

54. Meena, H.P.; Kumar, J.; Upadhyaya, H.D.; Bharadwaj, C.; Chauhan, S.K.; Verma, A.K.; Rizvi, A.H. Chickpea mini core germplasm collection as rich sources of diversity for crop improvement. *J. SAT Agric. Res.* **2010**, *8*, 1–5.
55. Ramamoorthy, P.; Lakshmanan, K.; Upadhyaya, H.D.; Vadez, V.; Varshney, R.K. Root traits confer grain yield advantages under terminal drought in chickpea (*Cicer arietinum* L.). *Field Crops Res.* **2017**, *201*, 146–161. [CrossRef]
56. Purdy, S.; Cunniff, J.; Maddison, A.; Jones, L.; Barraclough, T.; Castle, M.; Davey, C.; Jones, C.; Shield, I.; Gallagher, J.; et al. Seasonal Carbohydrate Dynamics and Climatic Regulation of Senescence in the Perennial Grass, Miscanthus. *Bioenergy Res.* **2014**, *8*, 28–41. [CrossRef]
57. Purdy, S.J.; Maddison, A.L.; Cunniff, J.; Donnison, I.; Clifton-Brown, J. Non-structural carbohydrate profiles and ratios between soluble sugars and starch serve as indicators of productivity for a bioenergy grass. *AoB Plants* **2015**, *7*, plv032. [CrossRef]
58. Merchant, A.; Richter, A.; Popp, M.; Adams, M. Targeted metabolite profiling provides a functional link among eucalypt taxonomy, physiology and evolution. *Phytochemistry* **2006**, *67*, 402–408. [CrossRef]
59. Conselvan, G.B.; Fuentes, D.; Merchant, A.; Peggion, C.; Francioso, O.; Carletti, P. Effects of humic substances and indole-3-acetic acid on Arabidopsis sugar and amino acid metabolic profile. *Plant Soil.* **2018**, *426*, 17–32. [CrossRef]
60. Jones, M.G.; Outlaw, W.H.; Lowry, O.H. Enzymic assay of 10 to 10 moles of sucrose in plant tissues. *Plant Physiol.* **1977**, *60*, 379–383. [CrossRef]
61. R Core Team. *R: A Language and Environment for Statistical Computing*; R Team: Austria, Vienna, 2021.
62. Kuhn, M. Building Predictive Models in R Using the caret Package. *J. Stat. Softw.* **2008**, *28*, 1–26. [CrossRef]
63. Akaike, H. A new look at the statistical model identification. *IEEE Trans. Autom. Control.* **1974**, *19*, 716–723. [CrossRef]
64. Aho, K.; Derryberry, D.; Peterson, T. Model selection for ecologists: The worldviews of AIC and BIC. *Ecology* **2014**, *95*, 631–636. [CrossRef] [PubMed]
65. Hastie, T.; Tibshiranix, R.; Friedman, J. *The Elements of Statistical Learning, Data Mining, Inference, and Prediction*, 2nd ed.; Springer Series in Statistics; Springer Nature: Cham, Switzerand, 2009.
66. Tripathi, A. Feature Selection Techniques in Regression Model. Available online: https://towardsdatascience.com/feature-selection-techniques-in-regression-model-26878fe0e24e (accessed on 30 November 2022).
67. Heinze, G.; Wallisch, C.; Dunkler, D. Variable selection—A review and recommendations for the practicing statistician. *Biom. J.* **2018**, *60*, 431–449. [CrossRef] [PubMed]
68. Ceusters, N.; Luca, S.; Feil, R.; Claes, J.E.; Lunn, J.E.; Van den Ende, W.; Ceusters, J. Hierarchical clustering reveals unique features in the diel dynamics of metabolites in the CAM orchid Phalaenopsis. *J. Exp. Bot.* **2019**, *70*, 3269–3281. [CrossRef]

*plants*

MDPI

*Article*

# Somaclonal Variation for Genetic Improvement of Starch Accumulation in Potato (*Solanum tuberosum*) Tubers

**Walaa M. R. M. Adly [1], Gniewko Niedbała [2], Mohammad E. EL-Denary [1], Mahasen A. Mohamed [1], Magdalena Piekutowska [3], Tomasz Wojciechowski [2], El-Sayed T. Abd El-Salam [4] and Ahmed S. Fouad [4,*]**

1 Horticulture Research Institute, Agriculture Research Center, Giza 12619, Egypt
2 Department of Biosystems Engineering, Faculty of Environmental and Mechanical Engineering, Poznań University of Life Sciences, Wojska Polskiego 50, 60-627 Poznań, Poland
3 Department of Geoecology and Geoinformation, Institute of Biology and Earth Sciences, Pomeranian University in Słupsk, Partyzantów 27, 76-200 Słupsk, Poland
4 Botany and Microbiology Department, Faculty of Science, Cairo University, Giza 12613, Egypt
* Correspondence: ahmedsfouad@yahoo.com or ahmedsfouad@sci.cu.edu.eg

**Abstract:** Starch content is one of the major quality criteria targeted by potato breeding programs. Traditional potato breeding is a laborious duty due to the tetraploid nature and immense heterozygosity of potato genomes. In addition, screening for functional genetic variations in wild relatives is slow and strenuous. Moreover, genetic diversity, which is the raw material for breeding programs, is limited due to vegetative propagation used in the potato industry. Somaclonal variation provides a time-efficient tool to breeders for obtaining genetic variability, which is essential for breeding programs, at a reasonable cost and independent of sophisticated technology. The present investigation aimed to create potato somaclones with an improved potential for starch accumulation. Based on the weight and starch content of tubers, the somaclonal variant *Ros 119*, among 105 callus-sourced clones, recorded a higher tuberization potential than the parent cv Lady Rosetta in a field experiment. Although this somaclone was similar to the parent in the number of tubers produced, it exhibited tubers with 42 and 61% higher fresh and dry weights, respectively. Additionally, this clone recorded 10 and 75% increases in starch content based on the dry weight and average content per plant, respectively. The enhanced starch accumulation was associated with the upregulation of six starch-synthesis-related genes, namely, the *AGPase, GBSS I, SBE I, SBE II, SS II* and *SS III* genes. AGPase affords the glycosyl moieties required for the synthesis of amylose and amylopectin. GBSS is required for amylose elongation, while SBE I, SBE II, SS II and SS III are responsible for amylopectin.

**Keywords:** potato; tissue culture; somaclonal variation; starch; gene expression

**Citation:** Adly, W.M.R.M.; Niedbała, G.; EL-Denary, M.E.; Mohamed, M.A.; Piekutowska, M.; Wojciechowski, T.; Abd El-Salam, E.-S.T.; Fouad, A.S. Somaclonal Variation for Genetic Improvement of Starch Accumulation in Potato (*Solanum tuberosum*) Tubers. *Plants* **2023**, *12*, 232. https://doi.org/10.3390/plants12020232

Academic Editors: Ranjith Pathirana and Francesco Carimi

Received: 29 November 2022
Revised: 25 December 2022
Accepted: 31 December 2022
Published: 4 January 2023



## 1. Introduction

Potato (*Solanum tuberosum*) is the most crucial non-graminaceous food crop cultivated in 17.34 million hectares yielding 370 million tons (FAO 2019 https://www.fao.org/faostat/ar/#data/QCL, accessed on 23 October 2022). Tubers are an important source for carbohydrates, proteins and other essential nutrients including minerals (iron, magnesium, phosphorus, potassium and zinc), vitamins (thiamin, niacin, pyridoxine, riboflavin, folate and ascorbic acid, pantothenic acid), and dietary fibers [1]. Potato stands first among food crops for production of energy, proteins, vitamins and minerals per unit of land area and time [2]. Alongside the indoor culinary purposes, tubers are utilized in many food products [3] that accumulate massive amounts of peel directed to bioethanol production [4].

Starch is the most abundant carbohydrate in potato tubers; it is a mixture of two polysaccharides. The first is amylose which is a linear, long α-glucan with few branches, containing about 99% α-(1,4) linkages and only 1% α-(1,6) linkages. The second is amylopectin characterized with heavily branched structure related to having about 5% α-(1,6) linkages [5]. Starch synthesis is hosted in amyloplasts and catalyzed with granular-bound starch synthase (GBSS), specialized for elongation of growing $\alpha-1,4$ linkage with exclusive or nearly exclusive activity in the soluble phase, and both starch branching enzyme (SBE) and starch synthases (SSs) to construct the branched chains [6]. The glycosyl moieties required for both amylose and amylopectin are provided as ADP glucose synthesized under catalysis of ADP glucose pyrophosphorylase (AGPase) [7,8]. The increased rate of starch biosynthesis during the early stages of tuber induction and the subsequent developmental stages of tuberization is underlined

Starch is an agriculturally important product with multiple food and nonfood valuable purposes. In human nourishment, starch plays an essential role in offering the metabolic energy necessary to perform biological functions; it is the primary source of energy for an enormous portion of the world's population. As an industrial material, starch has many applications including adhesion, coating, encapsulation, gelling and thickening [5]. Thus, starch content is one of the major quality criteria targeted by potato breeding programs [9] especially in this era characterized by suffering from climate change and growing gap in food supply.

Traditional potato breeding is a laborious duty due to the tetraploid nature and immense heterozygosity of the potato genomes. The former is the reason for intra-species incompatibilities and inbreeding depression, whereas the latter inhibits the introduction of novel traits through traditional breeding [10]. In addition, screening for functional genetic variation in wild relatives is slow and strenuous [11,12]. Adding to these difficulties, genetic diversity which is the raw material of breeding programs is hindered with vegetative propagation [13] applied for potato cultivation in most countries [14].

Based on genetic engineering and in vitro screening, biotechnology can overcome many of potato breeding obstacles [15,16]. However, the health and environmental issues about genetically modified foods [17] put in vitro screening as the first runner of biotechnological tools utilized in potato breeding. Generally, it depends on the genetic variations that arise during in vitro conditions, and these genetic variations are termed somaclonal variations [18]. Such variations can be attributed to point mutations, numerical and structural chromosomal variations as well as epigenetic variations including hypo- and hypermethylation of DNA [19]. Nevertheless, the molecular mechanisms underlying somaclonal variation require further elucidation [20].

In vitro screening of somaclones was successfully used to produce new potato lines with improved tolerance against drought [21], salinity [22], cadmium [15], early blight [23] and postharvest diseases [24]. In addition, several research groups introduced in vitro selected potato lines with improved starch content [25–28]. However, the genetic expression profile underlining the enhanced starch accumulation was not investigated. The transcription level of several genes can be rapidly and accurately estimated in the same extract using few chemicals, compared with assaying of the corresponding enzymes activities that may require several extraction methods and an arsenal of reagents. The increased rate of starch biosynthesis during the early stages of tuber induction and the subsequent developmental stages of tuberization is underlined with up-regulation of starch synthesis genes that was more pronounced when final tuber size was attained [29].

Generally, in vitro procedures expose plant material to oxidative stress and subsequent mutations [30]. However, although regeneration from preformed meristems (e.g., buds) does not normally produce variants, passing through a callus phase promotes somaclonal variation [31].

Therefore, the aim of the present investigation is to exploit somaclonal variation in callus-sourced regenerated plants in order to select new potato lines with a high starch content and to elucidate the gene expression profiles of starch-related genes, namely, *AGPase, GBSS I, SBE I, SBE II, SS II* and *SS III.*

## 2. Results

### 2.1. Tissue Culture, Acclimatization and Minituber Production

Potato in vitro plants (Figure 1A) were selected to provide internode explants. As a response to a callus induction medium, the internode explants swelled with a synchronized appearance of green nodular calli at the cut edges after a couple of weeks (Figure 1B). After being transferred to a shoot regeneration medium, the calli expanded to the whole explant surface.

**Figure 1.** Tissue culture and minituber formation of potato (*Solanum tuberosum*): four-week-old in vitro plants multiplicated via nodal cuttings on basal medium (**A**), callus induction on internode cuttings placed on basal medium fortified with 1-naphthaleneacetic acid (NAA) at 0.186 mg $L^{-1}$ and 6-benzylaminopurine (BAP) at 2.25 mg $L^{-1}$ (**B**), shoot regeneration from calli on basal medium containing BAP at 2.25 mg $L^{-1}$ (**C**), rooting of regenerated shoots on basal medium supplemented with 0.1 mg $L^{-1}$ indole-3-butyric acid (IBA) and 0.5 mg $L^{-1}$ indole-3-acetic acid (IAA) (**D**), acclimation of regenerated plants (**E**), four-week-old acclimated plants (**F**) and minituber formation of acclimated plants (**G**).

After eight weeks on the regeneration medium, an average of six shoots per explant (Figure 1C) were regenerated on about 60% of the explants, yielding 146 shoots, of which only 113 (77.3%) were able to survive following three successive nodal-cutting-based multiplication steps and produced corresponding clones. The obtained shoots were rooted successfully on a rooting medium to produce plants ready for acclimatization (Figure 1D). The rooted shoots of all the clones were acclimatized in 5 cm pots (Figure 1E) for two weeks in a greenhouse, and then they were transferred to 25 cm pots (Figure 1F). Fourteen weeks later, the produced $G_0$ minitubers were collected (Figure 1G).

The in vitro plants, used as a source of explants, were multiplied via nodal cuttings to establish control clones, and they were designated as meristem-derived (M-D) clone plants. Each of the *M-D* clone plants produced an average of five minitubers, weighing about 25 g. For the callus-sourced clones, a wide spectrum of tuberization potentials was recorded among the different clones. The variation among the different putative somaclones ranged from 2 to 15 for the number of minitubers per plant and 4–27 g for the fresh weight of tubers per plant.

*2.2. Tuberization and Starch Accumulation*

Compared with the $G_1$ tubers produced by the *M-D* clone, the primary screening of the $G_1$ tubers produced by the 105 callus-sourced clones failed to achieve better tuberization potentials in terms of the number of tubers, tuber weight and starch content, except for the clone named *Ros 119*. Although bearing the same number of tubers produced by the *M-D* clone (Figures 2 and 3), *Ros 119* exhibited tubers with 20 and 36% higher fresh and dry weights (Figure 4), in addition to 10 and 49% increases in starch content on the bases of tuber dry weight and average content per plant, respectively (Figure 5). The previous superiority of the *Ros 119* clone over the *M-D* one was observed to be intensified in the $G_2$ tubers. Compared with the *M-D* clone, the *Ros 119* tubers exhibited 38, 57, 11 and 71% increases in fresh weight, dry weight, starch content expressed in mg/g dry weight and starch content expressed in g per plant, respectively.

**Figure 2.** Potato plants carrying $G_2$ tubers: the commercial cultivar Lady Rosetta (**A**), *M-D* clone (**B**) and *Ros 119* clone (**C**).

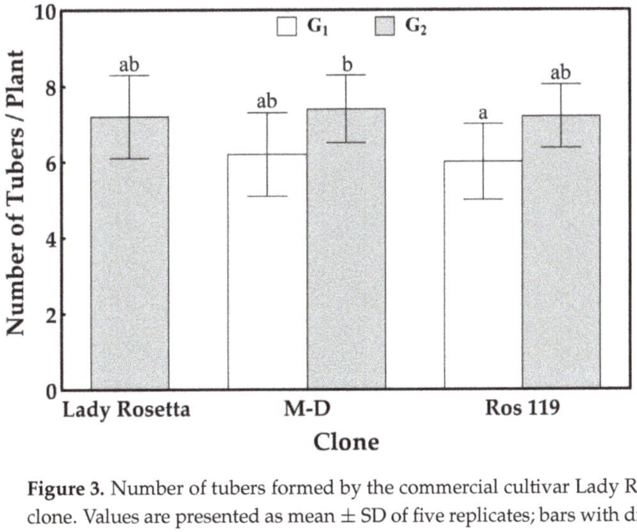


**Figure 3.** Number of tubers formed by the commercial cultivar Lady Rosetta, *M-D* clone and *Ros 119* clone. Values are presented as mean ± SD of five replicates; bars with different letters are significantly different, based on the LSD test, at $p < 0.05$.

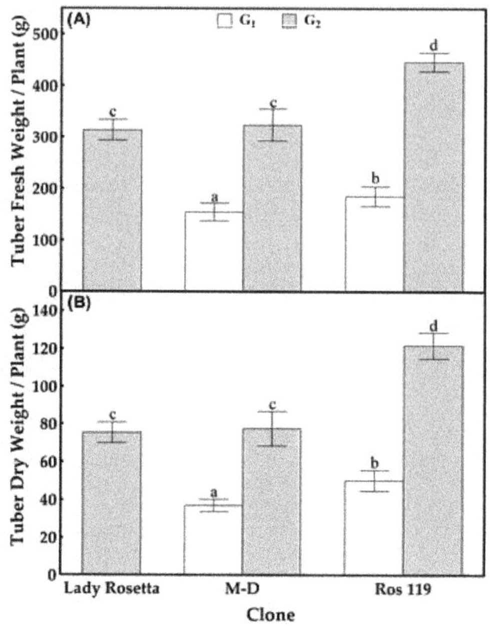


**Figure 4.** Fresh (**A**) and dry (**B**) weights of tubers formed by the commercial cultivar Lady Rosetta, *M-D* clone and *Ros 119* clone. Values are presented as mean ± SD of five replicates; bars with different letters are significantly different, based on the LSD test, at $p < 0.05$.

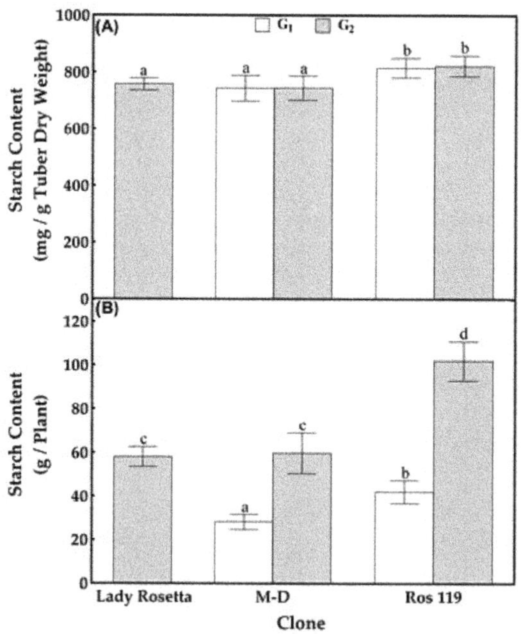


**Figure 5.** Tuber starch content on dry-weight (**A**) and per-plant (**B**) bases in tubers formed by the commercial cultivar Lady Rosetta, *M-D* clone and *Ros 119* clone. Values are presented as mean ± SD of five replicates; bars with different letters are significantly different, based on the LSD test, at $p < 0.05$.

Based on the results of the present study, the cv Lady Rosetta produced an average of 7.2 tubers per plant (Figures 2 and 3), weighing 314 and 75 g on the bases of fresh and dry weights, respectively (Figure 4). The starch determination in the tubers reflected their ability to accumulate about 756 mg of starch per g of dry weight, resulting in a total starch content of about 58 g per plant (Figure 5). Similar results were recorded when observing the G2 tubers produced by the M-D clone.

The primary data for the average weights of the $G_2$ tubers produced by the callus-sourced clones introduced *Ros 119* as a distinguished clone possessing fresh and dry weights that were about 42 and 61% higher, respectively, than those of the Lady Rosetta tubers. Similarly, the screening for starch accumulators among the callus-sourced clones reflected the ability of *Ros 119* to accumulate a starch content that was 10% higher (Figure 5Athan that of the cv Lady Rosetta tubers, without an accompanied significant variation in the average number of tubers per plant. The combined increases in tuber dry weight and starch content in *Ros 119* result in about a 75% increase in starch content per plant (Figure 5B).

### 2.3. Gene Expression Analysis

The monitoring of the gene expressions of the key enzymes involved in starch synthesis reflected an insignificant difference between the expression levels in the Lady Rosetta tubers and the corresponding gene expressions in the *M-D* clone tubers (Figure 6). However, the callus-derived clone *Ros 119* exhibited significantly elevated transcript abundances for the examined genes, compared with the corresponding genes in the Lady Rosetta tubers.

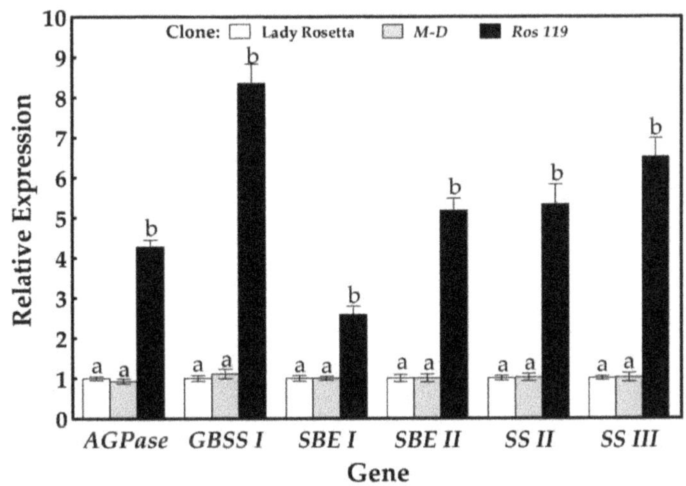


**Figure 6.** Relative expressions of starch-synthesis-related genes in tubers formed by the commercial cultivar Lady Rosetta, *M-D* clone and *Ros 119* clone. Values are presented as mean ± SD of five replicates; bars with different letters are significantly different, based on the LSD test, at $p < 0.05$.

In *Ros 119*, the expression level of the ADP glucose pyrophosphorylase (*AGPase*) gene responsible for the key reaction in starch content reached a 4.29-fold increase compared to the level quantified in the Lady Rosetta tuber. The transcript abundance of the granular-bound starch synthase (*GBSS*) gene responsible for amylose biosynthesis exhibited the most obvious enhancement among the examined genes, manifesting a 8.34-fold increase compared to the corresponding gene expression in the Lady Rosetta tubers. The expressions of the starch-branching enzyme (*SBE*) genes responsible for $\alpha-1.6$ linkages in amylopectin exhibited 2.57- and 5.17-fold increases compared to the expressions measured in the Lady Rosetta tubers for *SBE I* and *SBE II*, respectively, while the expressions of the starch synthase II (*SS II*) and starch synthase III (*SS III*) genes responsible for $\alpha-1.4$ linkages in the branched chains reached 5.31- and 6.5-fold increases, respectively, compared to the corresponding gene expressions.

## 3. Discussion

In the current study, callus was induced on internode explants on a medium supplemented with 2.25 mg $L^{-1}$ BAP and 0.186 mg $L^{-1}$ NAA, while shoot regeneration was achieved following auxin removal and the addition of $AgNO_3$ at 4 mg $L^{-1}$. The same media were utilized for callus induction followed by shoot regeneration from potato leaves and internode explants in a previous study [32]. Similarly, Kumlay and Ercisli [33] employed a medium supplemented with both cytokinin and auxin to induce callus formation on the leaves and internode explants of four potato cultivars, whereas shoot regeneration commenced following auxin omission. The same approach was implemented by Ghosh et al. [34], who started with leaf-sourced explants of three potato cultivars.

Auxins are important players in callus initiation [35]; they afford a narrow range for cell fate transition [36], which necessitates their withdrawal or at least the lowering of their concentration in regeneration media. However, cytokinins influence callus initiation [33]; they are the major participants in regeneration media, where shoot regeneration is the consequence of interconnections among cytokinin receptors, the development of shoot meristems and cell cycles [37]. The role played by $AgNO_3$ in shoot regeneration media can be attributed to the intrusion with ethylene perception, which mitigates the hindering influence of the accretion of the gas hormone on shoot regeneration [38,39]. In addition, $AgNO_3$ enhances the accumulation of polyamines [32], whose role in the improvement of morphogenesis has been previously recorded in potato shoot cultures [40].

In the present study, the rooting of the regenerated roots was achieved on a medium supplemented with 0.1 mg $L^{-1}$ IBA and 0.5 mg $L^{-1}$ IAA. A combination of the aforementioned auxins was utilized by Hajare et al. to initiate the rooting of potato-regenerated shoots [41]. IBA is generally employed to root potato-regenerated shoots, either alone [42,43] or in combination with another auxin [44,45].

The genetic variations that regularly arise during regeneration from callus cells are attributed to numerical and structural chromosomal variations; point mutations; and epigenetic variations, including the hyper- and hypo-methylation of DNA [19]. These variations, termed somaclonal variations [18], introduced *Ros 119* as a new clone. *Ros 119* is a starch-rich clone able to accumulate a significantly higher amount of starch in its tubers compared with the commercial cultivar Lady Rosetta. Similar to our results, Thieme and Griess [27], Rosenberg et al. [25] and Bayati et al. [24] used somaclonal variation to introduce potato lines with enhanced starch content. However, the introduced lines had different starch accumulation potentials and different frequencies in their occurrence, which could be attributed to the randomness of the variations responsible for the enhanced starch accumulation.

The increase in the starch accumulation recorded in the present study can be attributed to the increase in the expression of AGPase, SSs, GBSS and SBEs genes (Figure 7). The up-regulation of starch synthesis-related genes was documented during tuber formation of potato [29,46]. Based on microarray analysis, Kloosterman et al. [46] recorded up-regulated profile for the genes addressed in the present study during the early stages of tuberization, that was maintained until the final tuber size was reached. The authors attributed these results to the increased rate of starch biosynthesis. Supported with results of transcriptome analysis, Ferreira et al. [29] documented low expressions of starch-biosynthesis-related genes at early tuberization stages followed by upregulation at terminal stages.

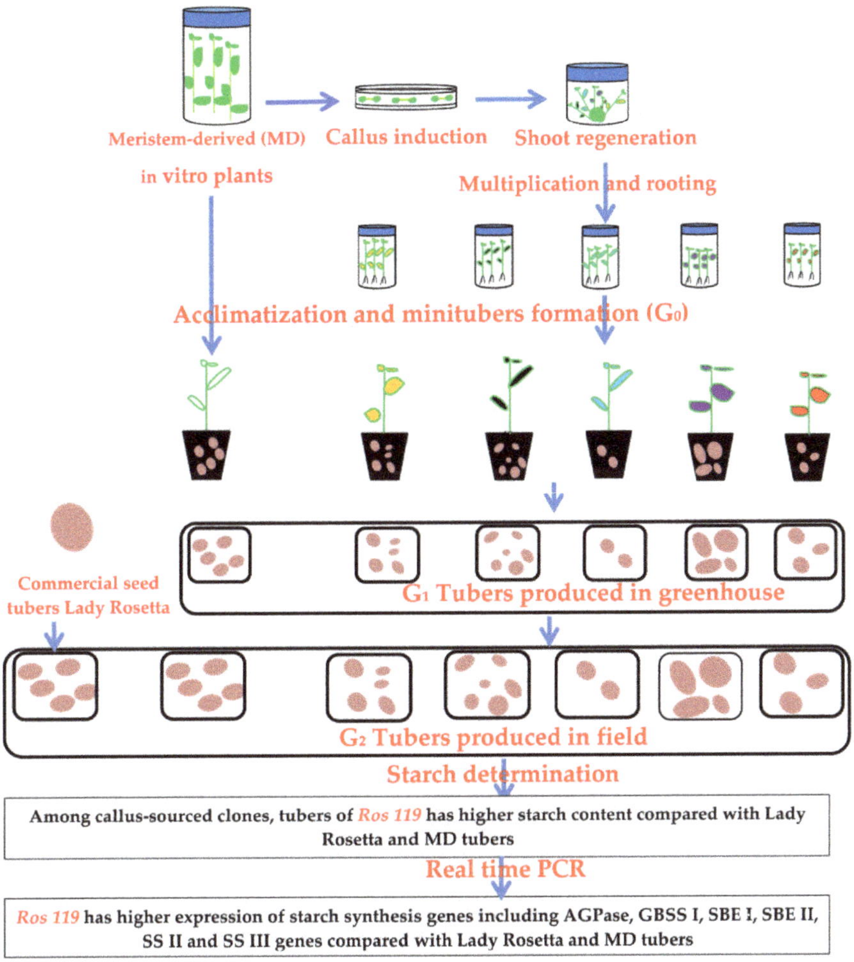


**Figure 7.** Graphical summary showing the production of potato somaclones with distinguished ability to accumulate starch based on upregulation of starch-synthesis-related genes.

Variations in gene expressions in tissue-culture-derived plants have also been documented in *Phalaenopsis* 'Wedding Promenade' [46] and rice [47]. When investigating the fruit transcriptome of cucumber somaclones, Pawe\lkowicz et al. [48] documented a different differential gene expression in each clone. The authors attributed the results to variations in the genic region and the interactions among molecular networks, which initiate specific pathways. Similar findings were recorded by López-Hernández and Cortés [49], who examined the transcriptomes of mint somaclones.

An increase in starch content accompanied by an increase in the expression of the *AGPase* gene was previously recorded in potato by Müller-Röber et al. [50,51] and Stark et al. [50,51]. The synchronization between the increase in starch content and the expressions of the *AGPase, SS, GBSS* and *SBE* genes has been recorded in potato [8], rice [52], wheat [53] and lanzhou lily bulb [54].

AGPase catalyzes the key step in starch biosynthesis; it produces ADP glucose, which provides the glycosyl moiety required for starch biosynthesis [7,8]. Thus, AGPase is the rate-limiting enzyme in starch accumulation in potato [55]. ADP-glucose is actively transferred through the activity of specialized transporters into amyloplasts [56,57] for the subsequent synthesis of amylose and amylopectin. Starch biosynthesis is catalyzed by a group of enzymes, including GBSS, SSs and SBEs. Both GBSS and SSs are responsible for chain elongation by catalyzing the formation $\alpha-1{,}4$ glucosidic bonds in amylose and amylopectin, respectively, while SBEs are responsible for the formation of $\alpha-1.6$ linkages at the branch points of amylopectin [6]. Several SSs have been characterized in plants; however, they are mainly referred to as SS II and SS III in potato [8].

## 4. Materials and Methods

### 4.1. Explant Preparation

Virus-free tubers of potato (*Solanum tuberosum*) cv Lady Rosetta were kindly provided by the Agricultural Research Center, Cairo, Egypt. The tubers were kept at 10 °C until sprouting; then, the sprouts were separated from the tubers and immersed in tap water mixed with a few drops of a commercial liquid detergent, agitated for 10 min and then thoroughly rinsed in running water for 30 min to remove the remaining detergent. The washed sprouts were placed in a 250 mL caped jar containing about 100 mL of 20% commercial Clorox (5% chlorine) (Clorox Egypt) containing a few drops of Tween 20, and then the jar was shaken for five minutes. In a laminar flow cabinet, the surface-sterilized sprouts were picked and rinsed thoroughly in sterile distilled water. The meristems were aseptically excised and transferred into sterile tubes (one explant/tube), each containing 10 mL sterile basal medium (MS medium [58] supplemented with 100 mg $L^{-1}$ myoinositol and 30 g $L^{-1}$ sucrose) supplemented with 0.01 mg $L^{-1}$ NAA, 0.1 mg $L^{-1}$ gibberellic acid ($GA_3$) and 2 mg $L^{-1}$ calcium pantothenate. The medium was solidified using 7 g $L^{-1}$ agar, and the pH was adjusted to 5.7 before autoclaving for 20 min at 121 °C. The cultures were incubated at 25 °C and at a light intensity of 15.8 Wat $m^{-2}$ with 16/8 h light/dark cycles under cool white fluorescent lamps. The same incubation conditions were implemented throughout the study. After six weeks, shoots that were about 5–7 cm were collected and cut into nodal cuttings for micropropagation in order to establish control clones designated as meristem-derived (*M-D*) clones, which were employed as a source of internode explants for callus initiation. The plant material was subcultured at three-week intervals in 400 mL glass jars (4–5 cuttings/jar), each containing approximately 50 mL basal medium.

### 4.2. Callus Induction, Shoot Regeneration and Rooting

Based on our previous publication [32], about 1 cm of each of the internode explants was utilized for callus induction on a basal medium augmented with 2.25 mg $L^{-1}$ BAP and 0.186 mg $L^{-1}$ NAA. Six weeks later, the explants carrying calli were transferred to a regeneration medium consisting of a basal medium supplemented with 2.25 mg $L^{-1}$ BAP and 4 mg $L^{-1}$ $AgNO_3$. After eight weeks, the regenerated shoots were separated from the remaining callus and each subjected to three cycles of subculture on a basal medium in order to produce a sufficient number of shoots considered as a putative clone that received a characteristic code. The shoots of each clone were transferred to a rooting medium consisting of a basal medium supplemented with 0.1 mg $L^{-1}$ IBA and 0.5 mg $L^{-1}$ IAA [41]. Four weeks later, the healthy rooted plants were harvested and acclimatized.

At the beginning of September 2019, the healthy rooted shoots of all regenerated clones, including those of the *M-D* clone plants, were carefully collected and rinsed thoroughly with tap water to remove the remaining culture media adhering to the roots. The washed plants were transplanted into 5 cm pots (one plant/pot) filled with an autoclaved soil mixture of sand: peat moss: vermiculite (1:1:1, *v*/*v*). The plants were covered with transparent plastic bags to retain a high humidity, irrigated regularly with sterilized water and maintained in a greenhouse. Two weeks later, the plastic bags were punctured with a paper punch and kept for another week; thereafter, the acclimated plantlets were transplanted in 25 cm pots. At the end of December, $G_0$ minitubers were harvested, washed and stored at 2 °C until they were used as seed tubers.

In the middle of February 2020, the $G_0$ minitubers of each clone were planted in $1 \times 1 \times 0.25$ m boxes containing peat moss:sand:vermiculite:perlite:foam (40:40:10:5:5) in a greenhouse following the recommendations of the Egyptian Ministry of Agriculture for agricultural practices concerning cultivation, fertilization, irrigation and pest and disease control. After 16 weeks, the $G_1$ tubers of each clone were harvested, washed, counted and weighed. The tuber samples of each clone were dried at 50 °C until constant weight and used for starch quantification, while the remaining tubers were stored as seed tubers at 4 °C. At the beginning of September 2020, the stored $G_1$ tubers were planted in an open field for field experiments, and $G_2$ tubers were collected 16 weeks later. The field experiments were carried out in Samannud, Gharbiya Governorate, Egypt. (30°52′54″ N 31°14′11″ E). The soil at the study area was loamy soil with pH 7.8 and an electrical conductivity (EC) of 2.96 EC/dsm$^{-1}$. The research site's weather was semi-arid with rainy winter (rainfall of 10 mm, average day/night temperature of 19/10 °C and relative humidity of 60%). Seven tubers of each clone were planted at a depth of 20 cm, with an in-row spacing of 25 cm and an inter-row spacing of 72 cm. The agricultural practices guidelines of the Egyptian Ministry of Agriculture were followed for cultivation, fertilization, irrigation and pest and disease control. The tubers of the commercial cultivar Lady Rosetta were planted under conditions the same as those of the control clones. After 16 weeks, five plants of each clone were randomly picked for sampling. The tubers of each plant were counted and manipulated separately, where they were washed with water to remove any remaining soil particles and dried with a clean towel. After the determination of the total fresh weight of the tubers of each plant, three sections of about 5 mm thickness from the bottom, middle and top of each tuber were collected using a sharp knife. The collected sections from the tubers of each plant were mixed and further cut into fine pieces before weighing out one g to be stored at −80 °C in order to represent the plant in a qPCR analysis. The remaining pieces were added to bulk tuber parts and dried at 50 °C until constant weight for dry weight determination and starch quantification.

### 4.3. Starch Determination

The sugar-free dry tuber tissue prepared via the repeated extraction in iso-propanol (80% *v*/*v*) was used for starch quantification. After an overnight drying step at 70 °C as described by Kumar et al. [59], the dried fragments were homogenized in perchloric acid (60% *v*/*v*) for starch hydrolysis. The liberated glucose was estimated using the anthrone method [60].

### 4.4. Real-Time Quantitative PCR

The stored tuber samples were crushed into a fine powder in liquid nitrogen and used for total RNA extraction using an RNeasy Plant Mini kit (Qiagen, Hilden, Germany), with the purification step using DNase according to the manufacturer's instructions. RNA purity and concentration were confirmed with a NanoDrop spectrophotometer (ND 2000c, Thermo Fisher Scientific, Wilmington, DE, USA). One µg of the purified RNA was converted to cDNA using a Sensi-FAST™ cDNA synthesis kit following the standard protocol from the manufacturer. Quantitative real-time PCR for AGPase, GBSS I, SBE I, SBE II, SS II and SS III cDNAs was carried out on a Mx3000P (Stratagene, CA, USA) qPCR system using specific primers (Table 1) [8]. The thermal profile of the real-time system was 95 °C for 10 min, followed by 40 cycles (95 °C for 15 s and 60 °C for 60 s).

The relative transcript abundance of tublin was used as an endogenous control, to which the transcription levels of the AGPase, GBSS I, SBE I, SBE II, SS II and SS III genes were normalized using the $2^{-DDCt}$ method [61]. The expression recorded in the tubers of the commercial cultivar Lady Rosetta was employed as a quantification unit.

**Table 1.** List of primers used for qRT-PCR (5′–3′).

| Gene | Identification ID | Primer Sequence |
|---|---|---|
| *Tublin* | LOC102577624 | 5′-GTCAGTCTGGTGCTGGTAATAA-3′<br>5′-TCTCAGCCTCCTTCCTTACA-3′ |
| *AGPase* | LOC102577790 | 5′-TT CCTT CCACCAACCAAGATAG-3′<br>5′-CACTATGG AGTGTT CCACAGAA-3′ |
| *GBSS I* | LOC102583115 | 5′-CTTGCGTTTGCTGAGATGATAAA-3′<br>5′-CAGAAGCTCCTAAGCCCAATAG-3′ |
| *SBE I* | LOC102596498 | 5′-GCGAACATGTGTGGCTTATTAC-3′<br>5′-TCTCGTCACTCTCCTCGATATT-3′ |
| *SBE II* | LOC102590711 | 5′-CTCTGGATAGACCGTCAACATC-3′<br>5′-AGGTACCCTT CTCCTCCTAATC-3′ |
| *SS II* | LOC102583115 | 5′-CAACAGGACCTACTTCAACAGA-3′<br>5′-CTACCACTCCCACCATCATAAG-3′ |
| *SS III* | LOC102577674 | 5′-GTCACCTGTTCGTGTATCATCT-3′<br>5′-CCACTCTCTT CCGATCTCTTTG-3′ |

### 4.5. Statistical Analysis

The results of all clones are presented as the mean of five replicates ± standard deviation (SD). One-way ANOVA with the LSD post hoc statistical test was utilized to compare the different clones on the basis of tuber yield per plant manifested in number of tubers, the fresh and dry weights of the tubers, the starch content on dry-weight and per-plant bases and the expressions of six starch-related genes at $p = 0.05$ using SPSS v. 14.

## 5. Conclusions

Somaclonal variation is an efficient breeding tool. It can provide breeders with new genotypes with favorable traits, which is essential for breeding programs targeting yield improvements. Somaclonal variation is a time-efficient alternative to conventional breeding, and it is able to provide potato clones with an outstanding potential for starch accumulation at a reasonable cost. It is an acceptable approach that is able to manipulate starch synthesis genes without issues arising regarding genetic transformation. However, the emerging clone requires genetic characterization to participate effectively in breeding programs targeting qualitative and quantitative yield improvements, which is the aim of our future work.

**Author Contributions:** Conceptualization, E.-S.T.A.E.-S., A.S.F., M.E.E.-D. and M.A.M.; methodology, W.M.R.M.A.; software, W.M.R.M.A., G.N., M.P., T.W. and A.S.F.; validation, E.-S.T.A.E.-S., A.S.F., M.E.E.-D. and M.A.M.; formal analysis, W.M.R.M.A.; investigation, W.M.R.M.A.; resources, M.E.E.-D. and M.A.M.; data curation, W.M.R.M.A., G.N., M.P. and T.W.; writing—original draft preparation, W.M.R.M.A., G.N., M.P. and T.W.; writing—review and editing E.-S.T.A.E.-S., A.S.F., M.E.E.-D., G.N., M.P., T.W. and M.A.M.; visualization, W.M.R.M.A., G.N., M.P., T.W. and A.S.F.; supervision, E.-S.T.A.E.-S., A.S.F., M.E.E.-D. and M.A.M.; project administration, M.E.E.-D. and M.A.M.; funding acquisition, M.E.E.-D., M.A.M. and W.M.R.M.A. All authors have read and agreed to the published version of the manuscript.

**Funding:** This research received no external funding.

**Acknowledgments:** The authors acknowledge Salah El-Din Ahmed Mohamedien, the director of the main vegetable crops and hybrid production project, the Egyptian Ministry of Agriculture and Land reclamation, for the facilities he provided for the completion of this work.

**Conflicts of Interest:** The authors declare no conflict of interest.

## References


1. Devaux, A.; Goffart, J.-P.; Kromann, P.; Andrade-Piedra, J.; Polar, V.; Hareau, G. The Potato of the Future: Opportunities and Challenges in Sustainable Agri-Food Systems. *Potato Res.* **2021**, *64*, 681–720. [CrossRef] [PubMed]
2. Zaheer, K.; Akhtar, M.H. Potato Production, Usage, and Nutrition—A Review. *Crit. Rev. Food Sci. Nutr.* **2016**, *56*, 711–721. [CrossRef]
3. Zhang, H.; Fen, X.; Yu, W.; HU, H.; DAI, X. Progress of Potato Staple Food Research and Industry Development in China. *J. Integr. Agric.* **2017**, *16*, 2924–2932. [CrossRef]
4. Maroufpour, B.; Rad, F.A.; Yazdanseta, S. Bioethanol Production as Biofuel from Potato Peel Using Saccharomyces Cerevisiae PTCC 5052 and Zymomonas Mobilis PTCC 1718. *Bioagro* **2019**, *31*, 177–184.
5. Dupuis, J.H.; Liu, Q. Potato Starch: A Review of Physicochemical, Functional and Nutritional Properties. *Am. J. Potato Res.* **2019**, *96*, 127–138. [CrossRef]
6. Nazarian-Firouzabadi, F.; Visser, R.G. Potato Starch Synthases: Functions and Relationships. *Biochem. Biophys. Rep.* **2017**, *10*, 7–16. [CrossRef] [PubMed]
7. Ferrero, D.M.; Piattoni, C.V.; Diez, M.D.A.; Rojas, B.E.; Hartman, M.D.; Ballicora, M.A.; Iglesias, A.A. Phosphorylation of ADP-Glucose Pyrophosphorylase during Wheat Seeds Development. *Front. Plant Sci.* **2020**, *11*, 1058. [CrossRef]
8. Su, W.; Ye, G.; Zhou, Y.; Wang, J. Starch Synthesis and Gelatinization Properties of Potato Tubers. *Ciência Rural* **2022**, *52*, e20210050. [CrossRef]
9. Kumari, M.; Kumar, M.; Solankey, S.S. Breeding Potato for Quality Improvement. *Potato Incas All Over World* **2018**, *37*, 71482.
10. Diambra, L.A. Genome Sequence and Analysis of the Tuber Crop Potato. *Nature* **2011**, *475*, 189–195.
11. Fouad, A.; Hafez, R.; Hosni, H. Authentication of Three Endemic Species of the Caryophyllaceae from Sinai Peninsula Using DNA Barcoding. *Egypt. J. Bot.* **2019**, *59*, 483–491. [CrossRef]
12. Hameed, A.; Zaidi, S.S.-A.; Shakir, S.; Mansoor, S. Applications of New Breeding Technologies for Potato Improvement. *Front. Plant Sci.* **2018**, *9*, 925. [CrossRef] [PubMed]
13. Bisognin, D.A. Breeding Vegetatively Propagated Horticultural Crops. *Crop Breed. Appl. Biotechnol.* **2011**, *11*, 35–43. [CrossRef]
14. Simmonds, N. A Review of Potato Propagation by Means of Seed, as Distinct from Clonal Propagation by Tubers. *Potato Res.* **1997**, *40*, 191–214. [CrossRef]
15. Ashrafzadeh, S.; Leung, D.W. Novel Potato Plants with Enhanced Cadmium Resistance and Antioxidative Defence Generated after in Vitro Cell Line Selection. *PLoS ONE* **2017**, *12*, e0185621. [CrossRef]
16. Ricroch, A.E.; Hénard-Damave, M.-C. Next Biotech Plants: New Traits, Crops, Developers and Technologies for Addressing Global Challenges. *Crit. Rev. Biotechnol.* **2016**, *36*, 675–690. [CrossRef]
17. Karalis, D.T.; Karalis, T.; Karalis, S.; Kleisiari, A.S. Genetically Modified Products, Perspectives and Challenges. *Cureus* **2020**, *12*, e7306. [CrossRef]
18. Larkin, P.J.; Scowcroft, W.R. Somaclonal Variation—A Novel Source of Variability from Cell Cultures for Plant Improvement. *Theor. Appl. Genet.* **1981**, *60*, 197–214. [CrossRef]
19. Krishna, H.; Alizadeh, M.; Singh, D.; Singh, U.; Chauhan, N.; Eftekhari, M.; Sadh, R.K. Somaclonal Variations and Their Applications in Horticultural Crops Improvement. *3 Biotech* **2016**, *6*, 54. [CrossRef]
20. Albiski, F.; Najla, S.; Sanoubar, R.; Alkabani, N.; Murshed, R. In Vitro Screening of Potato Lines for Drought Tolerance. *Physiol. Mol. Biol. Plants* **2012**, *18*, 315–321. [CrossRef]
21. Zeid, I.M.A.; Soliman, H.I.; Metwali, E.M. In Vitro Evaluation of Some High Yield Potato (*Solanum Tuberosum* L.) Cultivars under Imposition of Salinity at the Cellular and Organ Levels. *Saudi J. Biol. Sci.* **2022**, *29*, 2541–2551. [CrossRef]
22. Mirkarimi, H.R.; Abasi-moghadam, A.; Mozafari, J. In Vitro and Greenhouse Evaluation for Resistance to Early Blight of Potato Isolated from Alternaria Alternata. *Agric. Sci.* **2013**, *4*, 473–476.
23. Soliman, W.; Ibrahim, M.; El Baz, H. In Vitro Evaluation of *Syzygium Aromaticum* L. Ethanol Extract as Biocontrol Agent against Postharvest Tomato and Potato Diseases. *Egypt. J. Bot.* **2019**, *59*, 81–94.
24. Bayati, E.; Gomarian, M.; Mirzaie-Nodousha, H.; Changizi, M.; Khaghani, S. Producing a Superior Genotype from Agria Potato Cultivar Using Somaclonal Variation. *Nexo Rev. Científica* **2021**, *34*, 671–681. [CrossRef]
25. Rosenberg, V.; Kotkas, K.; Särekanno, M.; Ojarand, A.; Vasar, V. Variation of Potato Meristem Clones-New Facts for Science and Practice. *Agron. Vestis* **2008**, *11*, 129–135.
26. Rosenberg, V.; Särekanno, M.; Kotkas, K.; Vasar, V.; Ojarand, A. Variation of Agronomic Traits of Potato Somaclones Produced by Meristem Culture. *S. Afr. J. Plant Soil* **2007**, *24*, 95–99. [CrossRef]
27. Thieme, R.; Griess, H. Somaclonal Variation in Tuber Traits of Potato. *Potato Res.* **2005**, *48*, 153–165. [CrossRef]
28. Kloosterman, B.; Vorst, O.; Hall, R.D.; Visser, R.G.; Bachem, C.W. Tuber on a Chip: Differential Gene Expression during Potato Tuber Development. *Plant Biotechnol. J.* **2005**, *3*, 505–519. [CrossRef]
29. Ferreira, S.J.; Senning, M.; Sonnewald, S.; Keßling, P.-M.; Goldstein, R.; Sonnewald, U. Comparative Transcriptome Analysis Coupled to X-Ray CT Reveals Sucrose Supply and Growth Velocity as Major Determinants of Potato Tuber Starch Biosynthesis. *BMC Genom.* **2010**, *11*, 93. [CrossRef]
30. Krishna, H.; Sairam, R.; Singh, S.; Patel, V.; Sharma, R.; Grover, M.; Nain, L.; Sachdev, A. Mango Explant Browning: Effect of Ontogenic Age, Mycorrhization and Pre-Treatments. *Sci. Hortic.* **2008**, *118*, 132–138. [CrossRef]

31. Zayova, E.; Vassilevska-Ivanova, R.; Kraptchev, B.; Stoeva, D. Somaclonal Variations through Indirect Organogenesis in Eggplant (*Solanum Melongena* L.). *Biol. Divers. Conserv.* **2010**, *3*, 1–5.
32. Adly, W.M.; Mazrou, Y.S.; EL-Denary, M.E.; Mohamed, M.A.; El-Salam, A.; El-Sayed, T.; Fouad, A.S. Boosting Polyamines to Enhance Shoot Regeneration in Potato (*Solanum Tuberosum* L.) Using AgNO3. *Horticulturae* **2022**, *8*, 113. [CrossRef]
33. Kumlay, A.M.; Ercisli, S. Callus Induction, Shoot Proliferation and Root Regeneration of Potato (*Solanum Tuberosum* L.) Stem Node and Leaf Explants under Long-Day Conditions. *Biotechnol. Biotechnol. Equip.* **2015**, *29*, 1075–1084. [CrossRef]
34. Ghosh, S.; Majumdar, S.; Sarkar, D.; Datta, K. An Efficient Adventitious Shoot Regeneration System for Potato (*Solanum Tuberosum* L.) Using Leaf Discs. *J. Plant Biochem. Biotechnol.* **2015**, *24*, 298–304. [CrossRef]
35. Shin, S.Y.; Choi, Y.; Kim, S.-G.; Park, S.-J.; Park, J.-S.; Moon, K.-B.; Kim, H.-S.; Jeon, J.H.; Cho, H.S.; Lee, H.-J. Submergence Promotes Auxin-Induced Callus Formation through Ethylene-Mediated Post-Transcriptional Control of Auxin Receptors. *Mol. Plant* **2022**, *15*, 1947–1961. [CrossRef] [PubMed]
36. Shin, J.; Seo, P.J. Varying Auxin Levels Induce Distinct Pluripotent States in Callus Cells. *Front. Plant Sci.* **2018**, *9*, 1653. [CrossRef] [PubMed]
37. Bidabadi, S.S.; Jain, S.M. Cellular, Molecular, and Physiological Aspects of in Vitro Plant Regeneration. *Plants* **2020**, *9*, 702. [CrossRef]
38. Malik, W.A.; Mahmood, I.; Razzaq, A.; Afzal, M.; Shah, G.A.; Iqbal, A.; Zain, M.; Ditta, A.; Asad, S.A.; Ahmad, I.; et al. Exploring Potential of Copper and Silver Nano Particles to Establish Efficient Callogenesis and Regeneration System for Wheat (*Triticum Aestivum* L.). *GM Crops Food* **2021**, *12*, 564–585. [CrossRef]
39. Zhao, X.-C.; Qu, X.; Mathews, D.E.; Schaller, G.E. Effect of Ethylene Pathway Mutations upon Expression of the Ethylene Receptor ETR1 from Arabidopsis. *Plant Physiol.* **2002**, *130*, 1983–1991. [CrossRef]
40. Feray, A.; Hourmant, A.; Brun, A.; Penot, M. Effect of Polyamines on Morphogenesis of in Vitro Potato Plants (Solanum Tuberosum Cv. Bintje). *Comptes Rendus De L'académie Des Sci. Série 3 Sci. De La Vie* **1993**, *316*, 1446–1451.
41. Hajare, S.T.; Chauhan, N.M.; Kassa, G. Effect of Growth Regulators on in Vitro Micropropagation of Potato (*Solanum Tuberosum* L.) Gudiene and Belete Varieties from Ethiopia. *Sci. World J.* **2021**, *2021*, 5928769. [CrossRef] [PubMed]
42. Abeuova, L.S.; Kali, B.R.; Rakhimzhanova, A.O.; Bekkuzhina, S.S.; Manabayeva, S.A. High Frequency Direct Shoot Regeneration from Kazakh Commercial Potato Cultivars. *PeerJ* **2020**, *8*, e9447. [CrossRef] [PubMed]
43. Elaleem, K.; Modawi, R.S.; Khalafalla, M.M. Effect of Plant Growth Regulators on Callus Induction and Plant Regeneration in Tuber Segment Culture of Potato (*Solanum Tuberosum* L.) Cultivar Diamant. *Afr. J. Biotechnol.* **2009**, *8*, 2529–2534.
44. Aamir, I.; Ali, R.; Zahid, M.; Shahid, M.; Khalid, Z.M.; Shaheen, A. Establishment of an Efficient and Reproducible Regeneration System for Potato Cultivars Grown in Pakistan. *Pak. J. Bot.* **2016**, *48*, 285–290.
45. Genene, G.; Mekonin, W.; Meseret, C.; Manikandan, M.; Tigist, M. Protocol Optimization for in Vitro Propagation of Two Irish Potato (*Solanum Tuberosum* L.) Varieties through Lateral Bud Culture. *Afr. J. Plant Sci.* **2018**, *12*, 180–187. [CrossRef]
46. Lee, H.-J.; Yoon, Y.-J.; Paek, K.-Y.; Park, S.-Y. Endoreduplication and Gene Expression in Somaclonal Variants of Clonally Propagated Phalaenopsis 'Wedding Promenade. *Hortic.Environ. Biotechnol.* **2017**, *58*, 85–92. [CrossRef]
47. Wang, N.; Yu, Y.; Zhang, D.; Zhang, Z.; Wang, Z.; Xun, H.; Li, G.; Liu, B.; Zhang, J. Modification of Gene Expression, DNA Methylation and Small RNAs Expression in Rice Plants under In Vitro Culture. *Agronomy* **2022**, *12*, 1675. [CrossRef]
48. Pawe\lkowicz, M.E.; Skarzyńska, A.; Mróz, T.; Bystrzycki, E.; Pląder, W. Molecular Insight into Somaclonal Variation Phenomena from Transcriptome Profiling of Cucumber (*Cucumis Sativus* L.) Lines. *Plant Cell Tissue Organ Cult. (PCTOC)* **2021**, *145*, 239–259. [CrossRef]
49. López-Hernández, F.; Cortés, A.J. Whole Transcriptome Sequencing Unveils the Genomic Determinants of Putative Somaclonal Variation in Mint (*Mentha* L.). *Int. J. Mol. Sci.* **2022**, *23*, 5291. [CrossRef]
50. Müller-Röber, B.; Sonnewald, U.; Willmitzer, L. Inhibition of the ADP-Glucose Pyrophosphorylase in Transgenic Potatoes Leads to Sugar-Storing Tubers and Influences Tuber Formation and Expression of Tuber Storage Protein Genes. *EMBO J.* **1992**, *11*, 1229–1238. [CrossRef]
51. Stark, D.M.; Timmerman, K.P.; Barry, G.F.; Preiss, J.; Kishore, G.M. Regulation of the Amount of Starch in Plant Tissues by ADP Glucose Pyrophosphorylase. *Science* **1992**, *258*, 287–292. [CrossRef] [PubMed]
52. Zhu, G.; Ye, N.; Yang, J.; Peng, X.; Zhang, J. Regulation of Expression of Starch Synthesis Genes by Ethylene and ABA in Relation to the Development of Rice Inferior and Superior Spikelets. *J. Exp. Bot.* **2011**, *62*, 3907–3916. [CrossRef] [PubMed]
53. Gu, Y.; Han, S.; Chen, L.; Mu, J.; Duan, L.; Li, Y.; Yan, Y.; Li, X. Expression and Regulation of Genes Involved in the Reserve Starch Biosynthesis Pathway in Hexaploid Wheat (*Triticum Aestivum* L.). *Crop J.* **2021**, *9*, 440–455. [CrossRef]
54. Li, W.; Huang, D.; Wang, B.; Hou, X.; Zhang, R.; Yan, M.; Liao, W. Changes of Starch and Sucrose Content and Related Gene Expression during the Growth and Development of Lanzhou Lily Bulb. *PLoS ONE* **2022**, *17*, e0262506. [CrossRef] [PubMed]
55. Hovenkamp-Hermelink, J.; Jacobsen, E.; Ponstein, A.; Visser, R.; Vos-Scheperkeuter, G.; Bijmolt, E.; De Vries, J.; Witholt, B.; Feenstra, W. Isolation of an Amylose-Free Starch Mutant of the Potato (*Solanum Tuberosum* L.). *Theor. Appl. Genet.* **1987**, *75*, 217–221. [CrossRef]
56. Bowsher, C.G.; Scrase-Field, E.F.; Esposito, S.; Emes, M.J.; Tetlow, I.J. Characterization of ADP-Glucose Transport across the Cereal Endosperm Amyloplast Envelope. *J. Exp. Bot.* **2007**, *58*, 1321–1332. [CrossRef]
57. Kirchberger, S.; Leroch, M.; Huynen, M.A.; Wahl, M.; Neuhaus, H.E.; Tjaden, J. Molecular and Biochemical Analysis of the Plastidic ADP-Glucose Transporter (ZmBT1) from Zea Mays. *J. Biol. Chem.* **2007**, *282*, 22481–22491. [CrossRef]

58. Murashige, T.; Skoog, F. A Revised Medium for Rapid Growth and Bio Assays with Tobacco Tissue Cultures. *Physiol. Plant.* **1962**, *15*, 473–497. [CrossRef]
59. Kumar, D.; Ezekiel, R.; Singh, B.; Ahmed, I. Conversion Table for Specific Gravity, Dry Matter and Starch Content from under Water Weight of Potatoes Grown in North-Indian Plains. *Potato J.* **2005**, *32*, 79–84.
60. Sadasivam, S.; Manickam, A. *Biochemical Methods for Agricultural Sciences*; Wiley Eastern Limited: Hoboken, NJ, USA, 1992.
61. Livak, K.J.; Schmittgen, T.D. Analysis of Relative Gene Expression Data Using Real-Time Quantitative PCR and the 2- ΔΔCT Method. *Methods* **2001**, *25*, 402–408. [CrossRef]

*plants*

MDPI

*Article*

# Genotype × Environment Interaction and Stability Analysis of Selected Cassava Cultivars in South Africa

**Assefa B. Amelework [1,*], Michael W. Bairu [1,2], Roelene Marx [1], Mark Laing [3] and Sonja L. Venter [1]**

1 Agricultural Research Council, Vegetable, Industrial and Medicinal Plants, Private Bag X293, Pretoria 0001, South Africa; bairum@arc.agric.za (M.W.B.); pienaarr@arc.agric.za (R.M.); sventer@arc.agric.za (S.L.V.)

2 African Centre for Crop Improvement, School of Agriculture, Earth and Environmental Sciences, University of KwaZulu-Natal, Private Bag X01, Scottsville, Pietermaritzburg 3209, South Africa

3 School of Agricultural Sciences, Food Security and Safety Niche Area, Faculty of Natural & Agricultural Sciences, North-West University, Private Bag X2046, Mmabatho 2735, South Africa; laing@ukzn.ac.za

* Correspondence: assefaa@arc.agric.za

**Abstract:** Cassava (*Manihot esculenta* Crantz) is an important root crop worldwide. It is adapted to a wide range of environmental conditions, exhibiting differential genotypic responses to varying environmental conditions. The objectives of this study were: (1) to examine the effect of genotype, environment and genotype × environment interaction (GEI) on fresh root yield (FRY) and dry matter content (DMC); and (2) to identify superior genotypes that exhibit high performance for the traits of interest using the genetic tools of additive main effects and multiplicative interaction (AMMI) and genotype stability index (GSI) analysis. Eleven cassava genotypes were evaluated in a randomized complete block design at six trial sites in South Africa. The combined analysis of variance based on AMMI revealed significant genotype, environment and GEI for the traits. The percentage variation due to GEI was higher than the percentage variation due to genotype for FRY, reflecting differential genotypic responses across the experimental sites. The proportion of variance due to genotype variation was larger for DMC. Genotype stability index (GSI) showed that UKF3 (G6), 98/0002 (G2) and P4/10 (G5) were the highest yielding and most stable genotypes for FRY, and 98/0002 (G1), UKF3 (G6) and UKF9 (G11) were the highest yielding and most stable genotypes for DMC. Cultivars 98/0002 and UKF3 were identified as providing high stability with superior fresh root yield and DMC. These genotypes could be recommended to farmers for food, feed and industrial applications without the need for further breeding. The AMMI-2 model clustered the testing environments into three mega-environments based on the winning genotypes for FRY and DMC. Mabuyeni (KwaZulu-Natal), Shatale (Mpumalanga) and Mandlakazi (Limpopo) would be the best testing sites in future cassava-genotype evaluation and breeding programs. This study provides a baseline for a future study on the GEI of cassava varieties, using a larger set of genotypes, factoring in seasonal variation.

**Keywords:** AMMI model; genotype × environment interaction; *Manihot esculenta*; stability; breeding sites

**Citation:** Amelework, A.B.; Bairu, M.W.; Marx, R.; Laing, M.; Venter, S.L. Genotype × Environment Interaction and Stability Analysis of Selected Cassava Cultivars in South Africa. *Plants* **2023**, *12*, 2490. https://doi.org/10.3390/plants12132490

Academic Editors: Ranjith Pathirana and Francesco Carimi

Received: 2 May 2023
Revised: 19 June 2023
Accepted: 22 June 2023
Published: 29 June 2023



## 1. Introduction

With increasing climate variability, cassava (*Manihot esculenta* Crantz) has proven to be among the most resilient food security crops for more than 800 million people in Sub-Saharan Africa. Cassava produces 40% more starch than rice and 25% more than maize [1]. The potential for cassava to play a key role in food security, climate risk mitigation and import substitution for industrial starch, livestock feed and biofuel feedstock in South Africa's economy has been reviewed by Amelework et al. [2]. Farmers can grow and harvest cassava on marginal soils with minimal capital input and rainfall of less than 500 mm per annum. In South Africa, 2.5 million households practice small-scale subsistence farming on less than 15% of the available agricultural land. The majority of this land is not suitable for the production of maize or vegetable crops, being in low rainfall areas with poor soils. A low-input and rainfed cassava farming system would benefit rural resource-poor farmers in South Africa [3]. However, it has only been grown as a minor subsistence crop by smallholders in the North East regions bordering Mozambique. The ARC and research partners have been researching agronomically suitable cassava genotypes and appropriate production systems in three provinces in South Africa. As a result, many farmers have expressed an interest in farming cassava.

New genotypes of cassava can be developed in a specific research location through either hybridization or mutation. However, to select the best-performing genotypes, it is necessary to evaluate the advanced breeding lines in a wide range of environments. Breeding lines tested in different environments always exhibit significant variation in terms of phenotypic performance owing to environmental variation and different biotic and abiotic stresses [4]. The analysis of genotypic interactions with the environmental conditions of multiple sites helps to quantify the adaptability and stability of genotypes [5]. The differential response of genotypes to different environmental conditions is termed the genotype by environment interaction (GEI).

The presence of GEI is a challenge for breeders in evaluating lines in multilocational trials. Quantifying and minimizing the GEI remains one of the top priorities of any breeding program. GEI reduces the association between phenotypic and genotypic values, thereby hampering the genetic progress in plant breeding programs [6]. The use of statistical models such as additive main effects and multiplicative interaction (AMMI) and genotype (G) main effect plus genotype × environment (GE) interaction (GGE) models assist breeders in quantifying and understanding the patterns of GEI and in evaluating the performance of genotypes in various environmental conditions. This allows breeders to select stable and adaptable genotypes for a range of environments [7].

In Africa, Latin America and Asia, where cassava is grown as a subsistence and industrial crop, research has been conducted to enhance the genetic profile of cassava, resulting in many genotypes being released for improved yield, dry matter and starch content, and resistance or tolerance to major insect pests and disease. Genetic improvement begins with the collection and evaluation of diverse genetic resources [8]. In the past five years, the Agricultural Research Council of South Africa (ARC) has collected a number of cassava cultivars from national and international research institutes. However, the deployment of these newly introduced genotypes into new production areas requires a basic understanding of their performance in the new environments and to identify the most useful environments for future testing and characterization of cassava germplasm [4]. The aims of this study were to evaluate the yield and dry matter content of selected cassava genotypes across different environments, to study the patterns of GEI, to identify superior genotypes that exhibit high-performing and stable genotypes in support of establishing a cassava starch industry in South Africa, and to identify mega-environments for future germplasm evaluation trials.

## 2. Results

### 2.1. AMMI Analysis of G × E Interaction

The AMMI model analysis of variance (ANOVA) of eleven cassava genotypes measured in four environments showed that the mean square (MS) for genotype, environment and GEI were highly significant ($p < 0.001$) for fresh root yield and dry matter content (Table 1). GEI was further partitioned by principal component analysis, which showed that the first two IPCAs MS (IPCA1 and IPCA2) were significant ($p < 0.001$) for both fresh root yield and dry matter content. It was evident from the AMMI analysis that the GEI sum of squares (SS) accounted for a larger proportion of the treatment SS (45.6%) for FRY, whereas the SS of the genotypes constituted the largest proportion of the treatment SS (51.9%) for DMC. The genotype SS accounted for 18.1% of the treatment SS for FRY, whilst environment and GEI SS accounted for 36.4% and 45.6%, respectively (Table 1). Unlike FRY, environment and GEI effects SS accounted for a smaller proportion of the treatment

SS for DMC. The model with the first few IPCAs that captured most of the GEI variation was the best model for extracting and explaining the GEI pattern from the dataset. In this study, the percentage goodness of fit by the first two IPCAs was 74.7% for FRY and 77.4% for DMC, indicating the usefulness of the AMMI model for extracting and understanding the patterns of GEI. The results also showed that the six environments varied both in the main and interaction effects.

**Table 1.** Additive main effects and multiplicative interaction (AMMI) analysis of variance for fresh root yield (t.ha$^{-1}$) and dry matter content (%) of 11 cassava genotypes measured on six trial sites.

| Source | DF | Fresh Root Yield | | | Dry Matter Content | | |
|---|---|---|---|---|---|---|---|
| | | SS | MS | % SS Explained | SS | MS | % SS Explained |
| Treatments | 65 | 173,736 | 2673 | | 12,784 | 197 | |
| Genotypes | 10 | 31,367 | 3137 *** | 18.05 | 6640 | 664 *** | 51.94 |
| Environments | 5 | 63,219 | 12,644 *** | 36.39 | 2726 | 545 *** | 21.32 |
| Interactions (GEI) | 50 | 79,150 | 1583 *** | 45.56 | 3418 | 68 *** | 26.74 |
| IPCA 1 | 14 | 39,273 | 2805 *** | 49.62 | 1658 | 118 *** | 48.51 |
| IPCA 2 | 12 | 19,846 | 1654 *** | 25.07 | 988 | 82 *** | 28.91 |
| Residuals | 24 | 20,031 | 835 | | 772 | 32 | |
| Error | 120 | 16,722 | 139 | | 939 | 7.8 | |

DF = degree of freedom; IPCA1 = the first interaction principal component; IPCA2 = the second interaction principal component; *** significant at $p < 0.001$.

### 2.2. The GEI Patterns of Traits and Genotypes Based on GGE Biplot Analysis

To visualize the performance of different genotypes across different environments, biplots were used. The AMMI and GGE biplots are generally used to explain the genotype adaptation or stability across environments. If the PCA score for a genotype or environment is near zero, then there is a small interaction impact; however, if a genotype and environment have the same sign on the PCA axis, there is a positive interaction; alternatively, they have different signs on the PCA axis, there is a negative interaction. Environments with large PCA scores show high interaction between the environments and genotypes and are discriminatory, whereas environments with PCA scores near zero have little interaction with the genotypes and have a low discriminatory value.

### 2.3. Fresh Root Yield

The GEI IPCA1 scores were plotted against the mean performances of the genotypes and environments in the AMMI1 model (Figure 1A). The x-coordinate indicates the main effects (means), and the y-coordinate indicates the effects of the interaction (IPCA1). The genotype and environment overall mean was 70.4 tons ha$^{-1}$. The superior genotypes were G1 > G3 > G6 > G7 > G5 > G11 > G2, which were located on the right two quadrants (top and bottom right) of the biplot. Environments were distributed from low-yielding environments in the left two quadrants (top and bottom left) to the high-yielding environments in the right two quadrants (top and bottom right). E3 > E6 > E2 > E4 were identified as higher-yielding environments, whereas E1 and E3 were relatively low-yielding environments.

Values closer to the origin of the axis (IPCA1) provide a smaller contribution to the interaction than those that are further away. Hence, G9 and G11 were the most unstable genotypes, with mean fresh root yields of 65.1 ton ha$^{-1}$ and 71.4 ton ha$^{-1}$, respectively, which are close to the overall mean (Figure 1A), while G2 and G5 were relatively stable genotypes. On the contrary, the majority of the genotypes revealed intermediate stability and performance. However, among these genotypes, 98/0002 had the highest mean fresh root yield (86.8 ton ha$^{-1}$) combined with stability comparable to the other genotypes. Similarly, some environments, such as E2 and E3, stood out as making little or little contribution to the interaction, E6 made a small contribution, and E1, E5 and E4 made a large contribution to the interaction.

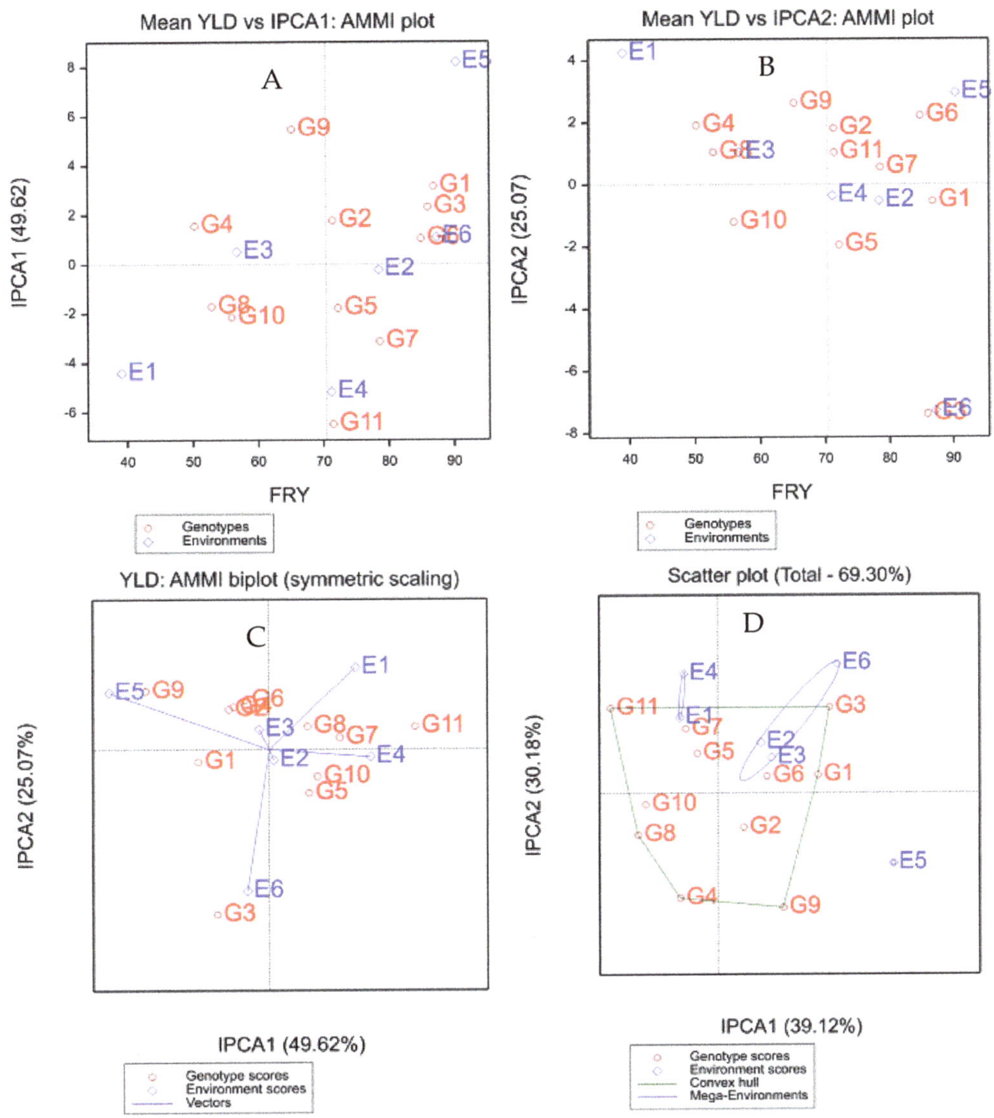


**Figure 1.** Biplot analysis for 11 cassava genotypes evaluated in six trial sites for fresh root yield (FRY). (**A**) = AMMI1 biplot showing the IPCA1 vs. main effect (means); (**B**) = AMMI1 biplot showing the IPCA2 vs. main effect; (**C**) = AMMI2 biplot showing the first two principal axes of interaction (IPCA2 vs. IPCA1) and (**D**) = GGE plot defining mega-environments using different winning genotypes tested. E1–E6 = the six testing locations, and G1–G11 = the eleven cassava genotypes used for this study.

The cumulative contribution from IPC1 and IPC2 included about 93% of the interaction MS (Figure 1C). According to the correlation between IPC1 and IPC2, the genotypes that were positioned near the origin had the least interaction, and the genotypes positioned near the axis had a stability that was more general. G1, G7 and G8 showed little or no interaction with the environments, while G2, G4, G5, G6 and G10 revealed a minimum interplay between genotype and environments, whereas G3, G9 and G11 were the most unstable. G1 had a PCA1 score of approximately zero on the IPCA1 axis, indicating that this genotype was the most stable across environments.

E2 and E3 were the largest contributors to the phenotypic stability of the genotypes, and these environments were among the lowest in mean FRY. However, E1, E5 and E6 contributed the most GEI. Any genotypes positioned closer to a certain environment have specific stability in that environment. Hence, G3 had a specific adaptation to E6, G9 to E5 and G5 and G10 to E4.

The partitioning of GGE showed that IPCA1 and IPCA2 accounted for 39.1% and 30.2% of the GGE sum of squares, respectively, explaining a total of 69.3% variation in FRY (Figure 1D). Based on the predicted means of FRY obtained from the AMMI2 model, three mega-environments were identified (Figure 1D). The first one contains E1 and E4, with G7 and G11 as the winner genotypes. The second mega environment constituted E2, E3 and E6, where G1 and G3 were the best genotypes at these sites. The last mega-environment was formed by one environment, E5, where genotype G9 was the winner (Figure 1D).

### 2.4. Dry Matter Content

The scatterplot of mean DMC vs. IPCA1 (Figure 2A) illustrates that G11 (49.9%) and G4 (49.8%) had the highest DMC, while G8 (27.5%) had the lowest. The vertical line that divides the horizontal axis into two parts is the mean DMC (43.3%). G11 > G4 > G6 > G1 > G5 > G2 > G9 had a higher DMC than the mean DMC, while G3, G7 and G8 had lower DMC values. In terms of IPCA1, G10, G2 and G9 had a maximum GEI and were the most unstable genotypes. The highest DMC was recorded at E4, followed by E5 and E6. However, in the rest of the locations, the genotypes performed below the mean DMC. E4 and E2 made large contributions to the GEI, while E1 made a smaller contribution (Figure 2A).

The cumulative percentage of the GEI that was captured by IPCA1 and IPCA2 was 77.4% (Figure 2C). According to the association between IPCA1 and IPCA2, G1, G3, G5 and G6 were the most stable genotypes with the least interaction, while G9, G2, G7, G4, G8 and G11 were the most unstable genotypes. The locations were ranked as E4 > E5 > E6 in terms of DMC. Furthermore, there was a positive interaction between E4 and G2, E2 and G4, E1 and G9.

The partitioning of GGE showed that IPCA1 and IPCA2 accounted for 68.1% and 16.5% of the GGE sum of squares, respectively, explaining 84.6% of the total variation (Figure 2D). Two mega environments were formed based on winning genotypes. The first mega environment constitutes E1, E2, E3, E5 and E6 with G1, G4, G5, G6 and G11 as winning genotypes, whereas the second environment contains E4 with G2, G9 and G10 as the winning genotypes.

### 2.5. Stability Analysis Using AMMI Model

AMMI stability value (ASV) was proposed by Purchase et al. [9] to quantify and rank genotypes according to their stability. The ranking of genotypes based on ASV for FRY and DMC are presented in Tables 2 and 3, respectively. The genotypes were ranked based on the ASV score, where low scores represent the most stable genotypes. Based on the ASV, the most stable genotype for fresh root yield were G6, G8, G4 and G2, with the lowest AS scores. With regard to DMC, G1, G5, G6 and G3 had the lowest ASV rank and most stable genotypes, while UKF8, G10, G2 and G4 were the least stable.

Another approach to determine yield stability is the use of the genotype stability index (GSI), calculated by ranking the mean performance of genotypes (RY) across environments. The YSI incorporates both mean yield and stability in a single criterion. The YSI ranked G6, G1 and G5 as the highest-yielding and most stable genotypes for FRY, whereas G9, G11 and G10 were the least stable for FRY. On the other hand, G1, G6 and G11 were ranked the highest and the most stable genotypes for DMC, whereas G10, G8 and G2 were ranked the least stable for DMC.

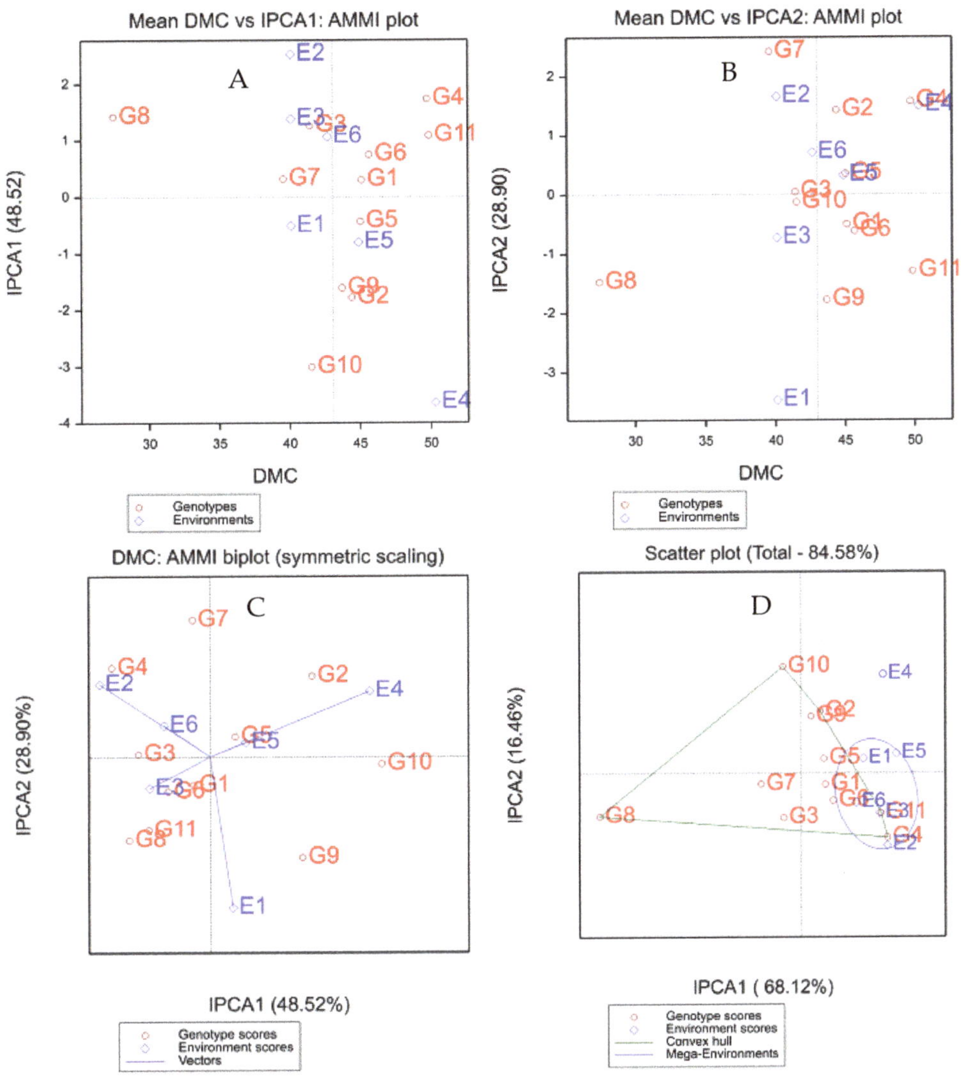


**Figure 2.** Biplot analysis for 11 cassava genotypes evaluated in six trial sites for dry matter content (DMC). (**A**) = AMMI1 biplot showing the IPCA1 vs. main effect (means); (**B**) = AMMI1 biplot showing the IPCA2 vs. main effect; (**C**) = AMMI2 biplot showing the first two principal axes of interaction (IPCA2 vs. IPCA1); and (**D**) = GGE plot defining mega-environments using different winning genotypes tested. E1–E6 = the six testing locations, and G1–G11 = the eleven cassava genotypes used for this study.

**Table 2.** Ranking of 11 cassava genotypes based on fresh root yield, Additive main effects and multiplicative interaction (AMMI) stability value (ASV) and genotype stability index (GSI) based on six trial sites.

| Genotype | Mean | RY | IPCA1 | IPCA2 | ASV | RASV | GSI | RGSI |
|---|---|---|---|---|---|---|---|---|
| G1 | 86.77 | 1 | 3.14 | −0.55 | 6.24 | 8 | 9 | 2 |
| G2 | 71.31 | 7 | 1.77 | 1.79 | 3.94 | 4 | 11 | 4 |

**Table 2.** *Cont.*

| Genotype | Mean | RY | IPCA1 | IPCA2 | ASV | RASV | GSI | RGSI |
|---|---|---|---|---|---|---|---|---|
| G3 | 85.91 | 2 | 2.31 | −7.40 | 8.70 | 9 | 11 | 6 |
| G4 | 50.22 | 11 | 1.56 | 1.91 | 3.63 | 3 | 14 | 8 |
| G5 | 72.13 | 5 | −1.78 | −1.95 | 4.03 | 5 | 10 | 3 |
| G6 | 84.84 | 3 | 1.04 | 2.21 | 3.02 | 1 | 4 | 1 |
| G7 | 78.51 | 4 | −3.14 | 0.54 | 6.23 | 7 | 11 | 5 |
| G8 | 52.8 | 10 | −1.72 | 1.03 | 3.56 | 2 | 12 | 7 |
| G9 | 65.11 | 8 | 5.47 | 2.62 | 11.13 | 10 | 18 | 11 |
| G10 | 55.91 | 9 | −2.15 | −1.21 | 4.43 | 6 | 15 | 9 |
| G11 | 71.37 | 6 | −6.49 | 1.02 | 12.88 | 11 | 17 | 10 |

RY = mean yield; IPCA1 = First interaction principal component; IPCA2 = Second interaction principal component; ASV = AMMI stability value; RASV = Rank of AMMI stability value; YSI = Genotype stability index; RYSI = Rank of genotype stability index.

**Table 3.** Ranking of 11 cassava genotypes based on dry matter content (DMC), additive main effects and multiplicative interaction (AMMI) stability value (ASV) and genotype stability index (GSI) based on their performance at six trial sites.

| Genotype | Mean | RY | IPCA1 | IPCA2 | ASV | RASV | GSI | RGSI |
|---|---|---|---|---|---|---|---|---|
| G1 | 45.13 | 4 | 0.30 | −0.50 | 0.71 | 1 | 5 | 1 |
| G2 | 44.41 | 6 | −1.78 | 1.43 | 3.31 | 10 | 16 | 9 |
| G3 | 41.44 | 9 | 1.26 | 0.04 | 2.11 | 4 | 13 | 6 |
| G4 | 49.75 | 2 | 1.73 | 1.57 | 3.30 | 9 | 11 | 5 |
| G5 | 45.03 | 5 | −0.44 | 0.35 | 0.81 | 2 | 7 | 4 |
| G6 | 45.66 | 3 | 0.74 | −0.62 | 1.39 | 3 | 6 | 2 |
| G7 | 39.56 | 10 | 0.32 | 2.41 | 2.47 | 6 | 16 | 8 |
| G8 | 27.46 | 11 | 1.42 | −1.48 | 2.80 | 7 | 18 | 10 |
| G9 | 43.71 | 7 | −1.62 | −1.78 | 3.25 | 8 | 15 | 7 |
| G10 | 41.54 | 8 | −3.01 | −0.13 | 5.05 | 11 | 19 | 11 |
| G11 | 49.88 | 1 | 1.08 | −1.30 | 2.23 | 5 | 6 | 3 |

RY = mean dry matter content; IPCA1 = First interaction principal component; IPCA2 = Second interaction principal component; ASV = AMMI stability value; RASV = Rank of AMMI stability value; YSI = Genotype stability index; RGSI = Rank of genotype stability index.

## 3. Discussion

Significant genotype × environment interactions reduce the progress of genotype selection because large interactions can reduce gains from selection and make the identification of superior genotypes difficult. Quantifying and understanding GEI is important when selecting genotypes adapted to a range of target environments that vary considerably [10]. Evaluation of genotypes across different environments is important to identify stable genotypes and high-yielding genotypes in specific environments and to identify sites that best represent the target environment [11]. In addition, genotypic stability and adaptability should be considered important aspects of yield trials [12]. An ideal genotype should have a superior and stable performance within and across environments. Several statistical approaches to the measurement of the stability of performance have been suggested to examine the stability of individual genotypes across environments [13]. Therefore, it is recommended that yield and stability are evaluated simultaneously in multi-site trials to reduce the effect of genotype by environment interaction and to make selection more precise [14].

Genotype effects were highly significant ($p < 0.001$) for FRY and DMC, indicating the presence of wide genetic variation among the genotypes for the traits. This variation suggests that the studied genotypes constituted diverse germplasm with sufficient genetic variation for breeding purposes, which could be improved by hybridization among the genotypes followed by selection. The significant environmental effect ($p < 0.001$) observed for all the agronomic traits signified the substantial influence of the environment on the

expression of the traits. This variation underlines the need to conduct multi-locational trials in order to identify genotypes with broad or specific adaption. The significant variation of the GEI effect ($p < 0.001$) found for the observed agronomic traits indicated that genotype and environment main effects were not sufficient to explain the observed phenotypic variation. This variation was the source of deviation in the performance of the genotypes across the different environments. Many researchers have reported similar findings [15–20]. The results confirmed that testing for GEI and assessing the stability of genotypes across environments is essential in breeding programs.

The AMMI analysis of variance revealed that GEI contributed the most variability (45.6%) of the total variation for the parameter FRY. However, the genotypic variance accounted for a large proportion of the observed phenotypic variance (51.9%) for DMC. Environmental variation contributed to 36.4% of the total variability in FRY and 21.3% in DMC. As such, only 18% of the variation was explained by the genetic contribution for the parameter FRY. Olayinka et al. [20] in Nigeria reported that in their study on cassava, more than 88% of the treatment SS was due to environmental variation for FRY. In this study, the variation in GEI constituted the larger proportion of the treatment SS for FRY. This finding agreed with many researchers [16–19,21–23], who reported the significant influence of the environment and GEI on the expression of yield and yield component traits. However, this finding was contrary to Benesi et al. [24], Olayinka et al. [20], Peprah et al. [25] and Tumuhimbise et al. [26], who observed non-significant GEI values for DMC, FRY and starch. The discrepancies could be due to the fact that genotypes might have similar responses across environments or that the testing environments that they used were similar in terms of spatial and temporal environmental conditions. The results of this particular study showed that the selected environments were adequately diverse to discriminate between genotypic performances under the different temporal and spatial environmental conditions used [27].

Stability analysis methods are often used to identify genotypes that have stable performance and respond positively to improvements in environmental conditions [28]. AMMI stability value (ASV) indicates the stability of genotypes in which genotypes with low ASVs are considered to be stable, whilst those with high ASV values are considered to be less stable genotypes [29]. Cultivars UKF3 (G6), UKF5 (G8) and P1/19 (G4) were the most stable for FRY and 98/0002 (G1), P4/10 (G5) and UKF3 (G6) were the most stable for DMC. Genotypic performance *per se* can be misleading due to the sensitivity of genotypes to environmental fluctuations. Similarly, stability alone does not ensure a high yield since a consistently low-yielding genotype can also be stable [30]. In some cases, the most stable genotypes do not have the best yield performances [17]. Hence, for breeding, agronomy and physiological studies, both performance and stability should be considered simultaneously to reduce the effect of GEI. Therefore, a high root yield is considered with stability in the estimation of the genotype stability index (GSI). Genotypes with lower GSI are desirable because they combine a high mean yield performance with stability [26]. Based on the GSI, UKF3 (G6), 98/0002 (G1) and P4/10 (G5) were identified as providing both high yield performance and stability for FRY, and 98/0002 (G1), UKF3 (G6) and UKF9 (G11) were identified as having high DMC and stability values.

The results of AMMI analysis indicated that the first two IPCAs were highly significant ($p < 0.001$) and contributed 74.7% and 77.4% of the total phenotypic variation for FRY and DMC, respectively. Gauch [31] proposed that the most accurate model for AMMI could be predicted using the first two IPCAs; hence the two IPCA scores were then used in the calculation of ASV, as postulated by Purchase et al. [9]. Genotypes with IPCA1 scores adjacent to the zero lines of the biplot indicated that these genotypes are suited to all environments, whereas IPCA1 vectors with the same sign and score but which are situated away from the zero lines of the biplot have genotypes that are adapted to a specific environment [32]. MSAF2 (G3) had IPCA1 scores of close to zero on the IPCA1 axis, which indicated that this genotype was suitable and stable across environments for FRY, while 98/0002 (G1), P4/10 (G5) and UKF4 (G7) were suitable and stable for DMC. The test sites Masibekela and Mabuyeni generated low interaction effects for FRY, while the sites Mandlakazi and Masibekela generated low interaction effects for DMC.

Studies on the suitability of the test sites to reflect the target environments is the next important step after determining GEI to quantify each of the test environments for their discriminative power, representativeness, inter-relatedness and redundancy among the test environments [33]. The concept of a mega-environment was introduced to subdivide a crop production region into several relatively homogeneous mega-environments, to breed and target adapted genotypes for each mega-environment, and to reduce research costs by eliminating redundant trial sites [34]. The first mega-environment recommended by the GGE plot includes Shatale (E4; Mpumalanga) and Nseleni (E1; KZN). Shatale (E4) would be a good environment for future preliminary screening and for breeding activities because it represents an intermediate-performance environment for FRY (71.1 ton $ha^{-1}$) and a high-performance environment for DMC (50.3%). The second mega-environment consisted of Masibekela (E3; Mpumalanga), Mutale (E6; Limpopo) and Mabuyeni (E2; KZN). Mabuyeni (E2) would be a good testing location because it provides a high-performance environment for FRY (78.3 ton-$ha^{-1}$) and an intermediate-performance environment for DMC (40.1%). The high FRY and DMC observed in Mutale could be attributed to the crop being harvested more than 2 months later than at the other sites. The third mega-environment only contained Mandlakazi (E5) (Limpopo). The three mega-environments represented the three provinces (Mpumalanga, KZN and Limpopo) in South Africa that are suitable for cassava cultivation.

The specific adaptability of a genotype to a particular environment could be assessed by analyzing the position of the genotypes with reference to the environmental vectors in the AMMI2 biplot. In addition, a "which-won-where" biplot was constructed for each trait to explore the possible existence of mega-environments within the studied environments and to identify winning genotypes in each mega-environment using GGE analysis. Scavo et al. [35], studying potato genotypes, and Khan [36], studying Bambara groundnut genotypes, used a similar approach. The results confirm the presence of distinct interactions between genotypes and environments for FRY and DMC. In the biplot, UKF9 (G11) and UKF4 (G7) were relatively far from the origin of the axes and close to Shatale (E4), indicating that they are well adapted for Shatale for FRY. On the other hand, 98/0002(G1), MSAF2 (G3) and UKF3 (G6) were found to be the best in the second mega-environment, Mutale (E6), with above-average values, while UKF7 (G9) was found to be well adapted to Mandlakazi. For DMC, only two mega-environments were identified, in which the first environment comprised all the environments except Mandlakazi with 98/0002 (G1), P1/19 (G4), P4/10 (G4), UKF3 (G6) and UKF9 (G11) as winning genotypes, and the second mega-environment consisted of only one location, Mandlakazi (E4), with 98/0505 (G2), UKF8 (G10) and UKF7 (G9) as the winning genotypes.

## 4. Materials and Methods

### 4.1. Trial Site Description

This research was conducted in three provinces that represent the tropical and subtropical agroecological zones in South Africa: KwaZulu-Natal (KZN), Mpumalanga and Limpopo. The data presented in this study were collected from 6 environments, namely Nseleni, Mabuyeni, Masibekela, Shatale, Mutale and Mandlakazi, during the 2019–2020 cropping season. Cassava yield and yield component traits were evaluated approximately 12 months after planting (MAP) at all locations except in Mutale (14 months). Detailed information on the location of each trial site and their GPS coordinates are presented in Table 4.

**Table 4.** Details and GPS coordinates of the six trial sites.

| Code | Location | District | Province | Soil Type | GPS Coordinate |
|---|---|---|---|---|---|
| E1 | Nseleni | Empangeni | KZN | Sandy | −28.634120, 31.912331 |
| E2 | Mabuyeni | King Cetshwayo | KZN | Silt | −28.853811, 31.961901 |
| E3 | Masibekela | Ehlanzeni | Mpumalanga | Sandy loam | −25.870814, 31.825738 |
| E4 | Shatale | Ehlanzeni | Mpumalanga | Sandy loam | −24.747785, 31.035320 |
| E5 | Mandlakazi | Mopani | Limpopo | Sandy loam | −23.801784, 30.377987 |
| E6 | Mutale | Vhembe | Limpopo | Clay | −22.721418, 30.572238 |

KZN: KwaZulu-Natal.

### 4.2. Planting Material and Experimental Design

The study evaluated 11 cassava genotypes acquired from the International Institute of Tropical Agriculture (IITA), the University of KwaZulu-Natal (UKZN) and the Agricultural Research Council (ARC; Table 5). Planting materials were multiplied using an in vitro tissue culture system, and the plantlets were acclimatized in a greenhouse before planting. The trial at each location was laid out in a randomized complete block design with 3 replications. Each genotype was planted in a plot size of 5 × 5 m comprising 5 rows of 5 plants with an inter- and intra-row spacing of 1 x 1 m, respectively. Plants were grown from disease-free, in vitro tissue cultured plantlets. The genotypes were grown under rainfed conditions. Neither fertilizers nor pesticides were applied. Standard agronomic practices were followed as recommended for cassava [37].

**Table 5.** Descriptions of cassava genotypes tested at six different trial sites.

| Code | | Type | Source | Trait |
|---|---|---|---|---|
| G1 | 98/0002 | Released cultivar | IITA | CMD resistance |
| G2 | 98/0505 | Released cultivar | IITA | CMD resistance |
| G3 | MSFA2 | Landrace | ARC | High FRY/Low *DMC* |
| G4 | P1/19 | Breeding line | IITA | High *DMC* |
| G5 | P4/10 | Breeding line | IITA | High *DMC* |
| G6 | UKF3 | Breeding line | Kenya | High SC |
| G7 | UKF4 | Breeding line | Kenya | High SC |
| G8 | UKF5 | Breeding line | Kenya | High SC |
| G9 | UKF7 | Breeding line | Kenya | High SC |
| G10 | UKF8 | Breeding line | Kenya | High SC |
| G11 | UKF9 | Breeding line | Kenya | High SC |

IITA = The International Institute of Tropical Agriculture; UKZN= University of KwaZulu-Natal; ARC= Agricultural Research Council; CMD = cassava mosaic disease; DMC = dry matter content; FRY = fresh root yield; SC = starch content. The UKF materials were originally from Kenya and were used for Ph.D. studies at the University of Kwazulu-Natal. UKZN is the custodian of the materials.

### 4.3. Data Collected and Preparation of Samples

Yield data was collected from five randomly selected plants per plot. The mean yield of the five plants was converted into ton ha$^{-1}$ using a plant density of 10,000 plants per hectare.

Dry matter content (*DMC*) was measured using 5 randomly selected storage roots. The roots were thoroughly cleaned with water and dried with a paper towel before being diced into 1 cm thick discs at 25%, 50% and 75% of the length from the base of the roots. The freshly-cut tuber discs were further sliced into smaller-sized cubes to facilitate oven drying. Five 100 g chopped cubes were taken from each sample and were oven-dried at 105 °C for 24 h. The dried cubes were weighed to obtain the dry matter content.

*DMC* was measured using the following equation:

$$Dry\ matter\ content\ (DMC) = \left(\frac{DW}{FW}\right) \times 100$$

### 4.4. Data Analysis

All the data generated were analyzed using GenStat statistical software version 19.1 [38]. The quality of the data was inspected for data logging errors, and outliers and extreme values were removed from the analysis. Data obtained from each location were analyzed separately by running a single location analysis of variance, and thereafter data from all four environments were pooled for analysis of variance (ANOVA) to perform the combined analysis of 11 cassava genotypes across the four environments to test the presence of significant genotype, environment and genotype × environment variation.

The pooled ANOVA was highly significant ($p < 0.001$) for genotype, environment and GEI components for FRY and *DMC*, justifying the use of Additive Main effect and Multiplicative Interaction (AMMI) and GGE biplot analyses to identify the stable genotypes. The AMMI model was used to determine the main effects and genotype × environment interactions. The AMMI model fits the additive effects for the genotypes and environments and the multiplicative term for interactions [39]. The AMMI model was as follows:

$$Y_{ij} = \mu + \alpha_i + \beta_j + \sum_{k=1}^{n} \lambda_k \gamma_{ik} \delta_{jk} + \varepsilon_{ij}$$

where $Y_{ij}$ = the yield of $i$th genotype in the $j$th environment over all replications, μ is the grand mean, $\alpha I$ is the $i$th genotype mean deviation, $\beta_j$ = the $j$th environment mean deviation, $\lambda_k$ is the singular value for IPC axis $k$, $\gamma_{ik}$ is the $i$th genotype eigenvector value for IPC axis $k$, $\delta_{jk}$ is the $j$th environment eigenvector value for IPC axis $k$ and $\varepsilon_{ij}$ is the error term.

Biplots were generated by plotting the first principal component axis (IPCA1) scores of the genotypes and the environments against their respective IPCA2 scores, resulting from the singular value decomposition of the environment or standardized G × E data [40]. The genotype main effect and genotype by environment interaction effect (GGE) biplot were generated based on 2 concepts, the biplot concept [41] and the GGE concept [42]. A GGE biplot analysis was applied for visual examination of the GEI pattern in the data set.

An AMMI stability value (*ASV*) was calculated for each genotype according to the relative contributions of the principal component axis scores (*IPCA1* and *IPCA2*) to the interaction sum of squares. The AMMI stability value (*ASV*), as described by Purchase et al. [9], was calculated as follows:

$$ASV = \sqrt{\left[\frac{IPCA1_{SS}}{IPCA2_{SS}}\,(IPCA_{score})\right]^2 (IPCA2_{score})^2}$$

The genotype stability index (YSI) was also calculated using the sum of the ranking based on yield and ranking based on the AMMI stability value.

$$GSI = RASV + RY$$

where $RASV$ = the rank of the genotypes based on the AMMI stability value; $RY$ = the rank of the genotypes based on yield across environments. YSI incorporates both mean yield and stability in a single criterion. Low values of both parameters show desirable genotypes with high mean yields and stability.

## 5. Conclusions

The aim of this study was to evaluate and assess the adaptability and stability of selected cassava genotypes based on their mean performance under a wide range of environments in order to select superior and stable genotypes. Selection of genotypes for stability is needed in most dryland environments, where the environment is variable and unpredictable. The stability and adaptability analysis, using AMMI biplots, ASV and GSI statistics, identified the cassava genotypes UKF3 (G6), 98/0002 (G1) and P4/10 (G5) as being the most stable genotypes with the highest root yields, while 98/0002(G1), UKF3 (G6) and UKF9 (G11) were found to be the most stable genotypes with high DMC. Cultivars 98/0002 and UKF3 were identified as combining high stability with superior FRY and

DMC. These genotypes could be recommended to farmers for food, feed and industrial application. In addition, genotypes identified as high performing and stable for both traits could be utilized as parental genotypes in future breeding programs. Similarly, the test sites of Mandlakazi (Limpopo), Mabuyeni (KZN) and Shatale (Mpumalanga) were identified as suitable and representative environments for all genotype evaluations and breeding for FRY and DMC. This study will serve as a baseline for further studies on GEI effects with a larger set of cassava genotypes and will factor in seasonal effects.

**Author Contributions:** Conceptualization, A.B.A., M.W.B., M.L. and S.L.V.; implementation, A.B.A. and R.M.; data analysis, A.B.A.; data collection, R.M.; writing of the manuscript, A.B.A., M.W.B., M.L. and S.L.V. All authors have read and agreed to the published version of the manuscript.

**Funding:** This research was funded by the Technology Innovation Agency (TIA), South Africa.

**Data Availability Statement:** Not applicable.

**Conflicts of Interest:** The authors declare no conflict of interest.

## References

1. Tonukari, N.J. Cassava and the future of starch. *Electron. J. Biotechnol.* **2004**, *7*, 5–8. [CrossRef]
2. Amelework, A.B.; Bairu, M.W.; Maema, O.; Venter, S.L.; Laing, M. Adoption and Promotion of Resilient Crops for Climate Risk Mitigation and Import Substitution: A Case Analysis of Cassava for South African Agriculture. *Front. Sustain. Food Syst.* **2021**, *5*, 617783. [CrossRef]
3. Giller, K.E.; Wilson, K.J. *Nitrogen Fixation in Tropical Cropping System*; CAB International: Wallingford, UK, 1991.
4. Fotso, A.K.; Hanna, R.; Kulakow, P.; Parkes, E.; Iluebbey, P.; Ngome, F.A.; Suh, C.; Massussi, J.; Choutnji, I.; Wirnkar, V.L. AMMI analysis of cassava response to contrasting environments: Case study of genotype by environment effect on pests and diseases, root yield, and carotenoids content in Cameroon. *Euphytica* **2018**, *214*, 155. [CrossRef]
5. Bustos-Korts, D.; Romagosa, I.; Borràs-Gelonch, G.; Casas, A.M.; Slafer, G.A.; van Eeuwijk, F. Genotype by environment interaction and adaptation. In *Encyclopaedia of Sustainability Science and Technology*; Meyers, R.A., Ed.; Springer Nature: Cham, Switzerland, 2018; pp. 1–44.
6. Fox, P.N.; Crossa, J.; Romagosa, I. Multi-environment testing and genotype × environment interaction. In *Statistical Methods for Plant Variety Evaluation*; Kempton, R.A., Fox, P.N., Cerezo, M., Eds.; Springer: Dordrecht, The Netherlands, 2012; pp. 117–138.
7. Ebem, E.C.; Afuape, S.O.; Chukwu, S.C.; Ubi, B.E. Genotype × Environment Interaction and Stability Analysis for Root Yield in Sweet Potato [*Ipomoea batatas* (L.) Lam]. *Front. Agron.* **2021**, *3*, 665564. [CrossRef]
8. Ceballos, H.; Iglesias, C.A.; Pérez, J.C.; Dixon, A.G. Cassava breeding: Opportunities and challenges. *Plant Mol. Biol.* **2004**, *56*, 503–516. [CrossRef] [PubMed]
9. Purchase, J.L.; Hatting, H.; van Deventer, C.S. Genotype × environment interaction of winter wheat (*Triticum aestivum* L.) in South Africa: II. Stability analysis of yield performance. *S. Afr. J. Plant Soil* **2000**, *17*, 101–107. [CrossRef]
10. Yan, W. GGE Biplot vs. AMMI Graphs for genotype-by-environment data analysis. *J. Ind. Soc. Agric. Stat.* **2011**, *65*, 181–193.
11. Nowosad, K.; Liersch, A.; Popławska, W.; Bocianowski, J. Genotype by environment interaction for seed yield in rapeseed (*Brassica napus* L.) using additive main effects and multiplicative interaction model. *Euphytica* **2016**, *208*, 187–194. [CrossRef]
12. Yan, W.; Kang, M.S.; Ma, B.; Woods, S.; Cornelius, P.L. GGE Biplot vs. AMMI Analysis of Genotype-by-Environment Data. *Crop Sci.* **2007**, *47*, 643–653. [CrossRef]
13. Annicchiarico, P. *Genotype × Environment Interactions: Challenges and Opportunities for Plant Breeding and Cultivar Recommendations*; FAO Plant Production and Protection Paper, 0259-2517, 174; Food and Agriculture Organization of the United Nations: Rome, Italy, 2022; pp. 1–126.
14. Jiwuba, L.; Danquah, A.; Asante, I.; Blay, E.; Onyeka, J.; Danquah, E.; Egesi, C. Genotype by Environment Interaction on Resistance to Cassava Green Mite Associated Traits and Effects on Yield Performance of Cassava Genotypes in Nigeria. *Front. Plant Sci.* **2020**, *11*, 572200. [CrossRef]
15. Chipeta, M.M.; Melis, R.; Shanahan, P.; Sibiya, J.; Benesi, I.R.M. Genotype × environment interaction and stability analysis of cassava genotypes at different harvest times. *J. Anim. Plant Sci.* **2017**, *27*, 901–919.
16. Egesi, C.N.; Ilona, P.; Ogbe, F.O.; Akoroda, M.; Dixon, A. Genetic Variation and Genotype × Environment Interaction for Yield and Other Agronomic Traits in Cassava in Nigeria. *Agron. J.* **2007**, *99*, 1137–1142. [CrossRef]
17. Akinwale, M.G.; Akinyele, B.O.; Dixon, A.G.O.; Odiyi, A.C. Genetic variability among forty-three cassava genotypes in there agro-ecological zones of Nigeria. *J. Plant Breed. Crop Sci.* **2010**, *2*, 104–109.
18. Adjebeng-Danquah, J.; Manu-Aduening, J.; Gracen, V.E.; Asante, I.K.; Offei, S.K. AMMI Stability Analysis and Estimation of Genetic Parameters for Growth and Yield Components in Cassava in the Forest and Guinea Savannah Ecologies of Ghana. *Int. J. Agron.* **2017**, *2017*, 8075846. [CrossRef]
19. Rotich, D.C.; Kiplagat, O.K.; Were, V.W. Genotype by environment interaction on yield components and stability analysis of elite cassava genotypes. *Int. J. Plant Breed. Crop Sci.* **2018**, *5*, 361–369.

20. Bakare, M.A.; Kayondo, S.I.; Aghogho, C.I.; Wolfe, M.D.; Parkes, E.Y.; Kulakow, P.; Egesi, C.; Rabbi, I.Y.; Jannink, J.-L. Exploring genotype by environment interaction on cassava yield and yield related traits using classical statistical methods. *PLoS ONE* **2022**, *17*, e0268189. [CrossRef] [PubMed]
21. Aina, O. Additive Main Effects and Multiplicative Interaction (AMMI) Analysis for Yield of Cassava in Nigeria. *J. Biol. Sci.* **2007**, *7*, 796–800. [CrossRef]
22. Ssemakula, G.; Dixon, A. Genotype × environment interaction, stability and agronomic performance of carotenoid-rich cassava clones. *Sci. Res. Essays* **2007**, *2*, 390–399.
23. Egesi, C.N.; Onyeka, T.J.; Asiedu, R. Environmental stability of resistance to anthracnose and virus diseases of water yam (*Dioscorea alata*). *Afr. J. Agric. Res.* **2009**, *4*, 113–118. [CrossRef]
24. Aina, O.O.; Dixon, A.G.O.; Paul, I.; Akinrinde, E.A. G × E interaction effects on yield and yield components of cassava (landraces and improved) genotypes in the savannah regions of Nigeria. *Afr. J. Biotechnol.* **2009**, *8*, 4933–4945.
25. Benesi, I.; Labuschagne, M.; Dixon, A.; Mahungu, N. Genotype × enviroment interaction effects on native cassava starch quality and potential for starch use in the commercial sector. *Afr. Crop. Sci. J.* **2004**, *12*, 205–216. [CrossRef]
26. Boakye, B.; Kwadwo, O.; Asante, I.; Parkes, E.Y. Performance of nine cassava (*Manihot esculanta* Crantz) clones across three environments. *J. Plant Breed. Crop Sci.* **2013**, *5*, 48–53. [CrossRef]
27. Tumuhimbise, R.; Melis, R.; Shanahan, P.; Kawuki, R. Genotype × environment interaction effects on early fresh storage root yield and related traits in cassava. *Crop. J.* **2014**, *2*, 329–337. [CrossRef]
28. Nduwumuremyi, A.; Melis, R.; Shanahan, P.; Theodore, A. Interaction of genotype and environment effects on important traits of cassava (*Manihot esculenta* Crantz). *Crop. J.* **2017**, *5*, 373–386. [CrossRef]
29. Farshadfar, E.; Mahmodi, N.; Yaghotipoor, A. AMMI stability value and simultaneous estimation of yield and yield stability in bread wheat (*Triticum aestivum* L.). *Aust. J. Crop. Sci.* **2011**, *5*, 1837–1844.
30. Hagos, H.G.; Abay, F. AMMI and GGE biplot analysis of bread wheat genotypes in the northern part of Ethiopia. *J. Plant Breed. Genet.* **2013**, *1*, 12–18.
31. Yan, W.; Tinker, N.A. Biplot analysis of multi-environment trial data: Principles and applications. *Can. J. Plant Sci.* **2006**, *86*, 623–645. [CrossRef]
32. Gauch, H.G. Statistical Analysis of Yield Trials by AMMI and GGE. *Crop. Sci.* **2006**, *46*, 1488–1500. [CrossRef]
33. Murphy, S.E.; Lee, E.A.; Woodrow, L.; Seguin, P.; Kumar, J.; Rajcan, I.; Ablett, G.R. Genotype × Environment Interaction and Stability for Isoflavone Content in Soybean. *Crop. Sci.* **2009**, *49*, 1313–1321. [CrossRef]
34. Khan, M.H.; Rafii, M.Y.; Ramlee, S.I.; Jusoh, M.; Mamun, A. AMMI and GGE biplot analysis for yield performance and stability assessment of selected Bambara groundnut (*Vigna subterranea* L. Verdc.) genotypes under the multi-environmental trials (METs). *Sci. Rep.* **2021**, *11*, 22791. [CrossRef]
35. Gauch, H.G., Jr.; Zobel, R.W. Identifying Mega-Environments and Targeting Genotypes. *Crop. Sci.* **1997**, *37*, 311–326. [CrossRef]
36. Scavo, A.; Mauromicale, G.; Ierna, A. Dissecting the Genotype × Environment Interaction for Potato Tuber Yield and Components. *Agronomy* **2023**, *13*, 101. [CrossRef]
37. Hauser, S.; Wairegi, L.; Asadu, C.L.A.; Asawalam, D.O.; Jokthan, G.; Ugbe, U. *Cassava System cropping guide. Africa Soil Health Consortium: Nairobi, Kenya*; CAB International: Wallingford, UK, 2014.
38. GenStat. *GenStat for Windows*, 19th ed.; VSN International Ltd.: Oxford, UK, 2020.
39. Jalata, Z. GGE-biplot Analysis of Multi-environment Yield Trials of Barley (*Hordeium vulgare* L.) Genotypes in Southeastern Ethiopia Highlands. *Int. J. Plant Breed. Genet.* **2011**, *5*, 59–75. [CrossRef]
40. Maroya, N.G.; Kulakow, P.; Dixon, A.G.O.; Maziya-Dixon, B.B. Genotype × Environment Interaction of Mosaic Disease, Root Yields and Total Carotene Concentration of Yellow-Fleshed Cassava in Nigeria. *Int. J. Agron.* **2012**, *2012*, 434675. [CrossRef]
41. Gabriel, K.R. The biplot graphic display of matrices with application to principal component analysis. *Biometrika* **1971**, *58*, 453–467. [CrossRef]
42. Yan, W. GGEbiplot—A Windows Application for Graphical Analysis of Multienvironment Trial Data and Other Types of Two-Way Data. *Agron. J.* **2002**, *93*, 1111–1118. [CrossRef]

*plants*

MDPI

*Article*

# Transcriptome Analysis Reveals Key Genes Involved in Weevil Resistance in the Hexaploid Sweetpotato

**Kanoko Nokihara [1,†], Yoshihiro Okada [2,†], Shinichiro Ohata [1] and Yuki Monden [1,*]**

1 Graduate School of Environmental and Life Science, Okayama University, Okayama, Okayama 700-8530, Japan; p97j1mjy@s.okayama-u.ac.jp (K.N.); ph563a4t@s.okayama-u.ac.jp (S.O.)
2 Kyushu Okinawa Agricultural Research Center, National Agriculture and Food Research Organization, Itoman, Okinawa 901-0336, Japan; yoc1973@affrc.go.jp
* Correspondence: y_monden@okayama-u.ac.jp
† These authors contributed equally to this work.

**Abstract:** Because weevils are the most damaging pests of sweetpotato, the development of cultivars resistant to weevil species is considered the most important aspect in sweetpotato breeding. However, the genes and the underlying molecular mechanisms related to weevil resistance are yet to be elucidated. In this study, we performed an RNA sequencing-based transcriptome analysis using the resistant Kyushu No. 166 (K166) and susceptible Tamayutaka cultivars. The weevil resistance test showed a significant difference between the two cultivars at 30 days after the inoculation, specifically in the weevil growth stage and the suppressed weevil pupation that was only observed in K166. Differential expression and gene ontology analyses revealed that the genes upregulated after inoculation in K166 were related to phosphorylation, metabolic, and cellular processes. Because the weevil resistance was considered to be related to the suppression of larval pupation, we investigated the juvenile hormone (JH)-related genes involved in the inhibition of insect metamorphosis. We found that the expression of some terpenoid-related genes, which are classified as plant-derived JHs, was significantly increased in K166. This is the first study involving a comprehensive gene expression analysis that provides new insights about the genes and mechanisms associated with weevil resistance in sweetpotato.

**Keywords:** transcriptome; RNA-seq; sweetpotato; weevil resistance; juvenile hormones; terpenes

**Citation:** Nokihara, K.; Okada, Y.; Ohata, S.; Monden, Y. Transcriptome Analysis Reveals Key Genes Involved in Weevil Resistance in the Hexaploid Sweetpotato. *Plants* **2021**, *10*, 1535. https://doi.org/10.3390/plants10081535

Academic Editors: Ranjith Pathirana and Francesco Carimi

Received: 17 June 2021
Accepted: 21 July 2021
Published: 27 July 2021



## 1. Introduction

Sweetpotato (*Ipomoea batatas* (L.) Lam.) is a member of the Convolvulaceae family that is widely cultivated in the tropical and temperate zones. As a valuable source of carbohydrates, vitamins, fiber, and minerals, sweetpotato is considered one of the most important crops in the world, with an annual production of over 100 million tons globally [1]. In recent years, the production of sweetpotato varieties with favorable cultivation-related traits, such as high added-value and resistance to diseases, pests, and low soil temperature, was conducted for the expansion of planting areas. However, sweetpotato is a typical non-model crop species and a hexaploid (2n = 6x = 90) with a complex mode of inheritance and a large genome (2–3 Gb); hence, breeding and genetic studies have been difficult. Furthermore, although some varieties can be self-fertile, most show self-incompatibility or mating incompatibility with a specific group of varieties. This reproductive pattern causes the sweetpotato genome to be highly heterozygous. Therefore, it is extremely difficult to identify the genes and the underlying molecular mechanisms associated with important agricultural traits in sweetpotato.

Two weevil species, namely *Cylas formicarius* (Fabricius) and West Indian sweetpotato weevil (*Euscepes postfasciatus* (Fairmaire)), are known as the most damaging pests of sweetpotato worldwide [2,3], causing economic damage to farmers in Central and South America and the South Pacific Islands [4]. In Japan, weevils are distributed in tropical and subtropical regions, including the Nansei and Ogasawara Islands [5]. Particularly in Okinawa, sweetpotato is an essential crop that supports the backbone of its tourism industry. However, the crop yield is 60% less than that of mainland Japan; this low yield is mainly due to the feeding damage caused by weevils. Therefore, the development of weevil-resistant cultivars is important to mitigate this problem. Both species can invade the root approximately 2 months after planting and then lay eggs. The hatched larvae move to the tuberous roots while feeding on the inside of the stem and eventually become pupae that hatch into adults, which break through the tuberous roots and escape to the outside, becoming a new generation of parent insects. Therefore, weevils spend most of their lifetime either in stems or roots, shielding them from insecticides and causing significant damage to sweetpotato plants. The tuberous roots damaged by feeding produce a phytoalexin called ipomeamarone [6]. Ipomeamarone is a secondary metabolite of sesquiterpenes that acts as a toxic substance to domestic animals. Because the production of phytoalexins makes the sweetpotato bitter, astringent, and even more toxic to animals, the tuberous roots damaged by weevils are not fit for consumption by humans or livestock [5,6].

To develop resistant varieties, the identification of resistance genes and the elucidation of the mechanisms underlying weevil resistance in sweetpotato are necessary. Previous studies have reported the differences in weevil susceptibility among sweetpotato varieties [7–9]. Yada and colleagues performed genetic analysis on an F1 mapping population derived from the resistant African landrace New Kawago and susceptible North American cultivar Beauregard and identified simple sequence repeat markers associated with weevil resistance [10]. From the pest side, one group performed a transcriptome analysis on sweetpotato weevil (*Cylas puncticollis* Boheman) and discovered the presence of a functional RNAi pathway that may be used as a new strategy for controlling this pest [11]. Using next-generation sequencing (NGS) technology, Okada et al. performed genome-wide association studies (GWAS) in sweetpotato and detected several genomic regions associated with weevil resistance [12]. However, there are no reports regarding the molecular mechanisms underlying weevil resistance in sweetpotato using transcriptome-based analysis. Recently, with the decreasing costs and increasing throughput of NGS technology, several groups have reported large-scale transcriptome studies in sweetpotato, revealing the key genes and a comprehensive knowledge of the mechanisms underlying important agricultural traits [13–20]. In addition, the whole genome sequence and functionally annotated genes of the closely related diploid species *Ipomoea trifida* have been previously released [21–23]. Because *I. trifida* is considered a model sweetpotato species, its high-quality genome and gene sequences can be utilized as a reference for the transcriptome analysis of other sweetpotato varieties.

In this study, we performed an RNA sequencing (RNA-seq)-based transcriptome analysis using resistant and susceptible sweetpotato cultivars to comprehensively analyze the differentially expressed genes (DEGs) that respond to the feeding damage caused by weevils. Specifically, we aimed to identify the related genes and to elucidate the molecular mechanisms associated with weevil resistance in sweetpotato.

## 2. Results

### 2.1. Phenotyping for Weevil Resistance

In the weevil resistance evaluation test, 10 adult West Indian sweetpotato weevils (*Euscepes postfasciatus* (Fairmaire)) (sex ratio, male/female = 1:1) were inoculated into the sweetpotato tubers of weevil-resistant Kyushu No. 166 (K166) and susceptible Tamayutaka cultivars. The degree of damage was investigated by counting the number of eggs on the surface of tuberous roots at 3 days after inoculation and the total number of insects at 15 and 30 days after inoculation. In addition, the growth stages of the insects were investigated at 30 days after inoculation. At 3 days after inoculation, the average number of eggs per tuberous root was 23.1 and 23.4 for K166 and Tamayutaka, respectively, and no significant difference was detected between the two (Figure 1a). At 15 days after inoculation, the

average number of insects per tuberous root was 2.7 and 3.0 for K166 and Tamayutaka, respectively, and again no significant difference was observed (Figure 1b). In contrast, the average number of insects at 30 days after inoculation was 4.0 and 13.5 for K166 and Tamayutaka, respectively, with a significant difference ($p < 0.05$) in the total number of insects between the two cultivars (Figure 1c). In K166, the total numbers of insects were six larvae (75.0%) and seven pupae (25.0%). On the other hand, of the total number of insects in Tamayutaka, the numbers of larvae and pupae were 7 (25.9%) and 20 (74.1%), respectively (Figure 1d). Thus, at 30 days after inoculation, there was a large difference in the number and growth state of insects between the two cultivars. These results suggest that weevils have no preferred spawning sites between the two cultivars, and that the resistant K166 suppressed weevil growth, especially during pupation.

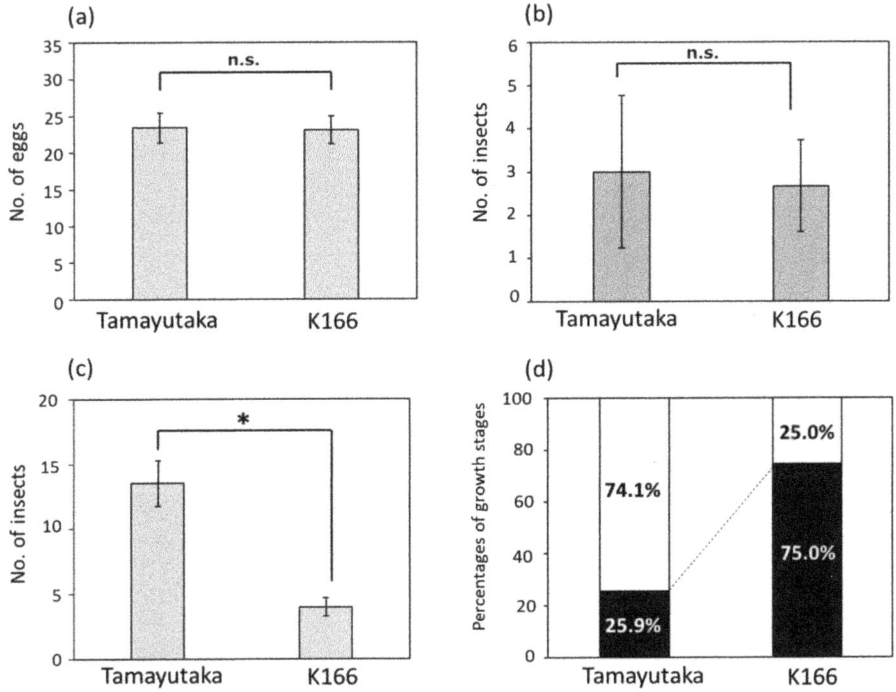


**Figure 1.** Results of the weevil resistance tests between the susceptible Tamayutaka and resistant Kyushu No. 166 (K166) cultivars. (**a**) The average number of eggs per tuberous root at 3 days after inoculation; (**b**) the average number of insects per tuberous root at 15 days after inoculation; (**c**) the average number of insects at 30 days after inoculation; (**d**) the growth state of insects at 30 days after inoculation. Black and white bars indicate the ratios of the larvae and pupae, respectively. * $p < 0.05$, n.s.: not significant.

### 2.2. RNA-Seq-Based Transcriptome Analysis

RNA-seq analysis was performed to identify the DEGs between the weevil-resistant and susceptible cultivars. Total RNA was extracted for RNA-seq library preparation using tuberous roots from two experimental plots (control and inoculation) with two replicates. Sequencing with HiSeqX produced a total of 219,712,549 reads for all samples (22,546,509–43,206,139). After preprocessing, a total of 209,834,637 reads (21,472,899–41,409,345) were obtained (Table S1). The average alignment rate was 72.79% (61.64–78.73%) after mapping the reads to the transcriptome sequences of *I. trifida* (Table S2). Using eXpress and edgeR, differential expression analysis between the weevil inoculation and control plots revealed 242 upregulated and 69 downregulated DEGs in K166 and 312 upregulated and 98 downregulated DEGs in Tamayutaka (Figure 2a,b, Supplementary Tables S3–S7), confirming the transcriptional response to the damage by weevil feeding in both cultivars. In contrast, 528 upregulated and 678 downregulated DEGs were detected between the control plots of K166 and Tamayutaka, indicating that there are many DEGs between the two cultivars (Figure 2c, Tables S3, S8 and S9). From the results of 1206 DEGs detected between cultivars, even in the control plot, the genetic backgrounds of these two cultivars were considered to be quite different. Moreover, 332 upregulated and 377 downregulated DEGs were detected between the weevil inoculation plots of K166 and Tamayutaka (Figure 2d, Tables S3, S10 and S11).

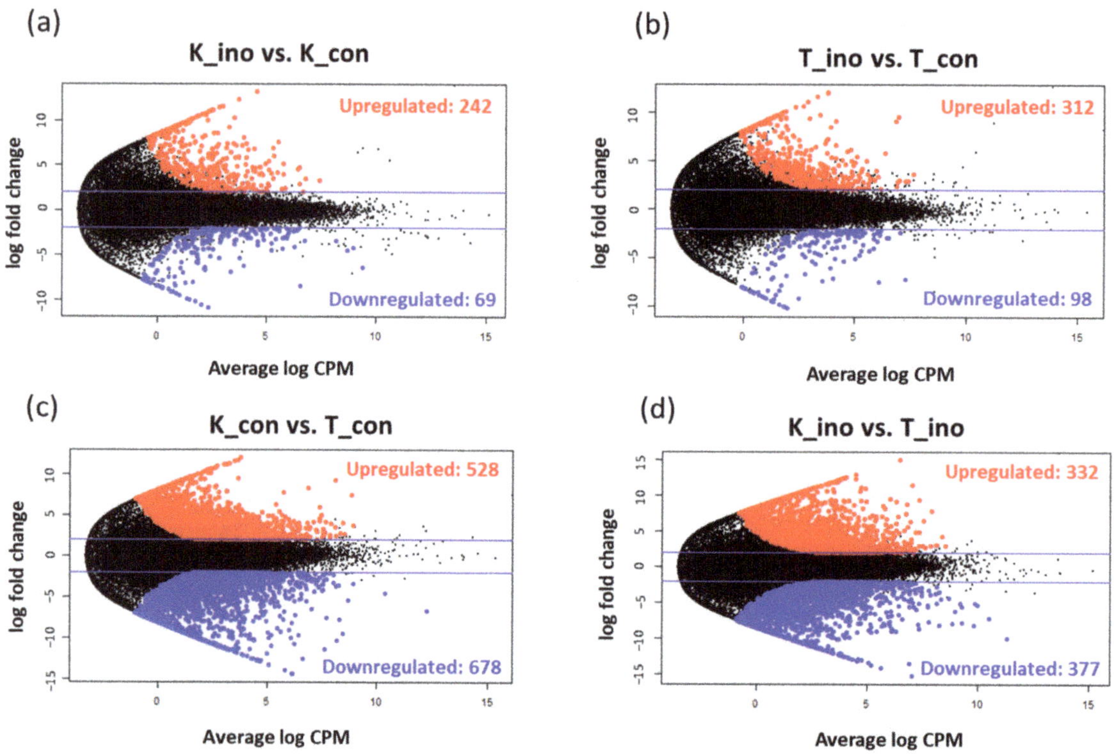


**Figure 2.** MA plots of the identified differentially expressed genes (DEGs). Red and blue dots represent the upregulated and downregulated DEGs, respectively. Black dots represent the non-differentially expressed genes. (**a**) DEGs between the control (K_con) and inoculation (K_ino) plots of Kyushu No. 166 (K166); (**b**) DEGs between control (T_con) and inoculation (T_ino) plots of Tamayutaka; (**c**) DEGs between the control plots of K166 and Tamayutaka; (**d**) DEGs between the inoculation plots of K166 and Tamayutaka.

In addition, we investigated the number of DEGs across four comparisons (K_ino vs. K_con, T_ino vs. T_con, K_con vs. T_con, and K_ino vs. T_ino) and the overlaps between each set of DEGs. Venn diagram analysis indicated that a total of 68 (55 + 3 + 3 + 7) DEGs were differentially expressed in both T_ino vs. T_con and K_ino vs. K_con conditions (Figure 3), suggesting that the expression levels of a relatively small number of these genes changed commonly in both varieties in response to weevil feeding damage. On the other hand, a total of 315 (303 + 5 + 3 + 4) DEGs were commonly detected between K_con vs. T_con and K_ino vs. T_ino comparisons, suggesting that these hundreds of DEGs showed different expression levels between cultivars regardless of control or inoculation plots. Interestingly, a total of 741 (36.9%) and 310 (15.4%) DEGs were detected that were present only in the control (comparison of K_con vs. T_con) and inoculation plots (comparison of

K_ino vs. T_ino), indicating that there are many genes with expression levels that changed specifically in each experimental plot.

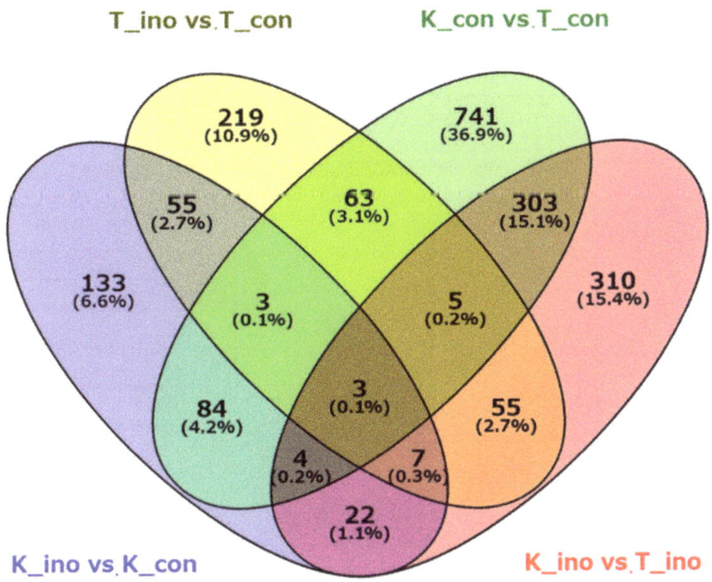


**Figure 3.** Venn diagram showing the number of DEGs and the overlaps among the four comparisons.

Gene ontology (GO) analysis was performed using OmicsBox for the functional annotation of the identified DEGs between K166 and Tamayutaka at 30 days after inoculation. In the biological process category, the top two enriched GO terms for the set of genes upregulated in K166 were "metabolic process" and "cellular process", followed by terms related to phosphorylation such as "phosphate-containing compound metabolic process", "phosphorylation", and "protein phosphorylation" (Figure 4a). In terms of molecular function, "binding" and "catalytic activity" were prominently represented. On the other hand, the top enriched GO terms for the upregulated genes in Tamayutaka were "oxidation-reduction process", "cellular anatomical entity", and "catalytic activity" in the biological process, cellular component, and molecular function categories, respectively (Figure 4b). In the control plot, the GO terms related to phosphorylation, metabolic, and cellular processes were not enriched in the upregulated DEGs of K166 (Figure S1). These results suggest that the upregulation of the genes involved in phosphorylation, metabolic, and cellular processes contributes to weevil resistance in K166.

### 2.3. *Juvenile Hormone (JH)-Related Genes*

Based on the results of the weevil inoculation test, we hypothesized that the resistance trait of K166 was likely to be due to the suppression of pupation during larval development. Hence, we analyzed the JH-related genes involved in the suppression of insect metamorphosis. Larval–pupal and pupal–adult transitions are controlled by the action of JHs and molting hormones in insects such as silkworms (*Bombyx mori* L.). JHs and JH analogues (JHAs or juvenoids) are known to prolong larval life by inhibiting the larval–pupal and pupal–adult transitions [24]. This mechanism is conserved in many insect species. In contrast, terpenes are a large and diverse class of organic compounds produced by a variety of plants. The biochemical actions of natural insect JHs and plant terpenes and terpenoid compounds are similar because terpenes mimic the action of insect JHs [24]. Therefore, it is possible that weevils may mistakenly recognize the JHs produced by sweetpotato, which may explain the suppression of weevil pupation in K166. To verify this hypothesis, we also investigated the expression levels of terpenoid-related genes and discovered 20 genes that were present in K166 and Tamayutaka (Figure 5). Five genes (*itf09g05600.t1*, *itf12g13950.t1*, *itf09g05580.t1*, *itf13g04680.t1*, and *itf12g14020.t1*) were highly expressed in K166 (Figure 5). Of these five genes, three (*itf09g05600.t1*, *itf09g05580.t1*, and *itf12g13950.t1*) had significantly increased expression in K166 (Figure S2, Tables S8 and S10), suggesting that these may be candidate genes that contribute to the inhibition of weevil pupation in K166. Interestingly, two (*itf09g05600.t1* and *itf09g05580.t1*) were found to be very closely located on chromosome 9, with a physical distance of approximately 2 kilobase (kb) from each other. The amino acid sequences of the two genes are also highly conserved (Figure S3). In addition, two functional domains (N-terminal and metal-binding) related to terpene synthase were present in both genes (Figure S4), suggesting their potential roles in terpene synthesis.

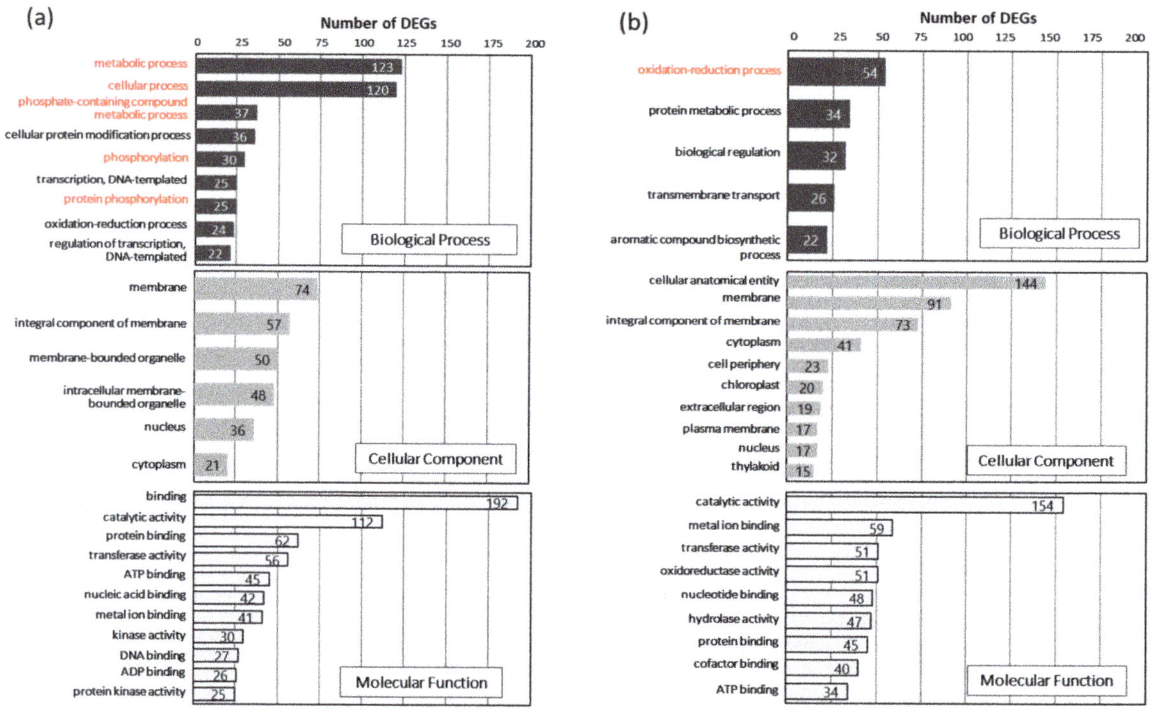


**Figure 4.** The enriched gene ontology (GO) terms of the identified differentially expressed genes (DEGs) between Kyushu No. 166 (K166) and Tamayutaka at 30 days after inoculation. The GO terms were classified in three categories: biological process, cellular component, and molecular function. (**a**) The top GO terms for the upregulated DEGs in K166; (**b**) the top GO terms for the upregulated DEGs in Tamayutaka.

In addition, we focused on the disease resistance-related genes and investigated their expression levels in both cultivars. By searching for genes that had resistance-related annotations and were highly expressed in the resistant K166 cultivar after inoculation, 30 genes were detected (Figure S5). There are many genes with annotations such as "NB-ARC domain-containing disease resistance protein", "TIR-NBS-LRR class", and "CC-NBS-LRR class". On the contrary, only nine genes were detected with resistance-related annotations and higher expression levels in Tamayutaka (data not shown). These results indicate that more disease-resistance genes were highly expressed in the resistant cultivar, which may contribute to resistance to some extent.

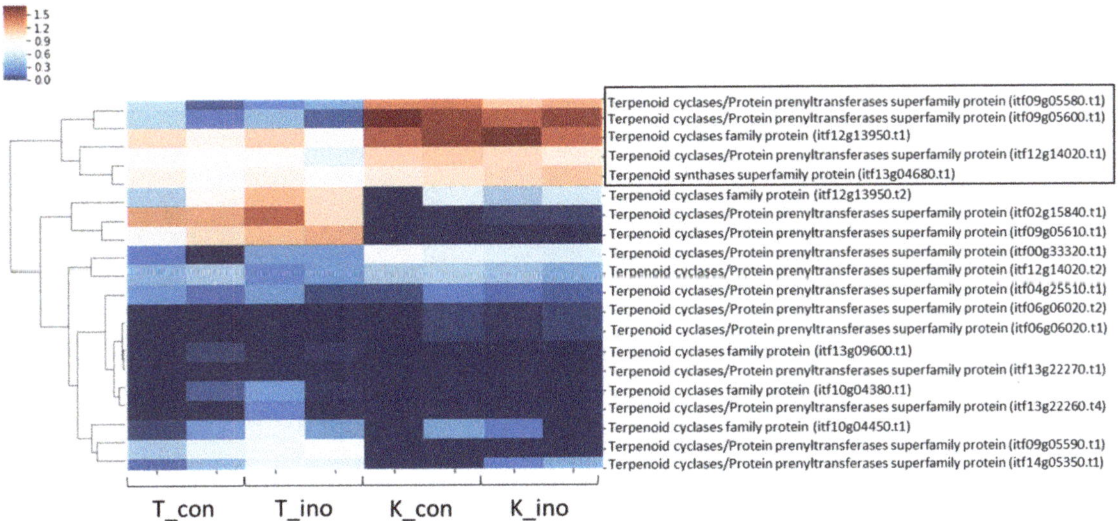


**Figure 5.** Hierarchical clustering and expression heatmap of 20 terpenoid-related genes. Black boxes indicate the genes that are highly differentially expressed between Kyushu No. 166 (K166) and Tamayutaka. The gene expression level (log2 fold change value) is represented by the blue (low) to red (high) color gradient. K_con: K166 samples in the control plot, K_ino: K166 samples in the inoculation plot, T_con: Tamayutaka samples in the control plot, T_ino: Tamayutaka samples in the inoculation plot.

## 3. Discussion

In this study, we performed a comprehensive transcriptome analysis of weevil resistance in sweetpotato. Although weevils are considered a serious pest worldwide, this insect species is distributed in specific regions only, such as the Nansei Islands in Japan. However, due to global warming, the distribution of weevils is expected to expand further, and weevil resistance may become the most desirable agricultural trait for sweetpotato cultivation in the future. To date, there have been no reports of NGS-based comprehensive gene expression analysis focused on investigating weevil resistance in sweetpotato. Therefore, our study provides novel insights into the transcripts that respond to damage by weevil feeding in sweetpotato.

RNA-seq analysis of weevil-resistant and susceptible cultivars revealed numerous DEGs between the two. Even in the control plots, many DEGs (1206) were detected between the two cultivars, reflecting the difference in their genetic backgrounds. Based on the pedigree information (Figure S6), the genetic backgrounds of the two cultivars are considered to be quite different. There were more upregulated than downregulated DEGs after weevil inoculation (98 downregulated and 312 upregulated DEGs in Tamayutaka, 69 downregulated and 242 upregulated DEGs in K166), indicating that transcriptional responses to the feeding damage by weevils occurred in both cultivars. Functional analysis of the DEGs also revealed that after weevil inoculation, the transcription of genes associated with metabolic processes, cellular processes, and phosphorylation were upregulated in the resistant K166 cultivar, suggesting that these genes may be critical for weevil resistance. On the other hand, it should be noted that while these DEGs may have contributed to weevil resistance, their expression levels may be the result of a response to feeding damage by weevils.

Plants possess a complex defense system against diverse pests and pathogens and a response system composed of pathogen detection, signal transduction, and defense response [25]. Plants can perceive certain elicitors in insect oral secretions that enter wounds during feeding and rapidly activate mitogen-activated protein kinase (MAPK) signaling [26]. MAPKs play critical roles in plant resistance against insect herbivores by regulating the herbivory-induced changes in phytohormones, the transcriptome activation of herbivore defense-related genes, and the accumulation of defensive metabolites. MAPKs consist of 11 domains that are found in all serine/threonine protein kinases [27], which are activated by the dual phosphorylation of the Thr and Tyr residues in the TxY motif of the activation loop (T-loop) located between subdomains VII and VIII. In the T-loop, activation occurs via MAPK kinases (MAPKKs), which are activated by MAPKK kinases (MAPKKKs) through the phosphorylation of conserved Ser and/or Thr residues. Activated MAPKs phosphorylate their substrates, including the transcription factors and enzymes that trigger downstream stress-related responses [26]. Thus, MAPK activation via phosphorylation may have immediately occurred after wounding and feeding by weevil larvae in K166, which subsequently induced defense reactions via phosphorylation of associated transcription factors and proteins.

In contrast, the expression of genes related to the oxidation-reduction process and oxidoreductase activity were upregulated in the susceptible Tamayutaka cultivar. Sweetpotato contains several phytoalexins, collectively known as furanoterpenoids, such as ipomeamarone and its precursor dehydroipomeamarone, ipomeanine, 1-ipomeanol, 4-ipomeanol, and 1,4-ipomeadiol [28–30]. Ipomeanine is produced by the oxidation of 4-ipomeanol, whereas 1,4-ipomeadiol and ipomeanol are produced by the reduction of ipomeanine [31]. After weevil inoculation, Tamayutaka was found to be more damaged than K166; thus, the production of phytoalexins such as ipomeamarone and ipomeanine was expected to be high in response to the damage. Consequently, a higher number of upregulated DEGs related to the oxidation-reduction process were detected in Tamayutaka than in K166. Furthermore, the upregulation of these genes was observed in the inoculation plots only (Figure S7), suggesting that both cultivars were damaged by weevil feeding and the resulting oxidation-reduction processes.

In the weevil inoculation test, there was no significant difference in the number of eggs at 3 days after inoculation or the total number of insects at 15 days after inoculation between the resistant and susceptible cultivars. In contrast, there was a large difference in the total number of insects and the growth stage at 30 days after inoculation. These results suggest that weevils have no preferred spawning sites between resistant and susceptible cultivars. However, the weevil growth, particularly pupation, was suppressed in K166, indicating that one possible mechanism of weevil resistance is the inhibition of weevil development and reproduction. We tested this hypothesis by investigating the JHs involved in the suppression of insect metamorphosis. A Krüppel homolog 1 gene (*Kr-h1*), which is induced by the JH via a JH receptor, plays a key role in the repression of insect metamorphosis [32]. The transcription factor Broad-Complex (*BR-C*) functions as a "pupal specifier" in the larval–pupal transition; JH-inducible Kr-h1 binds to the *BR-C* promoter region and represses its transcription, resulting in the inhibition of larval–pupal transition [33,34]. Therefore, JHs can inhibit larval–pupal and pupal–adult transitions in holometabolous insects. In such cases, the larva reaches the end of its life, and the reproduction of the next generation is halted. On the other hand, plant-derived metabolites are known to act as JHAs in insects, and the biochemical actions of insect JHs and plant terpenes and terpenoid compounds are similar [30]. In K166, the weevil may have misrecognized the JH produced by sweetpotato, resulting in the suppressed pupation of the larvae. Therefore, we also investigated the expression levels of terpenoid-related genes and discovered five genes that were upregulated in K166. In particular, three of the five genes had significantly increased expression levels in K166. Among the three genes, two possessed the N-terminal and metal-binding domains of terpene synthase, suggesting that the two genes may function in terpene synthesis and may be associated with the inhibited pupation of weevil larvae in K166. Interestingly, resistant sweetpotato cultivars may possess weevil-recognizing plant compounds that cause growth retardation in weevils. In addition, we revealed that more disease resistance-related genes were highly expressed in the resistant cultivar. These genes may also contribute to the expression of resistance in K166.

Therefore, future studies should investigate whether there is a difference in the amount of terpenes produced by weevil-resistant and susceptible cultivars and determine the correlation between the expression of terpenoid-related genes and the amount of terpenes produced. Furthermore, we are planning to perform additional genetic analyses such as quantitative trait loci (QTL) mapping and GWAS using the F1 populations derived from K166 and Tamayutaka. By investigating the DEGs in the selected QTL region, we can identify the candidate genes controlling weevil resistance in K166.

## 4. Materials and Methods

### *4.1. Resistance Evaluation Test by Weevil Inoculation*

The weevil-resistant K166 and susceptible Tamayutaka cultivars were chosen for this study. K166 is the progeny of a cross between Kyukei98160-1 and Murasakimasari. Tamayutaka is derived from a cross between Kanto No. 33 and Kuroshirazu (Figure S6). The breeding process and pedigree information are shown in Figure S6. For all plant samples, the tuberous roots produced in 2017 and cultivated at the National Agrobiological and Food Research Organization for Kyushu Okinawa Region (Miyakonojo City, Miyazaki Prefecture, Japan) were used. The weevil inoculation test and resistance evaluation were conducted at the Kyushu Okinawa Agricultural Research Center (Itoman City, Okinawa Prefecture, Japan). Adult West Indian sweetpotato weevils (*Euscepes postfasciatus* (Fairmaire)) were used for the inoculation test, and each sweetpotato sample was placed in a plastic case. We prepared two experimental plots (control and inoculation) with three biological replicates. In the inoculation plot, one sweetpotato tuber root was inoculated with 10 adults (sex ratio, male/female = 1:1). The degree of damage to the tuberous roots was investigated at 3, 15, and 30 days after inoculation. At 3 days after inoculation, the number of eggs on the surface of the tuberous root was counted. At 15 and 30 days after inoculation, the sweetpotato samples were dissected, and the degree of damage was investigated by counting the number of larvae, pupae, and adults in the tuberous roots.

### *4.2. RNA Extraction, Library Preparation, and RNA-Seq*

The tuberous roots from the control and inoculation groups with two replicates were collected. Total RNA was extracted using the phenol–chloroform method. Lithium chloride (LiCl) precipitation was performed to remove any impurities, and the extracted RNA was purified using RNeasy Plant Mini Kit (Qiagen, Hilden, Germany). The RNA yield (ng/μL) was measured using Qubit fluorometer (Thermo Fisher Scientific, Waltham, MA, USA), whereas the RNA quality was confirmed by agarose gel electrophoresis. The RNA-seq library was prepared using the KAPA mRNA HyperPrep Kit (KAPA Biosystems, Woburn, MA, USA). The library concentration of each sample was also measured using Qubit (Thermo Fisher Scientific). All samples were pooled in equal volumes to prepare the libraries for RNA-seq, and sequencing with HiSeqX (Illumina, San Diego, CA, USA) yielded 150-base pair (bp) paired-end reads.

### *4.3. Data Analysis*

The obtained RNA-seq data were analyzed using the following procedures. First, the quality of the paired-end reads was determined using FastQC [35]. Second, the adapter sequences and low-quality nucleotides were removed using Cutadapt [36]. The threshold value of the quality score was 30, and the minimum read length for trimming was 50 bp. After preprocessing, the reads were checked again by FastQC to confirm the quality. Third, using the transcriptome sequence of the publicly available sweetpotato diploid wild species *I. trifida* [22] as the reference, the preprocessed reads were aligned using Bowtie2 software [37]. Fourth, the gene expression levels were determined using eXpress (https://pachterlab.github.io/eXpress/index.html, accessed on 24 June 2019), and the DEGs were analyzed using edgeR [38]. Fifth, DEGs with false-discovery rate values < 0.05 and |log2 fold change| values > 2 were extracted for subsequent GO and enrichment analyses. Sixth, an in-house Python script was used to create a heatmap of the DEGs. Briefly, the transcripts per kilobase million (TPM) values from eXpress analysis were averaged within the iteration and then logarithmically converted and normalized to create a heatmap. The Venn diagram was generated using the online tool VENNY (v2.1) [39].

For the GO analysis, the FASTA sequences of the DEGs were imported into OmicsBox version 1.2 (BioBam) and aligned to the NCBI Viridiplantae NR database using blastx search (E-value $\leq 1.0 \times 10^{-3}$). Subsequent GO mapping was performed using the Blast2GO mapping against the latest version of the GO database to obtain the functional labels [40,41]. Then, the appropriate GO term was assigned to predict the function of the annotated sequences using an e-value cutoff of $1.0 \times 10^{-6}$ and an annotation cutoff of 55. Bar plots of the enriched GO terms were created for three categories: biological process, cellular component, and molecular function. For the candidate genes involved in weevil resistance, the homology of their amino acid sequences was confirmed using BioEdit. In addition, the functional gene domains were searched against several protein databases, including ProDom (http://prodom.prabi.fr, accessed on 5 November 2019) [42], Pfam (https://pfam.xfam.org, accessed on 5 November 2019) [43], SMART (http://smart.embl-heidelberg.de, accessed on 5 November 2019) [44], and PANTHER (http://pantherdb.org/, accessed on 5 November 2019) [45], using InterProScan [46].

**Supplementary Materials:** The following are available online at https://www.mdpi.com/article/10.3390/plants10081535/s1, Figure S1: The top GO terms for the upregulated DEGs in K166 in the control plot, Figure S2: The gene expression levels of the two candidate genes (*itf09g05600.t1* and *itf09g05580.t1*) based on the TPM values, Figure S3: The sequence homology of the amino acid sequences of the two genes (*itf09g05600.t1* and *itf09g05580.t1*), Figure S4: The gene structure and functional domains of the two genes, Figure S5: Hierarchical clustering and expression heatmap of 30 disease resistance-related genes, Figure S6: The breeding process and pedigree information of K166 and Tamayutaka, Figure S7: The enriched GO terms of the identified DEGs between the weevil inoculation and control plots, Table S1: The number of raw and preprocessed reads obtained through RNA sequencing, Table S2: Alignment rate of RNA sequencing reads, Table S3: Summary of the identified DEGs, Table S4: Upregulated genes between the weevil inoculation and control plots of Kyushu No. 166 (K166), Table S5: Upregulated genes between the weevil inoculation and control plots of Tamayutaka, Table S6: Downregulated genes between the weevil inoculation and control plots of Kyushu No. 166 (K166), Table S7: Downregulated genes between the weevil inoculation and control plots of Tamayutaka, Table S8: Upregulated genes between the control plots of Kyushu No. 166 (K166) and Tamayutaka, Table S9: Downregulated genes between the control plots of Kyushu No. 166 (K166) and Tamayutaka, Table S10: Upregulated genes between the weevil inoculation plots of Kyushu No. 166 (K166) and Tamayutaka, Table S11: Downregulated genes between the weevil inoculation plots of Kyushu No. 166 (K166) and Tamayutaka.

**Author Contributions:** Conceptualization, Y.O. and Y.M.; methodology, K.N. and Y.O.; data analysis, K.N. and S.O.; funding acquisition, Y.O.; writing and editing, K.N., Y.O. and Y.M. All authors have read and agreed to the published version of the manuscript.

**Funding:** This research was funded by NARO, grant number C20805.

**Institutional Review Board Statement:** Not applicable.

**Informed Consent Statement:** Not applicable.

**Data Availability Statement:** The RNA sequencing datasets were submitted to the DDBJ database. The accession number is DRA012193.

**Acknowledgments:** We thank Seizo Nakama, Nonoka Shinzato, and Chie Shimabukuro for their support in plant cultivation and investigation.

**Conflicts of Interest:** The authors declare no conflict of interest.

## References

1. Food and Agriculture Organization of the United Nations (FAO). Available online: http://www.fao.org/ (accessed on 12 July 2021).

2. Ames, T.; Smit, N.E.J.M.; Braun, A.R.; O'Sullivan, J.N.; Skoglun, L.G. *Sweetpotato: Major Pests, Diseases, and Nutritional Disorders*; International Potato Center (CIP): Lima, Peru, 1996; ISBN 92-9060-187-6.
3. Smith, T.P.; Hammond, A.M.; Story, R.N.; Burris, E. *Insect Pest Management in Louisiana Sweet Potatoes*; Pub. No. 2620; Louisiana State University Agricultural Center: Baton Rouge, LA, USA, 2008.
4. Sherman, M.; Tamahiro, M. The sweetpotato weevils in Hawaii: Their biology and control. *Hawaii Agric. Exp. Stn. Tech. Bull.* **1954**, *23*, 1–36.
5. Okada, Y.; Kobayashi, A.; Tabuchi, H.; Kuranouchi, T. Review of major sweetpotato pests in Japan, with information on resistance breeding programs. *Breed. Sci.* **2017**, *67*, 7382. [CrossRef] [PubMed]
6. Akazawa, T.; Uritani, L.; Kubota, H. Isolation of ipomeamarone and two coumarin derivatives from sweetpotato roots injured by the weevil *Cylas formicarius elegantulus*. *Arch. Biochem. Biophys.* **1960**, *88*, 150–156. [CrossRef]
7. Muyinza, H.; Talwana, H.L.; Mwanga, R.O.M.; Stevenson, P.C. Sweetpotato weevil (*Cylas* spp.) resistance in African sweetpotato germplasm. *Int. J. Pest Manag.* **2012**, *58*, 73–81. [CrossRef]
8. Korada, R.R.; Naskar, S.K.; Prasad, A.R.; Prasuna, A.L.; Jyothi, K.N. Differential volatile emission from sweet potato plant: Mechanism of resistance in sweet potato for weevil *Cylas formicarius* (Fab.). *Curr. Sci.* **2010**, *99*, 1597–1601.
9. Nottingham, S.F.; Son, K.C.; Wilson, D.D.; Severson, R.F.; Kays, S.J. Feeding and oviposition preferences of sweet potato weevil, *Cylas formicarius elegantulus* (Summers), on storage roots of sweet potato cultivars with differing surface chemistries. *J. Chem. Ecol.* **1989**, *15*, 895–903. [CrossRef]
10. Yada, B.; Alajo, A.; Ssemakula, G.N.; Brown-Guedira, G.; Otema, M.A.; Stevenson, P.C.; Mwanga, R.O.M.; Yencho, C. Identification of simple sequence repeat markers for sweetpotato weevil resistance. *Euphytica* **2017**, *213*, 129. [CrossRef]
11. Prentice, K.; Pertry, I.; Christiaens, O.; Bauters, L.; Bailey, A.; Niblett, C.; Ghislain, M.; Gheysen, G.; Smagghe, G. Transcriptome analysis and systemic RNAi response in the African sweetpotato weevil (*Cylas Puncticollis*, Coleoptera, Brentidae). *PLoS ONE* **2015**, *10*, e0115336. [CrossRef] [PubMed]
12. Okada, Y.; Monden, Y.; Nokihara, K.; Shirasawa, K.; Isobe, S.; Tahara, M. Genome-wide association studies (GWAS) for yield and weevil resistance in sweet potato (*Ipomoea batatas* (L.) Lam). *Plant Cell Rep.* **2019**, *38*, 1383–1392. [CrossRef] [PubMed]
13. Wang, Z.; Fang, B.; Chen, J.; Zhang, X.; Luo, Z.; Huang, L.; Chen, X.; Li, Y. *De novo* assembly and characterization of root transcriptome using Illumina paired-end sequencing and development of cSSR markers in sweet potato (*Ipomoea batatas*). *BMC Genom.* **2010**, *11*, 726. [CrossRef] [PubMed]
14. Xie, Z.; Wang, A.; Li, H.; Yu, J.; Jiang, J.; Tang, Z. High Throughput deep sequencing reveals the important roles of microRNAs during sweetpotato storage at chilling temperature. *Sci. Rep.* **2017**, *7*, 16578. [CrossRef] [PubMed]
15. Ponniah, S.K.; Thimmapuram, J.; Bhide, K.; Kalavacharla, V.K.; Manoharan, M. Comparative analysis of the root transcriptomes of cultivated sweetpotato (*Ipomoea batatas* [L.] Lam) and its wild ancestor (*Ipomoea trifida* [Kunth] G. Don). *BMC Plant Biol.* **2017**, *17*, 9. [CrossRef]
16. Ji, C.Y.; Chung, W.H.; Kim, H.S.; Jung, W.Y.; Kang, L.; Jeong, J.C.; Kwak, S.S. Transcriptome profiling of sweetpotato tuberous roots during low temperature storage. *Plant Physiol. Biochem.* **2017**, *112*, 97–108. [CrossRef]
17. Lau, K.H.; del Rosario Herrera, M.; Crisovan, E.; Wu, S.; Fei, Z.; Khan, M.A.; Buell, C.R.; Gemenet, D.C. Transcriptomic analysis of sweet potato under dehydration stress identifies candidate genes for drought tolerance. *Plant Direct* **2018**, *2*, e00092. [CrossRef] [PubMed]
18. Arisha, M.H.; Ahmad, M.Q.; Tang, W.; Liu, Y.; Yan, H.; Kou, M.; Wang, X.; Zhang, Y.; Li, Q. RNA-sequencing analysis revealed genes associated drought stress responses of different durations in hexaploid sweet potato. *Sci. Rep.* **2020**, *10*, 12573. [CrossRef] [PubMed]
19. Qin, Z.; Qin, Z.; Hou, F.; Hou, F.; Li, A.; Li, A.; Dong, S.; Dong, S.; Huang, C.; Wang, Q.; et al. Comparative analysis of full-length transcriptomes based on hybrid population reveals regulatory mechanisms of anthocyanin biosynthesis in sweet potato (*Ipomoea batatas* (L.) Lam). *BMC Plant Biol.* **2020**, *20*, 299. [CrossRef]
20. Suematsu, K.; Tanaka, M.; Kurata, R.; Kai, Y. Comparative transcriptome analysis implied a ZEP paralog was a key gene involved in carotenoid accumulation in yellow-fleshed sweetpotato. *Sci. Rep.* **2020**, *10*, 20607. [CrossRef] [PubMed]
21. Hirakawa, H.; Okada, Y.; Tabuchi, H.; Shirasawa, K.; Watanabe, A.; Tsuruoka, H.; Minami, C.; Nakayama, S.; Sasamoto, S.; Kohara, M.; et al. Survey of genome sequences in a wild sweet potato, *Ipomoea trifida* (H. B. K.) G. Don. *DNA Res.* **2015**, *22*, 171–179. [CrossRef]
22. Wu, S.; Lau, K.H.; Cao, Q.; Hamilton, J.P.; Sun, H.; Zhou, C.; Eserman, L.; Gemenet, D.C.; Olukolu, B.A.; Wang, H.; et al. Genome sequences of two diploid wild relatives of cultivated sweetpotato reveal targets for genetic improvement. *Nat. Commun.* **2018**, *9*, 4580. [CrossRef]
23. Li, M.; Yang, S.; Xu, W.; Pu, Z.; Feng, J.; Wang, Z.; Zhang, C.; Peng, M.; Du, C.; Lin, F.; et al. The wild sweetpotato (*Ipomoea trifida*) genome provides insights into storage root development. *BMC Plant Biol.* **2019**, *19*, 119. [CrossRef] [PubMed]
24. Doshi, S.S.; Shendage, A.N.; Khyade, V.B. The monoterpene compounds for juvenile hormone activity through changes in pattern of chitin deposition in the integument of fifth instar larvae of silkworm, *Bombyx mori* (L) (PM x CSR2). *World Sci. News* **2016**, *37*, 179–201.
25. Andersen, E.J.; Ali, S.; Byamukama, E.; Yen, Y.; Nepal, M.P. Disease resistance mechanisms in plants. *Genes* **2018**, *9*, 339. [CrossRef]
26. Hettenhausen, C.; Schuman, M.C.; Wu, J. MAPK signaling: A key element in plant defense response to insects. *Insect Sci.* **2015**, *22*, 157–164. [CrossRef] [PubMed]

27. Hanks, S.K.; Quinn, A.M.; Hunter, T. The protein kinase family: Conserved features and deduced phylogeny of the catalytic domains. *Science* **1988**, *241*, 42–52. [CrossRef] [PubMed]
28. Boyd, M.R.; Burka, L.T.; Harris, T.M.; Wilson, B.J. Lung-toxic furanoterpenoids produced by sweet potatoes (*Ipomoea batatas*) following microbial infection. *Biochem. Biophys. Acta* **1973**, *337*, 184–195. [CrossRef]
29. Oguni, I.; Uritani, I. Dehydroipomeamarone as an intermediate in the biosynthesis of Ipomeamarone, a phytoalexin from sweetpotato root infected with *Ceratocystis fimbriata*. *Plant Physiol.* **1974**, *53*, 649–652. [CrossRef]
30. Wilson, B.J.; Boyd, M.R.; Harris, T.M.; Yang, D.T.C. A lung oedema factor from mouldy sweet potatoes (*Ipomoea batatas*). *Nature* **1971**, *231*, 52–53. [CrossRef]
31. Chen, L.J.; DeRose, E.F.; Burka, L.T. Metabolism of furans in vitro: Ipomeanine and 4-ipomeanol. *Chem. Res. Toxicol.* **2006**, *19*, 1320–1329. [CrossRef]
32. Kayukawa, T.; Minakuchi, C.; Namiki, T.; Togawa, T.; Yoshiyama, M.; Kamimura, M.; Mita, K.; Imanishi, S.; Kiuchi, M.; Ishikawa, Y.; et al. Transcriptional regulation of juvenile hormone-mediated induction of Krüppel homolog 1, a repressor of insect metamorphosis. *Proc. Natl. Acad. Sci. USA* **2012**, *109*, 11729–11734. [CrossRef]
33. Kayukawa, T.; Nagamine, K.; Ito, Y.; Nishita, Y.; Ishikawa, Y.; Shinoda, T. Krüppel homolog 1 inhibits insect metamorphosis via direct transcriptional repression of broad-complex, a pupal specifier gene. *J. Biol. Chem.* **2016**, *291*, 1751–1762. [CrossRef]
34. Kayukawa, T.; Jouraku, A.; Ito, Y.; Shinoda, T. Molecular mechanism underlying juvenile hormone-mediated repression of precocious larval-adult metamorphosis. *Proc. Natl. Acad. Sci. USA* **2017**, *114*, 1057–1062. [CrossRef]
35. Andrews, S. FastQC: A Quality Control Tool for High Throughput Sequence Data. 2010. Available online: http://www.bioinformatics.babraham.ac.uk/projects/fastqc/ (accessed on 8 April 2019).
36. Martin, M. Cutadapt removes adapter sequences from high-throughput sequencing reads. *EMBnet J.* **2011**, *17*, 10–12. [CrossRef]
37. Langmead, B.; Salzberg, S.L. Fast gapped-read alignment with Bowtie 2. *Nat. Methods* **2012**, *9*, 357–359. [CrossRef]
38. Robinson, M.D.; McCarthy, D.J.; Smyth, G.K. edgeR: A Bioconductor package for differential expression analysis of digital gene expression data. *Bioinformatics* **2009**, *26*, 139–140. [CrossRef] [PubMed]
39. Oliveros, J.C. (2007–2015) Venny. An Interactive Tool for Comparing Lists with Venn's Diagrams. Available online: https://bioinfogp.cnb.csic.es/tools/venny/index.html (accessed on 3 July 2021).
40. Conesa, A.; Götz, S.; García-Gómez, J.M.; Terol, J.; Talón, M.; Robles, M. Blast2GO: A universal tool for annotation, visualization and analysis in functional genomics research. *Bioinformatics* **2005**, *21*, 3674–3676. [CrossRef]
41. Götz, S.; García-Gómez, J.M.; Terol, J.; Williams, T.D.; Nagaraj, S.H.; Nueda, M.J.; Robles, M.; Talón, M.; Dopazo, J.; Conesa, A. High-throughput functional annotation and data mining with the Blast2GO suite. *Nucleic Acids Res.* **2008**, *36*, 3420–3435. [CrossRef] [PubMed]
42. Servant, F.; Bru, C.; Carrère, S.; Courcelle, E.; Gouzy, J.; Peyruc, D.; Kahn, D. ProDom: Automated clustering of homologous domains. *Brief. Bioinform.* **2002**, *3*, 246–251. [CrossRef]
43. Mistry, J.; Chuguransky, S.; Williams, L.; Qureshi, M.; Salazar, G.A.; Sonnhammer, E.L.L.; Tosatto, S.C.E.; Paladin, L.; Raj, S.; Richardson, L.J.; et al. Pfam: The protein families database in 2021. *Nucleic Acids Res.* **2020**, *49*, D412–D419. [CrossRef]
44. Schultz, J.; Milpetz, F.; Bork, P.; Ponting, C.P. SMART, a simple modular architecture research tool: Identification of signaling domains. *Proc. Natl. Acad. Sci. USA* **1998**, *95*, 5857–5864. [CrossRef]
45. Mi, H.; Ebert, D.; Muruganujan, A.; Mills, C.; Albou, L.P.; Mushayamaha, T.; Thomas, P.D. PANTHER version 16: A revised family classification, tree-based classification tool, enhancer regions and extensive API. *Nucleic Acids Res.* **2020**, *49*, D394–D403. [CrossRef]
46. Zdobnov, E.M.; Apweiler, R. InterProScan—An integration platform for the signature-recognition methods in InterPro. *Bioinformatics* **2001**, *17*, 847–848. [CrossRef] [PubMed]

*plants*

MDPI

*Article*

# Analysis of Genetic Diversity and Resistance to Foliar Pathogens Based on Genotyping-by-Sequencing of a Para Rubber Diversity Panel and Progeny of an Interspecific Cross

**C. Bindu Roy [1,*,†], Shashi N. Goonetilleke [2,*,†], Limiya Joseph [1], Anu Krishnan [1], Thakurdas Saha [1], Andrzej Kilian [3] and Diane E. Mather [2]**

1 Rubber Research Institute of India, Kottayam 686 009, India
2 School of Agriculture, Food and Wine, Waite Research Institute, The University of Adelaide, Glen Osmond, SA 5064, Australia
3 Diversity Arrays Technology, Canberra, ACT 2617, Australia
* Correspondence: binduroy@rubberboard.org.in (C.B.R.); shashi.goonetilleke@adelaide.edu.au (S.N.G.)
† These authors contributed equally to this work.

**Abstract:** Para rubber trees (*Hevea brasiliensis*) are the largest major source of natural rubber in the world. Its major pathogens are *Phytophthora* spp., *Corynespora cassiicola*, and *Colletotrichum* spp. A rubber diversity panel of 116 clones using over 12,000 single nucleotide polymorphisms (SNPs) from DArTSeq genotyping revealed clear phylogenetic differences in clones that originated from different geographical regions of the world. An integrated linkage map constructed with an $F_1$ progeny of 86 from an interspecific cross between *H. brasiliensis* and *H. benthamiana* using 23,978 markers [10,323 SNPs and 13,655 SilicoDArTs] spanned 3947.83 cM with 0.83 cM average marker-interval. The genome scaffolds that were anchored to the linkage map, covering 1.44 Gb of *H. brasiliensis* reference genome, revealed a high level of collinearity between the genetic map and reference genome. Association analysis identified 12 SNPs significantly associated with the resistance against *Phytophthora*, *Corynespora*, and *Colletotrichum* in six linkage groups: 2, 6, 12, 14, 17, and 18. Kompetitive Allele-Specific PCR marker assays were developed for those 12 SNPs, screened with 178 individuals, and detected clear separation between two genotypes. Within the proximity to those SNPs, 41 potentially key genes that have previously been reported to associate with plant disease resistance were predicted with high confidence.

**Keywords:** disease resistance; genetic map; *Hevea brasiliensis*; KASP assay; marker-trait associations; phylogeny; potentially key disease resistance genes

**Citation:** Roy, C.B.; Goonetilleke, S.N.; Joseph, L.; Krishnan, A.; Saha, T.; Kilian, A.; Mather, D.E. Analysis of Genetic Diversity and Resistance to Foliar Pathogens Based on Genotyping-by-Sequencing of a Para Rubber Diversity Panel and Progeny of an Interspecific Cross. *Plants* **2022**, *11*, 3418. https://doi.org/10.3390/plants11243418

Academic Editors: Ranjith Pathirana and Francesco Carimi

Received: 13 October 2022
Accepted: 26 November 2022
Published: 7 December 2022



## 1. Introduction

*Hevea brasiliensis*, the Para rubber tree, is the only source of latex for the production of natural rubber, which is essential for aviation and other industries. It is a perennial, monoecious outcrossing tree species of the Euphorbiaceae family, with its center of origin in the Amazon forests. The species, which contains 36 chromosomes (2n = 36), is considered to be an amphidiploid that has stabilized its genome during the course of evolution [1]. Domestication of *H. brasiliensis* started in 1876 with Sir Henry Alexander Wickham transporting 70,000 rubber seeds from Brazil to the Royal Botanic Gardens, Kew, England. This collection, which came from a small area near Manaus on the banks of Rio Tapajos in the upper Amazon, represented a very small proportion of the gene pool available throughout the Amazon [2]. Only 2700 of these seeds germinated. Most of the seedlings were shipped to Asian countries: 1919 to Sri Lanka (of which 90% survived), 18 to Indonesia (of which two survived) and 50 to Singapore (none of which survived). Subsequently, 22 additional plants were sent to Singapore, all of which survived. The surviving plants from these shipments formed the basis of the rubber plantation industry in the East today [3,4]. This narrow genetic base was further constricted by the use of a limited number of high-yielding clones for propagation and by the use of cyclical generation-wise assortative mating and selection for yield in breeding [5].

High-yielding *Hevea* clones are highly susceptible to biotic and abiotic stresses, which can significantly affect latex production. Biotic stress is mainly due to three major pathogens: *Phytophthora* spp., which causes shoot rot, abnormal leaf fall, patch canker, and black stripe diseases; *Corynespora cassiicola*, which causes *Corynespora* leaf fall disease; *Colletotrichum* spp., which causes *Colletotrichum* leaf fall disease [6,7] (Supplementary Figure S1). Abiotic factors such as unfavorable climatic conditions, which adversely affect the growth and yield of rubber plants, act as predisposing factors for these diseases. Leaf fall caused by these diseases directly reduces plant growth and latex yield. Moderate to severe crop losses due to these diseases occur in all rubber-growing countries. Crop losses due to abnormal leaf fall disease in India, *Colletotrichum* leaf disease in Africa, and *Corynespora* leaf fall disease in China have been reported to range from 38–56% [8], from 7–45% [9], and from 20–25% [10], respectively. At present, the most extensively used management strategy for controlling these diseases is through the recurrent use of agrochemicals. While disease control achieved using genetics might provide sustainable intensification of crop production, introgression of disease resistance genes using backcrossing and phenotypic selection would be difficult for rubber because of its highly heterozygous nature, long breeding cycle, and the large area of land required for evaluation at each stage. Genetic mapping of resistance loci could enable the discovery of molecular markers that can be used to select resistant seedlings in breeding nurseries.

The first linkage map of *Hevea* was constructed for an interspecific cross between *H. brasiliensis* and *H. benthamiana*, using restriction fragment length polymorphism (RFLP) and amplified fragment length polymorphism (AFLP) markers [11]. Subsequently, simple sequence repeat (SSR) markers were mapped for several cross combinations [12–16]. Recently, genotyping-by-sequencing (GBS) and transcriptome sequencing have been used to construct high-density linkage maps for intraspecific crosses of *H. brasiliensis* [17,18]. Intraspecific crosses of *H. brasiliensis* have been used to map quantitative trait loci (QTLs) for growth-related traits [15,19] and for resistance against South American leaf blight [13]. No such mapping has been conducted for resistance against the other major diseases of rubber.

In the research that is presented here, we investigated genetic diversity and the genetic control of resistance to foliar diseases by applying DArTSeq genotyping-by-sequencing to a panel of *H. brasiliensis* clones and to progeny from an interspecific (*H. brasiliensis* × *H. benthamiana*) cross and conducted phylogenetic analysis and integrated genetic linkage mapping.

## 2. Results

### 2.1. Genotyping-by-Sequencing, Phylogenetic Analysis, and Linkage Mapping

DArTSeq genotyping-by-sequencing of a 116-member diversity panel (Supplementary Table S1) and a set of 86 interspecific progeny (derived from crosses between *H. brasiliensis* clone RRII 105 and *H. benthamiana* clone F4542) generated 14,315 single nucleotide polymorphisms (SNPs) and 34,000 presence-absence SilicoDArT markers (Supplementary Table S2). The 116-member diversity panel consisted of 110 *H. brasiliensis* clones, one each of five other *Hevea* species (*H. benthamiana* F4542, *H. pauciflora*, *H. nitida*, *H. spruceana*, *H. camargoana*) and an interspecific hybrid FX 516 (*H. benthamiana* x *H. brasiliensis* AVROS 363).

For the diversity panel, phylogenetic analysis was performed using data for 12,078 SNPs. Potential scale reduction factor values obtained using MrBayes Bayesian phylogenetic trees were between 1.0 and 1.1, and other examined parameters were within acceptable ranges. All *H. brasiliensis* clones were differentiated from individual clones of each of five other species (*H. benthamiana*, *H. camargoana*, *H. nitida*, *H. pauciflora*, *H. spruceana*) and from FX 516, an *H. benthamiana* × *H. brasiliensis* hybrid. Within *H. brasiliensis*, there were two main clades, one consisting of clones from Sri Lanka, Indonesia, and India, and the other consisting of clones from Malaysia and China. The clones from Malaysia showed the greatest diversity. The three clones from Indonesia were more similar to the clones from India than to the clones from Sri Lanka (Figure 1).

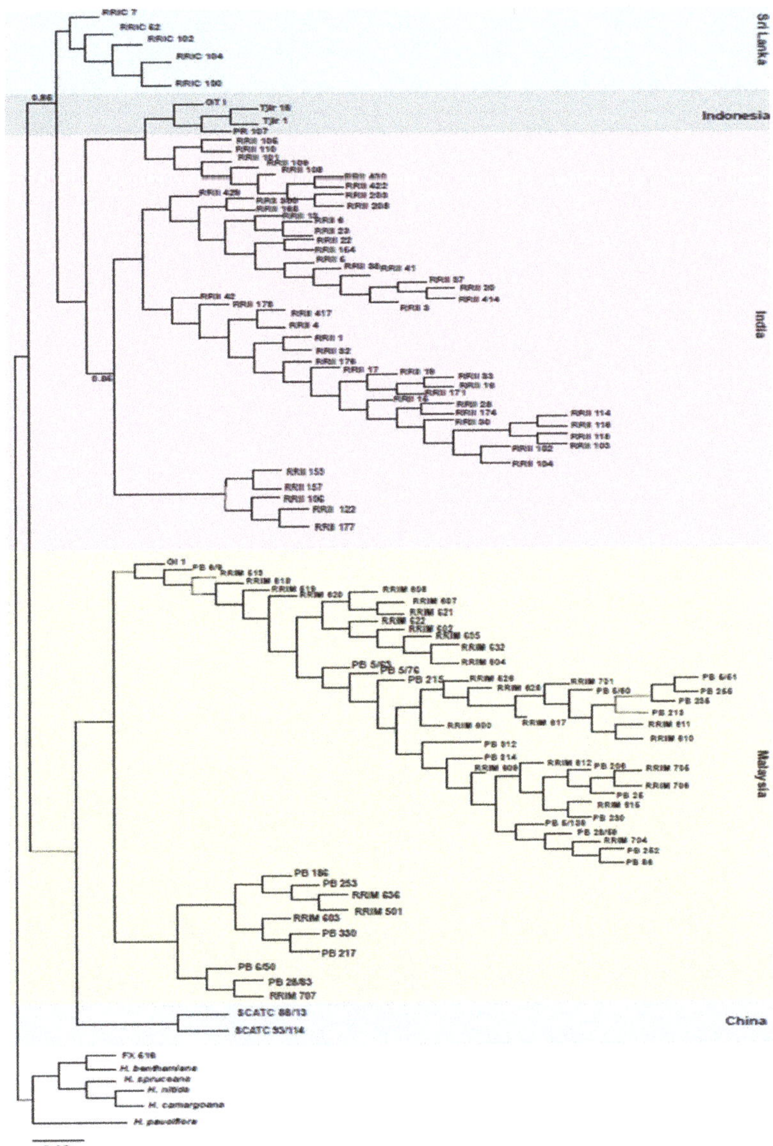


**Figure 1.** Phylogenetic tree constructed for 116 clones using SNP markers obtained from DArTSeq. Values indicated on the nodes are Bayesian posterior probabilities and are 1.0 unless otherwise indicated. The scale bar represents the probability of nucleotide substitutions per site.

Integrated linkage mapping performed with the software Lep-Map3 [20], using data for 23,978 markers (10,323 SNPs and 13,655 SilicoDArTs), resulted in a 3948 cM linkage map with 4757 loci across 18 linkage groups (Figure 2) (Supplementary Tables S3 and S4). Of the 23,978 mapped markers, 17,310 (72.2%) were successfully anchored to positions on pseudomolecules in the GT1 reference genome assembly for *H. brasiliensis* [21], and a further

451 (1.9%) were anchored to unplaced GT1 scaffolds (Supplementary Table S5). Among the remaining markers, 3532 were anchored to positions in an earlier draft assembly [22], bringing the total percentage of anchored markers to 88.8%.

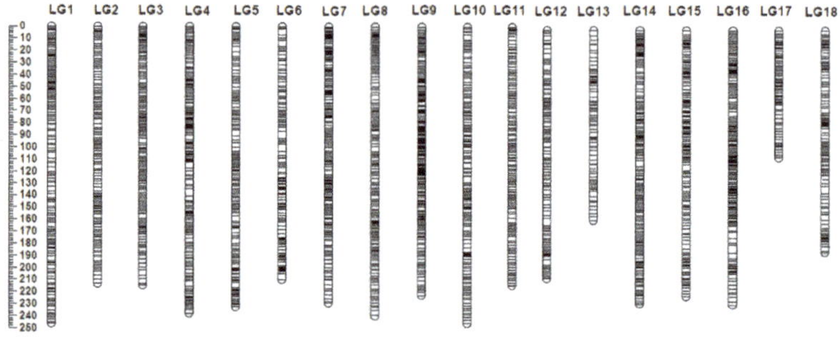


**Figure 2.** Positions at which markers were mapped on an 18-linkage-group integrated map constructed from genotyping-by-sequencing data for 178 RRII 105 × F4542 progeny. The scale to the left of the map shows genetic distances in cM.

A comparison of marker positions between the linkage map and the GT1 reference genome assembly indicated very high synteny and collinearity (Figures 3 and 4, Supplementary Table S6). For some chromosomes (e.g., LG5 in Figure 4), the relationship between genetic distances was highly linear. For some other chromosomes, this relationship was non-linear, indicating a non-uniform distribution of recombination along chromosomes. For example, the terminal regions of the chromosome LG7 are more recombinogenic than the central region of that chromosome, and the opposite is true for LG8 (Figure 4).

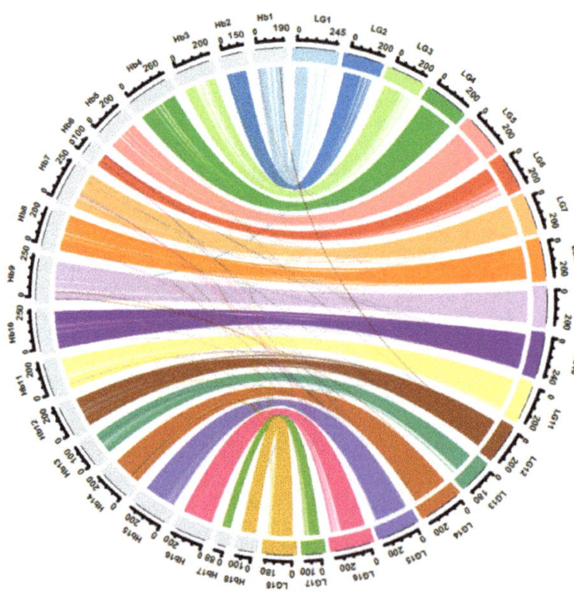

**Figure 3.** Synteny and collinearity of 17,310 markers between the RRII 105 × F4542 linkage map and the reference genome of *Hevea brasiliensis*. The outer circle indicates the genetic lengths of linkage groups LG1 through LG18 in cM and the physical lengths of pseudomolecules Hb1 through Hb18 in Mb.

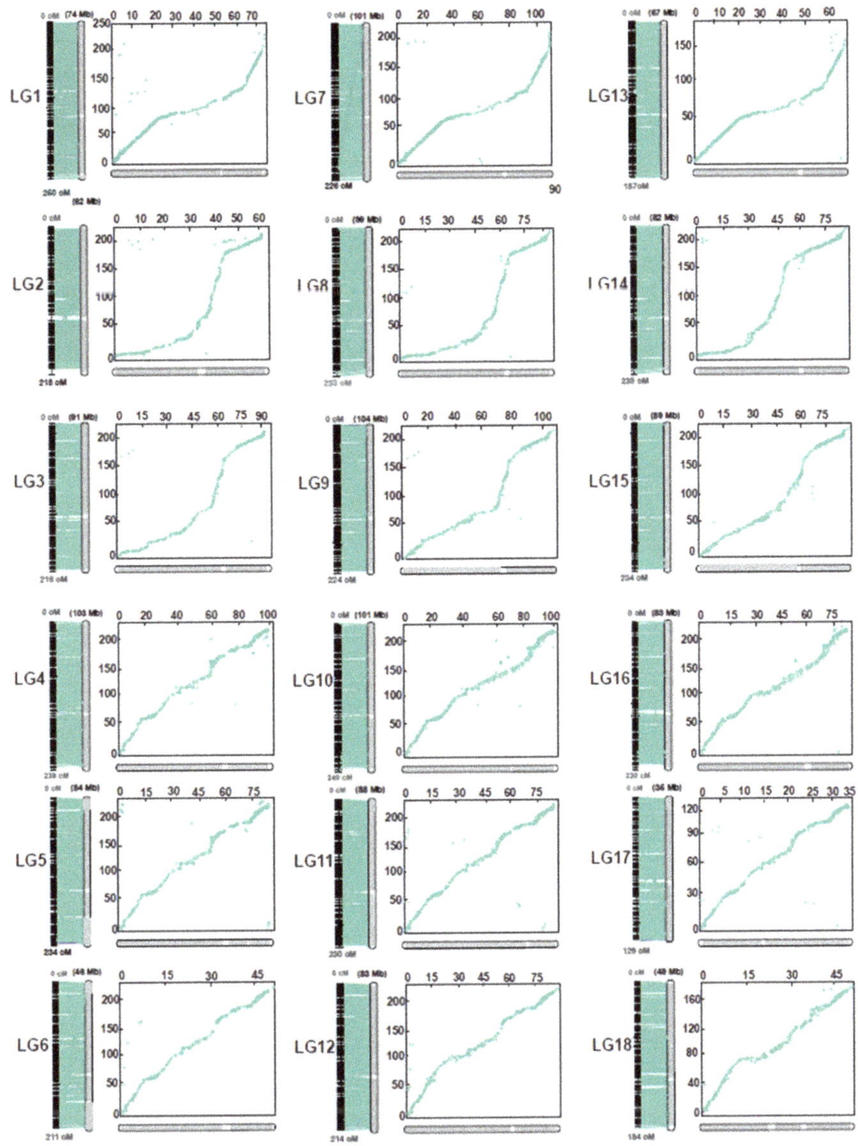


**Figure 4.** Relationship between the linkage map and the reference genome of *Hevea brasiliensis*. Each panel shows the genetic positions of markers on the linkage map (vertical axis, in cM) plotted against the physical positions of the same markers on the corresponding pseudomolecule of the reference genome (horizontal axis, in Mb). The right part represents the genetic and physical locations of the markers.

*2.2. Disease Resistance*

For *Phytophthora* spp., which infects rubber pods, petioles, and mature leaves, the *H. brasiliensis* parent RRII 105 was classified as susceptible (rated 4 on a 5-point scale), and the *H. benthamiana* parent F4542 was classified as resistant (rated 2). Among 85 RRII 105 × F4542 $F_1$ progeny assessed for resistance against *Phytophthora*, seven (8.2%) were highly resistant (rated 1), 22 (25.9%) were resistant (rated 2), 17 (20%) were moderately resistant (rated 3), 28 (32.9%) were susceptible (rated 4), and 11 (12.9%) were highly susceptible (rated 5) (Supplementary Figure S2).

For *Corynespora cassiicola* and *Colletotrichum acutatum*, which infect rubber tree leaves in the tender copper brown and light green stages, the *H. brasiliensis* parent RRII 105 was classified as highly susceptible (rated 5: leaf wilting always observed within 24 h of incubation with toxin extracts) and the *H. benthamiana* parent F4542 was classified as highly resistant (rated 1: no leaf wilting after 24 h of incubation with toxin extracts). Among 79 progeny tested with toxin extracts from *Corynespora cassiicola* and *Colletotrichum acutatum* extract, 43 (56.6%) were assigned intermediate ratings (2, 3, or 4) for both pathogens (Table 1).

**Table 1.** Numbers (and percentages) of *H. brasiliensis* x *H. benthamiana* $F_1$ progeny classified between 1 (highly resistant) and 5 (highly susceptible) for resistance against *Colletotrichum acutatum* and *Corynespora cassiicola*.

| *Colletotrichum acutatum* Resistance Classification | *Corynespora cassiicola* Resistance Classification | | | | | Total |
|---|---|---|---|---|---|---|
| | 1 | 2 | 3 | 4 | 5 | |
| 1 | 2 | 2 | 2 | 1 | 2 | 9 (11.3%) |
| 2 | 4 | 8 | 5 | 6 | 3 | 26 (32.9%) |
| 3 | 1 | 9 | 4 | 6 | 5 | 25 (31.6%) |
| 4 | 1 | 5 | 3 | 4 | 1 | 14 (17.8%) |
| 5 | 0 | 1 | 2 | 1 | 1 | 5 (6.4%) |
| **Total** | 8 (10.1%) | 25 (31.6%) | 16 (20.2%) | 18 (22.8%) | 12 (15.2%) | 79 (100%) |

Based on marker-trait association analysis, 12 SNPs were found to be significantly ($p < 1.0 \times 10^{-4}$) associated with one or more disease traits (Figure 5; Table 2): five (one on LG6 and four on LG18) for *Phytophthora* resistance, three (one on each of LGs 2, 4, and 18) for *Corynespora* resistance, and four (one on each of LGs 13, 14, and 17) for *Colletotrichum* resistance. In all cases, the favorable (resistance-associated) allele was from the *H. benthamiana* parent F4542. Of the 12 SNPs, 10 were anchored to the GT1 reference genome, and the other two were anchored to the rubber draft genome sequences. Except for one of the 12 markers, all were heterozygous in *H. benthamiana*.

**Table 2.** Markers that showed significant associations with disease traits in the integrated map of *Hevea brasiliensis* and *H. benthamiana*.

| Pathogen Associated with the Disease Trait | Linkage Group | Marker | Reference Genome Sequence | Physical Position (bp) | | $-Log_{10}$ $(p)$ | $R^2$ (%) (Variance Explained) | Allele Effect |
|---|---|---|---|---|---|---|---|---|
| | | | | Start | End | | | |
| *Phytophthora* | LG6 | 100057258 \| F \| 0–6:T > C | KB619684.1_ scaffold 218036 | 28,834 | 28,903 | 4.28 | 14 | 1.10 |
| | LG18 | 100033008 \| F \| 0–18:A > T | GT1 Reference genome (CM021243.1) | 18,128,717 | 18,128,649 | 4.52 | 18 | 3.21 |
| | | 100061510 \| F \| 0–34:C > A | GT1 Reference genome (CM021243.1) | 12,918,745 | 12,918,685 | 4.25 | 14 | 2.57 |
| | | 100053390 \| F \| 0–45:A > G | GT1 Reference genome (CM021243.1) | 15,754,360 | 15,754,413 | 4.25 | 14 | 2.57 |
| | | 100052000 \| F \| 0–31:C > T | GT1 Reference genome (CM021243.1) | 15,953,978 | 15,953,939 | 4.21 | 14 | 2.32 |

Table 2. Cont.

| Pathogen Associated with the Disease Trait | Linkage Group | Marker | Reference Genome Sequence | Physical Position (bp) Start | End | −Log10 (p) | $R^2$ (%) (Variance Explained) | Allele Effect |
|---|---|---|---|---|---|---|---|---|
| *Corynespora* | LG4 | 100104859 \| F \| 0–12:G > A | GT1 Reference genome (CM021229.1) | 34,907,019 | 34,906,974 | 4.21 | 14 | 2.30 |
| | LG6 | 100061575 \| F \| 0–46:G > A | GT1 Reference genome (CM021231.1) | 37,466,880 | 37,466,948 | 4.35 | 16 | 2.43 |
| | LG18 | 100074003 \| F \| 0–8:G > T | GT1 Reference genome (CM021243.1) | 44,256,460 | 44,256,528 | 4.35 | 16 | 2.41 |
| *Colletotrichum* | LG2 | 100085293 \| F \| 0–45:A > T | AJJZ010325919.1_ contig 401252 | 1511 | 1580 | 4.13 | 11 | 1.25 |
| | LG12 | 100044284 \| F \| 0–59:A > C | GT1 Reference genome (CM021237.1) | 73,830,891 | 73,830,823 | 4.26 | 14 | 1.56 |
| | LG14 | 100026178 \| F \| 0–30:G > T | GT1 Reference genome (CM021239.1) | 6,682,648 | 6,682,714 | 4.48 | 16 | 0.98 |
| | LG17 | 100113203 \| F \| 0–20:T > G | GT1 Reference genome (CM021242.1) | 2,385,929 | 2,385,861 | 4.21 | 16 | 0.97 |

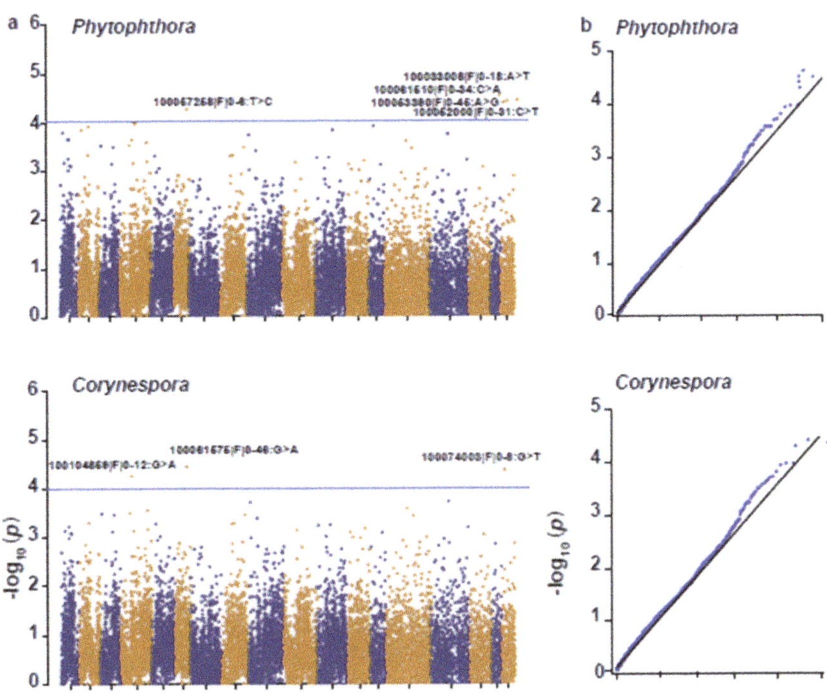


Figure 5. Cont.

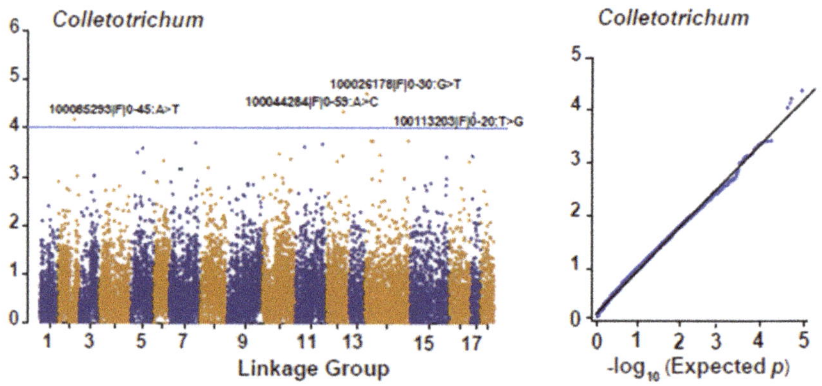


**Figure 5.** Manhattan plots and quantile-quantile plots from marker-trait association analysis for *Phytophthora, Corynespora,* and *Colletotrichum* resistance traits. In the Manhattan plots (**a**), each dot represents an SNP, showing the significance of its association with disease resistance ($-\log_{10}$ *p*-value) (vertical axis) plotted against its genetic position (horizontal axis), and the horizontal blue lines indicate the genome-wide significance threshold of $p = 1.0 \times 10^{-4}$. In the quantile-quantile plots (**b**), the black line represents the 95% confidence limit under the null hypothesis of no-marker-trait association, and the black dots represent *p*-values.

For the results of this work to be applied in marker-assisted rubber breeding, it could be useful to have simple assays for trait-associated SNPs. We designed Kompetitive Allele-Specific PCR (KASP) assays [23] for all 12 SNPs that showed significant associations with the disease traits (Supplementary Table S7). Those assays were then applied to the 178 RRII 105 × F4542 $F_1$ progeny, and genotypes were clearly separated from each other (Figure 6; Supplementary Figure S3). We looked at the two trait-associated SNPs on LG6 further, and for each of the SNPs, two genotypes (one heterozygous and one homozygous) had been observed in the DArTSeq results. With each of the KASP assays (WriKH1 for 100057258 | F | 0–6:T > C and WriHK7 for 100061575 | F | 0–46:G > A), both genotypes were observed and were clearly separated from each other (Figure 6).

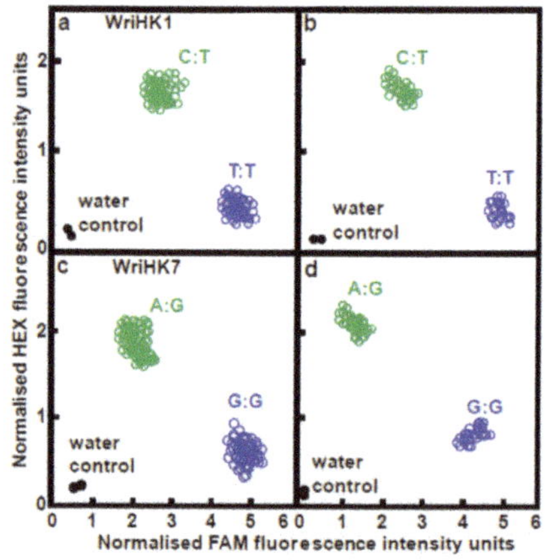


**Figure 6.** Results obtained with markers 100057258 | F | 0–6:T > C (WriKH1) and 100061575 | F | 0–46:G > A (WriHK7) on LG6. For both markers, left panels (**a**,**c**) show the marker results obtained for 178 *H. brasiliensis* × *H. benthamiana* progeny using KASP markers, and right panels (**b**,**d**) show the marker results obtained for 86 progeny using DArTSeq.

For all 12 SNPs, when we compared the genotypic scores obtained from the DArTSeq with the KASP assays, the discrepancies varied from two to three: two for the marker 100061575 | F | 0–46:G > A (WriHK1) and three for the marker 100057258 | F | 0–6:T > C (WriKH7) (Supplementary Table S8). For four of these, DArTSeq calls were homozygous, and KASP calls were heterozygous. For the other 10 SNPs, there were no discrepancies between the genotypic scores obtained from the DArTSeq and the KASP assays.

Of the 178 $F_1$ progeny studied, four of the F1 plants (9_26, 9_39, 9_76, and 13_1_34) carried the favorable alleles of *H. benthamiana* for all three disease traits. In addition, five of the F1 plants carried the favorable *H. benthamiana* alleles for resistance to *Phytophthora*, 22 carried the favorable alleles for *Corynespora*, and seven carried the favorable alleles for *Colletotrichum* (Supplementary Table S9).

Of the four progeny with favorable alleles at all 12 SNPs, three had been included in the subset that was phenotyped for all three disease resistance traits. All three were rated as highly resistant to *Phytophthora* (rated 1 with a lesion size < 0.69 cm) and highly resistant and/or resistant to *Corynespora* and *Colletotrichum* (rated 1 and/or 2) (Supplementary Table S9).

### 2.3. Significantly Associated Genes

The number of predicted genes near the trait-associated SNPs ranged from 67 to 36 (Supplementary Table S10). In the *Phytophthora*-associated region between 12 and 18.3 Mb on LG18, 55 genes were predicted with high confidence, including 12 with annotations related to innate immune response, host-pathogen relationships, or plant defense. Among these, XP_021650208.1 (disease resistance protein RPS2), GH714_038730 (BTB/POZ domain and ankyrin repeat-containing protein NPR1-like), XP_021650172.1 (heat shock factor protein HSF30-like), XP_021638503.1 (aquaporin NIP5-1), XP_021689264.1 (transcription factor BHLH089-like), XP_021640619.1 (zinc finger protein 6), XP_021665532.1 (cinnamoyl-CoA reductase 1) and GH714_038687 (WD repeat-containing protein VIP3) are similar to genes that have widely been reported to be associated with plant disease resistance and susceptibility. In the *Corynespora*-associated regions, between 44 and 45 Mb on LG18 and between 36.5 and 37.5 Mb on LG6, 67 and 26 genes were predicted, respectively. Among these, XP_021666909.1 (pathogenesis-related protein 5), XP_021647428.1 (WAT1-related protein At1g44800), XP_021660750.1 (ethylene-responsive transcription factor ERF014), and XP_021642270.1 (LRR receptor-like serine/threonine-protein kinase) are similar to genes that have been reported to be important in disease resistance in other plant species.

In the *Colletotrichum*-associated region between 6 and 6.7 Mb on LG14, 36 genes were predicted. Among these four had annotations associated with innate immune response (calcineurin B-like protein 7), biotic stress (O-acyltransferase WSD1-like), DNA methylation (DNA methylation 4-like), or systemic acquired resistance (laccase gene).

## 3. Discussion

In the research reported here, DArTSeq genotyping-by-sequencing was applied to a diversity panel and to a set of $F_1$ progeny from an interspecific *Hevea* cross. Numerous SNPs and presence-absence polymorphisms were discovered in both sets of materials.

For the diversity panel, the DArTSeq genotypic data were used for phylogenetic analysis. In the resulting phylogenetic tree, *H. brasiliensis* clones clustered according to the country from which they were obtained, despite all cultivated rubber having been derived from a narrow genetic base. The clear differentiation among clones from Sri Lanka, Indonesia, India, Malaysia, and China may involve founder effects (related to the particular seedlings dispersed from England in the 19th century and/or the effects of subsequent breeding efforts in individual countries. It may also indicate that there has been limited international germplasm exchange in rubber breeding. Despite Malaysia not having been the recipient of any of the seeds originally distributed from England, the clones obtained from Malaysia exhibited the greatest diversity, which may reflect the use of parents from multiple sources. Both the number of markers and the number of clones investigated here were much larger than in a previous phylogenetic analysis based on the application of 30 EST-SSR markers to 51 clones [24]. That analysis also showed a separation based on country of origin, with most clones from Malaysia readily differentiated from and less diverse than clones from South America.

The parents of the interspecific cross, the *H. brasiliensis* clone RRII 105 and the *H. benthamiana* clone F4542, were chosen based on a need to improve resistance against foliar diseases that are important in India, combined with an interest in understanding the genetic basis of this resistance. RRII 105 is the most widely cultivated clone in the traditional rubber-growing regions of India. It is high yielding but highly susceptible to *Corynespora cassiicola* and *Colletotrichum acutatum* and moderately susceptible to *Phytophthora meadii*. In contrast, F4542 is low-yielding but highly resistant to all three of these fungal pathogens. In Brazil, F4542 has been used as a parent in breeding for resistance to *Phytophthora* spp. and *Microcyclus* [25].

The integrated map constructed here is by far the most saturated linkage map available for rubber, with between eight and ten times more markers than prior maps [11–18]. This is due to a combination of factors: the use of an interspecific cross, the use of GBS to discover sequence polymorphisms, and the use of Lep-Map 3 software [20], which can generate robust maps from low-coverage datasets. The integrated map generated here exhibited very strong synteny and collinearity with a recently generated reference genome for *H. brasiliensis* [20], providing a solid basis for future forward and reverse approaches to identify causal genes.

KASP assays designed from the DArTSeq tags confirmed, with very few exceptions, the polymorphisms discovered by DArTSeq (Supplementary Table S8). In four of five cases where marker calls differed, KASP calls were heterozygous, and DArTSeq calls were homozygous. This probably indicates that just one of the alleles was sequenced in sufficient depth by DArTSeq genotyping. This has previously been reported in switchgrass [26] and almond [27].

This is the first report of significant marker-trait associations for resistance against foliar diseases caused by *Phytophthora* spp., *Corynespora cassiicola*, and *Colletotrichum acutatum* in rubber. Similar to what has been reported for resistance to several other fungal pathogens of rubber [28–30], continuous variation was observed, and multiple loci of small effect were detected. Among the mapping progeny, four clones carried favorable alleles at all 12 resistance-associated SNPs. Breeding for resistance could begin with the crossing of selected interspecific progeny with high-yielding clones of *H. brasiliensis* (to introgress resistance into productive backgrounds) and/or with intercrossing among selected interspecific progeny (to pyramid resistance alleles at multiple loci). In either case, the DArTSeq genotypes generated here can be used to select parents and design crosses, while molecular marker assays, such as the KASP assays developed here, could be used in early-stage selection for resistance.

Anchoring significantly associated marker sequences to the 6, 14, and 18 pseudomolecules of GT1 rubber reference genome sequence assembly enabled the identification of predicted genes near the trait-associated SNPs (Supplementary Table S10). Given that there is very little known about resistance mechanisms and host-pathogen relationships of rubber foliage diseases, none of these predicted genes can be excluded based on annotation. However, some of the genes detected in the candidate regions belong to gene families that have been reported to be associated with disease resistance, host-pathogen relationships, and innate immune responses in other plant species (Supplementary Table S10). Therefore, this study provides new resources for candidate genes for predicting resistance in rubber foliage diseases that might lead to improvement in the speed of breeding for multi-genic traits and elucidate the molecular mechanisms to combat the devastating foliage diseases in rubber.

## 4. Materials and Methods

### 4.1. Diversity Panel and H. brasiliensis (RRII 105) × H. benthamiana (F4542) Mapping Population

This research used a diversity panel consisting of 116 rubber clones (Supplementary Table S1) that belong to *H. brasiliensis* (54 from Malaysia, 51 from India, 5 from Sri Lanka, four from Indonesia, and 2 from China), each of one sample from *H. benthamiana*, *H. camargoana*, *H. pauciflora*, *H. spruceana*, and *H. nitida* and one sample from an interspecific hybrid between *H. benthamiana* and *H. brasiliensis* and a mapping population generated using an interspecific cross involving *H. brasiliensis* (RRII 105) and *H. benthamiana* (F4542) as the maternal and paternal parents, respectively. RRII 105 is the most popular clone in the traditional rubber-growing regions of India. It has a high yield but is moderately susceptible to *Phytophthora* spp. and highly susceptible to *Corynespora cassiicola* and *Colletotrichum* spp. F4542 has a low yield and is resistant to *Phytophthora* spp., *Corynespora cassiicola*, and *Colletotrichum* spp. The initial population was obtained in 2009 by performing hand pollination at the Rubber Research Institute of India (RRII). Due to the asynchronous flowering of the parents, pollination success was poor, and few progeny were obtained. Crossing was therefore repeated from 2011 to 2015 until sufficient $F_1$ progeny were obtained. The final mapping population consisted of 178 $F_1$ progeny, all of which are maintained in field nurseries using standard management practices.

### 4.2. DNA Extraction and DArTSeq Genotyping-By-Sequencing

Genomic DNA was extracted from 500 mg of lyophilized tender green leaf tissue following the CTAB method [31]. DNA quality was tested by electrophoresis on an ethidium bromide-stained 1% agarose gel, and quantity was measured using a NanoDrop™ spectrophotometer. Non-hybrids were eliminated based on the results of SSR marker *hmCT44* (GenBank Acc. No. AY962210) (Supplementary Figure S4). PCR reaction was carried out in a 10 µL final volume containing 20 ng of genomic DNA, 0.2 µM each of the forward and reverse primers (forward: 5′ TCTCATCCATGCAAGAACCCTA 3′ and reverse: 5′ GCGTTCCCAAATGCATACCT 3′), 200 µM dNTPs and 0.4 U of *Taq* DNA polymerase (GE Healthcare, UK). The thermocycling conditions were initial denaturation for 5 min at 95 °C followed by a touch-down PCR program for 7 cycles of 94 °C for 30 s, 63 °C for 1 min, Δ↓ 1 °C for 7 cycles, and 72 °C for 1 min. This was followed by normal cycling of 94 °C for 30 s, 56 °C for 1 min, 72 °C for 1 min for 23 cycles, and a final extension at 72 °C for 10 min. Once the PCR was completed, reactions were stopped immediately by the addition of 10 µL formamide loading buffer, and the amplification products were run on a 6% denaturing polyacrylamide gel containing 7 M urea using 0.6× TBE buffer at a constant power of 55 W. Gels were silver stained following the protocol described by Roy et al. [32] and samples that showed to be non-hybrids were removed from further analysis. The DArTSeq genotyping-by-sequencing technique (www.diversityarrays.com/dart-application-dartseq, accessed on 22 November 2022) was applied to DNA aliquots of parents and a randomly selected subset of 86 progeny at Diversity Arrays Technology (Bruce, ACT, Australia). In brief, DNA samples were digested with a combination of *Pst* I/*Mse* I restriction enzymes, and then the multiplexed reduced representation library was sequenced using single-end sequencing on an Illumina HiSeq 2500 with running 77 cycles. Data were provided by Diversity Arrays Technology, Australia, with polymorphisms scored as either co-dominant single nucleotide polymorphisms (SNPs) or as dominant (presence/absence) SilicoDArT sequence tags.

### 4.3. Phylogenetic Analysis

Phylogenetic analysis was performed using the DArTSeq data generated from the *Hevea* diversity panel of 116 rubber clones using MrBayes v3.2.6 [33] with the following parameters: a general time reversible model and a gamma-shaped distribution of rates across sites function were used. The Markov Chain Monte Carlo (MCMC) was set to three million generations with a sampling frequency of 100 and 250 burn-in. The analysis used all DArTSeq data from a VCF file with less than 5% missing data and for which the minor allele frequency was at least 5%. For the tree construction, diagnostic parameters such as the Potential Scale Reduction Factor (PSRF), shape of gamma distribution rate variation, and stationary rate frequency were considered to determine if the obtained tree is optimal and trustable. Chain convergence was checked in Tracer version 1.6 [34] by examining the log-likelihood plots, and the effective sample size values were ensured to be well above 200. The resulting phylogenetic tree was visualized with FigTree v.1.4.4 [31].

### 4.4. Construction of Integrated Genetic Map Using Lep-Map3 Software

For the construction of the integrated genetic map, all SNP markers that did not deviate from expected segregation patterns (1:1 or 1:2:1) and all SilicoDArT markers with less than 20% missing data points were selected. VCF tools [35] were used for filtering the VCF file, and SilicoDArT markers were coded as homozygote indels based on the parent, as mentioned in Lep-Map3. An integrated map was generated following the instructions in Lep-Map3 v0.2 [20] software, together with a pedigree file indicating the parents of the controlled cross. The modules in Lep-Map3 included several steps starting from ParentCall2 to remove erroneous or missing parental genotypes; Filtering2 to remove markers with high segregation distortion ($p$-value $< 1 \times 10^{-3}$); SeparateChromosome2 to assign markers into linkage groups by computing all pair-wise LOD scores between markers and to join markers with a user define LOD threshold; JoinSingles2All to assign singular markers to existing LGs by computing LOD scores between each single marker and markers from the existing LGs and OrderMarkers2 to order the markers within each LG by maximizing the likelihood of the data given in the order. To order markers within the linkage group, a LOD score of 5 was used. A total of iteration 100 was used to obtain a final map.

### 4.5. Genome Scaffold Anchoring and Comparative Mapping between the Integrated Map and the Rubber Reference Genome

All unique DArTSeq sequence reads mapped in the integrated map that was at least 64 bp long were aligned with the rubber reference genome sequence [36] using the BLAST+ tool version 2.12.0 (http://www.ncbi.nlm.nih.gov/blast, accessed on 22 November 2022). Each sequence read was considered to have been anchored to the rubber reference genome if it mapped to a unique site with >90% sequence similarity and an E-value $< 1 \times 10^{-15}$. Sequences that met these criteria were selected to compare the genetic positions in the integrated map with physical positions in 18 main pseudomolecules of the rubber reference genome. A circular plot was drawn using the Circlize R package version 0.4.15 [37].

### 4.6. Phenotypic Trait Measurement for Disease Resistance

Disease resistance was assessed for three major pathogens of rubber: *Phytophthora meadii*, *Corynespora cassiicola*, and *Colletotrichum acutatum* through in vitro studies. The parents (RRII 105 and F4542) and 86 $F_1$ progeny were phenotyped. To maintain uniformity in the leaf growth stage during inoculation processes, phenotyping was carried out using the initial population that was crossed in 2009. Due to the non-availability of the physiologically same stage of leaves with some progeny, *Phytophthora* resistance was assessed with 85 progeny, and resistance for *Corynespora* and *Colletotrichum* was assessed with 79 progeny.

#### 4.6.1. Screening for Phytophthora Resistance Using Zoospore Suspension

The plants were cut back for uniform growth and maturity. Mature leaves (two months old) were collected from each progeny and transported to the laboratory with the petiole dipped in water. Leaf discs of 3.5 cm diameter were taken from the leaves using a punching device. Four discs per leaf were taken, with a total of 32 discs per progeny. Leaf discs were placed with their abaxial surface upwards in Petri plates previously lined with three sterilized moist filter papers and kept in the inoculation room at 25 °C. The center of leaf discs was inoculated with 20 µL drops of spore suspension containing $2 \times 10^5$ zoospores/mL [38,39]. Following inoculation, the Petri plates were incubated at a temperature of 24 °C under alternate light and dark conditions. Leaf discs of RRIM 600, a highly susceptible clone, and FX 516, a tolerant clone to abnormal leaf fall (ALF) disease, were maintained as control. Disease severity was assessed by measuring the size of lesions developed on the leaf discs periodically from 72–144 h after inoculation. For *Phytophthora* resistance, the greatest differentiation was found at 96 h inoculation, and severity scores obtained at 96 h were therefore used for subsequent analysis. Screening was repeated three times, and the progeny were classified into five categories based on the necrotic area/lesion size: highly resistant (0.0–0.69 cm); resistant (0.7–1.39 cm); moderately resistant (1.4–2.09 cm); susceptible (2.1–2.79 cm), and highly susceptible (2.8–3.5 cm) (Supplementary Figure S5).

#### 4.6.2. Toxin-Based Screening for *Corynespora* and *Colletotrichum* Resistance

In order to screen for disease resistance of *Corynespora cassiicola* and *Colletotrichum acutatum*, toxin-based screening was employed using toxins extracted from the respective pathogen. Pure single spore cultures of *Corynespora cassiicola* and *Colletotrichum acutatum* were grown on potato dextrose agar medium. Twelve discs of 0.8 cm size from 10-day-old cultures were transferred aseptically to 100 mL of modified Czapek Dox liquid medium [40, 41] and incubated without agitation for 12–14 days in laboratory conditions (25 ± 2 °C). The culture filtrate was extracted using a vacuum filtration unit, and the crude toxin was used for screening. Glass vials of 15 mL capacity were used with 5 mL of crude toxin in a dilution of 1:6 in modified Czapek Dox liquid medium diluted in water. Healthy leaves collected in the morphogenetic stage C (limp, brownish to light green) [42] were excised underwater. The petioles of excised leaflets were immediately transferred to vials containing the diluted toxin. For each of the screening, similar stage leaves from each of the parents (*H. brasiliensis* is highly susceptible and *H. benthamiana* is highly tolerant) were used as controls in modified Czapek Dox liquid medium without crude toxin from pathogens. The leaflets were observed regularly for any signs of drooping [43]. Wilting intensity (water loss estimation) was visually assessed at 24 and 48 h following the treatment. Scoring for disease resistance was made on a scale of 1–5, with 1 indicating high resistance (with leaves remaining fresh even after 48 h of incubation in the toxin) and 5 indicating high susceptibility (with leaves completely wilted after just 24 h of incubation in the toxin) (Supplementary Figures S6 and S7). Ten replicates for each progeny were evaluated, and the experiment was repeated three times.

### 4.7. *Marker Trait Association Analysis Using Integrated Map*

Markers associated with each of these disease resistance traits were identified by performing association mapping analysis using TASSEL V5.2.73 software [44] using the Q-model (GLM Q-matrix as correction for population structure) implemented in General Linear Model (GLM) method. Marker alleles with $p$ values $1 \times 10^{-4}$ were declared significantly associated with each of the disease resistance, and a standard Bonferroni procedure was applied at $p < 0.000001$ [45].

### 4.8. *Development and Use of KASP Assays Where Significant Marker-Trait Association Were Detected*

For the SNPs where significant marker-trait associations were detected, DArTSeq tag sequences were selected for the development of allele-specific assays. First, these sequences were aligned with *Hevea* draft genomic contig sequences using the Map to Reference function in Geneious software version 10.1.8 [46] to obtain sequences of at least 100 bp. Based on these sequences, primer sets (each consisting of two allele-specific primers and one common primer) were designed using Kraken™ software (LGC Ltd., Teddington, UK). The primer sets were named using the prefixes WriHK, with Wri referring to the Waite Research Institute, H for *Hevea*, and K referring to KASP (Supplementary Table S7). Each KASP assay was tested with 178 (the 86 progeny that were used for DArTSeq analysis plus

96 additional progeny) of the *H. brasiliensis* × *H. benthamiana* $F_1$ progeny, with a no-template (water) sample included as a negative control.

For KASP assays, 1.972 μL of 1× KASP Master Mix (LGC Ltd., Teddington, UK) was added to 10 ng of DNA (5 μL of 2 ng/μL dried at 55 °C for 1 h). An aliquot of a primer mixture (0.028 μL, containing 12 μM of allele-specific forward and reverse primers and 30 μM of common primer) was added to each sample. The thermal cycling conditions for fragment amplification comprised two temperature steps in a Hydrocycler-16 PCR system (LGC Ltd., Teddington, UK). DNA was denatured at 94 °C for 15 min, followed by 10 cycles of 94 °C for 20 s, 61–55 °C for 60 s (dropping 0.6 °C per cycle), 26 cycles of 94 °C for 20 s, and 55 °C for 60 s. Fluorescence was detected using a Pherastar® Plus plate reader (BMG LABTECH, Ortenberg, Germany). Three further cycles (94 °C for 20 s and 57 °C for 60 s) were carried out, with fluorescence detected after each cycle. Data from the cycle that yielded the best separation among genotypic clusters were analyzed using Kraken™ software (LGC Ltd., Teddington, UK).

### 4.9. Candidate QTL Regions and Potential Key Genes

For each of these disease traits, the close proximity regions to the SNPs that showed significant marker-trait associations and which aligned with the GT1 rubber genome sequences were defined as the candidate QTL regions. Within those regions, the genes that have been reported to be associated with disease resistance, plant-pathogen relationships, and immune responses were selected as potential key genes for rubber foliage disease resistance traits.

**Supplementary Materials:** The supporting information can be downloaded at: https://www.mdpi.com/article/10.3390/plants11243418/s1. Figure S1: A healthy rubber plantation with dense canopy (a); mature trees infected with *Phytophthora* spp. causing abnormal leaf fall disease (b); young refoliating leaves showing typical symptoms of infection with *Corynespora cassiicola* causing *Corynespora* leaf fall disease (c) and shriveling of tender leaves infected with *Colletotrichum* spp. causing *Colletotrichum* leaf disease (d). Figure S2: *Phytophthora* disease severity measured by the size of lesion. Lesions caused by *Phytophthora meadii* were assessed with 85 progeny from a cross between *H. brasiliensis* and *H. benthamiana*. Five resistance categories are indicated as highly resistant (1), resistant (2), moderately resistant (3), susceptible (4), and highly susceptible (5). Disease severities of the resistant (*H. benthamiana*) and susceptible (*H. brasiliensis*) parents are indicated by triangle and circle, respectively. Figure S3: *Phytophthora* (a), *Corynespora* (b) and *Colletotrichum* (c) disease resistance of progeny based on KASP marker results. Level of resistance measured as size of lesion (a) rate of wilting in the leaf (b and c) caused by the resistant parent (*H. benthamiana*) is marked with a spotted line in black. The progeny (9_39) that had *H. brasiliensis* like genotype seems to carry resistance trait. Figure S4: Assessment for *Phytophthora* disease response in progeny population derived from a cross between *H. brasiliensis* and *H. benthamiana*. The degree of resistance is categorized as highly resistant (HR), resistant (R), moderately resistant (MR), susceptible (S) and highly susceptible (HS) based on size of lesion produced on leaf after 96 h of infection with zoospore suspension. Figure S5: Toxin-based screening for disease resistance to *Corynespora cassiicola*. Tender leaves that are 10 days old were excised and transferred to 5 mL of toxin fraction extracted from *C. cassiicola*. Disease resistance was scored from 1 to 5 based on the degree of wilting intensity of leaves caused by the toxin, where 1 indicates high level of resistance and 5 indicates high level of susceptibility. As controls, leaves from the resistant and susceptible parents *H. benthamiana* (Control 1) and *H. brasiliensis* (Control 2), respectively were placed on vials containing 5 mL of modified Czapek Dox liquid media without toxin from the pathogen. Figure S6: Toxin-based screening for disease resistance to *Colletotrichum acutatum*. Tender leaves that are 10 days old were excised and transferred to 5 mL of toxin fraction extracted from *C. acutatum*. Disease resistance was scored from 1 to 5 based on the degree of wilting intensity of leaves caused by the toxin, where 1 indicates high level of resistance and 5 indicates high level of susceptibility. As controls, leaves from the resistant and susceptible parents *H. benthamiana* (Control 1) and *H. brasiliensis* (Control 2), respectively were placed on vials containing 5 mL of modified Czapek Dox liquid media without toxin from the pathogen. Figure S7: Representative silver stained PAGE showing allelic profiles of the interspecific mapping population including the parents *H. brasiliensis* and *H. benthamiana* using a homozygous SSR marker hmCT44, which show polymorphism between the parents. Homozygous off types (P34, P63, P56, P14 and P108) could easily be detected from the heterozygous hybrid progeny. Table S1: Origin and Genealogies of 116 rubber clones and *Hevea* species used in the phylogenetic analysis. Table S2: Expected segregations based on the SNP configurations and observed SNP numbers for each parental configuration. Table S3: Description of the integrated map. Table S4: Integrated genetic linkage map of *Hevea brasiliensis* × *Hevea benthamiana*. Table S5: SNP and SilicoDArT tags aligned with the available rubber genome sequences (Draft genome and reference genome sequences of *H. brasiliensis*). Pseudomolecules with a code of "CM" belong to the reference genome sequences of *H. brasiliensis* and rest belong to the draft genome. Table S6: Markers used in Synteny analysis. Only markers aligned with the *H. brasiliensis* reference genome sequences were used in this analysis. Table S7: Allele-specific and common primer sequences for the KASP assays using trait-associated SNPs. Table S8: Comparison between KASP calls and DArTSeq calls for the markers 100057258 | F | 0–6:T > C (WriHK1) and 100061575 | F | 0-46:G>A (WriHK7). Table S9: Genotypic calls obtained with KASP assays for 178 progeny of *Hevea brasiliensis* × *Hevea benthamiana* and phynotypic data for the assayed disease traits. Table S10: Genes predicted with high confidence within a candidate regions on the pseudomolecules of 18, 6 and 14 in the *Hevea brasiliensis* GT1 reference genome.

**Author Contributions:** C.B.R., S.N.G., T.S. and D.E.M. designed the study, performed the experiments, and analyzed the data. A.K. (Anu Krishnan) and L.J. performed the phenotyping and DNA extraction. A.K. (Andrzej Kilian) performed DArT sequencing. C.B.R. and S.N.G. wrote the paper with input from all authors. All authors have read and agreed to the published version of the manuscript.

**Funding:** This research was funded by the Department of Science and Technology, Government of India (File No.: EMR/2016/001902) and the Australian Endeavour Fellowship of the Australian Government to CBR.

**Data Availability Statement:** The datasets generated and analysed during this study are available in the article and in its online supplementary materials.

**Acknowledgments:** The authors wish to thank Jayashree Madhavan for providing the interspecific hybrid progeny population used in the study. We gratefully acknowledge James Jacob, Former Director (Research), and Kavitha K. Mydin, Former Joint Director (Crop Improvement), Rubber Research Institute of India, for their support and encouragement throughout the work. Funding by the Department of Science and Technology, Government of India, and Australian Endeavour Fellowship of the Australian Government to CBR is gratefully acknowledged.

**Conflicts of Interest:** The authors declare no conflict of interest.

## References

1. Saha, T.; Priyadarshan, P.M. Genomics of *Hevea* rubber. In *Genomics of Tree Crops*; Schnell, R.J., Priyadarshan, P.M., Eds.; Springer: New York, NY, USA, 2012; pp. 261–298. [CrossRef]
2. Schultes, R.E. Wild *Hevea*-an untapped source of germplasm. *J. Rubber Res. Inst. Sri Lanka* **1977**, *54*, 1–31.
3. Djikman, M.J. *Hevea Thirty Years of Research in the Far East*; University of Miami Press: Corel Gables, FL, USA, 1953.
4. Baulkwill, W.J. The history of natural rubber production. In *Rubber*; Webster, C.C., Baulkwill, W.J., Eds.; England Longman Scientific and Technical: London, UK, 1989; pp. 1–56.
5. Simmonds, N.W. Rubber Breeding. In *Rubber*; Webster, C.C., Baulkwill, W.J., Eds.; England Longman Scientific and Technical: London, UK, 1989; pp. 85–124.
6. Sagaff, S.S.A.; Ali, S.N.; Mahyudin, M.M.; Wong, M.Y.; Yusop, M.R. Emerging and existing major leaf diseases of *Hevea brasiliensis* in Malaysia. *J. Curr. Opin. Crop Sci.* **2022**, *3*, 34–47.
7. Mazlan, S.; Jaafar, N.M.; Wahab, A.; Sulaiman, Z.; Rajandas, H.; Zulperi, D. Major diseases of rubber (*Hevea brasiliensis*) in Malaysia. *Pertanika J. Sch. Res. Rev.* **2019**, *5*, 1021.
8. Ramakrishnan, T.S. Experiments on the control of abnormal leaf fall on *Hevea* caused by *Phytophthora palmivora* in South India. In Proceedings of the Natural Rubber Conference, Kuala Lumpur, Malaysia, 26 September–1 October 1960; Rubber Research Institute of Malaya: Kuala Lumpur, Malaysia, 1960; pp. 454–466.
9. Ogbebor, N.O.; Adekunle, A.T.; Enobakhare, D.A. Inhibition of *Colletotrichum gloeosporioides* (Penz) Sac. causal organism of rubber (*Hevea brasiliensis* Muell. Arg.) leaf spot using plant extracts. *Afr. J. Biotechnol.* **2007**, *6*, 213–218.
10. Qu, W.; Li, Z.; Huang, G.; Lin, C.; Ni, W. The current and future potential geographic range of *Corynespora* Leaf Fall Disease in China. *Sens. Lett.* **2012**, *10*, 439–446. [CrossRef]

11. Lespinasse, D.; Rodier-Goud, M.; Grivet, L.; Leconte, A.; Legnate, H.; Seguin, M. A saturated genetic linkage map of rubber tree (*Hevea* spp.) based on RFLP, AFLP, microsatellite, and isozyme markers. *Theor. Appl. Genet.* **2000**, *100*, 127–138. [CrossRef]
12. Feng, S.P.; Wei-Guo, L.; Fei, Y.; Jing-Yi, W.; Yao-Ting, W. Construction of genetic linkage map for rubber tree (*Hevea brasiliensis*) based on SSR markers. *Hereditas* **2010**, *32*, 857–863.
13. Le Guen, V.; Garcia, D.; Doaré, F.; Mattos, C.R.R.; Condina, V.; Couturier, C.; Chambon, A.; Weber, C.; Espéout, S.; Seguin, M. A rubber tree's durable resistance to *Microcyclus ulei* is conferred by a qualitative gene and a major quantitative resistance factor. *Tree Genet. Genom.* **2011**, *7*, 877–889. [CrossRef]
14. Triwitayakorn, K.; Chatkulkawin, P.; Kanjanawattanawong, S.; Sraphet, S.; Yoocha, T.; Sangsrakru, D. Transcriptome sequencing of *Hevea brasiliensis* for development of microsatellite markers and construction of a genetic linkage map. *DNA Res.* **2011**, *18*, 471–482. [CrossRef]
15. Souza, L.M.; Mantello, C.C.; Suzuki, F.; Gazaffi, R.; Garcia, D.; Le Guen, V.; Garcia, A.A.F.; Souza, A.P. Development of a genetic linkage map of rubber tree (*Hevea braziliensis*) based on microsatellite markers. *BMC Proc.* **2011**, *5*, 39. [CrossRef]
16. Souza, L.M.; Gazaffi, R.; Mantello, C.C.; Silva, C.C.; Garcia, D.; Le Guen, V. QTL mapping of growth-related traits in a full-sib family of rubber tree (*Hevea brasiliensis*) evaluated in a sub-tropical climate. *PLoS ONE* **2013**, *8*, e61238. [CrossRef] [PubMed]
17. Pootakham, W.; Ruang-Areerate, P.; Jomchai, N.; Sonthirod, C.; Sangsrakru, D.; Yoocha, T.; Theerawattanasuk, K.; Nirapathpong-porn, K.; Romruensukharom, P.; Tragoonrung, S.; et al. Construction of a high-density integrated genetic linkage map of rubber tree (*Hevea brasiliensis*) using genotyping-by-sequencing (GBS). *Front. Plant Sci.* **2015**, *6*, 367. [CrossRef]
18. Conson, A.R.O.; Taniguti, C.H.; Amadeu, R.R.; Andreotti, I.A.A.; de Souza, L.M.; dos Santos, L.H.B.; Rosa, J.R.B.F.; Mantello, C.C.; da Silva, C.C.; Junior, E.J.S.; et al. High resolution genetic map and QTL analysis of growth-related traits of *Hevea brasiliensis* cultivated under suboptimal temperature and humidity conditions. *Front. Plant Sci.* **2018**, *9*, 1255. [CrossRef] [PubMed]
19. Francisco, F.R.; Aono, A.H.; da Silva, C.C.; Gonçalves, P.S.; Scaloppi Junior, E.J.; Le Guen, V.; Fritsche-Neto, R.; Souza, L.M.; Souza, A.P. Unravelling rubber tree growth by integrating GWAS and biological network-based approaches. *Front. Plant Sci.* **2021**, *12*, 768589. [CrossRef] [PubMed]
20. Rastas, P. Lep-MAP3: Robust linkage mapping even for low-coverage whole genome sequencing data. *Bioinformatics* **2017**, *33*, 3726–3732. [CrossRef]
21. Liu, J.; Shi, C.; Li, W.; Zhang, Q.; Zhang, Y. The chromosome-based rubber tree genome provides new insights into spurge genome evolution and rubber biosynthesis. *Mol. Plant* **2020**, *13*, 336–350. [CrossRef]
22. Pootakham, W.; Sonthirod, C.; Naktang, C.; Ruang-Areerate, P.; Yoocha, T.; Sangsrakru, D. *De novo* hybrid assembly of the rubber tree genome reveals evidence of paleotetraploidy in *Hevea* species. *Sci. Rep.* **2017**, *7*, 41457. [CrossRef]
23. He, C.; Holme, J.; Anthony, J. SNP genotyping: The KASP assay. In *Crop Breeding*; Humana Press: New York, NY, USA, 2014; pp. 75–86.
24. Perseguini, J.M.K.C.; Romão, L.R.C.; Briñez, B.; Junior, E.J.S.; Gonçalves, P.S.; Benchimol, L.L. Genetic diversity of cultivated accessions and wild species of rubber tree using EST-SSR markers. *Pesq. Agropec. Bras. Brasília* **2012**, *47*, 1087–1094. [CrossRef]
25. Narayanan, C.; Mydin, K.K. Breeding for disease resistance in *Hevea* spp.-Status, potential threats, and possible strategies. In Proceedings of the 4th International Workshop on Genetics of Host-Parasite Interactions in Forestry: Disease and Insect Resistance in Forest Trees, General Technical report PSW-GTR-240, Valley River Inn, Eugene, OR, USA, 31 July–5 August 2011; pp. 240–251.
26. Lu, F.; Lipka, A.E.; Glaubitz, J.; Elshire, R.; Cherney, J.H.; Casler, M.D.; Buckler, E.S.; Costich, D.E. Switchgrass genomic diversity, ploidy, and evolution: Novel insights from a network-based SNP discovery protocol. *PLoS Genet.* **2013**, *9*, e1003215. [CrossRef]
27. Goonetilleke, S.N.; March, T.J.; Wirthensohn, M.G.; Arús, P.; Walker, A.R.; Mather, D.E. Genotyping-by-sequencing in almond: SNP discovery, linkage mapping and marker design. *Genes Genomes Genet.* **2018**, *8*, 161–172. [CrossRef]
28. Wastie, R.L. Nursery screening of *Hevea* for resistance to *Gleosporium* leaf disease. *J. Rubber Res.* **1973**, *23*, 339.
29. Lim, T.M. A rapid laboratory method of assessing susceptibility of *Hevea* clones to *Oidium Hevea*. *Exp. Agric.* **1973**, *9*, 275. [CrossRef]
30. Tan, A.M.; Loo, T.P.; Vadivel, G.; Ros, M.; Bach, K.; Yoon, K.F. Survey of major leaf diseases of rubber in Peninsular Malaysia. *Plant. Bull. Rubber Res. Inst. Malays.* **1992**, *2*, 51.
31. Doyle, J.J.; Doyle, J.L. Isolation of plant DNA from fresh tissue. *Focus* **1990**, *12*, 13–15.
32. Roy, C.B.; Ravindran, M.; Saha, T. Efficient screening of AFLP primer combinations for evaluating genetic diversity among cultivated rubber (*Hevea brasiliensis*) clones. *Nat. Rubber Res.* **2012**, *25*, 21–30.
33. Rambaut, A. FigTree v1.4.2. A Graphical Viewer of Phylogenetic Trees. 2018. Available online: https://tree.bio.ed.ac.uk/software/figtree/ (accessed on 22 November 2022).
34. Milne, I.; Shaw, P.; Stephen, G.; Bayer, M.; Cardle, L.; Thomas, W.T.B.; Flavell, A.J.; Marshall, D. Flapjack-graphical genotype visualization. *Bioinformatics* **2010**, *26*, 3133–3134. [CrossRef] [PubMed]
35. Danecek, P.; Auton, A.; Abecasis, G.; Albers, C.A.; Banks, E. The variant call format and VCFtools. *Bioinformatics* **2011**, *27*, 2156–2158. [CrossRef]
36. Feng, L.Y.; Liu, J.; Gao, C.W.; Wu, H.B.; Gao, L.Z. Higher genomic variation in wild than cultivated rubber trees, *Hevea brasiliensis*, revealed by comparative analyses of chloroplast genomes. *Front. Ecol. Evol.* **2020**, *8*, 237–247. [CrossRef]
37. Gu, Z.; Gu, L.; Eils, R.; Schlesner, M.; Brors, B. Circlize implements and enhances circular visualization in R. *Bioinformatics* **2014**, *30*, 2811–2812. [CrossRef]
38. Rajalakshmy, V.K.; Joseph, A. Production of sporangia by *Phytophthora meadii*. *Indian Phytopathol.* **1986**, *39*, 470–471.

39. Joseph, A.; Mathew, F.; Mercy, M.A. Evaluation of wild Hevea germplasm against Phytophthora meadii causing abnormal leaf fall disease of rubber in India. In Proceedings of the International Rubber Workshop on Hevea Diseases in Africa, Benin City, Nigeria, 5–7 June 2012; pp. 55–65.
40. Borjesson, T.; Stollman, U.; Schnurer, T. Volatile metabolites and other indicators of *Penicillium aurantiogriseum* growth on different substrates. *Appl. Environ. Microbiol.* **1990**, *56*, 3705–3710. [CrossRef] [PubMed]
41. Breton, F.; Sanier, C.; D'Auzac, J. Role of cassiicolin, a host-selective toxin, in pathogenicity of *Corynespora cassiicola*, causal agent of a leaf fall disease of *Hevea*. *J. Rubber Res.* **2000**, *3*, 115–128.
42. Hallé, F.; Martin, R. Etude de la croissance rythmique chez l'hévéa (*Hevea brasiliensis*) Müll. Arg., (Euphorbiacées, crotonoïdées). *Adansonia* **1968**, *8*, 475–503.
43. Fernando, T.H.P.S.; Jayasinghe, C.K.; Wijesundera, R.L.C.; Silva, W.P.K.; Nishantha, E.A.D.N. Evaluation of screening methods against Corynespora leaf fall disease of rubber (*Hevea brasiliensis*). *J. Plant Dis. Prot.* **2010**, *117*, 24–29. [CrossRef]
44. Bradbury, P.J.; Zhang, Z.; Kroon, D.E.; Casstevens, T.M.; Ramdoss, Y.; Buckler, E.S. TASSEL: Software for association mapping of complex traits in diverse samples. *Bioinformatics* **2007**, *23*, 2633–2635. [CrossRef] [PubMed]
45. Schulze, T.G.; McMahon, F.J. Genetic association mapping at the crossroads: Which test and why? Overview and practical guidelines. *Am. J. Med. Genet.* **2002**, *114*, 1–11. [CrossRef] [PubMed]
46. Kearse, M.; Moir, R.; Wilson, A.; Stones-Havas, S.; Cheung, M.; Sturrock, S.; Buxton, S.; Cooper, A.; Markowitz, S.; Duran, C.; et al. Geneious Basic: An integrated and extendable desktop software platform for the organization and analysis of sequence data. *Bioinformatics* **2012**, *28*, 1647–1649. [CrossRef]

plants

MDPI

*Article*

# Increased Zygote-Derived Plantlet Formation through *In Vitro* Rescue of Immature Embryos of Highly Apomictic *Opuntia ficus-indica* (Cactaceae)

**Angela Carra [1], Caterina Catalano [1], Ranjith Pathirana [2,3], Maurizio Sajeva [4], Paolo Inglese [5], Antonio Motisi [1] and Francesco Carimi [1,*]**

[1] CNR—Istituto di Bioscienze e BioRisorse, Via Ugo La Malfa 153, 90146 Palermo, Italy; angela.carra@ibbr.cnr.it (A.C.); caterina.catalano@ibbr.cnr.it (C.C.); antonio.motisi@ibbr.cnr.it (A.M.)

[2] Plant & Food Research Australia Pty Ltd., #46 Plant Breeding, Waite Road, Urrbrae, SA 5064, Australia; ranjith.pathirana@adelaide.edu.au

[3] School of Agriculture, Food and Wine, University of Adelaide, Waite Road, Urrbrae, SA 5064, Australia

[4] Department of Biological, Chemical and Pharmaceutical Sciences and Technologies (STEBICEF), University of Palermo, Via Archirafi 18, 90123 Palermo, Italy; maurizio.sajeva@unipa.it

[5] Department of Agricultural Food and Forest Sciences, University of Palermo, Viale delle Scienze, 90128 Palermo, Italy; paolo.inglese@unipa.it

* Correspondence: francesco.carimi@ibbr.cnr.it

**Abstract:** *O. ficus-indica* (prickly pear cactus) is an important forage and food source in arid and semiarid ecosystems and is the most important cactus species in cultivation globally. The high degree of apomixis in the species is a hindrance in plant breeding programs where genetic segregation is sought for the selection of superior genotypes. To understand if *in ovulo* embryo rescue could increase the proportion of zygotic seedlings, we compared the mature seed-derived seedlings with those regenerated from *in vitro* embryo rescue at 20, 25, 30, 35, and 40 post-anthesis days (PADs) in four Italian cultivars. The seedlings were classified as apomictic or zygotic based on molecular marker analysis using inter-sequence single repeat (ISSR) primers. Multiple embryos were recovered from all the cultured immature ovules, and plantlets were regenerated and acclimatized to the field post hardening, with success rates ranging from 62% ('Senza spine') to 83% ('Gialla'). The level of polyembryony differed among cultivars and recovery dates, with the highest being 'Rossa', producing 4.8 embryos/ovule at 35 PADs, and 'Gialla', the lowest, with 2.7 at 40 PADs. The maximum number of embryos observed within a single ovule was 14 in 'Trunzara bianca'. ISSR analysis revealed that ovule culture at 35 PADs produced the highest percentage of zygotic seedlings in all the cultivars, from 51% ('Rossa') to 98% ('Gialla'), with a high genotype effect as well. Mature seeds produced much fewer seedlings per seed, ranging from 1.2 in 'Trunzara bianca' to 2.0 in 'Rossa' and a lower percentage of zygotic seedlings (from 14% in 'Rossa' to 63% in 'Gialla'). Our research opens a pathway to increase the availability of zygotic seedlings in *O. ficus-indica* breeding programs through *in ovulo* embryo culture.

**Keywords:** apomixis; embryo rescue; ovule culture; prickly pear; cactus pear; nucellar embryos; hybrid progeny

**Citation:** Carra, A.; Catalano, C.; Pathirana, R.; Sajeva, M.; Inglese, P.; Motisi, A.; Carimi, F. Increased Zygote-Derived Plantlet Formation through *In Vitro* Rescue of Immature Embryos of Highly Apomictic *Opuntia ficus-indica* (Cactaceae). *Plants* **2023**, *12*, 2758. https://doi.org/10.3390/plants12152758

Academic Editor: Andrew McCubbin

Received: 11 May 2023
Revised: 2 June 2023
Accepted: 9 June 2023
Published: 25 July 2023



## 1. Introduction

Prickly pear (*Opuntia ficus-indica* (L.) Mill.), also called cactus pear, Barbary fig, or nopal cactus, is a member of the *Cactaceae* family; it originated in central Mexico and is considered the most important cactus species in horticulture worldwide, with a global distribution [1,2]. Photosynthetic adaptation with Crassulacean acid metabolism (CAM), where carbon is fixed in the night when the air is cooler, allows *Opuntia* spp. to better adapt to conserve water in arid or semiarid environments than the C4 and C3 plants, with 3–5 times lower transpiration rates [2,3]. Its excellent adaptation to arid and semiarid climates makes its fruit and fresh stems (cladodes) an important source of human nutrition as well as a forage and fodder source for farm animals in such areas [1,2,4].

The fruits of *O. ficus-indica* cultivars, known as prickly pears or tunas, can be very sweet. They come in diverse colors and are highly appreciated in many cultures [1,5,6]. For a long time, Mexicans have used the tender young cladodes (nopalitos) as a source of green and fresh vegetables [2,7]. Although nopalitos are not a common food in industrialized countries [2,8], they are gaining popularity among European and US consumers for their health benefiting profile. Furthermore, the prickly pear has many medicinal properties and has been used in traditional Mexican medicine for the treatment of a variety of diseases, such as arteriosclerosis, diabetes, and gastritis [2,9,10]. Additionally, the large amount of biomass produced by the prickly pear due to its high shoot-to-root ratio combined with its high productivity makes it an ideal fodder and feed source for livestock in semiarid and arid climates. A five-year study comparing the response of three *Opuntia* species to irrigation in Logandale, Nevada, USA showed that *O. ficus-indica* is the preferred species in terms of biomass gain and fruit quality [11]. Global prickly pear production is dominated by three countries: Mexico (45%), Italy (12.2%), and South Africa (3.7%). In Mexico, the planted area is between 50,000 and 70,000 ha, with an annual production of 300,000–500,000 tons [5,9,12]. Cultivation in Italy is concentrated in Sicily with about 8600 hectares, ranking first among the Mediterranean regions for producing and exporting prickly pear fruits [13].

Thanks to a mix of reproductive strategies (sexual, apomictic, and other clonal strategies), reduced water loss due to succulent stems (including many other morphological and physiological features), and efficient photosynthesis through CAM and polyploid genomes, opuntioid genera have successfully established in many parts of arid and semi-arid ecologies, although they are endemic to the Americas [14–16]. For example, on the planet's driest continent, Australia, all opuntoid cacti present (*Austrocylindropuntia* spp., *Cylindropuntia* spp., and *Opuntia* spp.), except *O. ficus-indica*, were named as Weeds of National Significance in April 2012 [17]. *Opuntia* and *Cylindropuntia* are the most invasive among cacti in Australia, with tiger pear (*O. aurantiaca*), prickly pear cactus (*O. monacantha*), wheel cactus (*O. robusta*), white-spined prickly pear (*O. streptacantha*), common prickly pear (*O. stricta*), and velvet prickly pear (*O. tomentosa*) being the most widespread [18]. South Africa and Spain are also considered invasive hotspots for opuntoid cacti [14]. These facts illustrate the high degree of adaptation of *Opuntia* spp. to drier regions of the world where other crops struggle, and the need to exploit it in horticulture in these regions.

Exploiting apomixis is a natural way of cloning through seeds, as apomixis produces seedlings that are genetically identical to the mother plant without the involvement of male gametes [19,20]. Apomixis is a complex developmental process; it is historically subdivided into two categories, gametophytic and sporophytic, based on whether the embryo develops via a gametophyte (embryo sac) or directly from a diploid somatic (sporophytic) cell within the ovule [20]. Apomixis in angiosperms is rarely obligate; usually, apomictic plants produce asexual and sexual progeny within the same offspring generation, and asexuality is facultative. Therefore, a proportion of the offspring represents recombinants, but frequencies of sexuality vary a lot among genera, species, and different modes of apomixis [21,22]. Diplospory is a form of gametophytic apomixis in which an unreduced embryo sac forms from a megaspore mother cell with the circumvention of meiosis [20,23]. Apomixis has a genetic basis, but it is still a matter of question how it is regulated. The ability to produce genetically uniform progeny via seeds is of significant value for its potential in agriculture to fix complex favourable genotypes, particularly hybrids expressing heterosis or those obtained from wide crosses, to improve breeding programs' efficiency [19–21,23].

On the other hand, the apomictic embryos, which are genetically identical to their maternal parent, limit the range of genetic variability that can be observed in the progeny of a cross, and thus the possibility of finding new genotypes. In these cases, *in vitro* embryo rescue can be a very useful technique for breeding programs [24,25].

Apomixis frequently occurs in *Opuntia* spp., including *O. ficus-indica* [15,19]. It was initially described by Ganong in *O. vulgaris* as far back as 1898 [20]. The most common type of apomixis in *Opuntia* involves the development of adventitious embryos from nucellar tissue (sporophytic agamospermy) [21–23]. Or, as in *O. streptacantha*, embryos can develop from an unfertilized egg (diplospory parthenogenesis) [21]. More recently, Kaaniche-Elloumi et al. [24] reported that *O. ficus-indica* ovules showed both sporophytic and gametophytic embryogenesis.

The objectives of this research were to study apomixis in *O. ficus-indica* and to determine the incidence of sexual and apomictic embryos *in vitro* and *in vivo*. In order to determine whether the genotype influences the level of polyembryony, four different cultivars were used. We also analyzed the effect of the ovule isolation time on the proportion of sexual and apomictic offspring *in vitro*, with the aim of developing a protocol to increase the production of zygotic seedlings in crossbreeding programs.

## 2. Materials and Methods

### 2.1. Plant Material

To study polyembryony, four cultivars used for fresh fruit consumption were utilized: 'Rossa', 'Senza spine', 'Trunzara bianca', and 'Gialla'. All the genotypes analyzed are classified as facultative apomicts because they have the ability to reproduce both sexually and asexually through apomixis. Immature and mature fruits were collected from open pollinated adult plants growing at the germplasm repository for perennial plants at the Institute of Biosciences and BioResources of the National Research Council of Italy (CNR-IBBR), located in Collesano District (Province of Palermo), Italy (37°59′19.9″ N, 13°54′55.8″ E, 80 m a.s.l.). Fruit growth and development are shown in Figure 1 and Supplementary Figure S1.

**Figure 1.** Prickly pear fruits of four cultivars harvested at 5-day intervals from 20 to 40 post-anthesis days for *in vitro* embryo rescue and, at maturity, for *in vivo* germination. Bar = 2 cm.

### 2.2. Media and In Vitro Culture Methodology of Immature Ovules

Media preparation, culture conditions, and plant regeneration were conducted similarly to the methods described by Carimi et al. [25]. Fertilized ovules to be grown *in vitro* were excised from immature fruits collected at 5-day intervals from 20 to 40 post-anthesis days (PADs) during July and August. The immature fruits were rinsed with tap water and then surface-sterilized by immersion for 5 min in 70% ethanol and 30 min in 2% (*w*/*v*) sodium hypochlorite. Finally, the fruits were rinsed two times with sterile distilled water for 5 min under aseptic conditions (Figure 2A,E). After sterilization, one longitudinal cut was made immediately under the fruit epidermis, avoiding the core where fertilized ovules are embedded. After opening the immature fruits (Figure 2B,F), fertilized ovules were extracted under aseptic conditions (Figure 2C,G), and, by means of a longitudinal cut, the outer seed integument was removed under a stereoscopic microscope using a scalpel and forceps (Figure 2D,H).

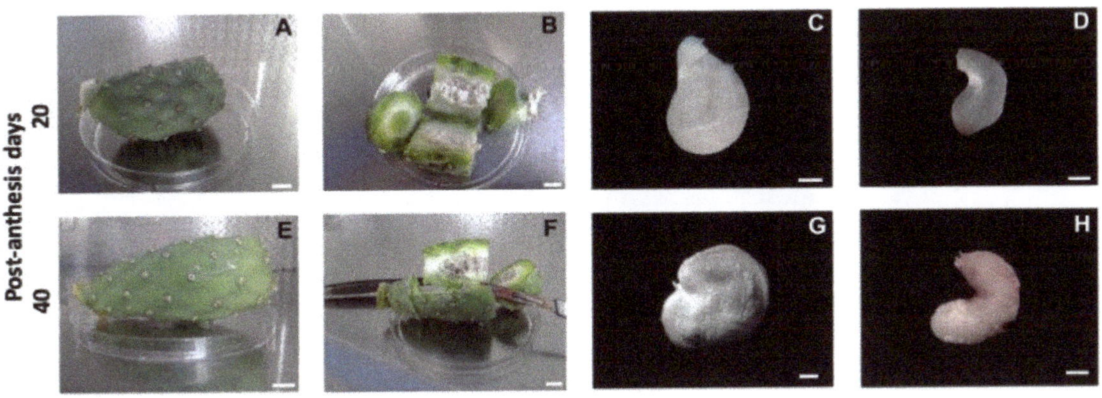


**Figure 2.** Procedure for ovule dissection under aseptic conditions. (**A**,**E**) Fruits harvested at different post-anthesis days (20 and 40, respectively) were surface sterilized in a laminar flow hood. Bar = 1 cm. (**B**,**F**) Fruits dissected in halves. Bar = 1 cm. (**C**,**G**) Immature ovules dissected from fruit. Bar = 1 mm. (**D**,**H**) The ovules after the outer integument was removed with a razor blade. Bar = 1 mm.

Immature ovules without integument were cultured on plant growth regulator-free Murashige and Skoog [26] (MS) basal medium (micro and macro salts and MS vitamins) supplemented with sucrose (50 g $L^{-1}$) and 500 mg $L^{-1}$ malt extract and solidified with 7 g $L^{-1}$ Plantagar (S 1.000, B&V, Italy). To induce embryo development and to determine the percentage of responsive ovules, each immature ovule was placed on 8 mL of medium in plastic Petri dishes (60 × 15 mm) sealed with Parafilm M (Figure 3A). Cultures were maintained in a climatic chamber at 26 °C with a 16 h photoperiod (40 μmol $m^{-2}$ $s^{-1}$ at shelf level provided by Osram Cool White 18 W fluorescent lamps).

Four weeks after incubation, each embryo sac was scored for the presence or absence of one or more embryos. About six weeks after incubation, the embryos generated from immature ovules were collected and transferred to solid MS medium prepared as previously described in Petri dishes (100 × 20 mm) and cultured for a further 4–6 weeks to allow plantlet development. Individual germinated somatic embryos (about 1–2 cm in length) were transferred to Magenta™ vessels to allow further growth (one embryo/Magenta™ vessel containing 50 mL of basal MS medium).

Once rooted, plantlets were transferred to autoclaved Jiffy® peat pellets and maintained for five weeks in a basal heating bench at 25 °C and at high relative humidity (95–98%). Subsequently, the plants were pricked into pots containing sterile soil, transferred to the greenhouse, and exposed to natural daylight conditions at 22/27 °C night/day.

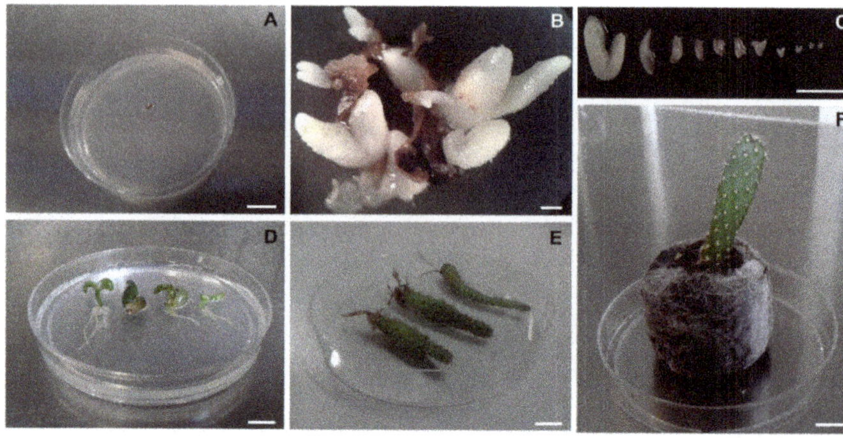


**Figure 3.** *In vitro* recovery procedure of *Opuntia ficus-indica* embryos and plantlets. (**A**) The fertilized immature ovule, without the outer integument, incubated on MS medium. Bar = 1 cm. (**B**) Embryos arising from fertilized immature ovule incubated *in vitro* on MS medium. Picture taken 3 weeks from the beginning of the experiment. Bar = 1 mm. (**C**) Embryos at different developmental stages dissected from a single fertilized ovule of 'Trunzara bianca' after 4 weeks of incubation. Bar = 5 mm. (**D**) Plantlets growing in Petri dish after 8 weeks from the beginning of the experiment. Bar = 1 cm. (**E**) Plants are removed from Magenta vessels and rinsed thoroughly in water to remove traces of medium; ready for transfer to Jiffy pots. Picture taken 14 weeks from embryo germination. Bar = 1 cm. (**F**) *Opuntia* plant acclimatized in a Jiffy pot after 19 weeks from the beginning of the experiment. Bar = 1 cm.

### 2.3. Seed Germination In Vivo

Fresh seeds were collected from mature fruits harvested in September stratified at 4 °C for 3 months in the dark and germinated into plastic pots (70 mm × 70 mm) containing sterile soil. The potted plants, covered with transparent polyethylene bags to maintain temperature and high humidity, were placed in a climate chamber at 25 ± 1 °C under the same culture conditions as described above. The percentage of germination and the number of plantlets produced per seed were evaluated four months after sowing.

### 2.4. DNA Extraction

DNA was extracted from young cladodes of the mother plants growing in the field and from young seedlings regenerated from ovules *in vitro* and from seeds *in vivo*. Seedlings from different cultivars were randomly selected from each different ovule isolation time for the analyses of genetic origin (zygotic or apomictic). All the samples were frozen in liquid nitrogen and stored at −80 °C. They were ground in a mortar with liquid nitrogen, and genomic DNA was extracted using the procedure described by Doyle and Doyle [27]. DNA was quantified by measuring $OD_{260}$, as described by Sambrook et al. [28].

### 2.5. Genetic Analysis

To assess the genetic origin of the progeny, mother plants and plantlets generated from ovules *in vitro* and from seeds *in vivo* were characterized by inter-simple sequence repeat polymorphic DNA (ISSR) marker analysis, as described by Siragusa et al. [29]. Briefly, a total of ten primers as reported by Fang and Roose [30], were used in preliminary experiments to assess the genetic origin of seedlings. Five of those primers, i.e., $(AC)_8YG$, $(AC)_8YA$, $(TCC)_5RY$, $(GA)_8YC$, and $(GA)_8YG$ were low informative and therefore were not included in the final study. The primers used in the final study were $(AG)_8YC$ [Annealing Temperature (Ta) 52.6 °C], $(AC)_8YT$ (Ta 50.3 °C), $(AG)_8YT$ (Ta 50.3 °C), $(GT)_8YG$ (Ta 52.1 °C), and $(CA)_8RG$ (Ta 51 °C). To distinguish apomictic from zygotic seedlings, a genetic analysis based on ISSR analysis was performed, as previously described in detail [29]. To confirm the reproducibility of the banding patterns, all analyses were repeated twice.

*2.6. Statistical Analysis*

The fruit growth pattern was evaluated by measuring fruit fresh weight, diameter, and length of 20 fruits for each cultivar at 20–40 PADs at 5-day intervals and at the ripening stage.

Each treatment *in vitro* and *in vivo* comprised 60 ovules or seeds, and experiments were performed in triplicate in a randomized complete block design. The effects of genotype and ovule developmental stage on the percentage of responsive ovules, the average number of plantlets generated per ovule, the percentage of monoembryonic ovules, and the percentage of ovules and seeds producing zygotic seedlings were tested by ANOVA ($p \leq 0.05$), and the differences among means were tested by Tukey's test. Prior to analysis, percentage data were arcsin square root transformed. Statistical analysis was performed using SigmaStat 3.5 for Windows.

## 3. Results

*3.1. Embryo Rescue In Vitro*

The first embryos emerged from the immature ovules cultured on MS medium (Figure 3A) about one week after culture initiation, and after 3–4 weeks, several embryos were visible on the surface of the ovule (Figure 3B). Fertilized ovules contained several embryos at different developmental stages (Figure 3C). The maximum number of viable embryos observed in a single fertilized ovule varied according to the cultivar: 'Senza spine' had a maximum of eight, 'Rossa' had a maximum of twelve, 'Gialla' had a maximum of twelve, and 'Trunzara bianca' had a maximum of fourteen.

A high percentage of embryos germinated *in vitro* (Figure 3D), and about 8–10 weeks after culture initiation, the plantlets grew normally (Figure 3E) with no significant differences found among the different genotypes. After about 3–4 months of culture *in vitro*, the quality of the roots was good, and plantlets were transferred to Jiffy peat pellets (Figure 3F). The percentage of acclimatized plantlets observed for the different cultivars was: 62%, 71%, 75%, and 83% for 'Senza spine', 'Trunzara bianca', 'Rossa', and 'Gialla', respectively.

Responsive ovules were collected from all genotypes at different PADs. The percentage of responsive explants ranged from 10% ('Gialla' collected at 40 PADs) to 97% ('Senza spine' collected at 35 PADs). The best result for all cultivars was obtained when collection was performed at 35 PADs (Figure 4).

The percentage of responsive ovules varied according to the collection time. The value increased when ovules were isolated in the period lasting from 20 to 35 PADs, while it decreased significantly at 40 PADs (Figure 4).

No significant differences were found among cultivars for the number of plantlets per ovule, while the number of plantlets per ovule was significantly lower for recovery at 40 PADs when compared to earlier periods of embryo rescue. Overall, data attest their value between 3.6 ('Gialla') and 4.6 ('Rossa') plantlets per ovule when genotype is considered, while data ranged between 3.5 (40 PADs) and 4.8 (35 PADs) for the number of plantlets regenerated per ovule according to time of recovery (Figure 5).

The percentage of monoembryonic ovules varied greatly in the experiment (Figure 6). The highest percentage was achieved with 'Trunzara bianca' collected at 40 PADs (22.9%) and the lowest percentage was recorded with the same cultivar collected at 20 PADs (5.0%). However, no significant differences were found among cultivars; percentages ranged from 13.4% ('Rossa') to 17.7% ('Trunzara bianca'). With regard to the time of recovery, no significant differences in percentages of monoembryonic ovules were observed.

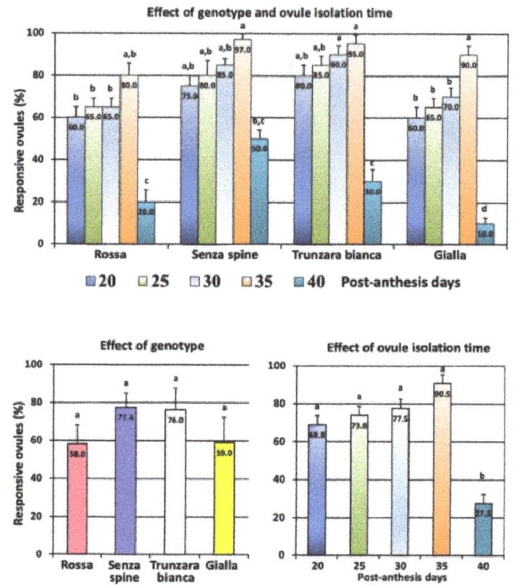


**Figure 4.** Effect of genotype and ovule isolation time on percentage of responsive ovules incubated *in vitro*. Different letters indicate significant differences (Tukey's test, $p < 0.05$, n = 60). Data represent values ± SE.

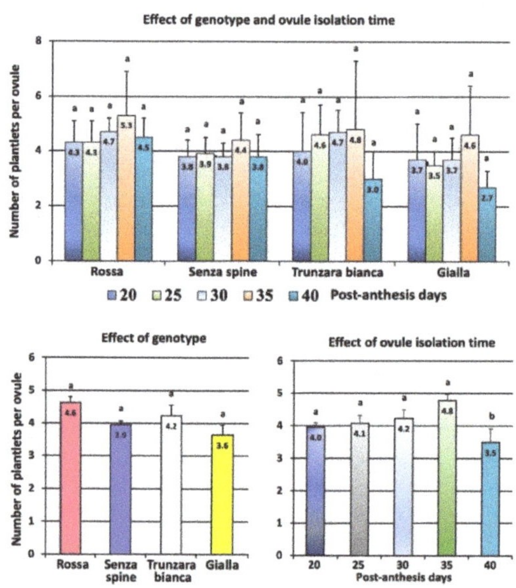


**Figure 5.** Effect of genotype and ovule isolation time on average number of plantlets generated per ovule. Different letters indicate significant differences (Tukey's test, $p < 0.05$ level, n = 60). Data represent values ± SE.

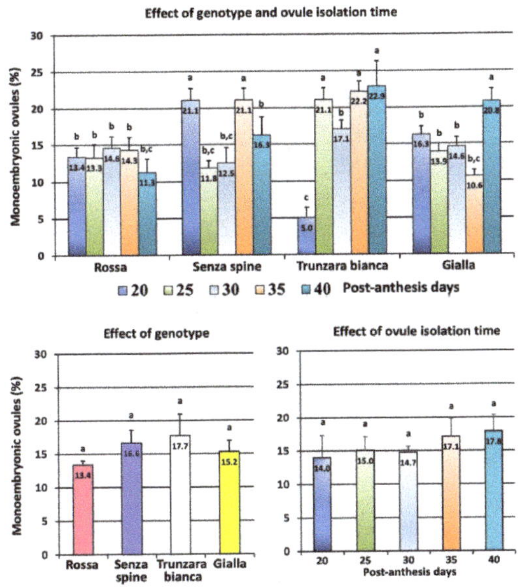


**Figure 6.** Effect of genotype and ovule isolation time on percentage of monoembryonic ovules. Different letters indicate significant differences (Tukey's test, $p < 0.05$ level, n = 60). Data represent values ± SE.

Ovules versus seeds: The number of plantlets generated per ovule *in vitro* and per seed *in vivo* varied greatly, and the number generated from ovules was significantly higher than those from seeds for all four cultivars (Figure 7). The average number of plantlets obtained per ovule ranged from 3.64 to 4.62 ('Gialla' and 'Rossa', respectively), with no significant differences among cultivars. Conversely, the average number of plantlets obtained from seeds *in vivo* was strongly reduced (Figure 7), ranging from 1.21 to 1.99 ('Trunzara bianca' and 'Rossa', respectively).

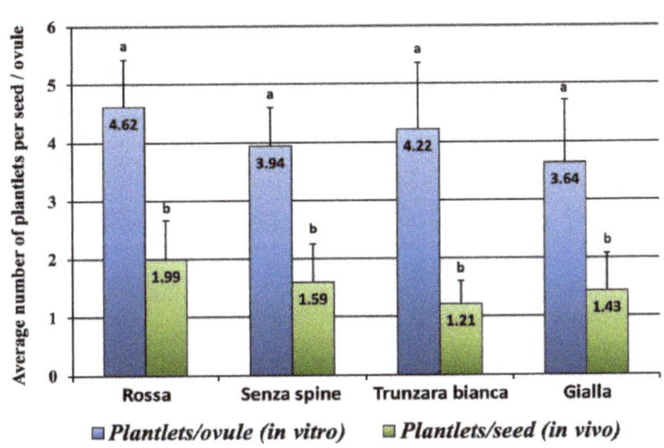


**Figure 7.** Average number of plantlets generated per ovule *in vitro* and per seed *in vivo*. Different letters indicate significant differences (Tukey's test, $p < 0.05$ level, n = 60). Bars correspond to mean values ± SE.

### 3.2. Genetic Analysis

ISSR primers were used to amplify the DNA of regenerants from each cultivar and to compare them to the respective mother plant. The presence of polymorphic bands allowed us to detect zygotic and apomictic seedlings (Figure 8).

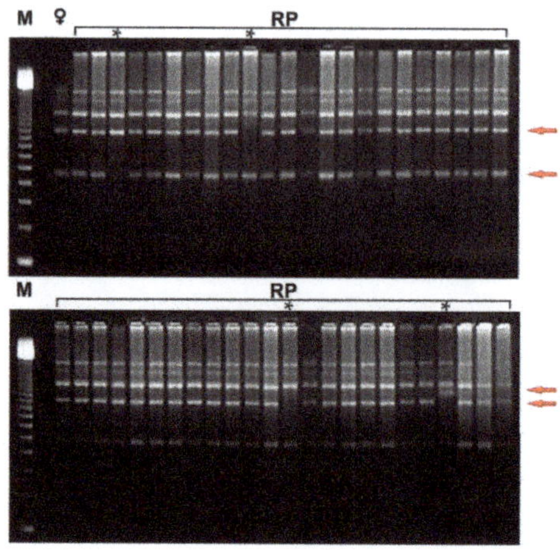


**Figure 8.** DNA analysis of plantlets recovered *in vitro* from immature ovule culture. Inter-simple sequence repeat polymorphic DNA (ISSR) profiles amplified from DNA extracted from 47 RP of 'Rossa' analysed using primer (GT)8 YG. M 100-bp DNA ladder; ♀ mother plant; RP 1–23 and 24–47 plantlets rescued *in vitro*. Arrows indicate polymorphic bands. The asterisk (*) indicates the profiles of the zygotic seedlings.

Screening using molecular markers revealed that a genotype had a significant effect on the percentage of ovules producing zygotic seedlings (Figure 9). The highest percentage was achieved with 'Gialla' collected at 35 PADs (98%) and the lowest percentage was recorded with 'Rossa' collected at 20 PADs (33%). Significant differences were also observed among cultivars; percentages ranged from 42.2% ('Rossa') to 92.4% ('Gialla'). However, no significant differences were found among different ovule isolation times; percentages ranged from 58.5% (20 PADs) to 74.5% (35 PADs).

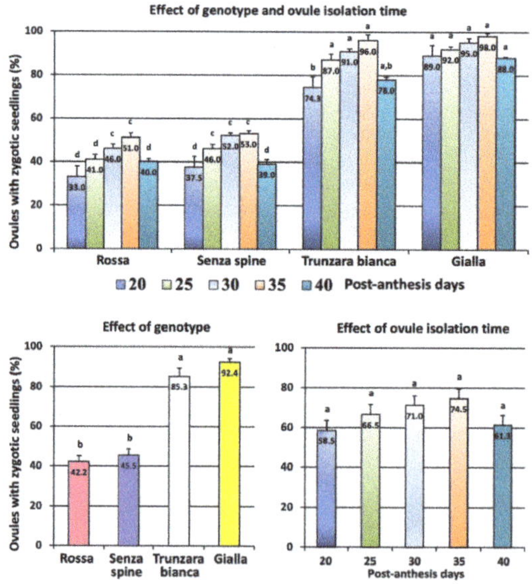


**Figure 9.** Percentage of ovules, collected at different days post anthesis, that yielded at least one zygotic seedling. Different letters indicate significant differences (Tukey's test, $p < 0.05$ level, n = 60). Bars correspond to mean percentage values ± SE.

Ovules versus seeds: From our results, it appears that the *in vitro* ovule culture procedure allows a more efficient recovery of zygotic embryos than the traditional *in vivo* seed germination procedure (Figure 10). The percentage of ovules with a zygotic seedling was higher than and significantly different from the values for seeds with a zygotic seedling in all the cultivars, 'Trunzara bianca' and 'Gialla' being the most responsive. The highest percentage was achieved with ovules of 'Gialla' (92.40%), and the lowest percentage was recorded with seeds of 'Rossa' (13.78%).

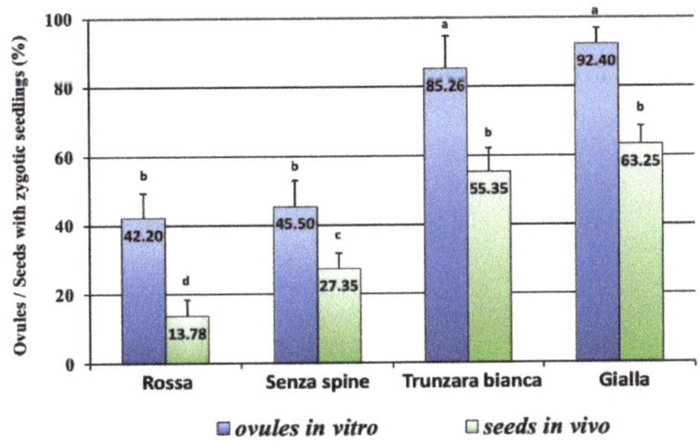


**Figure 10.** Percentage of ovules and seeds producing zygotic seedlings. Different letters indicate significant differences (Tukey's test, $p < 0.05$ level, n = 60). Bars correspond to mean percentage values ± SE.

## 4. Discussion

Under the rainfed conditions of the semiarid highlands of central Mexico, cactus pear (*O. ficus-indica*) is the main fruit crop, with more than 50,000–70,000 ha planted [5,22]. Cactus pear is also important in Italy and South Africa [12]; it is gaining importance in Chile [31], Brazil [32], and Egypt [10]; and it is becoming an important alternative crop for several countries in North Africa and other semiarid areas of the world [2–4,33,34]. Currently, in all countries with commercial plantations, the crop is produced from a few varieties that have either a direct origin in Mexico or have been derived from those [15,35]. The narrow nature of the germplasm base in Italy is also evident from our results of fruit characteristics, as there were no statistical differences in fruit length, diameter, or weight among the four studied cultivars. Producing varieties with better adaptation to the local environment, resistance to disease, and improved fruit or forage quality is an important objective in cactus pear breeding programs [15,35]. Climate change, while adding more opportunities for cactus cultivation in new areas, will require the achievement of other novel objectives in breeding programs [36,37]. Among the reproductive strategies evolved in *Opuntia* spp., apomixis and vegetative propagation by cladode detachment can be used for clonal propagation, and these are valuable tools for breeders and nurseries. This is the main reason for the lack of genetic diversity in *O. ficus-indica*. While the prevalence of apomixis in *O. ficus-indica* gives an additional tool for the nursery industry for vegetative propagation of elite genotypes, the identification of hybrids and progeny selection in crossbreeding programs becomes challenging, complicated, and inefficient because of apomixis [22].

Embryo rescue is a biotechnological approach used to overcome some technological difficulties encountered when using traditional plant breeding approaches. Early rescue of hybrid embryos allows the recovery of interspecific and intergeneric hybrids that are impossible to produce *in vivo* [38,39]. The method is also used to manipulate ploidy in cultivated species [38,40]. In this study, we explored another possible application of embryo rescue, i.e., to enhance the regeneration of zygotic embryos in *O. ficus-indica*. Previously, embryo rescue has been employed to increase the ratio of zygotic embryos to apomictic embryos in other apomictic species, such as citrus [25,41]. In cacti, Felker et al. [42] tested the progeny of a cross between *O. lindheimerii* and *O. ficus-indica* using randomly amplified polymorphic DNA (RAPD) markers and confirmed that four out of thirteen (30.8%) tested progeny were apomicts. In our study, we show conclusively that within *O. ficus-indica*, this ratio is genotype-dependent, with just 13.8% of the seedlings being zygotic in 'Rossa' against 63.3% in 'Gialla' in seedlings grown from mature seeds. While there are several mechanisms involved in apomixis, many studies have confirmed it to be controlled as a dominant trait [43,44], and the complexities unravelled in molecular studies can be attributed mainly to secondary factors resulting from the reproductive process [45,46]. Therefore, induced mutagenesis could be used to produce non-apomictic genotypes to help in crossbreeding programs [47,48]. However, a more urgent need is the development of methods to recover more zygotic seedlings from existing cultivars in crossbreeding programs. Our research was directed at a solution to solve this problem.

We attempted embryo rescue over five PADs periods from 20 to 40 days in four Italian cultivars with contrasting morphologies, and in all the cultivars, 35 PADs embryo rescue was the most successful in terms of the percentage of responsive ovules (80–95%) and the mean number of responsive embryos per ovule (4.4–5.3) in all four cultivars. In contrast, mature seeds produced very low numbers of seedlings per seed (1.2–2, or almost fourfold less). Our method of acclimation and hardening of immature embryo-derived seedlings was efficient, and the success rate was from 62% ('Senza spine') to 83% ('Gialla'). The next step in our research was to identify the origins of the seedlings, and we used ISSR markers, which are highly efficient and reliable [49–51]. Again, we found genotypic effects on the percentage of seedlings of zygotic origin, with 'Rossa' and 'Senza spine' producing significantly less (42 and 46%, respectively) than 'Trunzara bianca' and 'Gialla' (85 and 92%, respectively). Importantly, our method yielded a significantly higher percentage of zygotic embryo-derived seedlings than the counterpart mature seeds (14–63%) in all four cultivars. It should be noted that this higher percentage of zygotic embryos in our *in vitro* approach is from a fourfold higher ovule response compared to mature seeds, as already noted, thus making the yield of zygotic seedlings even greater. Of the four periods tested, 35 PADs recorded the highest yield of zygotic seedlings in all four cultivars, thus making our protocol easy to follow. It appears that for any species, the optimum period for the rescue of embryos needs to be identified, as previously recorded in apomictic sour orange (*Citrus aurantium*—125 PADs) [25] and in 'Shiranuhi' mandarin, a hybrid citrus [(*C. unshiu* × *C. sinensis*) × *C. reticulata*] (145 DAP) [41].

In preliminary experiments (data not presented), different combinations of plant growth regulators (PGR) were added to the culture medium to stimulate the *in vitro* development of the zygotic and apomictic embryos present in the immature ovules. Interestingly, we observed that in some combinations, PGR stimulated callus formation and the production of adventitious embryos from the different tissues of immature ovules. On the contrary, the PGR-free medium allowed the regular development of the zygotic and apomictic embryos already present in the immature ovules without the production of callus and adventitious embryos. Therefore, the immature ovules used in the present study were incubated on PGR-free medium to facilitate the recovery of zygotic embryos. The *in vitro* protocol used in our research is simple, as it consists of only MS media supplemented with malt extract and sucrose. For *in ovulo* embryo rescue of *Hylocereus* interspecific hybrids, another cactus of horticultural significance, Cisneros and Tel-Zur [52] used a combination of naphtheleneacetic acid, thidiazuron, and glutamine. In blueberry [38] and gentian [53] *in vitro* ovule culture, casein hydrolysate seems to be an essential ingredient. Thus, a reduced form of organic nitrogen seems to be essential for embryo growth *in vitro*, as also suggested by Sahijram et al. [54].

It is known that in *Opuntia* spp., apomixis can occur mainly through sporophytic agamospermy [21–23], where adventitious embryos develop from nucellar tissue. However, the development of embryos from unfertilised ovules (parthenogenesis) has also been observed in the genus *Opuntia* [21]. However, the exact reason for the abortion of zygotic embryos *in vivo* during the seed maturation processes is not known, and we are focusing on this aspect in our current research. Our hypothesis is that the numerous embryos of apomictic origin that are contained in the ovule (often, there are more than ten embryos per ovule) compete with the zygotic embryo by using the resources necessary for its development, causing its abortion. This hypothesis is substantiated by the increase in the proportion and number of zygotic embryos when the immature ovules are incubated on a culture medium providing sufficient nutrients, which enhances the chances of survival of zygotic embryos.

In conclusion, it can be stated that *in ovulo* embryo culture can increase the number of zygotic seedlings and their ratio to apomictic seedlings; therefore, this can play a significant role in crop improvement programs of apomictic *O. ficus-indica* involving hybridisation and selection in segregating populations.

**Supplementary Materials:** The following supporting information can be downloaded at: https://www.mdpi.com/article/10.3390/plants12152758/s1, Supplementary Figure S1: Immature fruit growth pattern of the four varieties of prickly pear used in the experiments.

**Author Contributions:** Conceptualization, A.C. and F.C.; data analysis, A.C., R.P. and F.C.; methodology, A.C., C.C., R.P., M.S., P.I., A.M. and F.C.; investigation, A.C., C.C., R.P., M.S., P.I., A.M. and F.C.; writing—original draft preparation, A.C., R.P. and F.C.; writing—review and editing, A.C., R.P. and F.C.; supervision, A.C. and F.C. All authors have read and agreed to the published version of the manuscript.

**Funding:** This research received no external funding.

**Data Availability Statement:** All data that were generated or analysed during this study have been included in this published article. Statistical data output is available upon request from the corresponding author.

**Acknowledgments:** We wish to thank Duncan Hedderley for valuable advice on the statistical analysis of the data.

**Conflicts of Interest:** The authors declare no conflict of interest.

## References

1. Arba, M. The potential of cactus pear (*Opuntia ficus-indica* (L.) Mill.) as food and forage crop. In *Emerging Research in Alternative Crops*; Springer: Cham, Switzerland, 2020; pp. 335–357.
2. Inglese, P.; Saenz, C.; Mondragon, C.; Nefzaoui, A.; Louhaichi, M. *Crop Ecology, Cultivation and Uses of Cactus Pear*; Food and Agriculture Organisation of the United Nations; International Center for Agricultural Research in the Dry Areas: Rome, Italy, 2017; p. 277.
3. Khodaeiaminjan, M.; Nassrallah, A.A.; Kamal, K.Y. Potential attribute of Crassulacean acid metabolism of *Opuntia* spp. production in water-limited conditions. In *Opuntia spp.: Chemistry, Bioactivity and Industrial Applications*; Ramadan, M.F., Ed.; Springer Nature: Cham, Switzerland, 2021; pp. 201–218.
4. Incedayi, B.; Dogan, N.; Copur, O.U. Assessment of cactus pear leather (pestil) as a new snack food. *J. Food Sci. Technol.* **2022**, *59*, 3158–3166. [CrossRef] [PubMed]
5. Ramírez-Rodríguez, Y.; Martínez-Huélamo, M.; Pedraza-Chaverri, J.; Ramírez, V.; Martínez-Tagüeña, N.; Trujillo, J. Ethnobotanical, nutritional and medicinal properties of Mexican drylands Cactaceae Fruits: Recent findings and research opportunities. *Food Chem.* **2019**, *312*, 126073. [CrossRef] [PubMed]
6. Barba, F.J.; Garcia, C.; Fessard, A.; Munekata, P.E.S.; Lorenzo, J.M.; Aboudia, A.; Ouadia, A.; Remize, F. *Opuntia Ficus Indica* Edible Parts: A Food and Nutritional Security Perspective. *Food Rev. Int.* **2020**, *38*, 930–952. [CrossRef]
7. Zimmermann, H.; Cuen, M.P.S.Y. Prickly Pear, The Other Face of Cactus Pear. *Acta Hortic.* **2006**, *728*, 289–296. [CrossRef]
8. Butera, D.; Tesoriere, L.; Di Gaudio, F.; Bongiorno, A.; Allegra, M.; Pintaudi, A.M.; Kohen, R.; Livrea, M.A. Antioxidant Activities of Sicilian Prickly Pear (*Opuntia ficus indica*) Fruit Extracts and Reducing Properties of Its Betalains: Betanin and Indicaxanthin. *J. Agric. Food Chem.* **2002**, *50*, 6895–6901. [CrossRef] [PubMed]
9. Daniloski, D.; D'Cunha, N.M.; Speer, H.; McKune, A.J.; Alexopoulos, N.; Panagiotakos, D.B.; Petkoska, A.T.; Naumovski, N. Recent developments on *Opuntia* spp., their bioactive composition, nutritional values, and health effects. *Food Biosci.* **2022**, *47*, 101665. [CrossRef]
10. Elella, F.M.A. Antioxidant and Anticancer Activities of Different Constituents Extracted from Egyptian Prickly Pear Cactus (*Opuntia ficus-indica*) Peel. *Biochem. Anal. Biochem.* **2014**, *158*, 1000158. [CrossRef]

11. Neupane, D.; Mayer, J.A.; Niechayev, N.A.; Bishop, C.D.; Cushman, J.C. Five-year field trial of the biomass productivity and water input response of cactus pear (*Opuntia* spp.) as a bioenergy feedstock for arid lands. *GCB Bioenergy* **2021**, *13*, 719–741. [CrossRef]
12. Palmieri, N.; Suardi, A.; Stefanoni, W.; Pari, L. *Opuntia ficus-indica* as an Ingredient in New Functional Pasta: Consumer Preferences in Italy. *Foods* **2021**, *10*, 803. [CrossRef]
13. ISTAT. Italian National Institute of Statistics Rome, 2021.
14. Novoa, A.; Le Roux, J.J.; Robertson, M.P.; Wilson, J.R.; Richardson, D.M. Introduced and invasive cactus species: A global review. *AoB Plants* **2015**, *7*, 1–14. [CrossRef]
15. Omar, A.A.; ElSayed, A.I.; Mohamed, A.H. Genetic diversity and ecotypes of *Opuntia* spp. In *Opuntia spp.: Chemistry, Bioactivity and Industrial Applications*; Ramadan, M.F., Ayoub, T.E.M., Rohn, S., Eds.; Springer International Publishing: Cham, Switzerland, 2021; pp. 181–199.
16. Rebman, J.P.; Pinkava, D.J. Opuntia Cacti of North America: An Overview. *Fla. Entomol.* **2001**, *84*, 474. [CrossRef]
17. Lloyd, S. *Situation Statement on Opuntioid cacti (Austrocylindropuntia spp., Cylindropuntia spp. and Opuntia spp.) in Western Australia*; Department of Agriculture and Food; Government of Western Australia: South Perth, Australia, 2014.
18. CISS. Weeds Australia Profiles—*Opuntia* spp. (Prickly Pears). Available online: https://profiles.ala.org.au/opus/weeds-australia/profile/Opuntia%20spp (accessed on 16 April 2023).
19. Mondragon-Jacobo, C. Preliminary identification of apomictic seedlings *Opuntia* spp. Breeding populations from central Mexico. *Acta Hortic.* **2002**, *581*, 119–124. [CrossRef]
20. Ganong, W.F. Upon Polyembryony and Its Morphology in *Opuntia vulgaris*. *Bot. Gaz.* **1898**, *25*, 221–228. [CrossRef]
21. Garcia Aguilar, M.; Pimienta Barrios, E. Cytological evidences of agamospermy in *Opuntia* (Cactaceae). *Haseltonia* **1996**, *4*, 39–42.
22. Jacobo, C.M. Verification of the apomictic origin of Cactus pear (*Opuntia* spp. Cactaceae) seedlings of open pollinated and crosses from central Mexico. *J. Prof. Assoc. Cactus Dev.* **2001**, *4*, 49–56.
23. Vélez-Gutierrez, C.; Rodríguez-Garay, B. Microscopic analysis of polyembryony in *Opuntia ficus-indica*. *J. Prof. Assoc. Cactus Dev.* **1996**, *1*, 39–48.
24. Kaaniche-Elloumi, N.; Jedidi, E.; Mahmoud, K.; Chakroun, A.; Jemmali, A. Gibberellic acid application and its incidence on in vitro somatic embryogenesis and fruit parthenocarpy in an apomictic cactus pear (*Opuntia ficus-indica* (L.) Mill.) clone. *Acta Hortic.* **2013**, *1067*, 225–230. [CrossRef]
25. Carimi, F.; De Pasquale, F.; Puglia, A.M. In vitro rescue of zygotic embryos of sour orange, *Citrus aurantium* L., and their detection based on RFLP analysis. *Plant Breed.* **1998**, *117*, 261–266. [CrossRef]
26. Murashige, T.; Skoog, F. A revised medium for rapid growth and bioassays with tobacco tissue cultures. *Physiol. Plant.* **1962**, *15*, 473–497. [CrossRef]
27. Doyle, J.J.; Doyle, J.L. A rapid DNA isolation procedure for small quantities of fresh leaf tissue. *Phytochem. Bull.* **1987**, *19*, 11–15.
28. Sambrook, J.; Fritsch, E.; Maniatis, T. *Molecular Cloning: A Laboratory Manual*, 2nd ed.; Cold Spring Harbor Laboratory Press: Cold Spring Harbor, NY, USA, 1989.
29. Siragusa, M.; Carra, A.; Salvia, L.; Puglia, A.M.; De Pasquale, F.; Carimi, F. Genetic instability in calamondin (*Citrus madurensis* Lour.) plants derived from somatic embryogenesis induced by diphenylurea derivatives. *Plant Cell Rep.* **2007**, *26*, 1289–1296. [CrossRef] [PubMed]
30. Fang, D.Q.; Roose, M.L. Identification of closely related citrus cultivars with inter-simple sequence repeat markers. *Theor. Appl. Genet.* **1997**, *95*, 408–417. [CrossRef]
31. Mora, M.; Matamala, P.; Sáenz, C. Characterization of Chilean consumers' attitudes and preferences toward different cactus pear colors. *Acta Hortic.* **2019**, *1241*, 143–148. [CrossRef]
32. Lenzi, M.; Orth, A.I. Mixed reproduction systems in *Opuntia monacantha* (Cactaceae) in Southern Brazil. *Braz. J. Bot.* **2012**, *35*, 49–58. [CrossRef]
33. Griffith, M.P. The origins of an important cactus crop, *Opuntia ficus-indica* (Cactaceae): New molecular evidence. *Am. J. Bot.* **2004**, *91*, 1915–1921. [CrossRef] [PubMed]
34. Mayer, J.A.; Cushman, J.C. Nutritional and mineral content of prickly pear cactus: A highly water-use efficient forage, fodder and food species. *J. Agron. Crop. Sci.* **2019**, *205*, 625–634. [CrossRef]
35. Jacoboa, C.M.; Chessab, I. Nopal (Opuntia spp.) genetic resources. In *Crop Ecology, Cultivation and Uses of Cactus Pear*; Inglese, P., Saenz, C., Mondragon, C., Nefzaoui, A., Louhaichi, M., Eds.; Food and Agriculture Organisation of the United Nations; International Center for Agricultural Research in the Dry Areas: Rome, Italy, 2017; pp. 43–50.
36. Pathirana, R.; Carimi, F. Management and Utilization of Plant Genetic Resources for a Sustainable Agriculture. *Plants* **2022**, *11*, 2038. [CrossRef]
37. Wijerathna-Yapa, A.; Pathirana, R. Sustainable Agro-Food Systems for Addressing Climate Change and Food Security. *Agriculture* **2022**, *12*, 1554. [CrossRef]
38. Pathirana, R.; Wiedow, C.; Pathirana, S.; Hedderley, D.; Morgan, E.; Scalzo, J.; Frew, T.; Timmerman-Vaughan, G. Ovule culture and embryo rescue facilitate interspecific hybridisation in blueberry (*Vaccinium* spp.). In Proceedings of the VIII International Symposium on in Vitro Culture and Horticultural Breeding, Coimbra, Portugal, 2–7 June 2015; Volume 1083, pp. 123–132. Available online: https://cdn.blueberriesconsulting.com/2017/12/Ovule-Culture-and-Embryo-Rescue-Facilitate-Interspecific-Hybridisation-in-BLUEBERRy.pdf (accessed on 10 May 2023).

39. Tan, M.-L.; Song, J.-K.; Deng, X.-X. Production of two mandarin × trifoliate orange hybrid populations via embryo rescue with verification by SSR analysis. *Euphytica* **2007**, *157*, 155–160. [CrossRef]
40. Xie, K.-D.; Yuan, D.-Y.; Wang, W.; Xia, Q.-M.; Wu, X.-M.; Chen, C.-W.; Chen, C.-L.; Grosser, J.W.; Guo, W.-W. Citrus triploid recovery based on 2x × 4x crosses via an optimized embryo rescue approach. *Sci. Hortic.* **2019**, *252*, 104–109. [CrossRef]
41. Kim, M.; Kim, S.H.; Kim, H.B.; Park, Y.C.; Song, K.J. Some factors affecting the efficiency of hybrid embryo rescue in the 'Shiranuhi' mandarin. *Hortic. Sci. Technol.* **2020**, *38*, 271–281. [CrossRef]
42. Felker, P.; Zapata, R.; Wang, X.; Medina, D.; Bunch, R.; Paterson, A. Fruit characters among apomicts and sexual progeny of a cross of the Texas native *Opuntia lindheimerii* (1250) with a commercial fruit type *Opuntia ficus–indica* (1281). *J. Prof. Assoc. Cactus Dev.* **2010**, *12*, 48–66.
43. Hand, M.L.; Koltunow, A.M.G. The Genetic Control of Apomixis: Asexual Seed Formation. *Genetics* **2014**, *197*, 441–450. [CrossRef] [PubMed]
44. Kaushal, P.; Dwivedi, K.K.; Radhakrishna, A.; Srivastava, M.K.; Kumar, V.; Roy, A.K.; Malaviya, D.R. Partitioning Apomixis Components to Understand and Utilize Gametophytic Apomixis. *Front. Plant Sci.* **2019**, *10*, 256. [CrossRef]
45. Grimanelli, D.; Leblanc, O.; Perotti, E.; Grossniklaus, U. Developmental genetics of gametophytic apomixis. *Trends Genet.* **2001**, *17*, 597–604. [CrossRef]
46. Ozias-Akins, P.; van Dijk, P.J. Mendelian Genetics of Apomixis in Plants. *Annu. Rev. Genet.* **2007**, *41*, 509–537. [CrossRef] [PubMed]
47. Ahloowalia, B.S.; Maluszynski, M. Induced mutations—A new paradigm in plant breeding. *Euphytica* **2001**, *118*, 167–173. [CrossRef]
48. Pathirana, R. Mutations in plant evolution, crop domestication and breeding. *Trop. Agric. Res. Ext.* **2021**, *24*, 124. [CrossRef]
49. Fernández, M.; Figueiras, A.; Benito, C. The use of ISSR and RAPD markers for detecting DNA polymorphism, genotype identification and genetic diversity among barley cultivars with known origin. *Theor. Appl. Genet.* **2002**, *104*, 845–851. [CrossRef]
50. Grativol, C.; Lira-Medeiros, C.D.F.; Hemerly, A.S.; Ferreira, P.C.G. High efficiency and reliability of inter-simple sequence repeats (ISSR) markers for evaluation of genetic diversity in Brazilian cultivated *Jatropha curcas* L. accessions. *Mol. Biol. Rep.* **2010**, *38*, 4245–4256. [CrossRef]
51. Raji, M.R.; Lotfi, M.; Tohidfar, M.; Zahedi, B.; Carra, A.; Abbate, L.; Carimi, F. Somatic embryogenesis of muskmelon (*Cucumis melo* L.) and genetic stability assessment of regenerants using flow cytometry and ISSR markers. *Protoplasma* **2017**, *255*, 873–883. [CrossRef] [PubMed]
52. Cisneros, A.; Tel-Zur, N. Embryo rescue and plant regeneration following interspecific crosses in the genus *Hylocereus* (Cactaceae). *Euphytica* **2010**, *174*, 73–82. [CrossRef]
53. Morgan, E.R. Use of *in ovulo* embryo culture to produce interspecific hybrids between *Gentiana triflora* and *Gentiana lutea*. *New Zealand J. Crop. Hortic. Sci.* **2004**, *32*, 343–347. [CrossRef]
54. Sahijram, L.; Soneji, J.R.; Naren, A.; Rao, B.M. Hybrid embryo rescue: A non-conventional breeding strategy in horticultural crops. *J. Hortic. Sci.* **2013**, *8*, 1–20.

*plants*

MDPI

*Article*

# In Vitro Polyploidization of *Brassolaeliocattleya* Hybrid Orchid

**Joe Abdul Vilcherrez-Atoche [1,2,*,†], Jéssica Coutinho Silva [3], Wellington Ronildo Clarindo [3], Mateus Mondin [4] and Jean Carlos Cardoso [1,*,†]**

1 Laboratory of Plant Physiology and Tissue Culture, Depart of Biotechnology, Plant and Animal Production, Center of Agricultural Sciences, Federal University of São Carlos (CCA/UFSCar), Rodovia Anhanguera, km 174, CP 153, Araras 13600-970, SP, Brazil

2 Master Science Graduate Program of Plant Production and Associated Bioprocesses, Center of Agricultural Sciences, Federal University of São Carlos, Araras 13600-970, SP, Brazil

3 Laboratório de Citogenética e Citometria, Departamento de Biologia Geral, Centro de Ciências Biológicas e da Saúde, Universidade Federal de Viçosa, Viçosa 36570-000, MG, Brazil

4 Departamento de Genética, Escola Superior de Agricultura "Luiz de Queiroz", Universidade de São Paulo, Piracicaba 13400-970, SP, Brazil

* Correspondence: joel210294@gmail.com (J.A.V.-A.); jeancardoso@ufscar.br (J.C.C.)

† These authors contributed equally to this work.

**Abstract:** The *Cattleya* (Orchidaceae–*Laeliinae* subtribe) intergeneric hybrids, such as *Brassolaeliocattleya* (*Blc.*), have great ornamental value, due to their compact-size, with large and high color diversity of flowers. Artificial induction of polyploidy brings agronomic, ornamental and genetic benefits to plants. Polyploidization efficiency depends on factors, such as the type of antimitotic, polyploidization method, concentrations, exposure times and type of explant. This study aimed to develop a protocol to polyploidize *Blc.* orchids, by testing two types of explants (seeds and protocorms), concentrations and exposure times to colchicine. The effects of colchicine on the in vitro development of explants were also investigated. The responses of explants to colchicine depended on the concentrations, exposure time and the interaction of these factors. Flow cytometric analysis evidenced high endopolyploidy and allowed the separation of polyploidized (4C, 8C and 16C peaks) from non-polyploidized (only 2C and 4C peaks) plants. The highest percentage of polyploid plants was regenerated from protocorms (16.4%) treated with colchicine instead of seeds (3.2%). Protocorms treated with colchicine at 500–750 μM for 18 h resulted in the best percentage of polyploidization. Additionally, in vitro natural polyploidization using protocorms was reported (11.5%). Cytological analyses allowed an estimation of the number of chromosomes of the parents (≡70), polyploidized (≡140) and non-polyploidized progeny (≡70).

**Keywords:** orchid hybridization; in vitro culture; seedling; colchicine; polyploids; flow cytometry

**Citation:** Vilcherrez-Atoche, J.A.; Silva, J.C.; Clarindo, W.R.; Mondin, M.; Cardoso, J.C. In Vitro Polyploidization of *Brassolaeliocattleya* Hybrid Orchid. *Plants* **2023**, *12*, 281. https://doi.org/10.3390/plants12020281

Academic Editors: Ranjith Pathirana and Francesco Carimi

Received: 22 November 2022
Revised: 1 January 2023
Accepted: 4 January 2023
Published: 7 January 2023



## 1. Introduction

Orchids of the genus *Cattleya* (Orchidaceae) are commonly called the "queens of orchids" [1], whose flowers are characterized by their large and wide petals in relation to the more elongated sepals, and have a lip with a great ornamental value [2]. In *Cattleya*, the currently commercialized plants are hybrids, obtained by interspecific or intergeneric hybridization, which has been the most widely breeding technique used to obtain a great diversity of commercial hybrids [3].

*Cattleya* hybrids show an excellent ornamental quality and acceptance in the international potted flower market [4] but the long juvenile period, the highly genotype-specific response to flowering and the short lifespan of flowers hampered the efficient development of large-scale production [3,5] similar to developed with other orchid genera, such as *Phalaenopsis*, *Dendrobium* and *Oncidium* hybrids, with greater commercial importance.

In orchid breeding programs, the addition of biotechnological tools to conventional hybridization, such as in vitro polyploidization, could be used to obtain superior genotypes with desirable characteristics that have not yet been obtained by conventional breeding programs [6].

Polyploidy is a phenomenon associated with organisms with more than two sets of chromosomes in their cells [7]. Polyploidy in plants can be naturally achieved either by endoreduplication [8] or by the fusion of unreduced gametes [9] but has widely been achieved by artificial induction using chemicals with antimitotic action. In plants, the most common antimitotic used is colchicine [10,11].

Colchicine inhibits microtubule formation in the chromatic spindle causing nondisjunction and subsequent duplication of chromosomes within plant cells, generating polyploid cells [12], which develop and regenerate into polyploid plants. Polyploidization produces interesting changes and new features in cultivated plants, ranging from morphological, horticultural and cellular changes that are affected by physiological, biochemical and genetic characteristics, with some improvements in relation to diploid organisms [12]. For example, polyploidization in flowers results in changes of development and architecture of plants and also in floral structures, as well as in the flowering season and number of flowers [13]. Polyploidization has been used in breeding programs in different ornamental plants, such as *Salvia coccinea* cv. Coral Nymph, *Gladiolus grandiflorus* and *Chrysanthemum carinatum*, in which polyploid plants showed larger flowers, with thicker petals and a longer shelf-life [14–16]. Additionally, in the orchid industry and trade, production of new polyploid cultivars usually results in superior ornamental characteristics compared to diploid cultivars [17].

However, the efficiency to induce chromosome duplication under in vitro conditions requires the development of a protocol including the methodology containing all the steps of in vitro cultivation, as well as the methods used for treating plant tissues with antimitotic chemicals. The factors affecting most the efficiency of polyploidization include the genotype, type of explant, culture medium and cultivation conditions, and those related to the antimitotic agent, such as molecule used and induction method, concentration and exposure time, and the method used to confirm polyploidization [10].

Different protocols for artificial induction of polyploid plants have already been developed for the main commercial genera of orchids, such as *Phalaenopsis* [18], *Cymbidium* [19], *Dendrobium* [20] and *Oncidium* [21]. In *Cattleya*, there are only two studies on the induction of polyploidy using only species, such as *Cattleya tigrina* [22] and *C. intermedia* [23]. In addition, there is limited knowledge about ploidy levels of interspecific and intergeneric hybrids as a result of hybridization, and efficient protocols aimed to achieve the polyploidization of commercial *Cattleya* hybrids, such as *Brassolaeliocattleya*. In this context, the objective of this study was to test different types of explants subjected to different exposure times and colchicine concentrations, aiming to develop a protocol for polyploidization and in vitro artificial chromosome duplication of a *Brassolaeliocattleya* hybrid. In addition, we also studied the different effects of colchicine on the in vitro development and growth of explants treated with this alkaloid.

## 2. Results

### 2.1. Colchicine Effects on Explant Development

(a) Seeds

The percentage of seeds containing embryos (total five repetitions of 100 seeds counted under microscopy) of the crossing between *Blc.* Haw Yuan Beauty × *Blc.* Goldenzelle "LC" was 81.25% [(number of seeds containing embryos/total number of seeds) * 100)].

Increases in colchicine concentration or exposure time caused a decrease in the percentage of germinated seeds (Table 1). The percentage of seeds that developed into seedlings, instead of protocorms, only increased using the highest concentration of 1000 μM (Table 1). On the contrary, increasing exposure time led to increases in protocorm development (54.99% to 68.36%), instead of seedlings with roots and shoots, which decreased from 45.01% to 31.64% (Table 1).

**Table 1.** Germination percentage (GP), percentages of seeds developed into protocorms (PrD) and into seedlings (SeD) and total fresh weight (TFW) of *Blc.* Haw Yuan Beauty x *Blc.* Goldenzelle "LC" seed progeny exposed to different times and concentrations of colchicine.

| | Seeds | | | |
|---|---|---|---|---|
| **Colchicine Treatment Time (h)** | **GP (%)** | **PrD (%)** | **SeD (%)** | **TFW (g)** |
| 6 | 44.86 a | 54.99 c | 45.01 a | 13.62 a |
| 12 | 31.53 c | 62.51 ab | 39.15 ab | 12.40 a |
| 18 | 29.94 c | 58.25 bc | 41.75 a | 12.26 a |
| 24 | 36.96 b | 68.36 a | 31.64 b | 13.42 a |
| **Colchicine concentrations (μM)** | | | | |
| 0.00 | 49.42 a | 62.88 a | 37.10 b | 14.52 a |
| 250 | 36.21 b | 60.38 ab | 39.62 ab | 12.76 ab |
| 500 | 34.86 b | 67.21 a | 34.86 b | 13.08 ab |
| 750 | 30.64 c | 61.75 a | 38.25 b | 12.50 ab |
| 1000 | 27.97 d | 52.91 b | 47.09 a | 11.77 b |
| F Colchicine treatment time (h) | 59.15 ** | 12.91 ** | 10.06 ** | 2.49 ns |
| F Colchicine concentrations (%) | 69.89 ** | 7.36 ** | 4.78 ** | 3.85 ** |
| F Interaction | 5.13 ** | 4.1 ** | 3.10 ** | 1.64 ns |
| CV (%) | 3.28 | 24.19 | 4.05 | 5.96 |

Mean values followed by different letters, in the same column, are significantly different by Tukey's test at 1% probability. ** significant at 1% probability ($p < 0.01$).

Both increases in colchicine concentration and exposure time had an effect on the total fresh weight, with a reduction of 2.75 g at 1000 μM colchicine (11.77 g/vial) in relation to the control treatment (14.52 g/vial) (Table 1).

(b) Protocorms

Increasing concentrations of colchicine resulted in a gradual decrease of protocorm survival (Table 2). The type of regeneration of protocorms, either by the proliferation of protocorm like-bodies (PLBs) or by the formation of new plantlets, were not affected by colchicine concentration (Table 2). In addition, increases in colchicine concentration gradually reduced the total fresh weight of tissues regenerated from protocorms, from 12.75 g (control) to 8.66 g (at 1000 μM), a reduction of 32% total fresh weight (Table 2).

Exposure time to colchicine had no effects on explant survival (Table 2). However, the best percentage of regeneration via plantlet formation, instead of PLBs proliferation, was reported at 12 h time exposure (Table 2). The largest difference occurred when treated for 6 h compared to longer exposure times, in which there was a reduction in the proliferation of PLBs and an increase in protocorms regenerated into plantlets (Table 2).

Colchicine also had important physiological effects on protocorms, resulting in the death of part of the protocorms and tissues exposed to colchicine. This was observed by the difference in color in part of the explants treated with the alkaloid, which showed a brown color, demonstrating partial or total death of the treated tissues (Figure 1A), compared to untreated protocorms (Figure 1B).

Although most protocorms showed a brown-color, after treatment with colchicine, the emergence of new points of regeneration of PLBs in tissues were observed, demonstrating that phytotoxic effects of colchicine did not completely affect the tissue, thus allowing their regeneration (Figure 2).

### 2.2. *Flow Cytometry and Cytogenetic Analysis*

Specific and high-quality peaks (2C, 4C, 8C or 16C) were obtained for the analyzed individuals (Figure 3). Flow cytometry confirmed the hybrid origin of the obtained seedlings (Figure 4). Nuclear 2C value showed that the female parent (*Blc.* Haw Yuan Beauty) has a higher DNA content (2C = 7.79 pg) than the male parent (*Blc.* Goldenzelle 'LC', 2C = 5.79 pg) (Figure 3A,B). All the progenies from the control (without colchicine treatment) showed intermediary peaks (Figures 3C and 4) and nuclear 2C value (2C = 6.86 pg). Flow cytometry also confirmed the polyploidized plants, which showed higher nuclear DNA content (2C = 14.37 pg) (Figure 3D).

**Table 2.** Percentages of survival (PS), PLBs proliferation (PLBP), regeneration into plantlets (RIP) and total fresh weight (TFW) of *Blc.* Haw Yuan Beauty × *Blc.* Goldenzelle "LC" progeny protocorms exposed to different times and concentrations of colchicine.

| | Protocorm | | | |
|---|---|---|---|---|
| **Colchicine Treatment Time (h)** | **PS (%)** | **PLBP (%)** | **RIP (%)** | **TFW (g)** |
| 6 | 64.67 a | 59.38 a | 40.62 b | 10.07 ab |
| 12 | 59.50 a | 36.94 b | 63.06 a | 11.70 ab |
| 18 | 63.34 a | 46.91 ab | 53.09 ab | 8.92 b |
| 24 | 61.67 a | 40.92 b | 59.08 a | 12.30 a |
| **Colchicine concentrations (μM)** | | | | |
| 0.00 | 72.92 a | 44.19 a | 55.80 a | 12.75 a |
| 250 | 71.25 a | 47.21 a | 52.79 a | 11.01 ab |
| 500 | 66.67 ab | 43.47 a | 56.53 a | 11.23 ab |
| 750 | 51.04 bc | 46.33 a | 53.67 a | 10.07 ab |
| 1000 | 49.58 c | 48.97 a | 51.03 a | 8.66 b |
| F treatment time | 0.49 ns | 5.0 ** | 5.25 ** | 4.07 * |
| F concentrations | 7.66 ** | 0.21 ns | 0.22 ns | 2.30 * |
| F Interaction | 1.92 * | 0.67 ns | 0.70 ns | 0.96 ns |
| CV (%) | 15.97 | 41.11 | 33.36 | 22.43 |

Mean values followed by different letters, in the same column, are significantly different by Tukey's test at 1% and 5% probability levels. ** significant at 1% probability ($p < 0.01$); * significant at 5% probability ($p < 0.05$).

**Figure 1.** Effects of colchicine on in vitro protocorm survival of *Blc.* Haw Yuan Beauty × *Blc.* Goldenzelle "LC": (**A**) protocorms cultured in MS medium for 120 days after treatment with colchicine at a concentration of 1000 μM and exposure time of 24 h, showing dead tissues, regeneration of PLBs ** and plantlets *; (**B**) control treatment with protocorms in plant regeneration (red arrow). Scale bar = 1 cm.

Flow cytometry histograms showed the presence of cells with different DNA ploidy levels, e.g., 2C, 4C, 8C and 16C cells, demonstrating that the leaves of progeny showed a high level of endopolyploidy (Figure 3C,D). Thus, the distinction between polyploid and non-polyploid plants was based on the following parameters: diploid plants were considered those showing only cells in 2C, 4C and sometimes 8C channels (Figure 3C), while polyploid plants showed cells only in 4C, 8C and 16C channels (Figure 3D).

**Figure 2.** Regeneration of progeny of *Blc.* Haw Yuan Beauty × *Blc.* Goldenzelle "LC" in response to colchicine treatment of protocorms: (**A**) plantlets regenerated from protocorms treated with colchicine; (**B**) plantlets regenerated from protocorms not treated with colchicine. Scale bar = 1 cm.

In general, polyploidized plants presented different morphology and architecture compared to non-polyploidized plants (Figure 5). Polyploid plants had oblong-shaped leaves with greater thickness, width and intensity of green color (Figure 5A), unlike non-polyploid plants, which showed lanceolate-shaped leaves with lower width, thickness and a light green color (Figure 5B).

Chromosome counting of the parents showed that the estimated number of chromosomes was approximately 70 (Figure 6B), which is similar to the F1 progenies (≡70) (Figure 6A). Root tips of polyploid plants, analyzed by flow cytometry, confirmed the occurrence of in vitro polyploidization (≡140 chromosomes), especially in treatments with colchicine (Figure 6C,D).

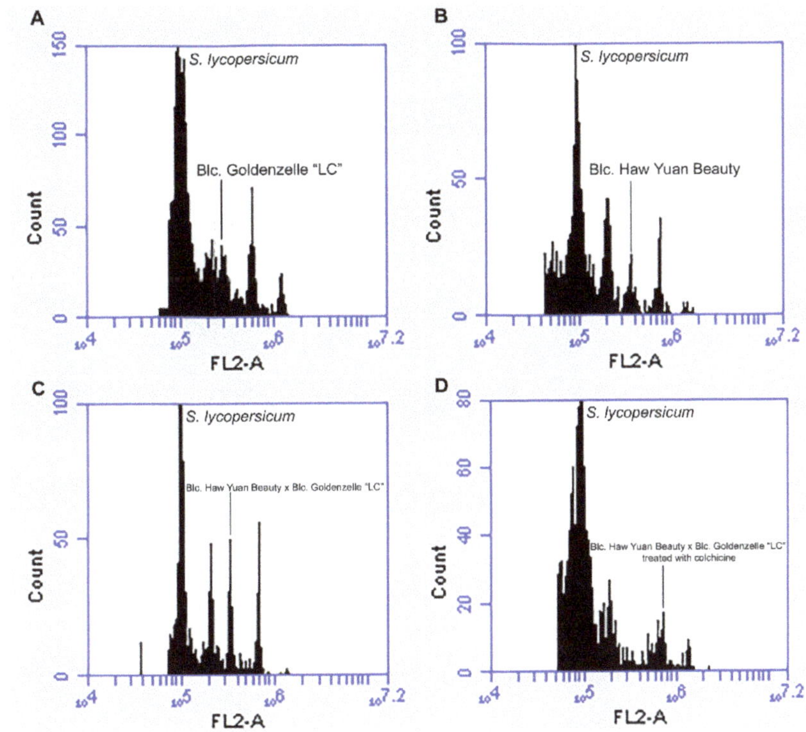


**Figure 3.** Flow cytometry histograms with nuclear 2C value measurements using the internal standard *S. lycopersicum* (2C = 2.00 pg): (**A**) male parent *Blc.* Goldenzelle "LC" with 2C = 5.79 pg; (**B**) female parent *Blc.* Haw Yuan Beauty with 2C = 7.79 pg; (**C**) *Blc.* Haw Yuan Beauty × *Blc.* Goldenzelle "LC" not treated with colchicine with 2C = 6.86 pg; (**D**) *Blc.* Haw Yuan Beauty × *Blc.* Goldenzelle "LC" treated with colchicine with 2C = 14.37 pg.

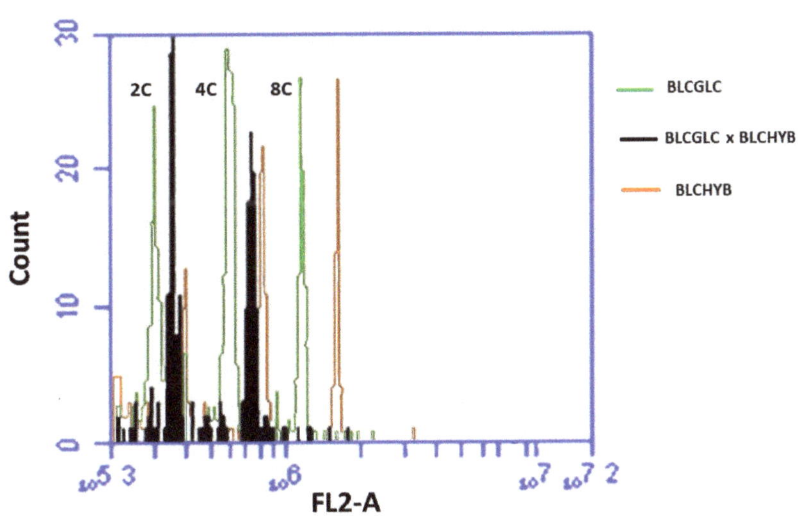


**Figure 4.** Comparison of the 2C, 4C and 8C peaks channels between *Blc.* Haw Yuan Beauty (BLCHYB) (orange line), *Blc.* Goldenzelle "LC" (BLCGLC) (green line) and the progeny of *Blc.* Haw Yuan Beauty × *Blc.* Goldenzelle "LC" not treated with colchicine (black line).

**Figure 5.** Development and in vitro rooting of plantlets regenerated from protocorm explants treated and not treated with colchicine: (**A**) vial containing only polyploid plants (P) of the progeny from *Blc.* Haw Yuan Beauty × *Blc.* Goldenzelle "LC"; (**B**) vial containing non-polyploidized plants (np) of the progeny from *Blc.* Haw Yuan Beauty × *Blc.* Goldenzelle "LC" and not treated with colchicine. Scale bar = 1 cm.

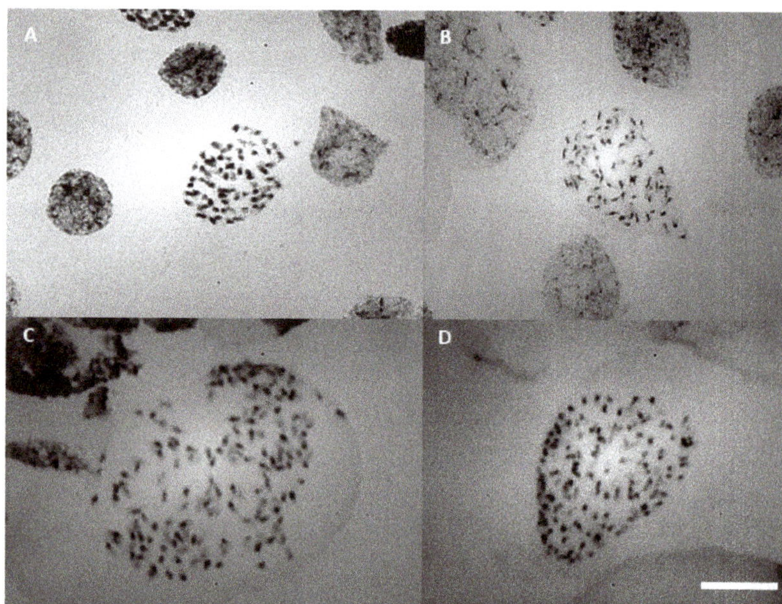


**Figure 6.** Prometaphase chromosome: (**A**) progeny of *Blc.* Haw Yuan Beauty × *Blc.* Goldenzelle "LC" without colchicine treatment; (**B**) *Blc.* Haw Yuan Beauty; (**C**,**D**) progeny of *Blc.* Haw Yuan Beauty × *Blc.* Goldenzelle "LC" treated with colchicine. Scale bar = 10 µm.

*2.3. Protocorms Resulted in the Best Polyploidization Rates Compared to Seeds*

Only five out of 154 progenies (3.21%) derived from seeds were polyploidized using colchicine (Figure 7A). Polyploid plants were derived only from seeds treated at 500, 750 and 1000 µM (Figure 7A). Differently and more effective than seeds, the use of protocorms treated with colchicine (Figure 7) resulted in a total of 35 polyploid plants out of the total of 213 analyzed (16.43% polyploidization efficiency). The best frequencies of polyploidization

(44 and 46%) were obtained using 1000 µM colchicine for 18 h and 500 µM colchicine for 6 h). Positive correlation between colchicine concentration and frequency of polyploidized seedlings was observed for colchicine treatment of protocorms (Figure 7B). The treatment time was also positively correlated with frequency of polyploidization until 18 h. The treatment for 24 h resulted in a reduction in polyploid frequency (14.6%) and was lower compared to shorter induction times (Figure 7C). In addition, natural polyploidization occurred in low frequency (11.5%) in water-treated protocorms (Figure 7A).

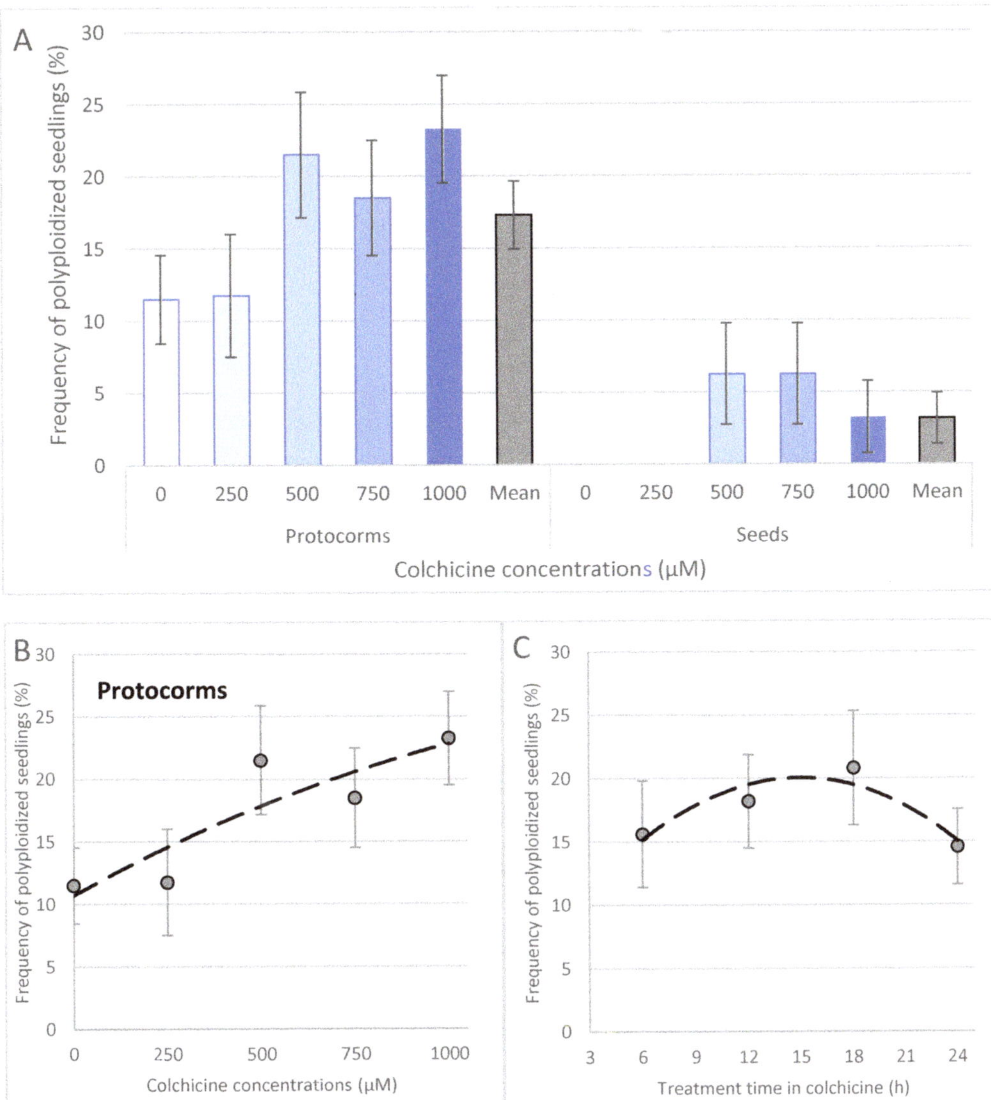


**Figure 7.** Polyploidy frequency in seedlings of *Blc.* hybrid. The type of the explant and concentration of colchicine in the frequency of polyploidization (**A**) and correlation with concentrations (**B**) and treatment time (**C**) in protocorms treated with colchicine. The Student's *t*-test showed significant values of coefficient of correlation (r) (0.05 *) for protocorms (treatment time and colchicine concentration) and for seeds (only for colchicine concentration). Equations and r values: (**B**) $y = -4 \times 10^{-6}x^2 + 0.0164x + 10.714$, $r = 0.882$ *; (**C**) $y = -0.0611x^2 + 1.8267x + 6.4$, $r = 0.913$ *.

## 3. Discussion

### 3.1. Colchicine Switched the In Vitro Development of Blc. Orchid

Colchicine has been the main substance used for chromosome duplication in plants. However, colchicine phytotoxicity was reported in different types of tissues used as explants, where the different levels of toxicity depend on the concentrations and exposure times used to treat the plant material [24].

In orchids, such as *Bletilla striata*, the treatment with the highest concentration of colchicine and longer exposure time (0.4% colchicine for 9 days) resulted in the lowest seed germination rate [25]. Chung et al. [13] also induced polyploidy using hybrid seeds of *Calanthe* (*C. discolor* × *C. sieboldii*) (Orchidaceae) and reported a slight gradual decrease in seed germination as a consequence of increasing concentration and exposure time of seeds to colchicine or oryzalin.

Lone et al. [26] observed a reduction in the survival rate of protocorms of *Cattleya tigrina* at the highest concentration of colchicine (1%), showing survival rates of 100%, 96% and 84%, with 24, 48 and 72 h of exposure, respectively, and compared to the control (100%). Similar results were reported for *Vanda* (Orchidaceae) protocorms treated with colchicine, aiming to induce polyploidy. The tissue response to colchicine resulted in greater accumulation of phenolic compounds, resulting in limited development of treated explants [27].

In the current experiment, lower survival percentages (between 40 and 80%) were observed after treating with colchicine. Reduction in growth of plant tissues treated with colchicine has been attributed to its negative effects on meristematic cells, producing abnormal cycles during cell division [28] and high concentrations of colchicine cause a decrease in mitotic division rates in treated tissues [29], which cause decreases in plant growth. These negative effects of colchicine were also observed in the present study with the hybrid progeny from *Blc.* Haw Yuan Beauty x *Blc.* Goldenzelle "LC".

### 3.2. Colchicine Is Efficient at Inducing Polyploidization in Blc. Orchid

Chung et al. [13], using seeds of *Calanthe* (*C. discolor* × *C. sieboldii*) as explants, reported an increase in the percentage of tetraploids as a function of time of treatment (from 3 to 7 days) and concentration (0.05% to 0.1%) of colchicine. *Calanthe* hybrid seeds treated with 0.1% colchicine for 3 and 7 days produced a high percentage of polyploids, 74% and 81%, respectively [13].

Among the differences between the present study with *Blc.*, using seeds as explants, and that conducted by Chung et al. [13], was the time of exposure of seeds to colchicine and the genotype. While the latter authors adopted 3–7 days, the present study with *Blc.* used 6–24 h (or 1 day). Despite this, in *Blc.* it was observed that the longest time of treatment with colchicine (24 h) was not the one with the highest frequency of polyploidized plants. Chung et al. [13] also concluded that the high percentage of polyploids in *Calanthe* is also due to its genome that allow the easy duplication of the number of chromosomes.

Even with a low rate of colchicine-treated seeds that developed into polyploidized plants in the present study on *Blc.*, the small seed size enabled the treatment of a high number of seeds using a reduced amount of colchicine solution, a reagent with high cost and risks associated with its manipulation [6].

Different from seeds used as explants, increases in the percentage of polyploids in colchicine-treated protocorms of *Blc.* are correlated with colchicine concentrations and exposure times and were superior or similar (38–46%) to those previously obtained with other orchid species and hybrids. For example, in protocorms of the hybrid *Cymbidium sinenthese* 'Lv mosu' × *Cym. hybridum* 'Shijieheping', the treatment with 0.03% colchicine for 72 h generated the best result, with 36% polyploids [30]. PLBs of *Dendrobium chrysotoxum* treated with 0.04% colchicine for 24 h produced a polyploid frequency of 47% [31]. Additionally, protocorms of *Phalaenopsis equestris*, *Phal. fasciata* and *Phal.* Betty Hausermann showed high-frequency of polyploidization, 46%, using the treatment with 50 mg $L^{-1}$ colchicine for 10 h [32].

Other interesting observation of our study are the presence of polyploidized seedlings from protocorms not treated with colchicine. Three main factors could explain such results: polyploid seedlings were generated during or as result of hybridization; the occurrence of natural endoreduplication in protocorm cells, which formed polyploidized PLBs and plantlets; the use of liquid medium and the presence of anaerobiosis combined with the time of treatment of the protocorms induces polyploidy.

Although polyploidization in response to hybridization has already been reported in orchids [33], this is not the case of natural polyploid plants from *Brassolaeliocattleya*, since flow cytometry of non-treated progenies showed intermediary nuclear DNA content between two parents. The hypothesis of endopolyploid cells inside protocorms that resulted in natural polyploid plants is strengthened by flow cytometry analysis that showed a high number of endopolyploid cells in tissues of protocorms of the studied hybrid.

Thus, the most accepted hypothesis is that the sections, manipulation and exposure of endopolyploid cells of the isolated protocorms resulted in the production of polyploid PLBs in the absence of colchicine. This result and conclusion were confirmed by Chen et al. [34], in which the sectioning of PLBs from *Phalaenopsis aphrodite* resulted in up to 34% polyploid PLBs. These authors [34] concluded that the main cause of polyploids regenerated from PLBs was derived from endopolyploid cells observed in their tissues. Furthermore, *Cattleya* is an orchid genus in which it has been observed endopolyploidy events in different tissues of several species, such as *C. trigina* [22], *C. trianae, C. grandis, C. guttata, C. labiata, C. cernua, C. tenius, C. elongata, C. crispata, C rupestres, C. aclandiae, C. amethystoglossa, C. pfisterii, C. rupestris, C. sincorana, C. loddigesii* and *C. granulosa* [35].

This spontaneous system for polyploidization of orchids using protocorms is extremely interesting from a practical point of view, as it does not require complex and additional procedures for using colchicine [18], a high-cost product that can pose health risks to the operator, whether in the preparation of solutions or in the treatments applied to plant tissues.

### 3.3. *Ploidy Levels and Chromosome Counting in Blc. Orchids*

Direct and indirect methods can be used for identification and confirmation of polyploid plants compared to diploid ones [36]. Flow cytometry is also considered a high-throughput system for early screening of polyploid plants in orchids [37].

The presence of endopolyploid cells, as reported for *Blc.* progenies in the present study, was also observed in other orchids, such as *Cattleya tigrina,* with ploidy levels of 2C and 4C and self-polyploidized seedlings resulting in cells with ploidy levels of 4C and 8C [22]. In our study, polyploidized plants showed only 4C and 8C peaks and at much lower frequencies, 16C, instead of 2C and 4C observed in most of non-colchicine treated protocorms.

As a consequence, seedlings of *Blc.* also showed changes in morphological features, and similar to observed in the leaves, such as thickness, length and intensity of green color in polyploids of *C. tigrina* [22], *Dendrobium formosum* [38], *Cymbidum lowianum* [39] *Phalaenopsis amabilis* and *Phal. amboinensis* [40]. The intensity of leaf color is used as a morphological marker for the identification of polyploidized plants [36], where it is believed that this effect is related to chromosomal duplication, causing an increase in the content of pigments and the enzyme production in polyploid plant cells [22].

Root tips of polyploid plants were analyzed by chromosome counting and confirmed the occurrence of in vitro polyploidization (≡140 chromosomes).

The combined results of cytogenetics with flow cytometry analysis confirmed that polyploidized plants contained twice the number of chromosomes and more than twice the DNA content of non-polyploidized plants. However, it was not possible to determine the exact number of chromosomes from the parents and progeny obtained from the cross between the two cultivars of *Brassolaeliocattleya* (*Blc.* Haw Yuan Beauty x *Blc.* Goldenzelle "LC") used in the present study, which have ornamental and horticultural characteristics from *C. briegeri, C. intermedia, C. forbesi, C. loddigesii, C. dowiana, C. trianae, C. tenebrosa* and

*C. bicolor* [41,42]. Cytological studies have shown that most *Cattleya* species mentioned above have 40 chromosomes [43,44], but there are also intraspecific chromosomal variations, such as in *C. trianae* and *C. bicolor* with 42 chromosomes [44,45]. Similarly, *Brassavola* and *Laelia* also belong to the subtribe *Laeliinae*, therefore they have 40 somatic chromosomes, in addition to some chromosomal variations of 42, 44 and 60 identified in the genus *Laelia* [45].

Molecular cytogenetic techniques allowed increasing the amount of information on the evolution of the karyotype in the subtribe Laeliinae [43,46]. A study on the evolutionary karyotype diversity in the subtribe *Laeliinae*, using molecular cytogenetics together with chromosome band analysis, demonstrated that *C. trianae* showed a fusion of a chromosome pair as a rearrangement mechanism. *Laelia gouldiana* presented a polyploid karyotype and *L. marginata* had a supernumerary chromosome [43] indicating the high degree of chromosomal variations common to this subtribe, to which the hybrid used in the present study belongs.

Difficulties encountered in obtaining the exact number of chromosomes in *Blc*. Haw Yuan Beauty × *Blc*. Goldenzelle "LC" were related to the high number of chromosomes found, in both parents and progeny, and the difficulty in finding perfect metaphases, as well as the low rate of cell divisions observed in root meristems during the in vitro culture of *Blc*. Haw Yuan Beauty x *Blc*. Goldenzelle "LC".

## 4. Material and Methods

### 4.1. Plant Material, In Vitro Establishment and Growth

Seeds from a mature capsule, eight months old, were obtained from the crossing between two hybrid cultivars: *Brassolaeliocattleya* 'Haw Yuan Beauty' and *Blc*. Goldenzelle "LC" (Orchidaceae) germplasm collection at UFSCar, Araras, Brazil. Seeds were dried at room temperature for 24 h before storage in 1.5 mL eppendorf type vials at 8 °C.

Disinfection and in vitro seeding were carried out using a solution containing one volume of bleach (2.0–2.5% active chlorine) and nine volumes of autoclaved deionized water. Seeds were immersed in this solution for 12 min under agitation, followed by three washes in autoclaved deionized water. Seeding was performed in 30 mL Murashige and Skoog culture medium [47] with the macronutrient concentration reduced by half, with 2% sucrose (Synth®, Diadema-SP, Brazil), 1.2 g $L^{-1}$ activated charcoal (Synth®), 0.1 g $L^{-1}$ inositol (Synth®) and pH adjusted to 5.7 before the addition of 6.4 g $L^{-1}$ agar (Agargel®, João Pessoa-PB, Brazil) inside glass flasks (240 mL capacity) covered with polypropylene caps. The culture media contained in the flasks was sterilized by autoclaving for 25 min at 121 °C and 1 atm.

### 4.2. Colchicine Treatment Procedures

Colchicine (Sigma-Aldrich®, Saint Louis, MO, USA) was prepared from a stock solution of colchicine, previously dissolved in a 1% (*v/v*) solution of dimethylsulfoxide (Synth®). Five concentrations were tested in this experiment: 0.00 µM; 250 µM; 500 µM; 750 µM and 1000 µM, combined with four immersion times: 6, 12, 18 and 24 h.

Treatments with colchicine were applied to seeds collected and stored from the *Blc*. hybrid and used as explants, and to protocorms also obtained from the germination of the same seeds, but after 90 days of in vitro cultivation in the culture medium described in Section 4.1. For seed treatment, 5 mg fresh mass of seeds per treatment and 15 protocorms per replication (60 per treatment) were used for this study (Figure 8). The graphical abstract with step-by-step experiment is also presented (Figure 9).

For the seeds, the different concentrations of colchicine were applied as pre-treatment, by the immersion of seeds in solutions containing different concentrations of colchicine, using a 250 mL Erlenmeyer flask kept in a horizontal rotary shaker at 60 rpm and in dark conditions at 25 ± 1 °C during exposure times. After exposure time, seeds were subjected to asepsis and inoculated in vitro in a germination culture medium, as described in Section 4.1.

**Figure 8.** Types of explants used for polyploidy induction: (**A**) macroscopic view of the seeds; (**B**) microscopic view (×40) of viable seeds (VS) containing embryos and non-viable seeds (NVS); (**C**) protocorms used in colchicine treatments. Scale bar = 1mm (**B**) and Scale bar = 1 cm (**C**).

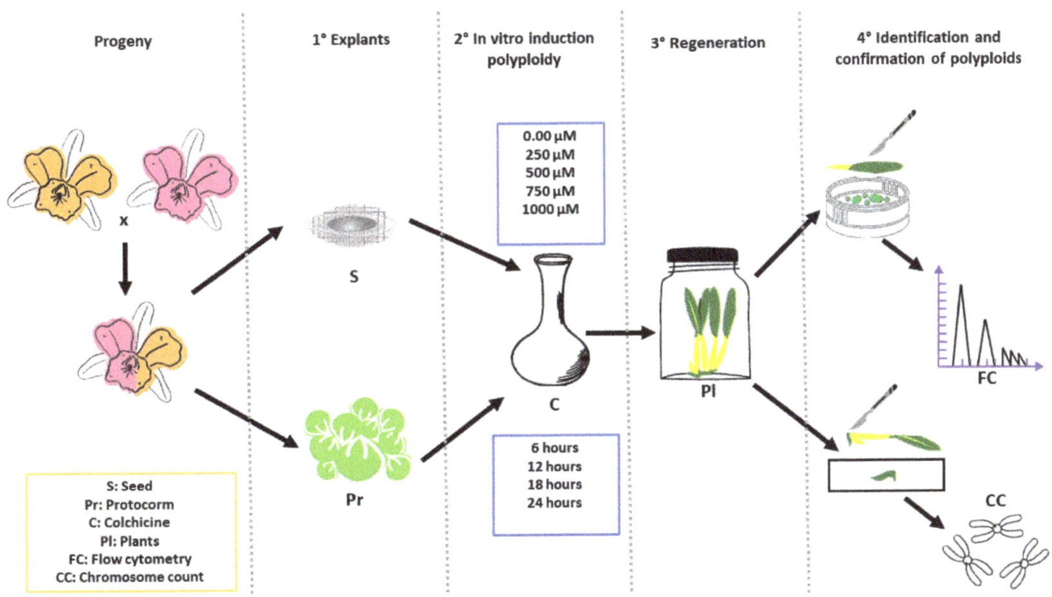


**Figure 9.** Graphical abstract of the main procedures, treatments and analysis of *Blc.* progeny treated with colchicine aiming polyploidization.

In vitro protocorms, obtained after 90 days of seed germination, were selected and subjected to different concentrations of colchicine and exposure times. The colchicine solution was filter-sterilized using a Millipore Millex™ EMD syringe filter (<0.22 µm) and poured into a 250 mL Erlenmeyer flask containing liquid MS medium (with no activated charcoal and agar). After preparing the different concentrations of colchicine, protocorms were immersed and maintained in horizontal rotary shakers at 60 rpm in dark conditions at 25 ± 1 °C during the exposure times.

For each explant, controls were made for each exposure time, with all detailed procedures maintained, except for the addition of colchicine in contact with seeds or protocorms. At the end of the colchicine exposure times, all explants were washed with sterile deionized water three times to remove colchicine from the explants, and then cultured again in MS culture medium containing charcoal and agar, as described in Section 4.1. There was no addition of plant growth regulators to the culture medium.

Cultivation of seeds or protocorms was carried out at 25 ± 2 °C with lighting provided by Light-Emitting Diodes (LEDs) in the red and blue wavelengths (3:1), with a Photosynthetically Photon Flux Density (PPFD) of ≡ 50 μmol $cm^{-2}$ $s^{-1}$ and photoperiod of 16 h.

For both explants, a 5 × 4 factorial completely randomized design was adopted, with five concentrations of colchicine and four exposure times. In total, four replications were used per treatment, consisting of 240 mL glass flasks containing 30 mL culture medium with the seeds (at least 5.0 mg/vial) or protocorms (15/vial).

### 4.3. Effects of Colchicine on Seeds and Protocorms

Evaluations of in vitro germination, regeneration and development were specific to each type of explant:

(a) Seeds: Using the counting method, the percentage of seeds was evaluated by the presence or absence of the embryo inside the testa with the aid of an optical microscope (Nikon Eclipse e200, Nikon Instruments, Japan) with a 4X objective lens (Figure 1B). This included the percentage of germination 180 days after seeding; the percentage of embryos that developed into protocorms and/or seedlings based on fresh mass calculation of each type of development after germination; the total fresh weight of plant tissue obtained per vial.

(b) Protocorms: After 180 days of culture following the treatment with colchicine, the percentages of survival and death of protocorms, the percentages of PLBs proliferation and the regeneration into plantlets and the total fresh weight obtained per vial were recorded.

### 4.4. Flow Cytometry Analysis

The nuclear 2C value was measured from eight seedlings (3–4 months old) of each treatment of seeds, and from 13 seedlings (7–8 months age) of each treatment of the protocorms in MS culture medium. The internal standard used for the analysis was *Solanum lycopersicum* L., 1753, 'Stupické' (2C = 2.00 pg) [25].

A leaf fragment of ~2 $cm^2$ from individuals of *Blc.* Haw Yuan Beauty × *Blc.* Goldenzelle "LC" and the internal standard were simultaneously chopped [48] for about 30 sec in a Petri dish containing 0.5 mL OTTO-I lysis buffer [49] supplemented with 50 μg $mL^{-1}$ RNAse (Sigma®) and 2 mM dithiothreitol (Sigma®) [50] and incubated for 3 min. A total of 0.5 mL of the same buffer was added and the suspension was filtered through a 30 μm diameter nylon mesh (Partec® Gmbh, Munster, Germany) in a 2.0 mL microtube. After centrifugation at 100× $g$ for 5 min, the supernatant was discarded and 100 μL of the same buffer was added to the precipitate; the material was vortexed and incubated for 10 min.

Subsequently, 0.5 mL modified OTTO-II staining buffer [49,50] (400 mM $Na_2HPO_4.H_2O$, 2 mM dithiothreitol (Sigma®), 50 μg $mL^{-1}$ RNAse (Sigma®) and 75 μg $mL^{-1}$ propidium iodide (PI (Sigma®), excitation/emission wavelengths: 480–575/550–740 nm) was added to the suspensions.

Suspensions were filtered through 20 μm nylon mesh (Partec®) into the reading tubes (Partec®) and kept for 30 min in the dark to stain the nuclei. Then, suspensions were analyzed in a flow cytometer (BD Accuri C6 flow cytometer, Accuri cytometers,

Belgium) equipped with a 488 nm laser to promote PI excitation and PI emission to the FL2 (615–670 nm) and FL3 (>670 nm).

Fluorescence peaks of the G0/G1 nuclei of each individual of *Blc.* Haw Yuan Beauty × *Blc.* Goldenzelle "LC" and the internal standard were analyzed from histograms using BD Accuri™ C6 software. G0/G1 peaks with a coefficient of variation (CV) less than 5% were considered to determine the level of DNA ploidy. The nuclear 2C value of each individual, in picograms (pg), was calculated using the formula below:

DNA content of each individual (pg) = [(mean channel of peak G0/G1 of each individual of *Blc.* Haw Yuan Beauty × *Blc.* Goldenzelle "LC") *2.00 pg *S. lycopersicum*]/(mean channel of peak G0/G1 from *S. lycopersicum*) *Blc.* Haw Yuan Beauty and *Blc.* Goldenzelle "LC" cultivars used as parents were also analyzed by flow cytometry.

### 4.5. Chromosome Counting

Root tips were collected from seedlings in in vitro conditions. Root tips (~2 cm) were pre-treated in a solution of 8-hydroxyquinoline (8-HQ) (Sigma®) at 300 ppm and cyclohexamine (Sigma®) at 25 ppm (19:1) for 24 h at a controlled temperature of 27 °C. The pre-treated roots were fixed in a 3:1 Carnoy solution for 24 h at a controlled room temperature of 27 °C.

The Feulgen method was used to stain the roots [51]. Roots were washed twice for five min with distilled water and hydrolyzed in a 5N HCl solution (Merck KGaA, Darmstadt, Germany) at 60 °C for 12 min. Afterwards, two washes were made with distilled water for five min and hydrolyzed roots were incubated in Schiff's reagent for 45 min in the dark.

Stained roots were treated twice in a 0.01 M citrate buffer solution for 5 min each, followed by enzymatic digestion using a mixed solution of cellulase (Serva 16420, Germany, final concentration of 1.4 U $mL^{-1}$) and pectinase (Calbiochem 515883, Germany, final concentration of 29.4 U $mL^{-1}$) (1:1) for one hour of incubation.

After digestion, roots were placed in a citrate buffer solution on ice until mounting the slides. The protocol described by Mondin and Aguiar-Perecin [51] was followed for preparation of cytological slides containing mitotic metaphases, in which the root was placed in a 45% acetic acid solution (Merck KGaA, Darmstadt, Germany) for ~2 min and then the root meristem was macerated on a slide with a drop of 1% acetic carmine. A coverslip was placed on the macerated tissue and heated with a lamp for later crushing/squashing. For chromosome counting, slides were analyzed using a Zeiss Axiophot 2 microscope using the appropriate filter. Images were acquired by the PCO CCD camera and digitized in the IKARUS software (Metasystems, Germany).

Images of cells containing mitotic metaphases were captured with a 100X objective lens. Image J software was used to analyze and count the chromosomes.

### 4.6. Statistical Analysis

All data were analyzed using the AgroEstat Online software (http://www.agroestat.com.br/), in which homogeneity and homoscedasticity tests, and analysis of variance (ANOVA) were run, and when a difference was detected, the means were compared by Tukey's test ($p < 0.05$ and $<0.01$). In addition, the frequency of polyploids from different concentrations and treatment time with colchicine were submitted to regression analysis and the coefficient of correlation (r values) were tested using Student's *t*-test.

## 5. Conclusions

This study reports the different effects of colchicine on in vitro growth and development of seeds and protocorms of the *Brassolaeliocattleya* hybrid, from the crossing between *Blc.* Haw Yuan Beauty × *Blc.* Goldenzelle "LC". In addition, we developed an efficient methodology for polyploidization of this orchid of high ornamental value. In addition, the potential of a colchicine-free polyploidization was demonstrated by using individualized protocorms as explants. Flow cytometry and cytological analysis were efficient in estimating ploidy and separating polyploid plants from non-polyploid plants.

**Author Contributions:** J.A.V.-A. and J.C.C. contributed with conceptualization, methodology, validation, formal analysis, investigation, resources, data curation, writing-original draft preparation and review and editing. J.C.C. also contributed with supervision and funding acquisition. M.M. contributed with methodology and formal analysis of cytogenetic studies. J.C.S. and W.R.C. contributed with methodology and formal analysis of flow cytometry analysis. All authors have read and agreed to the published version of the manuscript.

**Funding:** This study was partially financed by the Coordenação de Aperfeiçoamento de Pessoal de Nível Superior–Brazil (CAPES)–Finance Code 001 and supported by Conselho Nacional para o Desenvolvimento Científico e Tecnológico (CNPQ, Brazil), Process number 311083/2018-8.

**Data Availability Statement:** Data associated with this study may also be found at https://repositorio.ufscar.br/bitstream/handle/ufscar/16352/Escrita.DissertacaoFINAL04-07-22.pdf?sequence=1&isAllowed=y (accessed on 3 January 2023).

**Acknowledgments:** J.A.V.-A. thanks to the Carla Midori Iiyama (PPGPVBA-CCA/UFSCar) for collaboration in sample collection and to Silvia Cristina Menuzzo-Molina (Cytogenomics and Epigenetics Laboratory–ESALQ/USP) for help in sample preparation for chromosome counting. J.C.C. thanks to Conselho Nacional de Desenvolvimento Científico e Tecnológico for the Project number 311083/2018-8 and 317162/2021-7.

**Conflicts of Interest:** The authors declare no conflict of interest.

## References

1. AOS. The Showy Cattleya, Queen of the Orchids. 2013. Available online: http://www.aos.org/Default.aspx?id=%0A443 (accessed on 16 May 2022).
2. Cavero, M.; Collantes, B.; Patroni, C. *Orquídeas Del Peru*; Centro de Datos para la Conservación del Perú: La Molina, Peru, 1991.
3. Cardoso, J.C.; Martinelli, A.P.; da Silva, J.A.T. A novel approach for the selection of Cattleya hybrids for precocious and season-independent flowering. *Euphytica* **2016**, *210*, 143–150. [CrossRef]
4. Takane, R.J.; Yanagisawa, S.S.; Pivetta, K.F.L. *Cultivo Moderno de Orquídeas: Cattleya e Seus Híbridos*; UFC: Fortaleza, Brazil, 2010.
5. Cardoso, J.C.; Ono, E.O.; Rodrigues, J.D. Gibberellic acid and water regime in the flowering induction of Brassocattleya and Cattleya hybrid orchids. *Hortic. Bras.* **2010**, *28*, 395–398. [CrossRef]
6. Vilcherrez-Atoche, J.A.; Iiyama, C.M.; Cardoso, J.C. Polyploidization in Orchids: From Cellular Changes to Breeding Applications. *Plants* **2022**, *11*, 469. [CrossRef]
7. Soltis, P.S.; Marchant, D.B.; Van de Peer, Y.; Soltis, D.E. Polyploidy and Genome Evolution in Plants. *Curr. Opin. Genet. Dev.* **2015**, *35*, 119–125. [CrossRef] [PubMed]
8. Jolanta, M.; Kolano, B.; Sas-Nowosielska, H. Endopolyploidy in Plants. In *Plant Genome Diversity*; Springer: Vienna, Brazil, 2013; Volume 2, pp. 99–119.
9. Ranney, T.G. Polyploidy: From Evolution to New Plant Development. *Comb. Proc. Int. Plant Propagators' Soc.* **2006**, *56*, 137–142.
10. Dhooghe, E.; Van Laere, K.; Eeckhaut, T.; Leus, L.; Van Huylenbroeck, J. Mitotic chromosome doubling of plant tissues in vitro. *Plant Cell Tissue Organ Cult. (PCTOC)* **2011**, *104*, 359–373. [CrossRef]
11. Mondin, M.; Neto, A.D. Citogenética Vegetal Enfatizando a Família Orchidaceae. *Orchidstudium* **2006**, *4*, 24–54.
12. Manzoor, A.; Ahmad, T.; Bashir, M.A.; Hafiz, I.A.; Silvestri, C. Studies on Colchicine Induced Chromosome Doubling for Enhancement of Quality Traits in Ornamental Plants. *Plants* **2019**, *8*, 194. [CrossRef]
13. Chung, M.Y.; Kim, C.Y.; Min, J.S.; Lee, D.-J.; Naing, A.H.; Chung, J.D.; Kim, C.K. In Vitro Induction of Tetraploids in an Interspecific Hybrid of Calanthe (Calanthe Discolor × Calanthe Sieboldii) through Colchicine and Oryzalin Treatments. *Plant Biotechnol. Rep.* **2014**, *8*, 251–257. [CrossRef]
14. Kobayashi, N.; Yamashita, S.; Ohta, K.; Hosoki, T. Morphological Characteristics and Their Inheritance in Colchicine-induced Salvia Polyploids. *J. Jpn. Soc. Hortic. Sci.* **2008**, *77*, 186–191. [CrossRef]
15. Manzoor, A.; Ahmad, T.; Bashir, M.A.; Baig, M.M.Q.; Quresh, A.A.; Shah, M.K.N.; Hafiz, I.A. Induction and Identification of Colchicine Induced Polyploidy in Gladiolus *Grandiflorus 'White Prosperity'. Folia Hortic.* **2018**, *30*, 307–319. [CrossRef]
16. Kushwah, K.S.; Verma, R.C.; Patel, S.; Jain, N. Colchicine Induced Polyploidy in *Chrysanthemum carinatum* L. *J. Phylogenet. Evol. Biol.* **2018**, *6*, 2. [CrossRef]
17. Huy, N.P.; Tam, D.T.T.; Luan, V.Q.; Tung, H.T.; Hien, V.T.; Ngan, H.T.M.; Duy, P.N.; Nhut, D.T. In vitro polyploid induction of Paphiopedilum villosum using colchicine. *Sci. Hortic.* **2019**, *252*, 283–290. [CrossRef]
18. Chen, W.-H.; Tang, C.-Y. A Protocol for the Induction of Polyploids in Phalaenopsis Orchids by In Vitro Method Without Using Anti-Microtubule Agents. In *Orchid Propagation: From Laboratories to Greenhouses—Methods and Protocols*; Lee, Y.I., Yeung, E.T., Eds.; Humana Press: New York, NY, USA, 2018; pp. 317–330.
19. Ji, B.X.; Chen, D.W.; Zhang, C.C.; Min, D.; Huang, W.J.; Wang, Y. High Efficient Polyploid Induction of Cymbidium Hybridium. *Bull. Bot. Res.* **2011**, *31*, 558–562.

20. Vichiato, M.R.M.; Vichiato, M.; Pasqual, M.; Castro, D.M.; Dutra, L.F. Indução e Identificação de Tetraplóides Em Dendrobium Nobile Lindl.(Orchidaceae). *Rev. Ciência Agronômica* **2007**, *38*, 385–390.
21. Cui, G.R.; Zhang, Z.X.; Zhang, C.Y.; Hu, N.B.; Sui, Y.H.; Li, J.Q. Polyploid Induction and Identification of Oncidium. *Acta Prataculturae Sin.* **2010**, *19*, 184–190.
22. Menezes-Sá, T.S.A.; de Fátima Arrigoni-Blank, M.; da Costa, A.S.; de Almeida Santos-Serejo, J.; Blank, A.F.; Soares, C.A.; Moura, G.M.S. Chromosome Doubling in Cattleya Tigrina, A. Rich. *Sci. Plena* **2019**, *15*, 11. [CrossRef]
23. Silva, P.A.K.X.; Callegari-Jacques, S.; Bodanese-Zanettini, M.H. Induction and Identification of Polyploids in Cattleya Intermedia Lindl.(Orchidaceae) by in Vitro Techniques. *Ciência Rural* **2000**, *30*, 105–111. [CrossRef]
24. Wibisono, K.; Syarifah, I.A.; Waras, N.; Suhest, S. Sensitivity in Callus Tissue of Plectranthus Amboinicus (L.) through Mutation Induction with Colchicine. *AGRIVITA J. Agric. Sci.* **2021**, *44*, 82–95. [CrossRef]
25. Li, M.; Ding, B.; Huang, W.; Pan, J.; Ding, Z.; Jiang, F. Induction and Characterization of Tetraploids from Seeds of Bletilla Striata (Thunb.) Reichb. F. *BioMed Res. Int.* **2018**, *2018*, 3246398.
26. Lone, A.B.; Unemoto, L.K.; Takahashi, L.S.A.; Faria, R.T. Sobrevivência e desenvolvimento in vitro de Cattleya (Orchidaceae) submetida a tratamentos com colchicina. *Semin. Ciências Agrárias* **2010**, *31*, 1337. [CrossRef]
27. Mustika, T.; Indrianto, A. Improvement of Orchid Vanda Hybrid (Vanda Limbata Blume x Vanda Tricolor Lindl. Var. Suavis) by Colchicines Treatment in Vitro. *Modern. Appl. Sci.* **2016**, *10*, 83–89.
28. Thao, N.T.P.; Ureshino, K.; Miyajima, I.; Ozaki, Y.; Okubo, H. Induction of tetraploids in ornamental Alocasia through colchicine and oryzalin treatments. *Plant Cell Tissue Organ Cult. (PCTOC)* **2003**, *72*, 19–25. [CrossRef]
29. Heo, J.-Y.; Jeong, S.-H.; Choi, H.-R.; Park, S.-M. Polyploid Production in Lilium Leichtlinii Var. *Maximowiczii Using Colchicine. JAPS J. Anim. Plant Sci.* **2016**, *26*.
30. Song, L.; Yang, J.; Liu, D.; Li, Z.; Wang, Y. Polyploid Induction in Cymbidium Sinenthese'Lv Mosu'× Cymbidium Hybridum'Shijieheping'F1 Generation. *Guangxi Zhiwu/Guihaia* **2018**, *38*, 188–194.
31. Atichart, P. Polyploid Induction by Colchicine Treatments and Plant Regeneration of Dendrobium Chrysotoxum. *Thai J. Agric. Sci.* **2013**, *46*, 59–63.
32. Griesbach, R.J. Colchicine-induced polyploidy in phalaenopsis orchids. *Plant Cell Tissue Organ Cult. (PCTOC)* **1981**, *1*, 103–107. [CrossRef]
33. De, L.C.; Rao, A.N.; Rajeevan, P.K.; Pathak, P. ORCHID IMPROVEMENT-AN OVERVIEW. *J. Orchid. Soc. India* **2014**, *28*, 35–45.
34. Chen, W.H.; Tang, C.Y.; Kao, Y.L. Ploidy doubling by in vitro culture of excised protocorms or protocorm-like bodies in Phalaenopsis species. *Plant Cell Tissue Organ Cult. (PCTOC)* **2009**, *98*, 229–238. [CrossRef]
35. Souza, B.C.Q.; Souza, A.C.M.; Almeida, E.M.; Felix, L.P. Orchidaceae-IAPT/IOPB Chromosome Data 26. *Taxon* **2017**, *66*, 1487–1499.
36. Eng, W.-H.; Ho, W.-S. Polyploidization Using Colchicine in Horticultural Plants: A Review. *Sci. Hortic.* **2019**, *246*, 604–617. [CrossRef]
37. Grosso, V.; Farina, A.; Giorgi, D.; Nardi, L.; Diretto, G.; Lucretti, S. A high-throughput flow cytometry system for early screening of in vitro made polyploids in Dendrobium hybrids. *Plant Cell Tissue Organ Cult. (PCTOC)* **2018**, *132*, 57–70. [CrossRef]
38. Yenchon, S.; Te-Chato, S. Polyploidy Induction of Dendrobium Formosum by Colchicine Treatment in Vitro. *Acta Hortic.* **2014**, *1025*, 81–88. [CrossRef]
39. Xuejiao, L.; Zhilin, L.; Lipin, H. Induction and Identification of Polyploids in Wild Cymbidium Lowianum. *Chin. Agric. Sci. Bull.* **2010**, *13*, 60.
40. Rahayu, E.M.D.; Sukma, D.; Syukur, M.; Irawati, D. Polyploid Induction of *Phalaenopsis Amabilis* (L.) Blume and Phalaenopsis Amboinensis, J. J. Smith by Colchicine in In Vitro Culture. *Indones. J. Agron.* **2015**, *43*, 219–226. [CrossRef]
41. IOF. Orchid Hybrid: Brassolaeliocattleya Goldenzelle. ORCHIDS.ORG. 2022a. Available online: https://www.orchids.org/grexes/brassolaeliocattleya-goldenzelle (accessed on 25 January 2022).
42. IOF. Orchid Hybrid: Brassolaeliocattleya Haw Yuan Beauty. ORCHIDS.ORG. 2022b. Available online: https://www.orchids.org/grexes/brassolaeliocattleya-haw-yuan-beauty (accessed on 25 January 2022).
43. Querino, B.C.; Ferraz, M.E.; Mata-Sucre, Y.; Souza, G.; Felix, L.P. Cytomolecular diversity of the subtribe Laeliinae (Epidendroidae, Orchidaceae) suggests no relationship between genome size and heterochromatin abundance. *Plant Syst. Evol.* **2020**, *306*, 19. [CrossRef]
44. Souza, B.C.Q. Citogenética Da Subtribo Laeliinae (Orchidaceae: Epidendroideae): Regiões Heterocromáticas e Localização Do DNA Ribossomal. In *Dissertação de Mestrado*; Universidade Federal da Paraıba: Paraıba, Brazil, 2011.
45. Felix, L.P.; Guerra, M. Variation in chromosome number and the basic number of subfamily Epidendroideae (Orchidaceae). *Bot. J. Linn. Soc.* **2010**, *163*, 234–278. [CrossRef]
46. Moraes, A.P.; Chinaglia, M.; Palma-Silva, C.; Pinheiro, F. Interploidy Hybridization in Sympatric Zones: The Formation of E Pidendrum Fulgens× E. Puniceoluteum Hybrids (E Pidendroideae, O Rchidaceae). *Ecol. Evol.* **2013**, *3*, 3824–3837. [CrossRef]
47. Murashige, T.; Skoog, F. A Revised Medium for Rapid Growth and Bioassays with Tobacco Tissue Cultures. *Physiol. Plant.* **1962**, *15*, 473–497. [CrossRef]
48. Galbraith, D.W.; Harkins, K.R.; Maddox, J.M.; Ayres, N.M.; Sharma, D.P.; Firoozabady, E. Rapid Flow Cytometric Analysis of the Cell Cycle in Intact Plant. Tissues. *Science* **1983**, *220*, 1049–1051. [CrossRef]

49. Otto, F. Chapter 11 DAPI Staining of Fixed Cells for High-Resolution Flow Cytometry of Nuclear DNA. In *Methods in Cell Biology*; Academic Press: Cambridge, MA, USA, 1990; Volume 33, pp. 105–110. [CrossRef]
50. Praça-Fontes, M.M.; Carvalho, C.R.; Clarindo, W.R.; Cruz, C.D. Revisiting the DNA C-values of the genome size-standards used in plant flow cytometry to choose the "best primary standards". *Plant Cell Rep.* **2011**, *30*, 1183–1191. [CrossRef] [PubMed]
51. Mondin, M.; Aguiar-Perecin, M.L.R. Coloração Pelo Método de Feulgen. 2009. Available online: http://www.genetica.esalq.usp.br/citogenetica/protocolos/Feulgen.pdf (accessed on 10 October 2022).

plants

MDPI

*Article*

# Gamma Radiation Induced In-Vitro Mutagenesis and Isolation of Mutants for Early Flowering and Phytomorphological Variations in Dendrobium 'Emma White'

**Rubina Sherpa [1,2], Ramgopal Devadas [2,*], Sadashiv Narayan Bolbhat [1], Tukaram Dayaram Nikam [3] and Suprasanna Penna [4,5]**

1 Department of Botany, Annasaheb Awate College, Manchar 410503, India
2 Indian Council of Agricultural Research-National Research Centre on Orchids, Pakyong 737106, India
3 Department of Botany, Savitribai Phule Pune University, Pune 411007, India
4 Nuclear Agriculture and Biotechnology Division, Bhabha Atomic Research Centre, Mumbai 400085, India
5 Amity Institute of Biotechnology, Amity University Mumbai, Mumbai 410206, India
* Correspondence: ramgopal.devadas@icar.gov.in

**Abstract:** In vitro mutagenesis offers a feasible approach for developing new orchid cultivars through genetic manipulation. In the present study, protocorm-like bodies (PLBs) were exposed to gamma rays (10, 20, 40, 60, 80 Gy) to study in vitro growth responses and induction of mutants in Dendrobium 'Emma White'. Both proliferation and regeneration of PLBs decreased progressively with increasing doses, except for a significantly enhanced growth response at 10 Gy. The optimal dose of gamma radiation for mutagenesis was found in the range 10 to 25 Gy based on the growth reduction curve. Analysis using a high-throughput cell analyzer revealed a significant reduction in nuclear DNA content at > 40 Gy doses. At 10 Gy treatment, the growth attributes, such as root length, plant height and leaf number, were significantly increased by 36%, 26% and 20%, respectively, compared to the control. This increase was significant over other tested doses as well. Testing of random amplified polymorphic DNA markers revealed the presence of detectable polymorphism among gamma mutant plantlets with a polymorphism information content value at 0.41. The gamma-ray-induced earliness in flower development was observed within 294 days post ex vitro growth of 10 Gy mutant compared to the control plants flowered after 959 days. Our results highlight the significance of gamma radiation in inducing enhanced growth, morphological variations and early floral initiation in Dendrobium, providing a basic framework for mutation breeding and improvement of orchids.

**Keywords:** orchids; mutation; genetic markers; polymorphism; random amplified polymorphic DNA

**Citation:** Sherpa, R.; Devadas, R.; Bolbhat, S.N.; Nikam, T.D.; Penna, S. Gamma Radiation Induced In-Vitro Mutagenesis and Isolation of Mutants for Early Flowering and Phytomorphological Variations in Dendrobium 'Emma White'. *Plants* **2022**, *11*, 3168. https://doi.org/10.3390/plants11223168

Academic Editors: Ranjith Pathirana and Francesco Carimi

Received: 14 September 2022
Accepted: 15 November 2022
Published: 18 November 2022



## 1. Introduction

Dendrobium is the second largest genus after Bulbophyllum in the Orchidaceae family. Dendrobiums were used as rich medicinal plants in many old-world countries, including both China and India [1]. Dendrobiums also have potential demand all over southeast Asia and other tropical parts for exports due to the wide range of choices for flower color, shape, texture and longevity [2]. Worldwide Dendrobium marketing and trade occur broadly as cut flowers and potted plants. Dendrobium occupied among the top ten orchid taxa of commercially traded and propagated live plants at 2.3% (hybrids) and 3.4% (species) between 2006 and 2015 [3]. Thailand is the largest producer and exporter of Dendrobium, with 70% of the total 2.1 billion cut flowers exported globally [4]. However, only a few Dendrobium varieties dominate export trade from southeast Asian countries, and the majority of these varieties are genetically derived from *Dendrobium phalaenopsis* species [5]. 'Sonia' cultivar from Thailand and 'Uniwai' cultivar from Hawaii occupy 70% of total Dendrobium production [6,7], indicating the limited choice in varieties. Developing new genetic stocks with desirable traits will be helpful in meeting the demand for new Dendrobium varieties across international and domestic markets [8,9].

Development of a large number of new Dendrobium varieties is constrained by compatibility barriers within inter-sectional crossing in the genus and lack of advanced segregating progenies to construct genetic and linkage maps [8,10]. Further, linkage drag and negative linkage for promising characters, apart from prolonged juvenile phase [11] and mortality at ex vitro hardening [12], hinder the selection process. The absence of natural hybrids from the center of origin (Australia) also supports the crossability problems in Dendrobium [13], and even self-pollination leads to drastic adverse effects on several traits that were unresponsive to selection in this species [14]. Conventional breeding programs involving traditional hybridization to transfer desirable traits are laborious and take 12–13 years to develop new cultivars [15]. Hence, application of mutation breeding can offer quick and better solutions to overcome such inherent pre- and post-zygotic barriers to develop new vegetatively propagated Dendrobium cultivars. A wide range of trait-specific mutant varieties from plant structure to biotic and abiotic stress resistance with high yield have been successfully achieved through mutation breeding in other crops [16,17].

Among orchids, the initial studies on mutagenesis were conducted in Cattleya using gamma rays [18]. The changes in flower color pigmentation and flower size were reported in Dendrobium 'Sonia' cultivar through gamma radiation [19]. Induced mutations in orchid breeding for genetic improvement are restricted to a few genera; however, from a breeding point of view for developing new cultivars, it is essential to determine the correct mutagen, the optimum mutagen dose and the choice of developmental stage for treatment. Such studies will provide a baseline for other mutation breeding work for orchid varietal development. In the present study, we have studied the effect of gamma-radiation-induced in vitro and ex vitro growth responses and mutagenic changes at the cellular and genetic level in Dendrobium hybrid 'Emma White'.

## 2. Results

### 2.1. Growth Response of Protocorm Like Bodies (PLBs) to Different Doses of Gamma Radiation

The fresh weight of protocorm-like bodies (PLBs) of Dendrobium 'Emma White' measured a month after irradiation (MAI) ranged from 448.2 mg (10 Gy) to 209.1 mg (80 Gy), indicating a gradual decrease in tissue biomass with increasing doses of gamma irradiation as compared to the control (Table 1). However, the reduction by ~ 60 to 70% in tissue fresh weight was more evident at doses > 40 Gy. At 3 MAI, no significant differences were observed for proliferation among PLBs irradiated at 10 and 20 Gy doses (Figure 1A). In contrast, the proliferation was reduced drastically by 35 to 50% for PLBs treated with gamma ray doses of 40, 60 and 80 Gy compared to the control. At 6 MAI, a 28% increase in proliferation was observed at 10 Gy, in contrast to doses > 40 Gy, where the proliferation reduced significantly by 70 to 90% compared to the control (Figure 1A). A similar effect on the regeneration capacity of PLBs was observed at subsequent stages of growth and development. Initially, there was no evidence of PLB regeneration into shoots at higher doses, except for 10 and 20 Gy, where the regeneration percentage was comparable to the control (Figure 1B). At 6 MAI, PLBs exposed to 10 Gy radiation showed an increase (4%) in regeneration as compared with the control. However, a lower regeneration response at a gamma dose of 40 Gy (18.86%) with suppressed growth was observed at 60 and 80 Gy doses, indicating that PLBs are sensitive to higher doses of gamma radiation (Figure 1B). The average number of days required for initiation of leaf primordium from PLBs irradiated at 10 and 20 Gy was 36.2 days and 38.8 days compared to the control (32.2 days). On the other hand, a prolonged delay in the initiation of leaf primordium at a dose of 40 Gy (106.4 days) was observed, indicating an adverse effect of gamma irradiation on PLBs differentiation and the developmental process.

In the present study, all the PLBs of Dendrobium 'Emma white' irradiated with gamma ray doses of 10, 20, 40, 60 and 80 Gy survived with no browning or desiccation, making it difficult to estimate the lethal dose ($LD_{50}$). Hence, the radiation sensitivity of PLBs towards gamma rays was assessed based on the growth reduction dose ($GR_{50}$), at which the growth of PLBs reduced by 50% [20]. The $GR_{50}$ was estimated to be 25.52 Gy based on the initial regeneration response of irradiated PLBs (Figure 2), suggesting an optimal dose range of 10 to 25 Gy for irradiation of Dendrobium 'Emma White' PLBs.

**Table 1.** Fresh weight of protocorm-like bodies of Dendrobium 'Emma White' one month after irradiation with gamma rays.

| Treatment | Fresh Weight (mg) |
|---|---|
| Control | 648.6 ± 27.4 a |
| 10 Gy | 448.2 ± 16.6 b |
| 20 Gy | 294.4 ± 14.3 c |
| 40 Gy | 237.1 ± 4.9 d |
| 60 Gy | 232.0 ± 4.7 d |
| 80 Gy | 209.1 ± 5.9 d |

The data represent mean values ± standard error. Means with different lower-case letters (a, b, c, d) are significantly different at $p < 0.05$ (Duncan's Multiple Range Test).

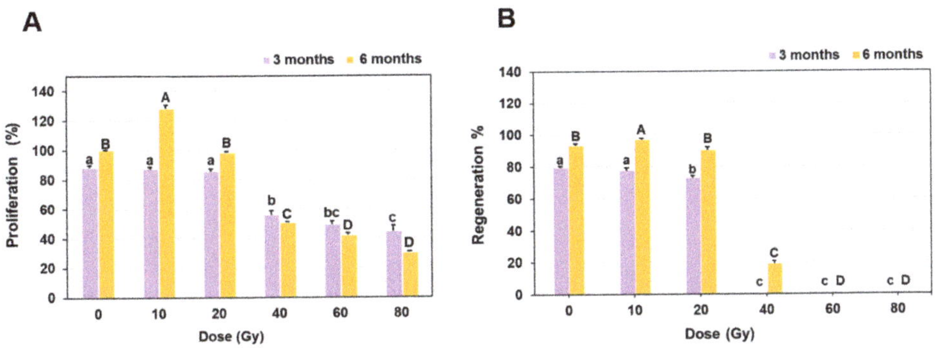


**Figure 1.** Growth response of protocorm-like bodies (PLBs) of Dendrobium after gamma irradiation. PLBs of Dendrobium 'Emma White' were irradiated with gamma ray doses of 10, 20, 40, 60 and 80 Gy and incubated at 22 ± 2 °C and 65–70% relative humidity with 16 h photoperiod. An unirradiated set of PLBs was maintained as control. After three and six months of irradiation, growth response of irradiated PLBs based on proliferation percentage (**A**) and regeneration percentage (**B**) was recorded for each treatment. Data represent mean values ± standard error ($n$ = 10). Means with different lower-case letters (a, b, c, d) are significantly different at $p < 0.05$ (Duncan's Multiple Range Test).

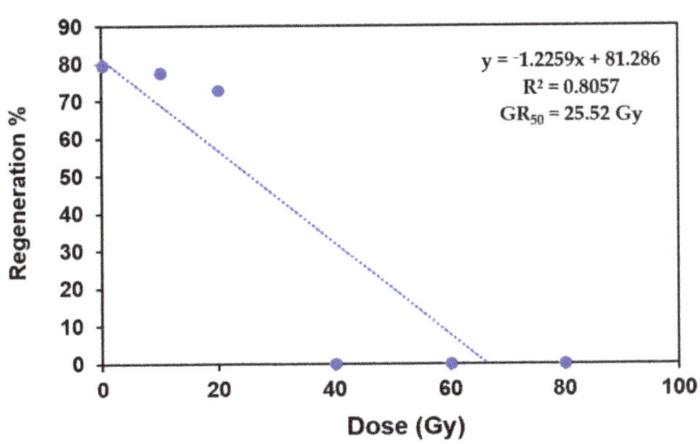


**Figure 2.** Linear regression analysis to estimate growth reduction dose ($GR_{50}$) of gamma-irradiated protocorm-like bodies of Dendrobium 'Emma White'. Shoot regeneration response of PLBs irradiated with gamma rays of 10, 20, 40, 60 and 80 Gy after three months of radiation treatment was used to estimate $GR_{50}$. The mean values of regeneration percentage were analyzed using linear regression equation (y = mx + c).

### 2.2. Nuclear DNA Content Estimation and Cell Cycle Analysis in Irradiated PLBs by High-Throughput Cell Analyser (HTCA)

The estimation of nuclear DNA content based on the average fluorescence intensity ranged from 101 to 85%, with a maximum of 101%, followed by 96% under 10 and 20 Gy doses, respectively (Figure 3A). In contrast, gamma-irradiated PLBs at higher doses showed a significant decrease in DNA content, with a maximum (15%) reduction at 80 Gy. Similarly, at 40 and 60 Gy, the DNA content of PLBs was reduced by ~7 to 9% compared to the control. The results indicated that gamma radiation exerted more effect in PLBs irradiated at higher doses, as was evident from the average fluorescence intensity of nuclear DNA generated by high-throughput cell analyzer (HTCA) (Supplementary Table S1). A histogram of the FL-3 fluorescence intensity peak distribution of nuclei isolated from gamma-irradiated PLBs after five months of irradiation is shown in Figure 3B. The percentage of cell count at the S + G2M phase in gamma-irradiated PLBs was decreased by ~ 14 to 50% with increasing irradiation doses of 20, 40, 60, and 80 Gy, indicating inhibition of cell cycle progression at the higher dose. Unlike other treatments, the cell count percentage in the 10 Gy treatment was increased by 2% at the S + G2M phase compared to the control treatment used as the benchmark value (Supplementary Table S1).

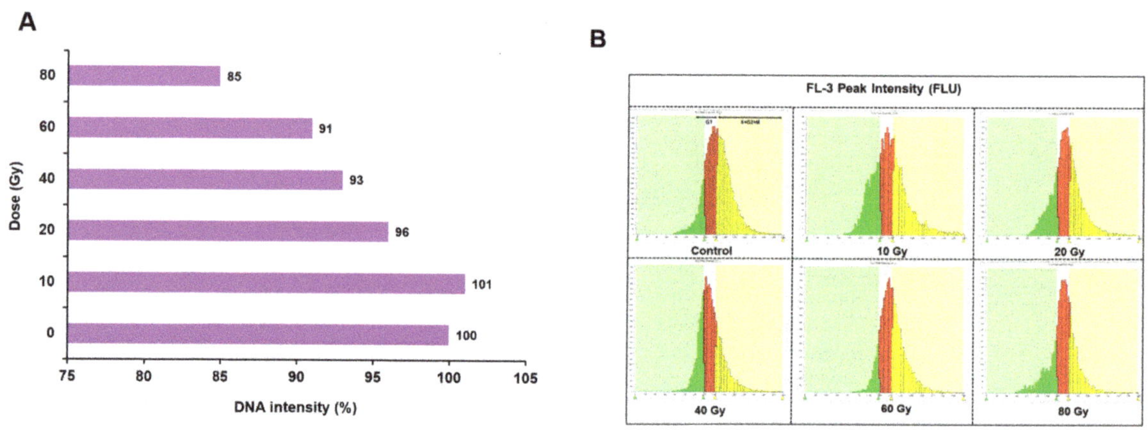


**Figure 3.** High-throughput-cell-analyser (HTCA)-based nuclear DNA content estimation and cell cycle analysis in protocorm-like bodies (PLBs) of Dendrobium under different gamma radiation treatments. PLBs of Dendrobium 'Emma White' were irradiated with gamma ray doses of 10, 20, 40, 60 and 80 Gy and incubated at 22 ± 2 °C and 65–70% relative humidity with 16 h photoperiod. After five months of treatment, nuclei were isolated from irradiated and control PLBs to estimate the nuclear DNA content based on the average fluorescence intensity of nuclei generated by HTCA, (**A**) Percentage of DNA intensity calculated based on fluorescence intensity of nuclei under different radiation treatments of 10, 20, 40, 60 and 80 Gy, including control using HTCA, (**B**) FL-3 fluorescence intensity histogram of nuclei isolated from PLBs, incubated with propidium iodide overnight at 4 °C in dark conditions. Gate between 0–1200 indicates dead cells (green), 1201–1600 live non-dividing cells in G1 phase (red) and 1601–2800 dividing cells in S, G2 and M phase (yellow) of cell cycle. The 488 nm laser-excited propidium iodide dye (FL-3) intercalates DNA and facilitates enumeration of DNA per nuclei number.

### 2.3. Frequency and Spectrum of Variation among Gamma-Irradiated Population of Dendrobium 'Emma White'

During the initial period of in vitro regeneration, plantlets developed from gamma-irradiated PLBs did not exhibit any morphological variation. However, after seven to eight months of irradiation, variation in the leaf shape and shooting pattern was observed. Alterations in leaf morphology (yellowing, asymmetrical and oval to cordate shapes) was observed at 10, 20 and 40 Gy doses (Figure 4). However, no shoot regeneration was observed from gamma-irradiated PLBs at 60 and 80 Gy. The highest frequency of variation (51.6%) was observed in the gamma-irradiated population of 40 Gy (Table 2), followed by the lower frequencies in PLBs irradiated at 20 Gy (9.5%) and 10 Gy (7%). The maximum spectrum of variation (10) was observed at 40 Gy compared to the spectrum of variation of 7 at lower doses (10, 20 Gy). The number of plantlets regenerated was reduced by 80% at 40 Gy compared to the highest number of regenerated plantlets at 10 Gy and the control. Among the mutagenic changes observed for morphological traits in the gamma-irradiated population, the variations induced in leaf margin (serration) were higher in proportion than other leaf variations (leaf margin > leaf apex > leaf vein > leaf shape) (Table 2).

**Figure 4.** Morphological variations among in vitro plantlets regenerated from gamma-irradiated protocorm-like bodies (PLBs) of Dendrobium 'Emma White'. PLBs of Dendrobium 'Emma White' were irradiated with gamma ray doses of 10, 20, 40, 60 and 80 Gy and incubated at 22 ± 2 °C and 65–70% relative humidity with 16 h photoperiod. Seven to eight months post irradiation, alterations in growth and structure of shooting pattern were observed at 10, 20 and 40 Gy doses. (**A**) Control unirradiated plantlet, (**B–D**) leaf variations at 10 Gy, (**E–H**) leaf variations at 20 Gy, (**I–M**) Leaf variations at 40 Gy. Morphological variation observed in leaf distribution pattern (**B**,**C**,**E**); leaf margin, midrib, apex, shape, and color (**D**,**M**); leaf margin (**G**,**J**,**K**,**L**); leaf color (**I**); leaf texture (**H**); multiple shoots formation (**F**).

**Table 2.** Variation frequency and variation spectrum of leaf mutants derived from gamma-irradiated protocorms-like bodies (PLBs) of Dendrobium 'Emma White'.

| Gamma Dose | Leaf Margin | Leaf Vein | Leaf Texture | Leaf Apex | Leaf Shape | | | | Leaf Distribution | | Leaf Color | Plantlets Regenerated | Total No. of Mutants | Variation Frequency (%) | Variation Spectrum |
|---|---|---|---|---|---|---|---|---|---|---|---|---|---|---|---|
| | T/S | DM | T/R | DN | Lobed | Linear | Ovate | Distorted | LC | T/R | YLWM | | | | |
| Control | - | - | - | - | - | - | - | - | - | - | - | 300 | 0 | - | - |
| 10 Gy | 4 | 4 | - | 6 | - | 2 | 2 | - | 1 | 2 | - | 300 | 21 | 7.0 | 7 |
| 20 Gy | 5 | 2 | 3 | 3 | - | - | 1 | 5 | 3 | - | - | 230 | 22 | 9.5 | 7 |
| 40 Gy | 8 | 3 | 2 | 5 | 1 | - | 1 | 3 | 2 | 1 | 5 | 60 | 31 | 51.6 | 10 |

T/S toothed/serrated; DM double midrib; T/R thick/rough; DN deep notch; LC leaf clump; T/R twisted/rolled; YLWM yellow leaves with a white margin. Variation frequency (%) calculated based on morphological variations observed to the total regenerated plantlets.

### *2.4. Phenotypic Variation among In Vitro Plantlets Derived from Gamma-Irradiated PLBs*

A significant effect of gamma radiation was observed on plant growth attributes viz., plant height, leaf length, leaf number, leaf width, root number and root length (Figure 5). Except for the 10 Gy treatment, all the other gamma irradiation doses showed a dose-dependent decrease in plant growth. The lowest radiation dose induced a significant difference in growth of plantlets when compared to the control. Compared with the control, the root length was increased by 36%, plant height by 26% and leaf number by 20% in the 10 Gy treatment. However, no significant differences were observed for root width in all treatments. The overall decrease in plant growth attributes at dose 20 Gy ranged from a minimum reduction in leaf width by 2% to a maximum reduction in leaf number by 9% compared to the control. At a higher dose of 40 Gy, significant differences were observed in plant height, leaf number and length, root number and length. The plant height was reduced by 33%, followed by a 30% reduction in leaf length at a 40 Gy dose compared to the control.

### *2.5. Genetic Analysis of Regenerated Mutants Based on Random Amplified Polymorphic DNA Markers*

The genetic divergence among gamma-irradiated mutants of 'Emma White' with respect to control plants was analyzed based on random amplified polymorphic DNA (RAPD) markers (Figure 6). A total of 33 bands were generated using six RAPD primers, and the number of bands ranged from four to eight, with an average of 5.50 bands per primer, of which four bands were monomorphic and the remaining twenty-nine bands were polymorphic (Table 3). The percentage of polymorphism was found to be 100% for all the primers except for OPAW13, 33.33%, and an average polymorphism of 88.89%. The polymorphism information content (PIC) value of different RAPD primers ranged from 0.15 to 0.5, with an average of 0.41 (Table 3). Among all six RAPD primers, two primers (OPB5 and OPF1) were found with a PIC value at 0.5, indicating a highly informative nature for the study of gamma-ray-induced mutants. OPAW13 showed the minimum PIC value at 0.15. NEI72 dissimilarity-coefficient-based genetic distance ranged from 0.07 to 0.89, with an average distance of 0.37, indicating a greater range of genetic diversity among gamma-irradiated mutants of 'Emma White' (Table 4). A maximum genetic distance of 0.89 was observed between mutants generated at 40 Gy_2 and 10 Gy_2. In contrast, the lowest distance of 0.07 was obtained between 20 Gy_3 and Control_2, followed by the distance of 0.09 between two control plants. The average genetic distance between the mutants and the control was found to be in the order of 0.16 (20 Gy) < 0.26 (40 Gy) and 0.37 (10 Gy), indicating the most distant genetic divergence among the 10 Gy mutants with respect to the control. A dendrogram constructed by the UPGMA-based clustering method showed three major clusters (I, II, III) with a genetic dissimilarity coefficient of 0.40 and with a cophenetic correlation coefficient of 0.82 [21]. Cluster I included mutants of 20 Gy and 40 Gy, including controls of 'Emma White'; cluster II included mutants of 10 Gy and 40 Gy, whereas cluster III included three mutants of 10 Gy (Figure 7).

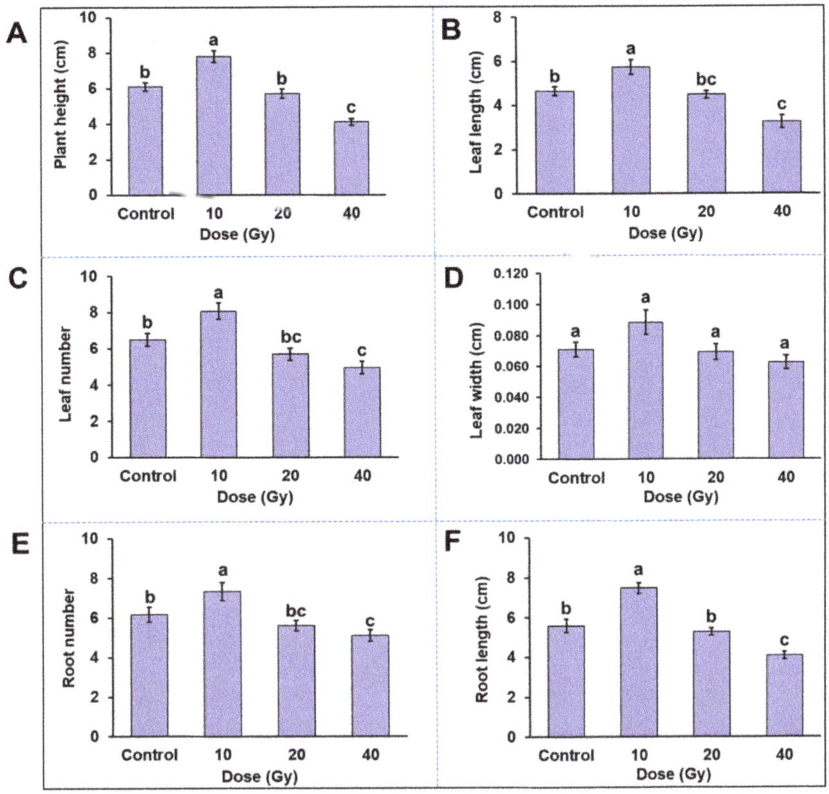


**Figure 5.** Variations in phenotypic characters of in vitro plantlets at 10, 20 and 40 Gy gamma radiation treatments. Protocorm-like bodies (PLBs) of Dendrobium 'Emma White' were irradiated with gamma rays of 10, 20, 40, 60 and 80 Gy and incubated at 22 ± 2 °C and 65–70% relative humidity with 16 h photoperiod. After 11–12 months of irradiation, differential phenotype of regenerated plantlets at 10, 20 and 40 Gy doses were recorded concerning plant growth attributes (**A**) plant height, (**B**) leaf length, (**C**) leaf number, (**D**) leaf width, (**E**) root number, (**F**) root length and root width. No significant difference in root width was observed under all the responsive treatments. Data represent mean value ± standard deviation ($n$ = 10). Mean values with different lower-case letters (a, b, c, d) are significantly different at $p < 0.05$ (Duncan's Multiple Range Test).

**Table 3.** Random amplified polymorphic DNA (RAPD) banding pattern analysis in gamma-ray-induced mutant plantlets at doses 10, 20 and 40 Gy.

| SI. No. | Primer | Total No. of Bands | Monomorphic | Polymorphic | Polymorphism (%) | PIC |
|---|---|---|---|---|---|---|
| 1 | OPB12 | 4 | 0 | 4 | 100 | 0.48 |
| 2 | OPA4 | 4 | 0 | 4 | 100 | 0.41 |
| 3 | OPAW13 | 6 | 4 | 2 | 33.33 | 0.15 |
| 4 | OPAW17 | 6 | 0 | 6 | 100 | 0.41 |
| 5 | OPB5 | 8 | 0 | 8 | 100 | 0.5 |
| 6 | OPF1 | 5 | 0 | 5 | 100 | 0.5 |
| 7 | Total | 33 | 4 | 29 | - | - |
| 8 | Mean | 5.50 | 0.67 | 4.83 | 88.89 | 0.41 |

**Table 4.** Genetic dissimilarity matrix from random amplified polymorphic DNA binary data of gamma-ray-induced mutants of Dendrobium 'Emma White'.

| | C_1 | C_2 | 10 Gy_1 | 10 Gy_2 | 10 Gy_3 | 10 Gy_4 | 20 Gy_1 | 20 Gy_2 | 20 Gy_3 | 40 Gy_1 | 40 Gy_2 | 40 Gy_3 |
|---|---|---|---|---|---|---|---|---|---|---|---|---|
| C_1 | 0 | | | | | | | | | | | |
| C_2 | 0.09 | 0 | | | | | | | | | | |
| 10 Gy_1 | 0.37 | 0.33 | 0 | | | | | | | | | |
| 10 Gy_2 | 0.37 | 0.33 | 0.33 | 0 | | | | | | | | |
| 10 Gy_3 | 0.48 | 0.36 | 0.38 | 0.56 | 0 | | | | | | | |
| 10 Gy_4 | 0.48 | 0.36 | 0.46 | 0.15 | 0.44 | 0 | | | | | | |
| 20 Gy_1 | 0.13 | 0.13 | 0.55 | 0.55 | 0.61 | 0.61 | 0 | | | | | |
| 20 Gy_2 | 0.24 | 0.27 | 0.46 | 0.38 | 0.61 | 0.43 | 0.20 | 0 | | | | |
| 20 Gy_3 | 0.10 | 0.07 | 0.32 | 0.39 | 0.43 | 0.50 | 0.10 | 0.26 | 0 | | | |
| 40 Gy_1 | 0.26 | 0.19 | 0.66 | 0.57 | 0.37 | 0.54 | 0.28 | 0.38 | 0.17 | 0 | | |
| 40 Gy_2 | 0.45 | 0.33 | 0.63 | 0.89 | 0.16 | 0.61 | 0.41 | 0.49 | 0.33 | 0.20 | 0 | |
| 40 Gy_3 | 0.13 | 0.17 | 0.51 | 0.51 | 0.56 | 0.48 | 0.09 | 0.15 | 0.15 | 0.30 | 0.37 | 0 |

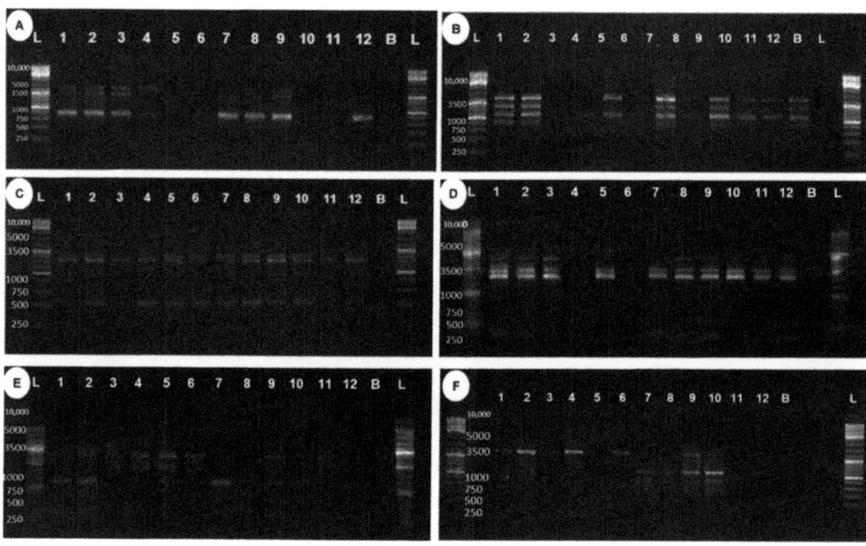


**Figure 6.** Random amplified polymorphic DNA (RAPD) polymorphism in gamma-ray-induced mutants of Dendrobium 'Emma White'. Total genomic DNA was isolated from in vitro young leaves of control and mutant plantlets using a modified CTAB method for RAPD marker level study to analyze the radiation-induced genetic variability in 'Emma White'. Twenty-eight decamer RAPD primers were used, out of which six primers (OPB 12, OPA 4, OPAW 13, OPAW 17, OPB 5, OPF 1) that produced clear and scorable bands were used to generate RAPD binary matrix. Bands were scored as the presence (1) or the absence (0) for control and mutant DNA samples. DNA ladder (lane L); control 1 (lane 1); control 2 (lane 2); 10 Gy_1 (lane 3); 10 Gy_2 (lane 4); 10 Gy_3 (lane 5); 10 Gy_4 (lane 6); 20 Gy_1 (lane 7); 20 Gy_2 (lane 8); 20 Gy_3 (lane 9); 40 Gy_1 (lane 10); 40 Gy_2 (lane 11); 40 Gy_3 (lane 12); blank without DNA to check cross-contamination (lane B). Primers: (**A**) OPB12; (**B**) OPA4; (**C**) OPAW13; (**D**) OPAW17; (**E**) OPB5; (**F**) OPF1. GeneRuler1 Kb DNA Ladder.

### 2.6. Leaf Variation among 10 Gy Mutant Population

After 7–8 months of ex vitro growth and development, morphological variations were observed among plants regenerated from 10 Gy-irradiated PLBs compared to the control (Figure 8). The differences were more apparent in leaf structure and arrangement of individual shoots. These variations ranged from double apexed leaves, fused leaves, multiple and raised midribs, serrated leaf tips, bilobed leaves, twisted leaves and broad/elliptic/ovate/triangular/linear-shaped leaves. Under a 10 Gy irradiation dose, 13 leaf mutants (10/4, 10/5, 10/16, 10/17, 10/21, 10/29, 10/33, 10/35, 10/37, 10/41, 10/79, 10/85, 10/111) were observed with morphological traits distinct from the control (Figure 8).

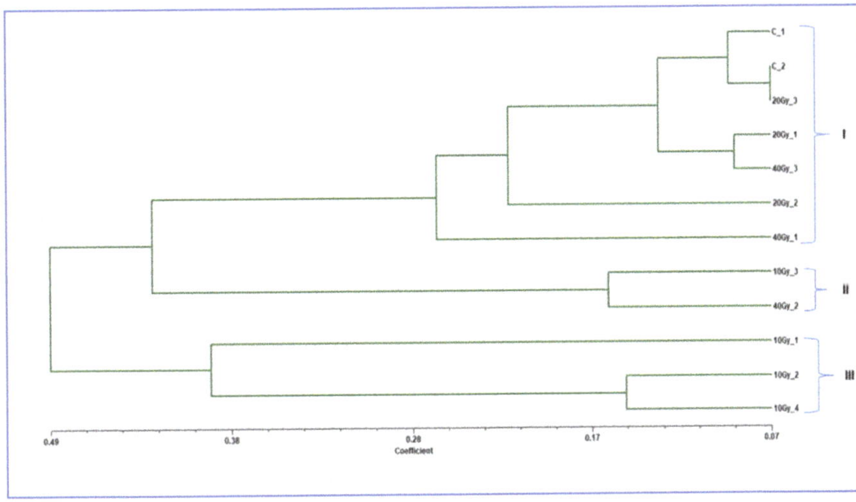


**Figure 7.** Dendrogram based on random amplified polymorphic DNA (RAPD) binary data. Dendrogram constructed using Nei's dissimilarity coefficient by UPGMA-based clustering method showed three major clusters (I, II, III), with a cophenetic correlation coefficient of 0.89. Cluster I included mutants of 20 Gy_1, 2, 3 and 40 Gy_1, 3, including two controls of 'Emma White' C_1, 2; cluster II included mutants of 10 Gy_3 and 40 Gy_2, whereas cluster III included three mutants of 10 Gy_1, 2, 4.

**Figure 8.** Morphological variations in 10 Gy mutant plants of Dendrobium 'Emma White' after 7–8 months of ex vitro transfer of well grown in vitro plantlets. (**A**) Control ((**B**):10/29) Leaves twisted in an anti-clockwise direction from the base. ((**C**):10/5) Closed leaf with marginal serrations around apex. ((**D**):10/85) Deformed leaf emerged from base of another fused leaf. ((**E**):10/35) Bilobed leaf with a deep notch, leaves fused to form two separate midribs. ((**F**):10/17) Asymmetric ovate leaf with two midribs,

two middle leaves emerged from fourth leaf. ((**G**):10/4) Broad leaf with two midribs, notch at apex, folded base. ((**H**):10/37) Deltate (triangular leaf) with uneven leaf growth. ((**I**):10/33) Third and fourth leaf fused with other leaves. ((**J**):10/35) Two leaves emerged from the base of another bilobed leaf. ((**K**):10/111) Closed leaf with marginal tooth or serrations around apex. ((**L**):10/21) Oval-shaped leaf with two midribs. ((**M**):10/79) Two midribs with slight elevation, small pseudo-bulb like appearance at the base. ((**N**):10/41) Long needle-shaped leaf (12.4 × 1.3 cm) opened at top. ((**O**):10/16) Leaf with two midribs, twisted, three pointed apexes.

### 2.7. Isolation of Early Flowering Mutant among 10 Gy Mutant Population

The first early flowering mutant (10/46) was isolated within 294 days of ex vitro transfer compared to several years required for flower development in control plants. This mutant plant was observed with a plant height of 7.2 cm, with the growth appearing stunted and with no change in flower color. In addition, alteration in the structure of the flower was observed as dorsal and lateral sepals fused with no petals. The lip of the early mutant flower appeared slightly rounded with wavy edges and longer side lobes compared to the flower of the mother plant (Supplementary Figure S1). The flower bud formation was initiated at 224 days of ex vitro growth, with a potted vase life of 50 days, as indicated by the number of days to withering. Among the gamma-ray-induced population, five other early flowering mutant lines (10/27, 10/39, 10/60, 10/118 and 10/7) were recovered within 457 to 678 days to first flowering as compared to the control plants flowered after 959.14 days (Figure 9, Table 5). Our results present a workflow for optimizing in vitro growth parameters and irradiation doses to isolate different morphological variations and desirable mutants, such as the early flowering mutant. Figure 10 depicts a complete workflow of the gamma-irradiation-induced mutagenesis system in Dendrobium 'Emma White'.

**Figure 9.** Gamma-radiation-induced early flowering mutant lines recovered from gamma-irradiated protocorm-like bodies of Dendrobium 'Emma White' at dose 10 Gy (**A–C**) and control (**D**).

**Table 5.** Performance of gamma-radiation-induced early flowering mutants recovered from irradiated protocorm-like bodies of Dendrobium 'Emma White' at dose 10 Gy compared with control plants (C).

| 10 Gy Mutant and Control Lines | DFBI | DTFF | DTW |
|---|---|---|---|
| 10/46 | 224 | 294 | 50 |
| 10/27 | 405 | 457 | 107 |
| 10/39 | 463 | 512 | 79 |
| 10/60 | 595 | 645 | 86 |
| 10/118 | 574 | 669 | 53 |
| 10/7 | 613 | 678 | 81 |
| C/4 | 906 | 954 | 85 |
| C/5(1) | 886 | 923 | 126 |
| C/5(2) | 886 | 923 | 76 |
| C/8 | 910 | 969 | 87 |
| C/13 | 790 | 862 | 101 |
| C/11 | 870 | 961 | 52 |
| C/6 | 1067 | 1112 | 79 |

C: control un-irradiated plants; values within bracket depict flower spike number of the same flower. DFBI: number of days to first bud initiation. DTFF: number of days to first flowering. DTW: number of days to withering.

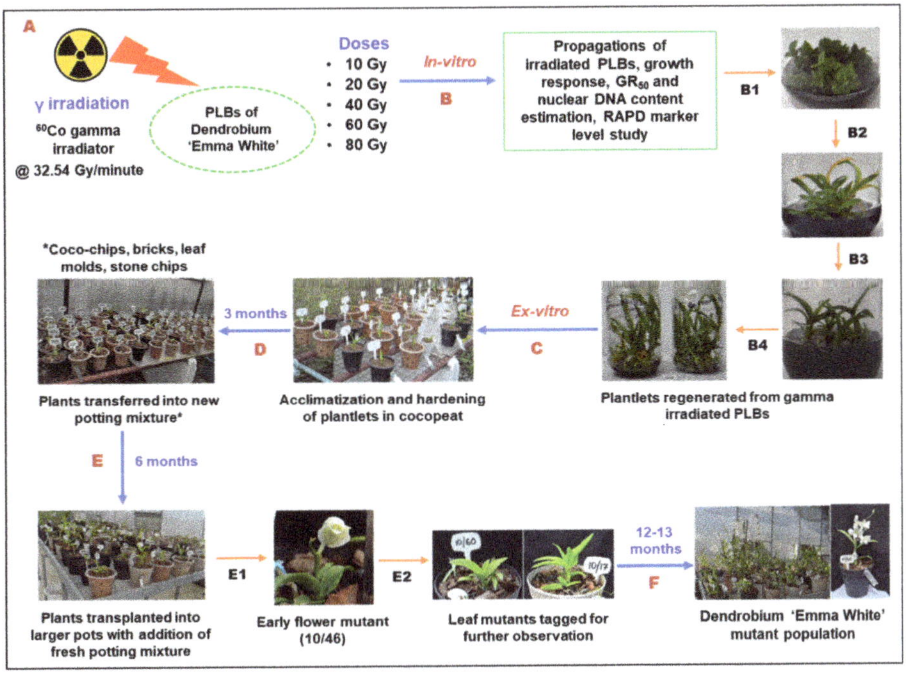


**Figure 10.** Workflow of gamma irradiation-induced mutagenesis in Dendrobium 'Emma White'. (**A**) Protocorm-like bodies (PLBs) generated from the shoot-tip of 'Emma White' was irradiated with different doses of gamma rays (10, 20, 40, 60, 80 Gy); (**B**) Irradiated PLBs were transferred onto the fresh culture medium and incubated in a culture room at 22 ± 2 °C and 65–70% relative humidity with 16 h of photoperiod. After three and six months of gamma irradiation, growth response of PLBs was recorded based on proliferation and regeneration percentage; nuclei isolated from PLBs irradiated with gamma rays were used for estimation of nuclear DNA content based on fluorescence intensity using high throughput cell analyser (HTCA); genetic diversity of in-vitro plantlets regenerated from gamma irradiated PLBs at dose 10, 20, and 40 Gy were analysed using random amplified polymorphic DNA (RAPD) markers; (**C**) Ex-vitro acclimatization and hardening of well rooted mutant and control plantlets in coco-peat; (**D**) After three months, plants were re-potted into a mixture of media containing coco-chips, brick pieces, leaf molds, and stone chips in 1:2:2:1 ratio; (**E**) After six months, plants were transplanted into larger pots with fresh potting mixture; (**E1**) First flowering in mutant 10/46 was observed within 294 days of ex-vitro growth; (**E2**) Morphological variations were observed in leaves of 10 Gy mutant plants as compared to control; (**F**) After twelve to thirteen months, well-grown Dendrobium 'Emma White' mutant population were established with early flowering at dose 10 Gy.

## 3. Discussion

The present study describes gamma-radiation-induced mutagenesis and its effect on various stages of in vitro and ex vitro plant growth in Dendrobium 'Emma White'. Compared to shoots and plantlets, PLBs have been proposed as the most suitable explant to induce variations in different orchid species due to their higher sensitivity toward gamma rays [22,23]. Thus, PLBs of Dendrobium hybrid Emma White were used as explants in the present study. Gamma rays are suitable for obtaining mutants with minor radiation damage [24]. The results revealed a significant effect of gamma irradiation with respect to the overall growth and diverse morphological variations, including early flowering, in 'Emma White'. The growth response was inversely proportional to increasing radiation doses. However, it is interesting that a lower radiation dose showed enhancement in proliferation and regeneration of PLBs after the radiation treatment (Figure 1). In addition, the results obtained from the study of differential phenotype of in vitro regenerated plantlets indicated the stimulatory effect of low-dose gamma radiation, which is in accordance with the previous report of enhanced growth responses at lower doses of gamma rays [25–29]. This effect may probably be attributed to the induced physiological and hormonal changes resulting in increased growth and development [30]. The effect of higher doses of gamma radiation indicated minimal proliferation and regeneration responses and reduced plant growth, suggesting radiation-induced cellular damage. The result corresponds to a previous report of growth inhibition at gamma doses > 40 Gy in two cultivars of Cymbidium hybrid [31]. A similar decrease in relative growth rate was reported in the growth of PLBs in Cymbidium at doses > 80 Gy [32]. Such a reduction in growth following gamma exposure could be attributed to significant oxidative damage resulting in altered chloroplast structure [25], membrane injuries [32] and a substantial decrease in nucleic acid and soluble protein levels [33], inhibiting metabolism and plant growth [34]. The higher doses alter stomatal morphology, resulting in inadequate gaseous exchange and, hence, lower plant viability [35]. Thus, irradiation at a lower dose is highly recommended in mutation breeding in orchids due to the adverse effects of higher doses on plant growth and survival [35]. In general, explant mortality after irradiation has been observed in many orchid cultivars with higher doses [22,23,36,37], including few other ornamentals [38–40].

Nuclear DNA content is suggested as an index of radiation damage in plants [24]. To understand the effect of radiation at the cellular level, nuclear DNA content was estimated based on fluorescence intensity of isolated nuclei using HTCA, which allows rapid high-throughput content screening in a short period over traditional flow cytometry [41]. The DNA content decreased significantly at doses more than 40 Gy, with more pronounced reduction at 80 Gy, suggesting that PLBs are more sensitive to higher doses of gamma rays. Nuclear DNA estimation can be used to measure radiation damage in mutation breeding experiments [35,42,43]. The decrease in DNA content has been related to signal-transduction-induced cell cycle arrest at the G2M phase of cell division [44]. The present study assessed the gamma-radiation-induced frequency of 7%, 9.5% and 51.6% based on morphological variations in regenerants recovered after radiation at 10, 20 and 40 Gy, respectively (Table 2). Such variations are not common in progenies arising from hybridized or self-pollinated orchids [45]. The previous study reported a mutation frequency of 3% induced by ion-bean (C6+) irradiation in two Paphiopedilum species, with no detectable variations by gamma rays [22]. The variegated chlorophyll leaf color mutants (0.4 Gy) and leaf shape mutants (2 Gy) were identified in two Dendrobium species with ion-beam (C6+), respectively [46]. To reduce the occurrence of undesired severe alterations in mutation breeding, the ideal doses that provide high frequency and a spectrum of desirable mutations can be chosen over radiation doses with the highest mutation frequency [47]. Due to their higher heterozygosity, orchids have a high mutation rate and different mutation types in a short cycle [48]. In the case of ornamental plants, single trait changes are generally observed, including harmful and unpredictable changes [49].

Molecular markers are considered an important tool for assessing plant genetic diversity in breeding programs [50]. Among various molecular markers, RAPD is the most common, inexpensive and reliable method for evaluating genetic variability [51], especially for plants such as orchids, where the availability of specific primers is limited. RAPD has been successfully employed in genetic diversity studies of many orchid cultivars [52–55]. The present study assessed the polymorphism in genomic DNA of mutant plantlets regenerated from gamma-irradiated PLBs of 'Emma White' using RAPD markers. A total of thirty-three scorable bands were generated using six RAPD primers, of which twenty-nine bands were polymorphic, with an average polymorphism of 88.89% (Table 3). The results suggest the effectiveness of gamma radiation in inducing higher polymorphism among mutants compared to the control. A previous study reported polymorphism of 46.5% in chemical mutagenesis of Dendrobium 'Earsakul' using ISSR markers [56]. As evident from the RAPD banding pattern, the absence of bands could be attributed to various kinds of DNA damage induced by gamma irradiation treatment resulting in generation of genetic diversity among mutants and between mutants and the control. Except for OPAW13, all the primers used in this study showed PIC values > 0.4, indicating informativeness in evaluating and quantifying polymorphism among the mutant population. In addition, the high cophenetic correlation coefficient of 0.82 was observed using the dissimilarity matrix and clustering method, which revealed that the dendrogram precisely preserved the pairwise distances between the original data points, making the dendrogram generated from the UPGMA-based clustering method viable for genetic diversity studies. Cluster analysis delineated mutants of different doses and the control into three distinct clusters, one comprising the control and mutants under 20 and 40 Gy treatment, while the third cluster included mutants under 10 Gy, indicating their potential genetic distinctness from other treatments and the control (Figure 7).

Morphological variations were detected from the in vitro differentiation stage to the ex-vitro stage, with altered leaf structure, multiple serrations, deep notches and asymmetric leaf arrangement in the mutant population (Figures 4 and 8). In a study using gamma-ray-induced mutants of rice, it has been shown that an increase in leaf vein density could result in enhanced photosynthetic efficiency, which indicates that trait alterations can be useful in higher plant productivity [57]. A pre-flowering period in orchids generally requires several years of vegetative growth depending on genera, species and habitat [11,58–61]. In the case of Dendrobium, it requires a minimum juvenile period of 2 to 5 years for floral induction [62]. In the present study, we isolated an early flowering mutant (10/46) among the 10 Gy-irradiated mutant population after 294 days of ex vitro transfer of in vitro plantlets. Generally, MADS-box genes are involved in floral organ expression and patterning during development in all angiosperms [63]. Low-dose gamma irradiation could have triggered the upregulation of these genes or some other signaling pathways involved in flower development, promoting early flowering. The isolated early flowering mutants are good candidates to further study flowering at the molecular level. A previous study reported overexpression of MADS-box genes (OMADS4 and OMADS1) in transgenic *Arabidopsis* and *Oncidium* cultivars, enabling early flowering [60]. In contrast, delayed flower bud formation to full bloom at 5 Gy gamma radiation treatment was reported in chrysanthemums [64]. We have also observed changes in the floral morphology in terms of in fused sepals and missing petals in the early flower mutant line (10/46) (Supplementary Figure S1). Similar morphological changes due to gamma irradiation influencing the shape of petals, sepals and the lip were also observed with PLBs of 'Sonia Kai' hybrid of Dendrobium [65], which resulted in recovery of four commercial mutants [19]. These changes could be due to a significant decrease in stomata size and cellular damage by radiation [36], which can influence the pattern of both growth and proliferation. However, early flowering in orchids after gamma treatment has not been reported, except for *Phalaenopsis aphrodite* treated at 15 Gy [66]. Our results suggest that mutagenesis can be used to isolate morphological mutants and mutants with desirable attributes for further improvement of valuable orchid hybrids such as Dendrobium 'Emma White'.

## 4. Materials and Methods

### *4.1. Plant Material for Gamma Irradiation Treatment*

PLBs of Dendrobium 'Emma White' (D. 'Singapore White' x D. 'Joan Kushima') were used as explant for mutagenesis, which is a complex hybrid derived from five Dendrobium species [9]. The PLBs generated from the shoot tip of 'Emma White' were developed previously at the institute (unpublished work). Before gamma radiation treatment, the established PLBs were maintained on Gamborg basal medium [67] supplemented with 2% sucrose, 0.15% activated charcoal (AC), 0.2 mg/L naphthalene acetic acid (NAA), gelled with 0.7% agar and adjusted at pH 5.8. After two weeks, the PLBs were irradiated with five doses of gamma rays (10, 20, 40, 60 and 80 Gy at a dose rate of 32.54 Gy/min) using $^{60}$Co gamma irradiator (Gamma Chamber 5000, Bhabha Atomic Research Centre, Trombay, Mumbai, India) as per the standard protocols [68]. All irradiated PLBs were transferred onto the fresh basal medium with the same supplementation. Gamborg basal medium was used as a culture medium throughout the experiment. An un-irradiated set of PLBs were maintained as control. The cultures were incubated at 22 ± 2 °C and 65–70% relative humidity with 16 h photoperiod provided by white fluorescent tube lights (Philips, India, 40 w).

### 4.2. Evaluation of Growth Response of Irradiated PLBs

To assess the immediate effect of gamma ray exposure in 'Emma White' the fresh weight of PLBs was measured after a month of irradiation treatment. In addition, growth responses to different irradiation doses were examined based on the survival, proliferation and shoot regeneration rate of PLBs, recorded at three and six months after irradiation. Subsequent development of PLBs after irradiation was investigated by recording the total number of days required for PLBs differentiation in terms of leaf primordium initiation. All growth parameters, including survival, proliferation and regeneration, were calculated based on the percentage of the number of responded PLBs to the total number of PLBs cultured. Accordingly, the optimal dose for mutagenesis was evaluated considering the PLBs growth response after irradiation. After every three months interval, PLBs were transferred into the fresh culture medium, allowing continuous growth and development. The experiment was performed with ten biological replicates and 10–12 PLBs per replicate.

### 4.3. High-Throughput Cell Analysis and Estimation of Nuclear DNA Content of PLBs

After five months of gamma irradiation, nuclei were isolated from irradiated and control PLBs for high-throughput cell analysis and estimation of nuclear DNA following the procedure using Tris-$MgCl_2$ buffer [69]. The isolated nuclei pellet was re-suspended in 600 µL of propidium iodide staining buffer overnight at 4 °C in dark conditions. The following day, 200 µL of stained nuclei solution was pipetted into a flat bottom corning 96-well plate and centrifuged at 3000 rpm for 6 min. The plate with loaded samples was placed inside the high-throughput cell analyzer (HTCA) (TTP Labtech's acumen® Cellista, Melbourn, UK), and flow data were analyzed using acumen Cellista software (version 4.2.5.0.69208). The un-irradiated set of PLBs was processed similarly and taken as control. Based on DNA-intercalating fluorescent dye (propidium iodide), FL3 was selected as the standard fluorescence filer (488 nm excitation and 620 ± 30 nm emission) for the present study. The average fluorescence intensity of nuclei generated by this laser scanning imaging cytometer was used to estimate nuclear DNA content. Based on the selected fluorescence gate, nuclei population defined as G1 (gate 1201–1600) and S + G2M (gate 1601–2800) were used to estimate the average percentage of nuclei in the S + G2M phase of the cell cycle. The experiment was performed with six biological replicates (~25 mg of PLBs per replicate), where each replicate was further divided into two technical replicates. Thus, a total of 12 replicates were used for each irradiation treatment.

### 4.4. Random Amplified Polymorphic DNA (RAPD)-Based Divergence Analysis of In Vitro Gamma Mutants

Total genomic DNA was isolated from in vitro young leaves of control and mutant plantlets using a modified cetyl trimethylammonium bromide (CTAB) method [70]. The isolated DNA was checked for its concentration and purity level using Nanodrop spectrophotometer. A preliminary marker level study was conducted using RAPD to analyze radiation-induced genetic diversity in 'Emma White' mutants. A total of 28 decamer RAPD primers were selected based on the previous studies [52,71–74], of which six primers (OPB 12, OPA 4, OPAW 13, OPAW 17, OPB 5, OPF 1) that produced strong and scorable bands were considered for further analysis. The details of the primers are listed in Supplementary Table S2. The 10 µL of polymerase chain reaction (PCR) mix was prepared containing 2x GoTaq® Green Master Mix (5 µL), 10µM primer (1 µL), nuclease free water (2 µL) and DNA sample (50 ng). PCR amplification was performed using Applied Biosystem Veriti 96-well Thermal Cycler with the first cycle of initial denaturation at 94 °C for 5 min, followed by 35 cycles of denaturation at 94 °C for 45 s, annealing temperature (Tm) specific to individual primers (refer Supplementary Table S2) for 1 min, extension at 72 °C for 2 min and a final cycle of extension at 72 °C for 5 min, followed by hold at 4 °C. The DNA amplicons were analyzed using 1.8% agarose gel, and bands were visualized and photographed using gel documentation system UV trans illuminator ECX-F20.M (GeNei).

### 4.5. *Morphological Variation Analysis of Putative Mutants at In Vitro and Ex Vitro Stage*

The gamma-irradiated regenerated PLBs were continuously monitored for any detectable morphological changes compared to control. Sub-culture was conducted at a regular interval of two months, allowing continuous growth and development of plantlets. Variation frequency was calculated as the percentage of mutants to the total regenerant PLBs and variation spectrum as the total mutant number [31,75]. After 12–13 months of post irradiation in vitro growth, the expression of differential phenotype was recorded in terms of plant height, leaf number, leaf length, leaf width, root number, root length and root width. The well-rooted plantlets were transferred ex vitro for acclimatization and hardening. A separate set of un-irradiated plantlets were maintained as the control for observation. The plantlets were washed with water to remove excess agar attached to roots, rinsed in 1% systemic fungicides (Carbendazin 50% WP, 02 min) and air-dried (10–15 min) and planted in small pots (10 × 7 cm) with coco-peat. The potted plants were kept inside a polyhouse under controlled conditions. After three months, plants were re-potted into a mixture of media containing coco-chips, brick pieces, leaf molds and stone chips (1:2:2:1 ratio). After six months, plants were transplanted into larger pots (15 × 16 cm) with fresh potting mixture. Each hardened plant was numbered accordingly as radiation dose followed by plant number, e.g., 10/1, 10/2, 10/3, 10/4, 10/5, etc. The morphological variations among the putative mutant populations were regularly monitored and flowering traits, such as days to flower bud initiation (DFBI), days to first flowering (DTFF) and days to withering (DTW), were recorded.

### 4.6. *Statistical Analysis*

The experiments were conducted in a completely randomized design (CRD). The statistical data were analyzed using R version 4.1.2 (Accessed on 1 November 2021). The influence of gamma irradiation on growth response of PLBs was analyzed using one-way ANOVA, followed by Duncan's Multiple Range Test (DMRT)-based post hoc analysis to determine the significant differences between irradiation treatments. The optimal gamma dose for mutagenesis of Dendrobium cultivar was analyzed using linear regression equation ($y = mx + c$) in Microsoft Excel, where y is the dependent variable (proliferation and regeneration rate), x is the independent variable (gamma radiation dose), m is the slope and c is the y-intercept, respectively. For RAPD data analysis, amplified bands were scored for the presence (1) or the absence (0) in a binary matrix. NEI72 coefficient-based genetic distance was calculated to generate dissimilarity matrix, and a dendrogram was constructed using unweighted pair group method with arithmetic mean (UPGMA) clustering method in NTSYSpc software, version 2.10 e [76]. The percentage of polymorphism was calculated as the number of polymorphic bands divided by the total number of bands [77]. The polymorphism information content (PIC) value of a marker was calculated using Gene-Calc bioinformatic tools [78].

## 5. Conclusions

The work described here establishes an optimized in vitro mutagenesis method for isolating gamma-radiation-induced mutants in Dendrobium 'Emma White' (Figure 10). It will reduce the time required for radiation dose optimization to generate a mutant population with desired traits, especially in orchids. Furthermore, a low radiation dose of 10 Gy showed a significantly profound stimulatory effect on overall growth and early flower development, indicating the usefulness of low doses in mutation breeding. The isolated mutants with economically valuable traits can be used in plant improvement and for further research into functional genes and related signaling pathways that influence early flowering in mutants.

**Supplementary Materials:** The following supporting information can be downloaded at: https://www.mdpi.com/article/10.3390/plants11223168/s1, Figure S1: Comparison of early flower mutant recovered from irradiated PLBs at 10 Gy dose with the flower of 'Emma White' mother plant; Table S1: Mean DNA intensity and nuclei count of gamma-irradiated PLBs after five months of irradiation treatment; Table S2: RAPD primers used for genetic study of gamma-irradiated mutants.

**Author Contributions:** Conceptualization, R.D.; methodology, R.D. and R.S.; software, R.D. and R.S.; validation, R.D., R.S. and S.P.; formal analysis, R.D. and R.S.; investigation, R.S., R.D. and S.P.; resources, R.S. and R.D.; data curation, R.D. and R.S.; writing—original draft preparation, R.S. and R.D.; writing—review and editing, R.D., R.S. and S.P.; visualization, R.D. and S.P.; supervision, S.N.B. and T.D.N.; project administration, R.D.; funding acquisition, R.D. All authors have read and agreed to the published version of the manuscript.

**Funding:** This study was funded by the Department of Atomic Energy (DAE), Board of Research in Nuclear Sciences (BRNS), Government of India (Project code: 35/14/22/2016-BRNS/35061).

**Data Availability Statement:** All data generated or analyzed during this study are included in this published article (and its Supplementary Materials). Some additional data on de novo whole genome sequence of the gamma-irradiated mutant Dendrobium hybrid cultivar 'Emma White' (10 Gy) were deposited with the NCBI with SRA accession number SRR16008784 and Genbank assembly accession GCA_021234465.1, available in the public domain via BioProject ID PRJNA763052. Additional data are available in the GigaScience GigaDB repository.

**Acknowledgments:** We are thankful to Bhabha Atomic Research Centre (BARC), Mumbai, India for providing gamma irradiation facility and lab infrastructure to conduct high-throughput cell analysis work. We would also like to thank Director, ICAR-National Research Centre on Orchids, Sikkim, India for providing institutional support. We are grateful for the project funding grant by the Department of Atomic Energy (DAE), Board of Research in Nuclear Sciences (BRNS), Government of India (project code: 35/14/22/2016-BRNS/35061).

**Conflicts of Interest:** The authors declare that they have no conflict of interest.

## References

1. Cakova, V.; Bonte, F.; Lobstein, A. *Dendrobium*: Sources of Active Ingredients to Treat Age-Related Pathologies. *Aging Dis.* **2017**, *8*, 827. [CrossRef] [PubMed]
2. Moudi, M.; Go, R.; Yong, C.Y.S.; Saleh, M.N. A Review on Molecular Systematic of the Genus Dendrobium Sw. *Acta Biol. Malays.* **2013**, *2*, 71–78. [CrossRef]
3. Hinsley, A.; de Boer, H.J.; Fay, M.F.; Gale, S.W.; Gardiner, L.M.; Gunasekara, R.S.; Kumar, P.; Masters, S.; Metusala, D.; Roberts, D.L.; et al. A Review of the Trade in Orchids and Its Implications for Conservation. *Bot. J. Linn. Soc.* **2018**, *186*, 435–455. [CrossRef]
4. Thammasiri, K. Current Status of Orchid Production in Thailand. *Acta Hortic.* **2015**, *1078*, 25–33. [CrossRef]
5. Baker, M.L.; Baker, C.O. *Orchid Species Culture: Dendrobium*; Timber Press: Portland, OR, USA, 1996.
6. Wanitprapha, K.; Kevin, M.; Yokoyama, K.M. Dendrobium Economic Fact Sheet. Available online: https://scholarspace.manoa.hawaii.edu/bitstream/10125/54681/CtahrpsEconFact13Dendro.pdf (accessed on 30 October 2022).
7. Nor Hazlina, M.S.; Wahba, L.E.; Fadelah, A.; Wickneswari, R. Genetic Relationships among 81 *Dendrobium* Accessions from Malaysia. *Malays. Appl. Biol.* **2013**, *42*, 35–40.
8. Devadas, R.; Pattanayak, S.L.; Singh, D.R. Studies on Cross Compatibility in *Dendrobium* Species and Hybrids. *Indian J. Genet. Plant Breed.* **2016**, *76*, 344. [CrossRef]
9. Devadas, R.; Khatiwara, P.; Barman, D.; Das, S.P. Breeding Dendrobium Phalaenopsis-Cane Type Hybrid in India: NRCO- 42 (Den. "Emma White" × Den. "Pompadour"). *Indian J. Genet. Plant Breed.* **2009**, *69*, 237–242.
10. Wilfret, G.J.; Kamemoto, H. Genome and Karyotype Relationships in the Genus *Dendrobium* (*Orchidaceae*). I. Crossability. *Am. J. Bot.* **1969**, *56*, 521–526. [CrossRef]
11. Zhang, S.; Yang, Y.; Li, J.; Qin, J.; Zhang, W.; Huang, W.; Hu, H. Physiological Diversity of Orchids. *Plant Divers.* **2018**, *40*, 196–208. [CrossRef]
12. Teixeira da Silva, J.A.; Jin, X.; Dobránszki, J.; Lu, J.; Wang, H.; Zotz, G.; Cardoso, J.C.; Zeng, S. Advances in *Dendrobium* Molecular Research: Applications in Genetic Variation, Identification and Breeding. *Mol. Phylogenet. Evol.* **2016**, *95*, 196–216. [CrossRef]
13. Adams, P.B. Systematics of *Dendrobiinae* (*Orchidaceae*), with Special Reference to Australian Taxa. *Bot. J. Linn. Soc.* **2011**, *166*, 105–126. [CrossRef]

14. Luyt, R.; Johnson, S.D. Hawkmoth Pollination of the African Epiphytic Orchid *Mystacidium Venosum*, with Special Reference to Flower and Pollen Longevity. *Plant Syst. Evol.* **2001**, *228*, 49–62. [CrossRef]
15. Teoh, S.B. A Breeding Program for the Development of Seed Propagated Uniform Hybrid Allopolyploid Aranda Cultivars for Cut Flowers. *Malay. Orchid Rev. Singap.* **1986**, *20*, 42–45.
16. Suprasanna, P.; Mirajkar, S.J.; Bhagwat, S.G. Induced Mutations and Crop Improvement. In *Plant Biology and Biotechnology*; Bahadur, B., Venkat Rajam, M., Sahijram, L., Krishnamurthy, K.V., Eds.; Springer: New Delhi, India, 2015; pp. 593–617. [CrossRef]
17. International Atomic Energy Agency Home Page. Available online: Https://Nucleus.Iaea.Org/Sites/Mvd/SitePages/Home.Aspx (accessed on 10 May 2021).
18. Thammasiri, K. Effects of gamma irradiation on protocorm-like bodies of *Cattleya alliances*. In Proceedings of the 6th Nuclear Science and Technology Conference, Bangkok, Thailand, 2–4 December 1996; p. 761.
19. Ariffin, S.; Mohamad, A.; Hassan, A.A.; Ahmad, Z.; Basiran, M.N. Flower Morphology of Dendrobium Sonia Mutants. In Proceedings of the Research and Development Seminar, Bangi, Malaysia, 12–15 October 2010; p. 4.
20. Álvarez-Holguín, A.; Morales-Nieto, C.R.; Avendaño-Arrazate, C.H.; Corrales-Lerma, R.; Villarreal-Guerrero, F.; Santellano-Estrada, E.; Gómez-Simuta, Y. Mean lethal dose (LD50) and growth reduction (GR50) due to gamma radiation in Wilman Lovegrass (*Eragrostis Superba*). *Rev. Mex. Cienc. Pecu.* **2019**, *10*, 227–238. [CrossRef]
21. Sokal, R.R.; Rohlf, F.J. The Comparison of Dendrograms by Objective Methods. *Taxon* **1962**, *11*, 33–40. [CrossRef]
22. Luan, L.Q.; Uyen, N.H.P.; Ha, V.T.T. In Vitro Mutation Breeding of *Paphiopedilum* by Ionization Radiation. *Sci. Hortic.* **2012**, *144*, 1–9. [CrossRef]
23. Xo, D.H.; Luan, L.Q. In Vitro Mutagenesis of Cymbidium La Bell "Anna Belle" by $\gamma$-Rays Irradiation and Oligochitosan Interaction. *Int. J. Environ. Agric. Biotechnol.* **2017**, *2*, 1540–1550. [CrossRef]
24. Yamaguchi, H.; Shimizu, A.; Hase, Y.; Tanaka, A.; Shikazono, N.; Degi, K.; Morishita, T. Effects of Ion Beam Irradiation on Mutation Induction and Nuclear DNA Content in *Chrysanthemum*. *Breed. Sci.* **2010**, *60*, 398–404. [CrossRef]
25. Wi, S.G.; Chung, B.Y.; Kim, J.-H.; Baek, M.-H.; Yang, D.H.; Lee, J.-W.; Kim, J.-S. Ultrastructural Changes of Cell Organelles in *Arabidopsis* Stems after Gamma Irradiation. *J. Plant Biol.* **2005**, *48*, 195–200. [CrossRef]
26. Kim, J.-H.; Chung, B.Y.; Kim, J.-S.; Wi, S.G. Effects of Planta Gamma-Irradiation on Growth, Photosynthesis, and Antioxidative Capacity of Red Pepper (*Capsicum Annuum* L.) Plants. *J. Plant Biol.* **2005**, *48*, 47–56. [CrossRef]
27. Minisi, F.; El-Mahrouk, M.; Rida, M.F.; Nasr, M. Effects of Gamma Radiation on Germination, Growth Characteristics and Morphological Variations of *Moluccella Laevis* L. *Am. Eurasian J. Agric. Environ. Sci.* **2013**, *13*, 696–704. [CrossRef]
28. Fathin, T.S.; Hartati, S.; Yunus, A. Diversity Induction with Gamma Ray Irradiation on *Dendrobium Odoardi* Orchid. *IOP Conf. Ser. Earth Environ. Sci.* **2021**, *637*, 012035. [CrossRef]
29. Pathirana, R. Plant Mutation Breeding in Agriculture. *CABI Rev.* **2011**, *2011*, 1–20. [CrossRef]
30. Kozlowska-Kalisz, J. The Influence of Ionizing Radiation on Biological Activity of Endogenous Growth Regulators in Orchids/*Cymbidium* in Tissue Culture. *Acta Hortic.* **1979**, *91*, 261–268. [CrossRef]
31. Kim, S.H.; Kim, S.W.; Ahn, J.-W.; Ryu, J.; Kwon, S.-J.; Kang, B.-C.; Kim, J.-B. Frequency, Spectrum, and Stability of Leaf Mutants Induced by Diverse $\gamma$-Ray Treatments in Two *Cymbidium* Hybrids. *Plants* **2020**, *9*, 546. [CrossRef]
32. Lee, Y.M.; Lee, H.J.; Kim, Y.S.; Kang, S.Y.; Kim, D.S.; Kim, J.B.; Ahn, J.W.; Ha, B.K.; Kim, S.H. Evaluation of the Sensitivity to Ionising $\gamma$ -Radiation of a *Cymbidium* Hybrid. *J. Hortic. Sci. Biotechnol.* **2016**, *91*, 109–116. [CrossRef]
33. Bajaj, Y.P.S. Effect of Gamma-Irradiation on Growth, RNA, Protein, and Nitrogen Contents of Bean Callus Cultures. *Ann. Bot.* **1970**, *34*, 1089–1096. [CrossRef]
34. Liu, N.; Lin, Z.; Guan, L.; Gaughan, G.; Lin, G. Antioxidant Enzymes Regulate Reactive Oxygen Species during Pod Elongation in *Pisum Sativum* and *Brassica Chinensis*. *PLoS ONE* **2014**, *9*, e87588. [CrossRef]
35. Bajpay, A.; Dwivedi, D. Gamma Ray Induced Foliage Variegation and Anatomical Aberrations in *Chrysanthemum* (*Dendranthema Grandiflora T.*) Cv. Maghi. *J. Pharm. Phytochem.* **2019**, *8*, 871–874.
36. Dehgahi, R.; Joniyasa, A. Gamma Irradiation-Induced Variation in Dendrobium Sonia-28 Orchid Protocorm-Like Bodies (PLBs). *Fungal Genom. Biol.* **2017**, *7*, 151. [CrossRef]
37. Billore, V.; Mirajkar, S.J.; Suprasanna, P.; Jain, M. Gamma Irradiation Induced Effects on in Vitro Shoot Cultures and Influence of Monochromatic Light Regimes on Irradiated Shoot Cultures of Dendrobium Sonia Orchid. *Biotechnol. Rep.* **2019**, *22*, e00343. [CrossRef]
38. Abdullah, T.L.; Endan, J.; Nazir, B.M. Changes in Flower Development, Chlorophyll Mutation and Alteration in Plant Morphology of *Curcuma Alismatifolia* by Gamma Irradiation. *Am. J. Appl. Sci.* **2009**, *6*, 1436–1439. [CrossRef]
39. Huang, Y.-L.; Yuan, S.-C.; Chang, K.-W.; Chen, F.-C. Gamma Irradiation Mutagenesis in *Monstera Deliciosa*. *Acta Hortic.* **2017**, *1167*, 213–216. [CrossRef]
40. Hosoguchi, T.; Uchiyama, Y.; Komazawa, H.; Yahata, M.; Shimokawa, T.; Tominaga, A. Effect of Three Types of Ion Beam Irradiation on Gerbera (*Gerbera Hybrida*) In Vitro Shoots with Mutagenesis Efficiency. *Plants* **2021**, *10*, 1480. [CrossRef]
41. Wylie, P.G.; Onley, D.J.; Hammerstein, A.F.; Bowen, W.P. Advances in Laser Scanning Imaging Cytometry for High-Content Screening. *ASSAY Drug Dev. Technol.* **2015**, *13*, 66–78. [CrossRef]
42. Yamaguchi, H.; Shimizu, A.; Degi, K.; Morishita, T. Effects of Dose and Dose Rate of Gamma Ray Irradiation on Mutation Induction and Nuclear DNA Content in *Chrysanthemum*. *Breed. Sci.* **2008**, *58*, 331–335. [CrossRef]

43. Oražem, P.; Štajner, N.; Bohanec, B. Effect of X-Ray Irradiation on Olive Shoot Culture Evaluated by Morphological Measurements, Nuclear DNA Content and SSR and AFLP Markers. *Trees* **2013**, *27*, 1587–1595. [CrossRef]
44. Preuss, S.B.; Britt, A.B. A DNA-Damage-Induced Cell Cycle Checkpoint in *Arabidopsis*. *Genetics* **2003**, *164*, 323–334. [CrossRef]
45. Madhusudan, S.; Wilson, D.M., III. Radiation Induced DNA Damage, Repair and Therapeutics. In *DNA Repair and Cancer*; CRC Press: Boca Raton, FL, USA, 2013. [CrossRef]
46. Abu Hassan, A.; Ariffin, S.; Ahmad, Z.; Basiran, M.N.; Oono, Y.; Hase, Y.; Shikazono, N.; Narumi, I.; Tanaka, A. *Mutation Induction of Orchid Plants by Ion Beams*; Japan Atomic Energy Agency–Review: Ibaraki, Japan, 2016. [CrossRef]
47. Brunner, H. Radiation Induced Mutations for Plant Selection. *Appl. Radiat. Isot.* **1995**, *46*, 589–594. [CrossRef]
48. Li, C.; Dong, N.; Zhao, Y.; Wu, S.; Liu, Z.; Zhai, J. A Review for the Breeding of Orchids: Current Achievements and Prospects. *Hortic. Plant J.* **2021**, *7*, 380–392. [CrossRef]
49. Su, J.; Jiang, J.; Zhang, F.; Liu, Y.; Ding, L.; Chen, S.; Chen, F. Current Achievements and Future Prospects in the Genetic Breeding of *Chrysanthemum*: A Review. *Hortic. Res.* **2019**, *6*, 109. [CrossRef]
50. Dhakshanamoorthy, D.; Selvaraj, R.; Chidambaram, A. Utility of RAPD Marker for Genetic Diversity Analysis in Gamma Rays and Ethyl Methane Sulphonate (EMS)-Treated Jatropha Curcas Plants. *Comptes Rendus Biol.* **2015**, *338*, 75–82. [CrossRef] [PubMed]
51. Ong, T. Detection of Genomic Instability in Lung Cancer Tissues by Random Amplified Polymorphic DNA Analysis. *Carcinogenesis* **1998**, *19*, 233–235. [CrossRef] [PubMed]
52. Khosravi, A.R.; Kadir, M.A.; Kadzemin, S.B.; Zaman, F.Q.; De Silva, A.E. RAPD Analysis of Colchicine Induced Variation of the Dendrobium Serdang Beauty. *Afr. J. Biotechnol.* **2009**, *8*, 1455–1465.
53. Mallón, R.; Rodríguez-Oubiña, J.; González, M.L. In Vitro Propagation of the Endangered Plant *Centaurea Ultreiae*: Assessment of Genetic Stability by Cytological Studies, Flow Cytometry and RAPD Analysis. *Plant Cell Tissue Organ Cult.* **2010**, *101*, 31–39. [CrossRef]
54. Srivastava, D.; Mc, G.; Sarangi, S. In Vitro Mutagenesis and Characterization of Mutants through Morphological and Genetic Analysis in Orchid *Aerides Crispa* Lindl. *Indian J. Exp. Biol.* **2018**, *56*, 385–394.
55. Hartati, S.; Muliawati, E.S. Short Communication: Genetic Variation of *Coelogyne pandurate*, *C. rumphii* and Their Hybrids Based on RAPD Markers. *Biodiversitas J. Biol. Divers.* **2020**, *21*, 4709–4713. [CrossRef]
56. Tantasawat, T.P.; Khairum, A.; Tharapreuksapong, A.; Poolsawat, O.; Tantasawat, P.A. Molecular Characterization of Dendrobium 'Earsakul' Mutants from in Vitro Selection for Black Rot Resistance. *J. Appl. Hortic.* **2017**, *19*, 130–134. [CrossRef]
57. Feldman, A.B.; Leung, H.; Baraoidan, M.; Elmido-Mabilangan, A.; Canicosa, I.; Quick, W.P.; Sheehy, J.; Murchie, E.H. Increasing Leaf Vein Density via Mutagenesis in Rice Results in an Enhanced Rate of Photosynthesis, Smaller Cell Sizes and Can Reduce Interveinal Mesophyll Cell Number. *Front. Plant Sci.* **2017**, *8*, 1883. [CrossRef]
58. Kull, T. *Cypripedium Calceolus* L. *J. Ecol.* **1999**, *87*, 913–924. [CrossRef]
59. Schmidt, G.; Zotz, G. Inherently Slow Growth in Two Caribbean Epiphytic Species: A Demographic Approach. *J. Veg. Sci.* **2002**, *13*, 527–534. [CrossRef]
60. Hsiao, Y.-Y.; Pan, Z.-J.; Hsu, C.-C.; Yang, Y.-P.; Hsu, Y.-C.; Chuang, Y.-C.; Shih, H.-H.; Chen, W.-H.; Tsai, W.-C.; Chen, H.-H. Research on Orchid Biology and Biotechnology. *Plant Cell Physiol.* **2011**, *52*, 1467–1486. [CrossRef]
61. Teixeira da Silva, J.A.; Hossain, M.M.; Sharma, M.; Dobránszki, J.; Cardoso, J.C.; Zeng, S. Acclimatization of in Vitro -Derived Dendrobium. *Hortic. Plant J.* **2017**, *3*, 110–124. [CrossRef]
62. Hee, K.H.; Loh, C.S.; Yeoh, H.H. Early in Vitro Flowering and Seed Production in Culture in Dendrobium Chao Praya Smile (*Orchidaceae*). *Plant Cell Rep.* **2007**, *26*, 2055–2062. [CrossRef]
63. Alvarez-Buylla, E.R.; Pelaz, S.; Liljegren, S.J.; Gold, S.E.; Burgeff, C.; Ditta, G.S.; Ribas de Pouplana, L.; Martínez-Castilla, L.; Yanofsky, M.F. An Ancestral MADS-Box Gene Duplication Occurred before the Divergence of Plants and Animals. *Proc. Natl. Acad. Sci. USA* **2000**, *97*, 5328–5333. [CrossRef]
64. Patil, U.; Karale, A.; Katwate, S. Mutation Breeding in *Chrysanthemum* (*Dendranthema Grandiflora* T.). *J. Pharmacogn. Phytochem.* **2017**, *6*, 230–232.
65. Ariffin, S.; Basiran, M.N. Induced Mutation of Dendrobium Orchid. In Proceedings of the Research and Development Seminar, Institute for Nuclear Technology Research, MINT, Bangi, Malaysia, 17–19 October 2000; p. 8.
66. Magdalita, P.M.; Pascual, A.O.S.; Villareal, R.L. Characterization and Flowering Behavior of Eleven Philippine Native *Phalaenopsis* Species and Gamma Irradiation Effects on *Phalaenopsis* Aphrodite. *Philipp. J. Sci.* **2019**, *149*, 1–10. [CrossRef]
67. Gamborg, O.L.; Murashige, T.; Thorpe, T.A.; Vasil, I.K. Plant Tissue Culture Media. *In Vitro* **1976**, *12*, 473–478. [CrossRef]
68. deAlmeida, C.E.; Ochoa, R.; de Lima, M.C.; David, M.G.; Pires, E.J.; Peixoto, J.G.; Salata, C.; Bernal, M.A. A Feasibility Study of Fricke Dosimetry as an Absorbed Dose to Water Standard for 192Ir HDR Sources. *PLoS ONE* **2014**, *9*, e115155. [CrossRef]
69. Sahoo, S.A.; Jha, Z.; Verulkar, S.B.; Srivastava, A.K.; Suprasanna, P. High-Throughput Cell Analysis Based Protocol for Ploidy Determination in Anther-Derived Rice Callus. *Plant Cell Tissue Organ Cult.* **2019**, *137*, 187–192. [CrossRef]
70. Doyle, J.J.; Doyle, J.L. Isolation of Plant DNA from Fresh Tissue. *Focus* **1990**, *12*, 13–15.
71. Goh, M.W.K.; Kumar, P.P.; Lim, S.H.; Tan, H.T.W. Random Amplified Polymorphic DNA Analysis of the Moth Orchids, *Phalaenopsis* (*Epidendroideae: Orchidaceae*). *Euphytica* **2005**, *141*, 11–22. [CrossRef]
72. Inthawong, S.; Bundithya, W.; Kuanprasert, N.; Apavatjrut, P. Analysis of Intersectional Hybrids of Dendrobium by RAPD Technique. *Kasetsart J. Nat. Sci.* **2006**, *40*, 456–461.

73. Niknejad, A.; Kadir, M.A.; Kadzimin, S.B.; Abdullah, N.A.P.; Sorkheh, K. Molecular Characterization and Phylogenetic Relationships among and within Species of *Phalaenopsis* (*Epidendroideae: Orchidaceae*) Based on RAPD Analysis. *Afr. J. Biotechnol.* **2009**, *8*, 5225–5240. [CrossRef]
74. Abbas, B.; Dailami, M.; Listyorini, F.H.; Munarti. Genetic Variations and Relationships of Papua's Endemic Orchids Based on RAPD Markers. *Nat. Sci.* **2017**, *9*, 377–385. [CrossRef]
75. Zalewska, M.; Jerzy, M. Mutation Spectrum in *Dendranthema Grandiflora* Tzvelev After In Vivo and In Vitro Regeneration of Plants from Irradiated Leaves. *Acta Hortic.* **1997**, *16*, 615–618. [CrossRef]
76. Jensen, R.J. *NTSYS-Pc: Numerical Taxonomy and Multivariate Analysis System*; Version 2.2; Exeter Publishing Setauke: New York, NY, USA, 2000.
77. Powell, W.; Morgante, M.; Andre, C.; Hanafey, M.; Vogel, J.; Tingey, S.; Rafalski, A. The Comparison of RFLP, RAPD, AFLP and SSR (Microsatellite) Markers for Germplasm Analysis. *Mol. Breed.* **1996**, *2*, 225–238. [CrossRef]
78. Nagy, S.; Poczai, P.; Cernák, I.; Gorji, A.M.; Hegedűs, G.; Taller, J. PICcalc: An Online Program to Calculate Polymorphic Information Content for Molecular Genetic Studies. *Biochem. Genet.* **2012**, *50*, 670–672. [CrossRef]

plants

MDPI

Article

# Chili Pepper *AN2* (*CaAN2*): A Visible Selection Marker for Nondestructive Monitoring of Transgenic Plants

**Sun-Hyung Lim [1,2,*], Da-Hye Kim [1,2], Myeong-Cheoul Cho [3] and Jong-Yeol Lee [4,*]**

1 Division of Horticultural Biotechnology, School of Biotechnology, Hankyong National University, Anseong 17579, Korea; kimdh143@naver.com
2 Research Institute of International Technology and Information, Hankyong National University, Anseong 17579, Korea
3 National Institute of Horticultural and Herbal Science, Rural Development Administration, Wanju 55365, Korea; chomc@korea.kr
4 National Institute of Agricultural Sciences, Rural Development Administration, Jeonju 54874, Korea
* Correspondence: limsh2@hknu.ac.kr (S.-H.L.); jy0820@korea.kr (J.-Y.L.); Tel.: +82-31-670-5105 (S.-H.L.); +82-63-238-4616 (J.-Y.L.)

**Abstract:** Selecting transformed plants is generally time consuming and laborious. To develop a method for transgenic plant selection without the need for antibiotics or herbicides, we evaluated the suitability of the R2R3 MYB transcription factor gene *CaAN2* from purple chili pepper (*Capsicum annuum*) for use as a visible selection marker. *CaAN2* positively regulates anthocyanin biosynthesis. Transient expression assays in tobacco (*Nicotiana tabacum*) leaves revealed that *CaAN2* actively induced sufficient pigment accumulation for easy detection without the need for a basic helix-loop-helix (bHLH) protein as a cofactor; similar results were obtained for tobacco leaves transiently co-expressing the anthocyanin biosynthesis regulators bHLH B-Peru from maize and R2R3 MYB mPAP1D from Arabidopsis. Tobacco plants harboring *CaAN2* were readily selected based on their red color at the shoot regeneration stage due to anthocyanin accumulation without the need to impose selective pressure from herbicides. Transgenic tobacco plants harboring *CaAN2* showed strong pigment accumulation throughout the plant body. The ectopic expression of *CaAN2* dramatically promoted the transcription of anthocyanin biosynthetic genes as well as regulators of this process. The red coloration of tobacco plants harboring *CaAN2* was stably transferred to the next generation. Therefore, anthocyanin accumulation due to *CaAN2* expression is a useful visible trait for stable transformation, representing an excellent alternative selection system for transgenic plants.

**Keywords:** alternative selection method; anthocyanin; *CaAN2*; *Capsicum annuum*; transformation; visible marker

**Citation:** Lim, S.-H.; Kim, D.-H.; Cho, M.-C.; Lee, J.-Y. Chili Pepper *AN2* (*CaAN2*): A Visible Selection Marker for Nondestructive Monitoring of Transgenic Plants. *Plants* **2022**, *11*, 820. https://doi.org/10.3390/plants11060820

Academic Editors: Ranjith Pathirana and Francesco Carimi

Received: 5 February 2022
Accepted: 18 March 2022
Published: 19 March 2022



## 1. Introduction

Pepper (*Capsicum annuum*) is an economically important vegetable that provides antioxidant compounds (with anti-inflammatory and antimicrobial effects) for the human diet. Engineering the pepper genome to produce novel and useful agronomic traits requires development of stable transformation methods and accurate selectable markers. Various selectable markers have been used for crop transformation, with their transformation efficiencies being strongly affected by the type of marker chosen [1,2]. The use of antibiotic and herbicide resistance genes as positive selection markers has prompted biosafety concerns about human health and the environment [3,4]. To address these concerns, β-glucuronidase and fluorescent proteins are generally used for the identification of transformed cells. However, these systems have several limitations, such as the need for destructive GUS staining methods and for expensive equipment to detect fluorescent signals [5,6]. Additionally, the process of genetic transformation and regeneration is time consuming and labor intensive, in terms of selection and characterization of transformed cells and occasionally can result in chimerism (a single plant tissue containing transformed and non-transformed sections), thus requiring additional experiments for gene transfer to subsequent generations.

Anthocyanins, a large class of secondary metabolites, are widely distributed in various plant tissues, including flowers, stems, leaves and fruits, with colors ranging from red to blue [7]. Therefore, anthocyanins could potentially be used as a selectable marker for the visual identification of transformed cells during in vitro culture. The anthocyanin biosynthetic pathway involves a multienzyme complex and is controlled by key transcription factors, including R2R3-MYB, basic helix-loop-helix (bHLH) and WD40 proteins, as well as MBW complexes [8,9]. The R2R3-MYB transcription factors, belonging to subgroup 5 (SG5) and SG6, activate anthocyanin biosynthesis. The ectopic expression of the R2R3-MYB transcription factor genes in apple, barrelclover, radish and strawberry leads to red or purple coloration in various tissues, including calli, root tips and leaves, by upregulating the expression of anthocyanin biosynthetic genes [10–12].

Most pepper plants have green stems and leaves, white flowers and fruits that turn from green to red at maturity; others have purple stems, leaves, flowers and fruits at the immature stage and red fruits at the ripe stage. *CaAN2* (*ANTHOCYANIN2*), an ortholog of petunia *PhAN2*, was isolated from purple pepper and shown to be responsible for the skin color of purple pepper fruits [13]. In purple-fruited pepper, variation in the *CaAN2* promoter region can enhance the expression level of this gene in various tissues, resulting in the accumulation of anthocyanin pigments [14].

In this study, to investigate the potential use of *CaAN2* as a visible selectable marker, a transient assay and stable transformation of tobacco (*Nicotiana tabacum*) plants was performed. The ectopic expression of *CaAN2* induced anthocyanin accumulation in tobacco. In addition, tobacco plants that were stably transformed with *CaAN2* showed strong pigment accumulation, which was steadily transferred to the next generation. These results indicate that *CaAN2* could be utilized as an alternative visible selectable marker to facilitate transgenic plant identification.

## 2. Results

### 2.1. *Anthocyanin Accumulation Determines the Green and Purple Coloration of Chili Pepper*

To investigate the mechanisms controlling anthocyanin biosynthesis in chili pepper, two pepper cultivars with different pigmentation patterns in fruits were analyzed. The green cultivar (G) has green leaves, green stems and white flowers, and its fruits are green at the mature green fruit stage 1 (FS1) and gradually become red at the red ripe stage (FS3). The purple cultivar (P) has green leaves, purple stems and purple flowers and its fruits are green at FS1, turn purple at the breaker stage (FS2) and are red at FS3 (Figure 1).

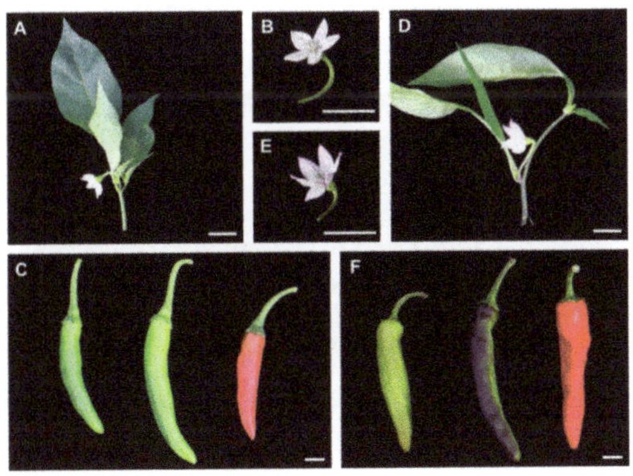


**Figure 1.** Comparison of the phenotypes of green (cv. AG188) and purple chili pepper (cv. 20GP15-2). (**A**) Green pepper plant; (**B**) green pepper flower; (**C**) green pepper fruits at different stages of development: mature green (left), breaker (middle) and red-ripe (right). (**D**) Purple pepper plant; (**E**) purple pepper flower; and (**F**) purple pepper fruits at different stages of development. Bars = 1 cm.

Anthocyanin contents were quantified in the leaves, stems, flowers and fruits at each stage in these two pepper cultivars (Figure 2). The anthocyanin contents were essentially consistent with the visible pigmentation patterns: anthocyanin levels were high in the stems, flowers and FS2 stage fruits of the P cultivar. These results suggest that the anthocyanin contents of stems, flowers and fruits are responsible for the differences in the purple coloration of the pepper cultivars.

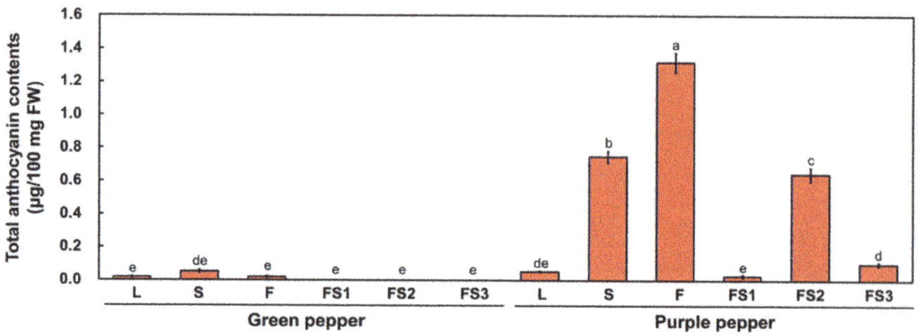


**Figure 2.** Anthocyanin contents of various green and purple pepper tissues. L, leaves; S, stems; F, flowers; FS1, mature green stage fruit; FS2, breaker stage fruit; and FS3, red-ripe stage fruit. Different letters indicate significantly different values ($p < 0.01$), as determined by two-way ANOVA followed by Duncan's multiple range tests.

### 2.2. *CaAN2 and CaTT8 Are Highly Expressed in Purple Chili Pepper Fruit*

To examine the expression of genes encoding two regulators of anthocyanin biosynthesis, the R2R3 MYB transcription factor gene *CaAN2* and the basic helix-loop-helix (bHLH) transcription factor gene *TRANSPARENT TESTA8* (*CaTT8*), quantitative reverse-transcription PCR (qRT-PCR) was performed with various tissues including leaves, stems, flowers and fruits at different developmental stages in both cultivars. *CaAN2* and *CaTT8* were expressed at higher levels in all tissues of the P vs. G cultivar and their expression levels reflected the extent of pigment accumulation in these tissues (Figure 3A). Specifically, *CaAN2* and *CaTT8* transcript levels were highest in flowers and purple fruit at breaker stage (FS2) in the P cultivar. Indeed, anthocyanin content was correlated with the simultaneous expression of *CaAN2* and *CaTT8*. These results suggest that *CaAN2* and CaTT8 cooperatively regulate anthocyanin biosynthesis in various tissues of chili pepper.

As shown in Figure 3B, the general phenylpropanoid biosynthetic genes *phenylalanine ammonia-lyase* (*CaPAL*), *cinnamate 4-hydroxylase* (*CaC4H*) and *4-coumarate coenzyme A:ligase* (*Ca4CL*) were highly expressed in the leaves of the P cultivar. However, early biosynthetic genes, including *chalcone synthase* (*CaCHS*), *chalcone isomerase* (*CaCHI*) and *flavanone hydroxylase* (*CaF3H*), were highly expressed in flowers of the P cultivar. The late biosynthetic genes, including *dihydroflavonol 4-reductase* (*CaDFR*), *anthocyanidin synthase* (*CaANS*) and *UDP-flavonoid glucosyl transferase* (*CaUFGT*), were highly upregulated in flowers and fruits (at stage FS2) of the P cultivar, which showed high anthocyanin contents. Taken together, these results confirm that anthocyanin accumulation reflects the expression levels of flavonoid biosynthetic genes across various tissues of different cultivars and that this expression occurs in tissues co-expressing *CaAN2* and *CaTT8*.

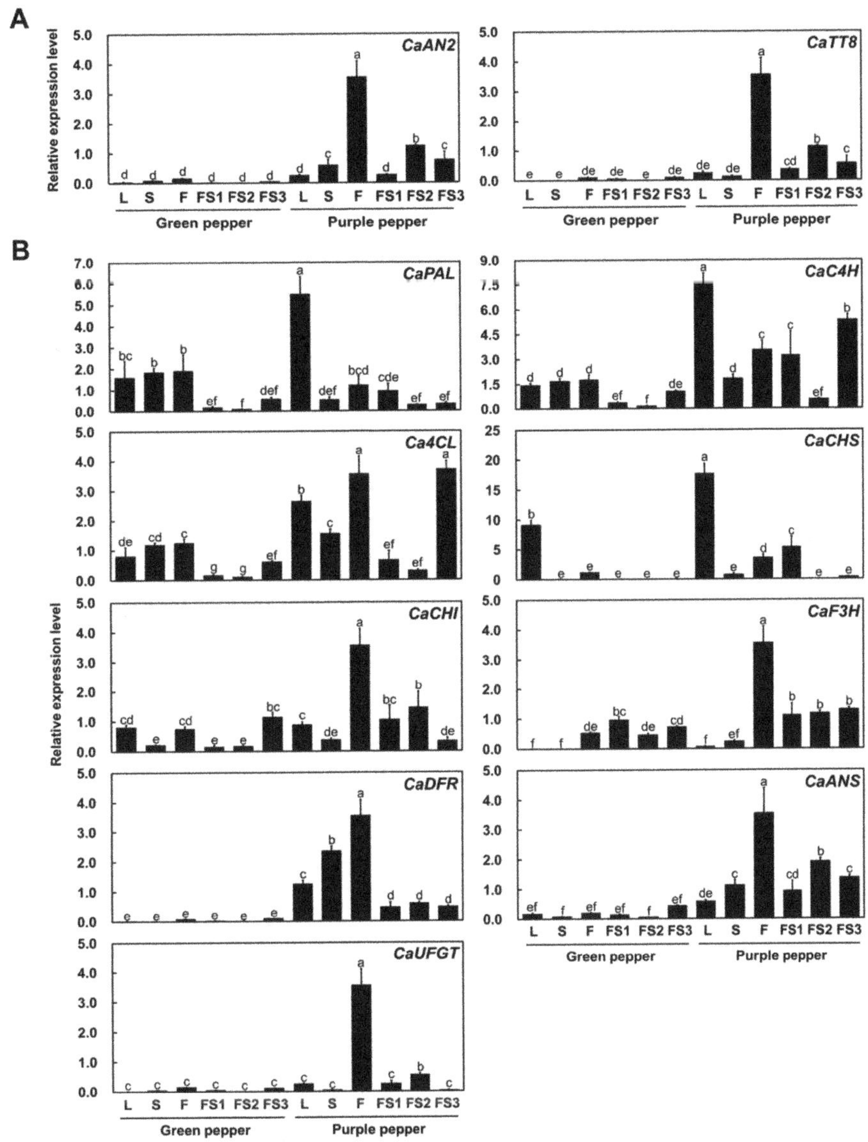


**Figure 3.** Relative transcript levels of *CaAN2* and other anthocyanin biosynthetic genes from green and purple peppers. Expression levels of anthocyanin regulatory genes (**A**) and anthocyanin biosynthetic genes (**B**). Results represent means ± SD from three independent biological replicates. *CaACTIN* was used as the reference gene. Different letters indicate significantly different values ($p < 0.01$), as determined by two-way ANOVA followed by Duncan's multiple range tests.

### 2.3. Isolation of *CaAN2* cDNA and Phylogenetic Analysis

To investigate the mechanism regulating anthocyanin biosynthesis in purple chili pepper, the R2R3-type transcription factor gene *CaAN2* was cloned from P cultivar leaves by PCR. The cDNA sequence of *CaAN2* was 100% identical to the previously reported sequence for *CaAN2*, comprising a 789-bp coding sequence encoding a predicted protein of 262 amino acids (GenBank accession number NP_001311547.1)

Flavonoid-related MYB transcription factors belonging to SG5 and SG6 regulate anthocyanin biosynthesis in different tissues of various plants, including leaves, fruits and seeds [11,15,16]. In the phylogenetic tree generated from R2R3-MYB proteins from various plant species (Figure 4A), *CaAN2* falls into the SG6 clade, together with PhAN2, AtPAP1 and MdMYB10: these eudicot MYB transcription factors actively regulate anthocyanin biosynthesis [16–19]. Sequence alignments showed that all SG5 and SG6 R2R3-MYBs share the highly conserved motif [D/E]Lx2[R/K]x3Lx6Lx3R in the R3 domain, which is functionally important for interactions between MYB and R/B-like bHLH proteins [11,16,19]. Additionally, the R2R3 domain of *CaAN2* contains five conserved tryptophans that are important for forming the helix-loop-helix protein architecture at the N terminus. At the end of the R3 domain, *CaAN2* harbors the conserved ANDV motif, a characteristic feature of SG6 R2R3-MYBs (Figure 4B). While *CaAN2* contains this conserved SG6 motif ([K/R]P[Q/R]P[Q/R]TF), it also harbors a highly variable C-terminal region compared to those of other anthocyanin-activating R2R3 MYB transcription factors [11,15,17] (Figure 4C). These results suggest that *CaAN2* activates anthocyanin biosynthesis in chili pepper.

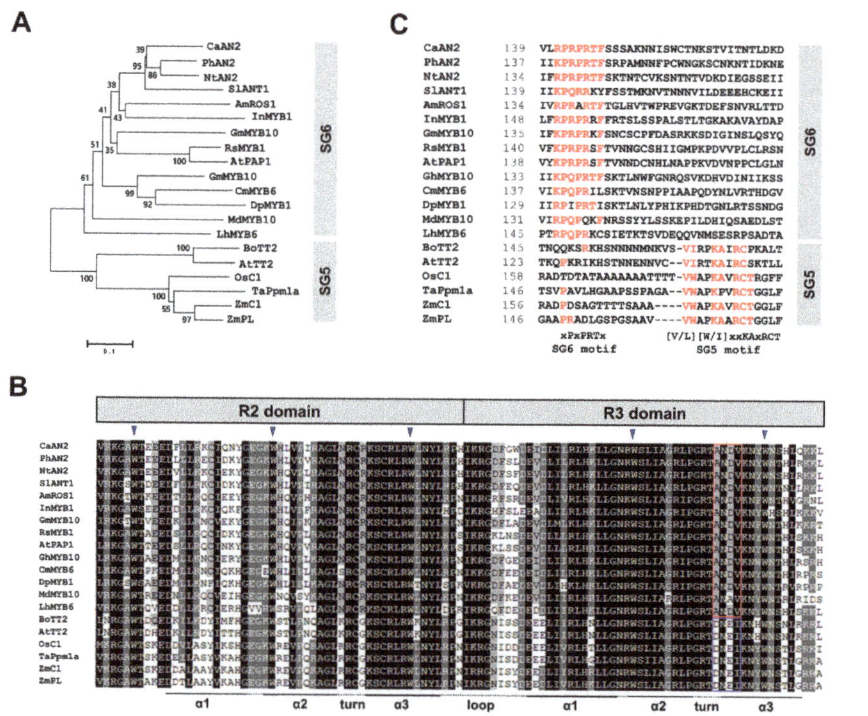


**Figure 4.** Phylogenetic analysis and multiple sequence alignment of anthocyanin-activating R2R3 MYBs. (**A**) Phylogenetic tree of pepper *CaAN2* and R2R3 MYB proteins from other plants. The phylogenetic tree was constructed using the neighbor-joining method with MEGA6 software. The GenBank accession numbers of species used in this study are listed in Supplementary Table S1. (**B**) Multiple sequence alignment of the R2 and R3 domains across the R2R3 MYB proteins shown in (**A**). The conserved residues ANDV and DNEI are represented by red and blue boxes, respectively. Inverted blue triangles indicate the conserved residues forming the inner hydrophobic core of the R2 and R3 domains. (**C**) Multiple sequence alignment of parts of the C-terminal regions of R2R3-MYB sequences, showing the SG5 and SG6 motifs. Amino acids matching either the SG6 or SG5 motif are indicated in red. The starting amino acid position of each sequence is given in the second column.

### 2.4. *CaAN2 Is an Active Regulator of Anthocyanin Biosynthesis*

To evaluate the role of *CaAN2* in anthocyanin biosynthesis, various combinations of diverse transcription factor genes, including genes encoding a bHLH *B-peru* from (*ZmB-peru*), R2R3 MYB *mPAP1D* from Arabidopsis (*AtmPAP1D*) and *CaAN2*, were transiently expressed in tobacco leaves (Figure 5). Individual infiltration with the *ZmB-peru AtmPAP1D* gene did not induce anthocyanin production, whereas co-infiltration with both genes induced anthocyanin accumulation in tobacco leaves. By contrast, the transient overexpression of the full-length genomic sequence or cDNA of *CaAN2* induced visible pigment accumulation in the absence of the bHLH transcription factor gene *ZmB-Peru*, as did the simultaneous expression of *CaAN2* and *ZmB-peru*. These results indicate that *CaAN2* is a positive regulator of anthocyanin biosynthesis.

**Figure 5.** Anthocyanin accumulation in a tobacco leaf transiently infiltrated with *Agrobacterium* expressing *CaAN2* and other anthocyanin biosynthesis regulatory genes. Tobacco leaves were transiently infiltrated with *Agrobacterium* cultures carrying empty vector or the indicated combinations of constructs harboring *ZmB-peru*, *AtmPAP1* and *CaAN2*. A representative photograph of a transiently infiltrated tobacco leaf at 5 days after agroinfiltration is shown.

### 2.5. *CaAN2 Is a Good Candidate Visible Selectable Marker Gene*

As shown in the transient assay, pigment accumulation was readily detected by the expression of *CaAN2*. To verify the utility of this gene as a visible marker, we transformed tobacco leaf disks with *Agrobacterium* containing *pB2GW7-CaAN2*, which includes the *bar* gene for phosphinothricin (PPT) resistance and cultured the resulting explants with or without PPT. Green shoots were detected more on PPT-free medium than on medium containing PPT (Figure 6). Additionally, purple shoots were observed on both PPT and PPT-free medium. Regenerated plantlets harboring the *CaAN2* gene showed distinct red coloration and were easily detected from the callus phase to plant regeneration. These results suggest that pigment accumulation via the expression of an anthocyanin activating regulator can be used to select transgenic plants without the need for selective pressure from antibiotics or herbicides.

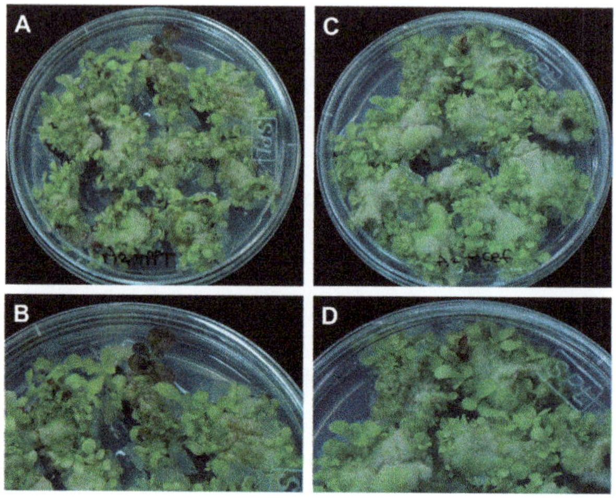


**Figure 6.** *CaAN2* expression in tobacco leaf explants enables the easy visual screening of transgenic tobacco plants. Purple coloration caused by anthocyanin accumulation can be observed in leaves grown on medium with (**A**,**B**) or without PPT (**C**,**D**).

## 2.6. Ectopic Expression of CaAN2 Strongly Promotes Anthocyanin Biosynthesis

Transgenic tobacco plants harboring *CaAN2* showed easily distinguishable pigmentation throughout the plant body, including leaves, stems and flowers (Figure 7A). The transgenic tobacco plants showed high anthocyanin contents, whereas the non-transgenic (NT) tobacco plants did not, indicating that *CaAN2* strongly increased anthocyanin production and conferred an intense red-purple color due to strong anthocyanin accumulation (Figure 7B). Taken together, these results demonstrate that heterologous expression of *CaAN2* strongly enhances anthocyanin accumulation and confers an intense red-purple color in tobacco.

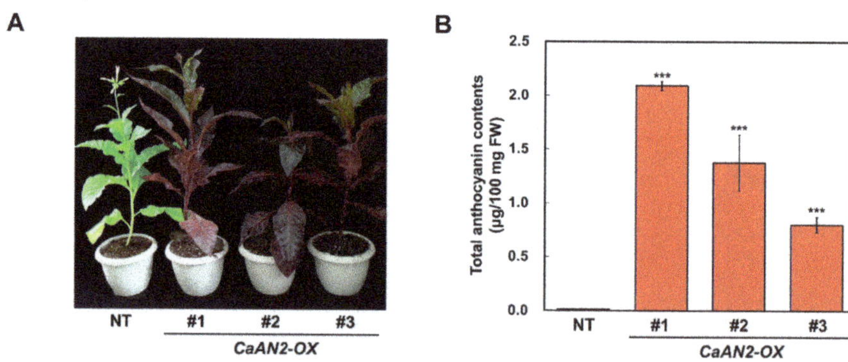


**Figure 7.** Phenotypes and relative anthocyanin contents of transgenic tobacco plants containing *CaAN2*. (**A**) Phenotypes of transgenic plants. (**B**) Total anthocyanin contents. NT, nontransgenic tobacco plant; *CaAN2-OX*, transgenic tobacco plants ectopically expressing *CaAN2*. All results represent mean values ± SD from three independent biological replicates. Asterisks indicate values that differ significantly from NT at $p < 0.001$ according to a Student's paired *t*-test.

qRT-PCR was performed to investigate the expression of anthocyanin biosynthetic genes and related transcriptional regulators in transgenic tobacco plants harboring *CaAN2*. To examine the relationship between leaf color and the transcript levels of anthocyanin biosynthetic pathway genes, the expression of nine structural genes, including the upstream genes *NtPAL* and *Nt4CL*; the early biosynthetic genes *NtCHS*, *NtCHI*, *NtF3H* and *flavonoid 3′-hydroxylase* (*NtF3′H*); and the late biosynthetic genes *NtDFR*, *NtANS* and *NtUFGT* was measured. The expression of anthocyanin biosynthesis regulatory genes, including the R2R3-MYB activator gene *NtAN2*, the bHLH activator gene *NtAN1*, the R2R3-MYB repressor gene *NtMYB3* and the R3 repressor gene *NtETC1* was also analyzed. The expression levels of all structural genes except *NtPAL* and *Nt4CL* were higher in transgenic vs. NT plants (Figure 8A). The transcript levels of these upregulated genes were similar to those of *CaAN2*. As expected, *CaAN2* transcripts were only present in transgenic plants but not in NT plants. In addition, the transcript levels of *NtAN1* and *NtAN2*, encoding endogenous anthocyanin biosynthesis activators, were high in transgenic plants but not in NT plants. Additionally, the transcript levels of the R2R3-MYB type repressor *NtMYB3* and the R3-MYB type repressor *NtETC1* were high in transgenic tobacco plants but not in NT plants (Figure 8B). These results indicate that the ectopic expression of *CaAN2* promotes the transcription of anthocyanin-biosynthesis-related regulators and biosynthetic genes, resulting in anthocyanin accumulation in tobacco leaves.

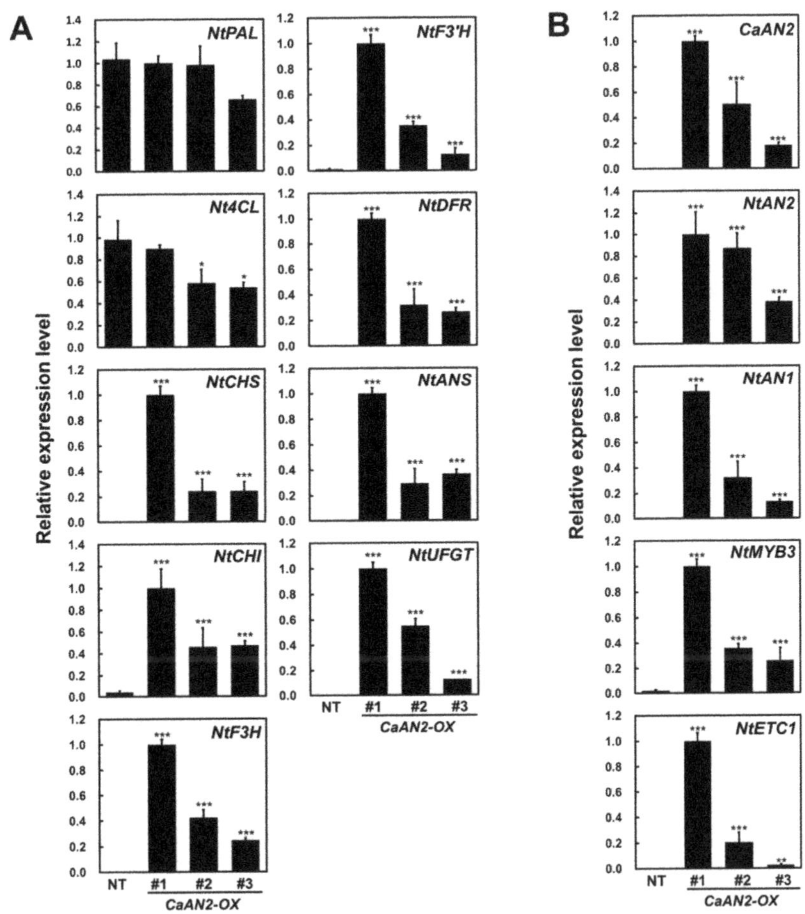


**Figure 8.** Expression profiles of anthocyanin biosynthetic genes and endogenous transcription factor genes in the leaves of nontransgenic tobacco (NT) and three independent transgenic tobacco lines harboring *CaAN2*. The relative transcript levels of anthocyanin biosynthetic genes (**A**) and regulatory genes (**B**) were measured by qRT-PCR, with *NtGAPDH* used as a reference gene. The biosynthetic

pathway genes evaluated include those encoding *chalcone synthase* (*NtCHS*), *chalcone isomerase* (*NtCHI*), *flavanone 3-hydroxylase* (*NtF3H*), *flavonoid 3′-hydroxylase* (*NtF3′H*), *dihydroflavonol 4-reductase* (*NtDFR*), *anthocyanidin synthase* (*NtANS*) and *UDP-flavonoid glucosyltransferase* (*NtUFGT*) as well as the upstream enzyme *phenylalanine ammonia-lyase* (*NtPAL*) and *4-coumarate-CoA ligase* (*Nt4CL*). Results are means ± SD from three independent biological replicates. *, ** and *** indicate values that differ significantly from NT at $p < 0.05$, $p < 0.01$ and $p < 0.001$, respectively, according to a Student's paired *t*-test.

### 2.7. Plant Coloration Facilitates Detection of and Selection against Chimerism

Chimerism is a fairly common occurrence during plant transformation and is a prime factor in the failure to transfer genes to subsequent generations [20]. Therefore, minimizing chimerism is indispensable for establishing an efficient and reliable transformation system. Here, transformed shoots with overall uniform red coloration were selected and progressed these shoots to subsequent generations. After the next several generations, *CaAN2* was stably transmitted, resulting in strong pigmentation due to anthocyanin accumulation. This process will be useful for reducing the occurrence of chimerism during transformation using *CaAN2* as a visible marker gene.

## 3. Discussion

The molecular genetic improvement of crops is strongly dependent on the selection of transformants with the desired traits. Selectable markers for antibiotic or herbicide resistance are commonly used for the screening of transformants. However, public concerns have been raised about the presence of selectable markers due to possible risks for human health and the environment [3]. Therefore, several alternative methods have been developed to generate marker-free transgenic plants, such as co-transformation and segregation, site-specific recombinase-mediated excision and intrachromosomal homologous recombination, but these techniques are costly, time-consuming and/or inefficient [21]. Additionally, plant selection systems consist of two components: chemical agents and selectable marker genes. During the tissue culture process, chemicals such as herbicides or antibiotics must be added to the tissue culture medium.

Anthocyanins, a group of flavonoid metabolites derived from phenylpropanoid compounds, are widely present in various plant tissues, including leaves, stems, flowers and fruits [7]. Additionally, high anthocyanin content in foods is generally considered beneficial to human health due to their strong antioxidant properties. Tissues containing anthocyanins can easily be discerned with the naked eye without the need for additional treatments. To utilize anthocyanin as a visible marker, it is essential to identify key genes for anthocyanin biosynthesis. In this study, the role of *CaAN2* from purple fruited chili pepper in anthocyanin pigment accumulation was verified. The high expression of *CaAN2* activated the transcription of anthocyanin biosynthetic genes from pepper as well as tobacco, resulting in pigment accumulation (Figures 2 and 5). Some anthocyanin-activating R2R3 regulators indispensably require bHLH transcription factors to induce anthocyanin accumulation [22]. Mangosteen (*Garcinia mangostana*) GmMYB10 did not activate the *AtAFR* or *GmDFR* promoter when expressed alone in a transient expression assay, but it activated these genes when co-expressed with AtbHLH2 [23]. In addition, transgenic Lisianthus containing snapdragon *AmROSEA1* (R2R3 MYB) showed anthocyanin accumulation only in sepals, which strongly express a bHLH cofactor gene [24]. Transgenic tobacco plants simultaneously expressing the maize *ZmB-peru* gene and the Arabidopsis *AtmPAP1D* gene displayed notable color changes compared to plants individually expressing *ZmB-peru* or *AtmPAP1D* [22]. Here, we determined that *CaAN2* alone is sufficient to activate anthocyanin biosynthesis in tobacco and pepper; therefore, it can be used as a visible reporter gene for plant transformation. Evaluation of the effect of *CaAN2* in a heterologous tobacco system confirmed that this gene can be useful for identifying transgenic plants without the need for the expensive equipment required for GFP detection or the chemical staining required for GUS detection.

The use of the genome-editing tool CRISPR-Cas9 for the molecular breeding of crops is on the rise. In addition, molecular characterization of transgene-free gene-edited plants is required and this is a time-consuming and labor-intensive process. The *OsC1* reporter gene is a valuable tool to aid in the visible screening of transformants at high efficiency [25]. The combination of Cas9 protein and *OsC1* represents a powerful selection system for transgenic rice plants, allowing transgene-free, gene-edited plants to be easily selected on the basis of a color change. Advanced methods based on anthocyanin accumulation can enable the robust, rapid selection of plants with gene-edited target traits.

Here, we successfully used *CaAN2* to monitor the transient transformation of tobacco and successfully selected transformed cells. The selection of transformed plants is generally labor- and time-intensive [26]. However, the expression of *CaAN2* can easily be monitored in transgenic plants based on anthocyanin accumulation. The application of *CaAN2* for CRISPR-Cas9-mediated genome editing is feasible for the selection of transgenic or transgenic-free genome-edited plants. Using this method, transgenic plants could be screened without the need for additional treatments.

## 4. Materials and Methods

### 4.1. Plant Materials

Two pepper (*Capsicum annuum*) cultivars, the green cultivar 'AG188' and the purple cultivar '20GP15-2', were used in this study; these cultivars are referred to as G and P, respectively. The seeds were obtained from the National Institute of Horticultural and Herbal Science (Wanju, Korea) and cultivated in the greenhouse. Tobacco (*Nicotiana tabacum* cv. Xanthi) plants were grown in greenhouses and growth chambers at Hankyong University (Anseong, Korea) under natural light at 26 $\pm$ 2 °C and used for transient *Agrobacterium* (*Agrobacterium tumefaciens*)-mediated infiltration assays and stable transformation.

To analyze anthocyanin contents and the transcript levels of anthocyanin biosynthetic genes, all samples were rapidly frozen in liquid nitrogen and stored at −80 °C. Each sample was ground to a powder and split into two aliquots: one for RNA extraction and the other to measure anthocyanin contents.

### 4.2. RNA Extraction, cDNA Synthesis and Isolation of Genomic DNA

Total RNA was extracted from various tissues of both pepper cultivars and from tobacco leaves using TRIzol reagent (Invitrogen, Carlsbad, CA, USA) and purified using a FavorPrep Plant Total RNA Mini Kit (Favorgen, Changzhi, Taiwan) according to the manufacturer's instructions. First-strand cDNA was synthesized from 2 μg total RNA using amfiRivert cDNA Synthesis Platinum Master Mix (GenDEPOT, Barker, TX, USA) for qRT-PCR analysis. Genomic DNA was extracted from the samples with a DNeasy Plant Mini Kit (Qiagen, Valencia, CA, USA) according to the manufacturer's instructions.

### 4.3. Measurement of Total Anthocyanin Contents

Total anthocyanin contents in frozen tissue samples were measured as previously described [27]. Each 100 mg (fresh weight) tissue sample was incubated in 600 μL extraction buffer (methanol containing 1% [*v/v*] HCl) for 6 h at 4 °C with moderate agitation. After the addition of 200 μL water and 200 μL chloroform, the sample was centrifuged at 14,000× *g* for 5 min at 4 °C to precipitate the plant debris. The absorbance of the supernatant was recorded at 530 nm ($A_{530}$) and 657 nm ($A_{657}$) using a microplate reader. Anthocyanin contents were determined according to the formula $A_{530} - (0.25 \times A_{657})$. Each sample was extracted and examined using three independent experiments.

### 4.4. Quantitative Reverse-Transcription PCR (qRT-PCR) Analysis

Transcript levels were measured by qRT-PCR using AccuPower 2x Greenstar qPCR Master Mix (Bioneer, Daejun, Korea) and a Bio-Rad CFX96 Detection System (Bio-Rad Laboratories, Hercules, CA, USA) according to the manufacturer's instructions. Gene expression levels were normalized to *actin* (*CaACTIN*) and *glyceraldehyde 3-phosphate dehydrogenase*

(*NtGAPDH*) for pepper and tobacco, respectively, as the reference gene. Three independent biological replicates were performed per sample. The primers used for RT-qPCR analysis are listed in Supplementary Table S2.

### 4.5. Gene Isolation and Sequence Analysis

The full-length coding sequence of *CaAN2* was amplified from cDNA and genomic DNA from purple pepper by PCR with PrimeSTAR HS DNA Polymerase (Takara, Otsu, Japan) using the primer pair *CaAN2* F/R (Supplementary Table S1). All amplicons were subcloned into the pENTR/D-TOPO vector (Invitrogen) for validation by sequencing. Multiple sequence alignments were generated using the CLUSTALW program (https://www.genome.jp/tools-bin/clustalw). A phylogenetic tree was constructed using the neighbor-joining method [28] with MEGA version 6 software [29].

### 4.6. In Planta Assays of the Anthocyanin Biosynthesis Activity of CaAN2

Amplified DNA products of *CaAN2* were cloned into the Gateway entry vector pDONR221 (Invitrogen) using PCR-specific primer sets (Supplementary Table S1) and incorporated into the Gateway destination vector pB2GW7 (VIB-Ghent University, Ghent, Belgium) via several Gateway cloning steps. The resulting constructs were introduced into *Agrobacterium* strain GV3101 for the transient infiltration assay and stable transformation.

To perform the tobacco agroinfiltration assay, *Agrobacterium* cultures harboring pB2GW7-*CaAN2*, pB2GW7-*ZmB-peru* and pB2GW7-*AtmPAP1* were grown in LB medium at 28 °C with shaking until the optical density at 600 nm reached 1.2. The bacteria were pelleted and resuspended in infiltration solution containing 10 mM $MgCl_2$ and 0.1 mM acetosyringone. Equal amounts of *Agrobacterium* suspensions harboring each construct were infiltrated into the abaxial surfaces of the expanded leaves of 6-week-old tobacco plants as described [28]. Photographs of infiltrated leaves were taken at 5 days after infiltration.

### 4.7. Plant Regeneration

Transgenic tobacco (*N. tabacum* cv. Xanthi) plants were generated by transformation with *Agrobacterium* containing the pB2GW7-*CaAN2* construct using the leaf disc method [22]. Briefly, tobacco seeds were surface sterilized and grown on solidified half-strength Murashige and Skoog (MS) medium (Duchefa, Haarlem, Netherlands). The plants were grown in a growth chamber under a 16 h light/8 h dark cycle at 26 ± 1 °C for 2 months. Leaf discs were obtained from the cultured plants and submerged in *Agrobacterium* mixture. To identify transgenic events, explants were cultured on shoot-inducing medium with or without 10 mg/L phosphinothricin (PPT, Duchefa). The regenerated shoots were subsequently transferred to MS medium to enable rooting and cultivated in a greenhouse to maturity. Three representative tobacco lines were selected for further analysis. Transgenic $T_3$ lines were developed by successive self-pollination of $T_0$ plants.

## 5. Conclusions

In this study, we characterized *CaAN2* as a common key regulator of anthocyanin pigmentation in purple chili pepper, cooperatively expressed with *CaTT8*. Through the transient assay and stable tobacco transformation, it confirmed that individual expression of *CaAN2* sufficiently induced the anthocyanin accumulation in a heterologous system. Additionally, anthocyanin accumulated phenotypes were stably inherited into the next generation without chimerism. Taken together, it indicates that *CaAN2* is useful as an alternative selection system for transgenic plants, as a visible selective marker gene.

**Supplementary Materials:** The following supporting information can be downloaded at: https://www.mdpi.com/article/10.3390/plants11060820/s1, Table S1: List of primers used in this study. Table S2: List of primers used in this study.

**Author Contributions:** Conceptualization, S.-H.L.; methodology, S.-H.L. and J.-Y.L.; software, S.-H.L. and D.-H.K.; validation, S.-H.L. and D.-H.K.; investigation, M.-C.C. and S.-H.L.; resources, M.-C.C.; data curation, S.-H.L. and J.-Y.L.; writing—original draft preparation, S.-H.L. and J.-Y.L.; writing—review and editing, S.-H.L.; supervision, S.-H.L.; project administration, S.-H.L.; funding acquisition, S.-H.L., S.-H.L. and J.-Y.L. devised the study and wrote the manuscript; J.-Y.L. and D.-H.K. generated the constructs and performed the molecular experiments. All authors contributed to the article and approved the submitted version. All authors have read and agreed to the published version of the manuscript.

**Funding:** This work was supported by a research grant from Hankyong National University in 2020.

**Institutional Review Board Statement:** Not applicable.

**Informed Consent Statement:** Not applicable.

**Data Availability Statement:** The data presented in this study, are available on request from the corresponding authors. The data are not publicly available because their elaboration are all reported in the manuscript.

**Conflicts of Interest:** The authors declare no conflict of interest.

## References

1. Long, D.; Wu, X.; Yang, Z.; Lenk, I.; Nielsen, K.K.; Gao, C. Comparison of three selectable marker genes for transformation of tall fescue (*Festuca arundinacea* Schreb.) plants by particle bombardment. *In Vitro Cell. Dev. Biol. Plant* **2011**, *47*, 658–666. [CrossRef]
2. Izawati, A.M.; Masani, M.Y.; Ismanizan, I.; Parveez, G.K. Evaluation on the effectiveness of 2-deoxyglucose-6-phosphate phosphatase (*$DOG^R1$*) gene as a selectable marker for oil palm (*Elaeis guineensis* Jacq.) embryogenic calli transformation mediated by *Agrobacterium tumefaciens*. *Front. Plant Sci.* **2015**, *6*, 727. [CrossRef] [PubMed]
3. Lucht, J.M. Public Acceptance of Plant Biotechnology and GM Crops. *Viruses* **2015**, *7*, 4254–4281. [CrossRef] [PubMed]
4. Lal, M.; Bhardwaj, E.; Chahar, N.; Dangwal, M.; Das, S. (Trans) Gene Flow: Mechanisms, Biosafety Concerns and Mitigation for Containment. In *Reproductive Ecology of Flowering Plants: Patterns and Processes*; Springer: Berlin/Heidelberg, Germany, 2020; pp. 335–394.
5. Jefferson, R.A.; Kavanagh, T.A.; Bevan, M.W. GUS fusions: β-glucuronidase as a sensitive and versatile gene fusion marker in higher plants. *EMBO J.* **1987**, *6*, 3901–3907. [CrossRef] [PubMed]
6. Lin, M.H.; Gresshoff, P.M.; Indrasumunar, A.; Ferguson, B.J. pHairyRed: A Novel Binary Vector Containing the *DsRed2* Reporter Gene for Visual Selection of Transgenic Hairy Roots. *Mol. Plant* **2011**, *4*, 537–545. [CrossRef] [PubMed]
7. Saigo, T.; Wang, T.; Watanabe, M.; Tohge, T. Diversity of anthocyanin and proanthocyanin biosynthesis in land plants. *Curr. Opin. Plant Biol.* **2020**, *55*, 93–99. [CrossRef] [PubMed]
8. Hichri, I.; Barrieu, F.; Bogs, J.; Kappel, C.; Delrot, S.; Lauvergeat, V. Recent advances in the transcriptional regulation of the flavonoid biosynthetic pathway. *J. Exp. Bot.* **2011**, *62*, 2465–2483. [CrossRef] [PubMed]
9. Petroni, K.; Tonelli, C. Recent advances on the regulation of anthocyanin synthesis in reproductive organs. *Plant Sci.* **2011**, *181*, 219–229. [CrossRef] [PubMed]
10. Kortstee, A.J.; Khan, S.A.; Helderman, C.; Trindade, L.M.; Wu, Y.; Visser, R.G.; Brendolise, C.; Allan, A.; Schouten, H.J.; Jacobsen, E. Anthocyanin production as a potential visual selection marker during plant transformation. *Transgenic Res.* **2011**, *20*, 1253–1264. [CrossRef] [PubMed]
11. Kim, D.H.; Lee, J.; Rhee, J.; Lee, J.Y.; Lim, S.H. Loss of the R2R3 MYB Transcription Factor RsMYB1 Shapes Anthocyanin Biosynthesis and Accumulation in *Raphanus sativus*. *Int. J. Mol. Sci.* **2021**, *22*, 10927. [CrossRef] [PubMed]
12. Ruan, Y.; Chen, K.; Su, Y.; Jiang, S.; Xu, P.; Murray, J.D. A Root Tip-Specific Expressing Anthocyanin Marker for Direct Identification of Transgenic Tissues by the Naked Eye in Symbiotic Studies. *Plants* **2021**, *10*, 605. [CrossRef] [PubMed]
13. Borovsky, Y.; Oren-Shamir, M.; Ovadia, R.; De Jong, W.; Paran, I. The A locus that controls anthocyanin accumulation in pepper encodes a MYB transcription factor homologous to *Anthocyanin2* of Petunia. *Theor. Appl. Genet.* **2004**, *109*, 23–29. [CrossRef] [PubMed]
14. Jung, S.; Venkatesh, J.; Kang, M.Y.; Kwon, J.K.; Kang, B.C. A non-LTR retrotransposon activates anthocyanin biosynthesis by regulating a MYB transcription factor in *Capsicum annuum*. *Plant Sci.* **2019**, *287*, 110181. [CrossRef] [PubMed]
15. Stracke, R.; Werber, M.; Weisshaar, B. The R2R3-MYB gene family in Arabidopsis thaliana. *Curr. Opin. Plant Biol.* **2001**, *4*, 447–456. [CrossRef]
16. Kim, D.H.; Yang, J.; Ha, S.H.; Kim, J.K.; Lee, J.Y.; Lim, S.H. An OsKala3, R2R3 MYB TF, Is a Common Key Player for Black Rice Pericarp as Main Partner of an OsKala4, bHLH TF. *Front. Plant Sci.* **2021**, *12*, 765049. [CrossRef] [PubMed]
17. Espley, R.V.; Brendolise, C.; Chagneì, D.; Kutty-Amma, S.; Green, S.; Volz, R.; Putterill, J.; Schouten, H.J.; Gardiner, S.E.; Hellens, R.P.; et al. Multiple repeats of a promoter segment causes transcription factor autoregulation in red apples. *Plant Cell* **2009**, *21*, 168–183. [CrossRef] [PubMed]

18. Borevitz, J.O.; Xia, Y.; Blount, J.; Dixon, R.A.; Lamb, C. Activation tagging identifies a conserved MYB regulator of phenylpropanoid biosynthesis. *Plant Cell* **2000**, *12*, 2383–2394. [CrossRef] [PubMed]
19. Yamagishi, M.; Shimoyamada, Y.; Nakatsuka, T.; Masuda, K. Two *R2R3-MYB* genes, homologs of petunia *AN2*, regulate anthocyanin biosyntheses in flower tepals, tepal spots and leaves of Asiatic hybrid lily. *Plant Cell Physiol.* **2010**, *51*, 463–474. [CrossRef] [PubMed]
20. Das Bhowmik, S.S.; Cheng, A.Y.; Long, H.; Tan, G.Z.H.; Hoang, T.M.L.; Karbaschi, M.R.; Williams, B.; Higgins, T.J.V.; Mundree, S.G. Robust Genetic Transformation System to Obtain Non-chimeric Transgenic Chickpea. *Front. Plant Sci.* **2019**, *10*, 524. [CrossRef]
21. Yau, Y.Y.; Stewart, C.N., Jr. Less is more: Strategies to remove marker genes from transgenic plants. *BMC Biotechnol.* **2013**, *13*, 36. [CrossRef]
22. Lim, S.H.; Sohn, S.H.; Kim, D.H.; Kim, J.K.; Lee, J.Y.; Kim, Y.M.; Ha, S.H. Use of an anthocyanin production phenotype as a visible selection marker system in transgenic tobacco plant. *Plant Biotechnol. Rep.* **2012**, *6*, 203–211. [CrossRef]
23. Palapol, Y.; Ketsa, S.; Lin-Wang, K.; Ferguson, I.B.; Allan, A.C. A MYB transcription factor regulates anthocyanin biosynthesis in mangosteen (*Garcinia mangostana* L.) fruit during ripening. *Planta* **2009**, *229*, 1323–1334. [CrossRef] [PubMed]
24. Schwinn, K.E.; Boase, M.R.; Bradley, J.M.; Lewis, D.H.; Deroles, S.C.; Martin, C.R.; Davies, K.M. MYB and bHLH transcription factor transgenes increase anthocyanin pigmentation in petunia and lisianthus plants, and the petunia phenotypes are strongly enhanced under field conditions. *Front. Plant Sci.* **2014**, *5*, 603. [CrossRef] [PubMed]
25. He, Y.; Zhu, M.; Wu, J.; Ouyang, L.; Wang, R.; Sun, H.; Yan, L.; Wang, L.; Xu, M.; Zhan, H.; et al. Repurposing of Anthocyanin Biosynthesis for Plant Transformation and Genome Editing. *Front. Genome Ed.* **2020**, *2*, 607982. [CrossRef]
26. Gao, X.; Chen, J.; Dai, X.; Zhang, D.; Zhao, Y. An effective strategy for reliably isolating heritable and Cas9-free arabidopsis mutants generated by CRISPR/Cas9-mediated genome editing. *Plant Physiol.* **2016**, *171*, 1794–1800. [CrossRef] [PubMed]
27. Lim, S.H.; Kim, D.H.; Jung, J.A.; Lee, J.Y. Alternative Splicing of the Basic Helix–Loop–Helix Transcription Factor Gene CmbHLH2 Affects Anthocyanin Biosynthesis in Ray Florets of Chrysanthemum (*Chrysanthemum morifolium*). *Front. Plant Sci.* **2021**, *12*, 669315. [CrossRef] [PubMed]
28. Saitou, N.; Nei, M. The neighbor-joining method: A new method for reconstructing phylogenetic trees. *Mol. Biol.* **1987**, *4*, 406–425. [CrossRef]
29. Tamura, K.; Stecher, G.; Peterson, D.; Filipski, A.; Kumar, S. MEGA6: Molecular Evolutionary Genetics Analysis version 6.0. *Mol. Biol.* **2013**, *30*, 2725–2729. [CrossRef]

MDPI
St. Alban-Anlage 66
4052 Basel
Switzerland
www.mdpi.com

*Plants* Editorial Office
E-mail: plants@mdpi.com
www.mdpi.com/journal/plants

www.ingramcontent.com/pod-product-compliance
Lightning Source LLC
Chambersburg PA
CBHW061618220926
51850CB00005B/540
* 9 7 8 3 7 2 5 8 1 3 0 7 0 *